How to Build
Everything

**Written by
Hannah Dolan**

**Models designed
and created by
Jessica Farrell and Nate Dias**

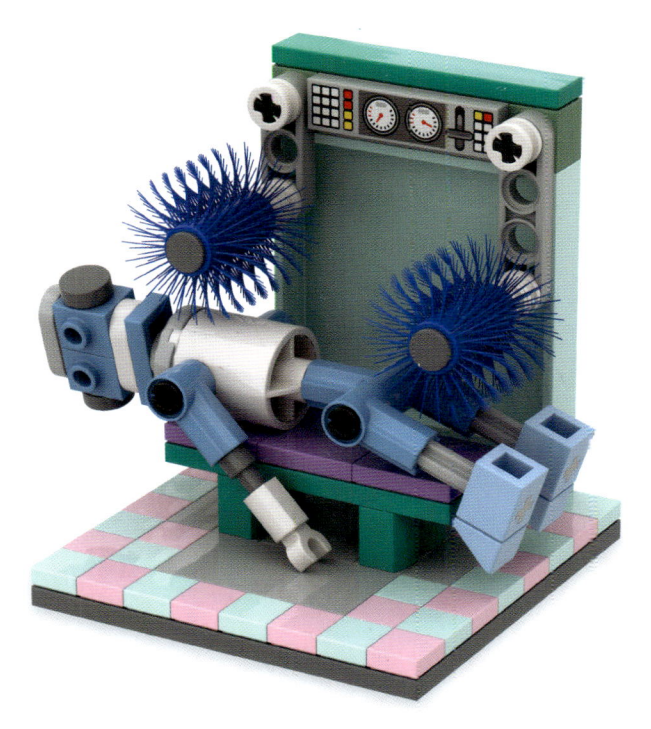

LEGO

CONTENTS

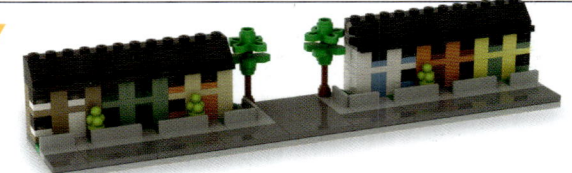

SUBURBAN STREET, PAGE 34

How to begin 6

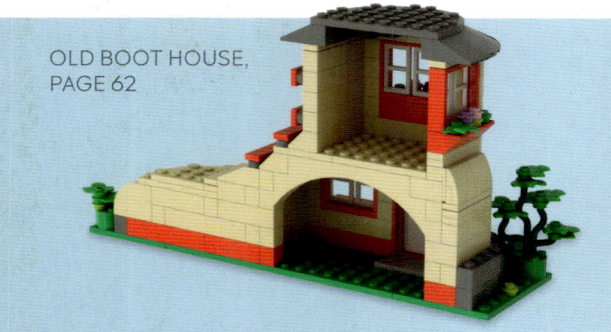

OLD BOOT HOUSE, PAGE 62

Chapter one: Houses 8
Simple house 10
Log cabin 12
Colors and textures 14
Beach house 16
Country cottage 18
Bungalow 20
Lighthouse 22
Fairy house 24
Festive house 26
Viking longhouse 28
Ski chalet 30
Modern house 32
Suburban street 34
Rainbow house 36
Townhouse 38
Model mash-up 40
Gingerbread house 42
Modular apartments 44
Hillside village 46

Playhouse 48
Snail house 50
Thatched cottage 52
Hamburger house 54
Extensions and extras 56
Underwater house 58
Flying house 60
Old boot house 62
Jungle hut 64
Fantasy castle 66
Medieval inn 68
Tree house 70
Furniture 72
Medieval castle 74
Haunted house 76

HIPPOPOTAMUS, PAGE 130

Chapter two: Animals 78
Giraffe 80
Piglet 82

Angelfish	84
Mouse	86
Fur and feathers	88
Parrot	90
Camel	92
Octopus	94
Model mash-up	96
Sheep	98
Turtle	100
Chameleon	102
Reindeer	104
Koi carp	106
Owl	108
Cat	110
Koala	112
Puppy	114
Imaginary critters	116
Sloth	118
Rabbit	120
Orca	122
Rooster	124
Fox	126
Peacock	128
Hippopotamus	130
Orangutan	132
Lion	134
Raccoon	136
Feeding time	138
Polar bear	140
Wolf	142
Panda	144

YELLOW TAXI, PAGE 198

Chapter three: Cars	146
Off-roader	148
Rocket car	150
Dragster	152
Engines	154
Animal cars	156
Bumper car	158
Family car	160
Auto rickshaw	162
Vintage car	164
City car	166
Moon buggy	168
Four wheeler	170
Golf cart	172
Race car	174
Hot-dog car	176
Black cab	178
Pickup truck	180
Switched-up pickups	182
Brick car	184
Sports car	186
Underwater car	188
Driverless car	190

LIMOUSINE, PAGE 214

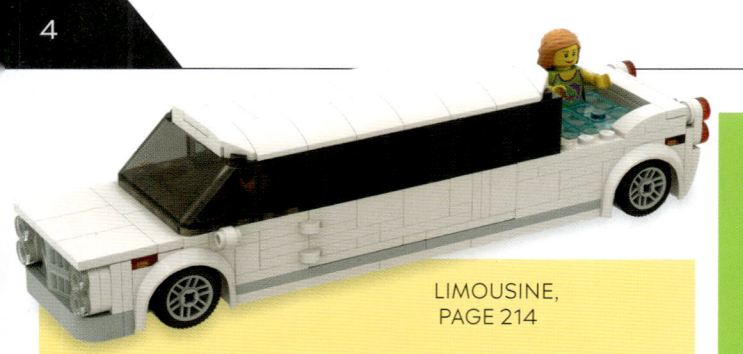

Cloud car	192
Hot rod	194
Pirate car	196
Yellow taxi	198
Street furniture	200
Gingerbread car	202
Camper	204
Ice-cream truck	206
Royal car	208
Monster truck	210
Model mash-up	212
Limousine	214

Chapter four: Dinosaurs	**216**
Brachiosaurus	218
Longisquama	220
Pteranodon	222
Compsognathus	224
Dimorphodon	226
Melanorosaurus	228
Claws and teeth	230
Velociraptor	232
Parasaurolophus	234
Cynognathus	236
Protoceratops	238
Micro dinosaurs	240
Doyouthinkhesaurus	242
Baryonyx	244
Pachycephalosaurus	246
Stegosaurus	248
Happyosaurus	250
Allosaurus	252
Habitats	254
Caudipteryx	256
Deinosuchus	258
Dimetrodon	260
Ichthyosaurus	262
Plesiosaurus	264
Iguanodon	266
Woolly mammoth	268
Ankylosaurus	270
Model mash-up	272
Diplodocus	274
Smilodon	276

STEGOSAURUS, PAGE 248

Kentrosaurus	278
Triceratops	280
Tyrannosaurus	282

ROBOT HORSE, PAGE 344

Chapter five: Robots	284
Boxy Bot	286
Robot Helper	288
Robo Recycler	290
Details and greebles	292
Chef Bot	294
Shopper Bot	296
Robo Puppy	298
Monster Mechanic	300
Delivery Bot	302
Dinobot Transporter	304
Hip-Hop Bot	306
Tidy Bot	308
Gardener Robot	310
Driverless Hovercar	312

Alien Robot	314
Model mash-up	316
Bath Time Bot	318
Robo-On-The-Go	320
Supersonic Robot	322
Underwater Explorer	324
Space Probe	326
Birthday Bot	328
Bots in the community	330
One-Bot Band	332
Mecha Bird	334
Quackbot	336
Odd Bot	338
Junk Robot	340
Warrior Mech	342
Robot Horse	344
Robo Cashier	346
Robot Spa	348
Android	350
Plant Cyborg	352

Extra bits and pieces	354
Acknowledgments	356

ROBO RECYCLER, PAGE 290

HOW TO BEGIN

Got my bricks. What's next?

Is your LEGO® collection on hand? Do you have some space to build? Then you're ready to begin. All that's left to do is decide which of the mountains of models in this book you want to create first. Will it be a cheery rainbow house, a fluffy sheep, a soaring *Dimorphodon*, a cool camper, or a handy robot helper? The models in the themed chapters start off easy and get harder as you continue, making each chapter an imaginative building journey. Let's go!

BREAKING DOWN BUILDING

Each model is broken down into three to five important building stages. You might not have all the bricks you need, but you don't have to copy the models brick by brick. The breakdowns show techniques to inspire your own amazing ideas. There are also three ideas galleries in each chapter that focus on particular details of models or ways to expand your building ideas. Look for a fun model mash-up in each chapter too.

The first picture is always the finished model

The dinosaur chapter notes the time period

Each model is classed as easy, medium, or hard

Some parts of models are broken down even more in these circles

DINOSAUR MODEL

The last step is one of the final stages

Smaller parts of models or extra build ideas

HOUSE IDEAS GALLERY

BRICK LISTS

If you'd like to see all of the bricks used in a particular model, go to

www.dk.com/LEGOhouses
www.dk.com/LEGOcars
www.dk.com/LEGOdinosaurs
www.dk.com/LEGOanimals
www.dk.com/us/information/how-to-build-lego-robots

HOW TO BEGIN

TECHNICAL TIPS

These notes for builders will help you understand some of the LEGO words and terms that are used a lot in this book.

Get to know the nuts and bolts of LEGO building!

LEGO DICTIONARY

2x3 brick

1x1 brick

Knobs are the round, raised bumps on top of bricks and plates. They fit into "tubes" on the bottom of pieces.

1x2 brick

Bricks are found in most LEGO models. They are named according to how many knobs they have on top.

1x3 plate

Plates are similar to bricks because they have knobs on top and tubes on the bottom, but they are much thinner.

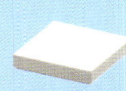

2x2 tile

Tiles are thin, like plates, but they have no knobs on top.

1x2 brick with hole

Holes inside bricks and other pieces can hold connectors such as pins, bars, and axles.

WAYS TO BUILD

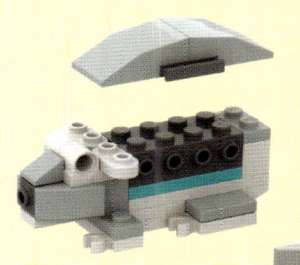

Upward
The "easy" models in this book show a lot of pieces stacked on top of each other, as shown in the middle of this mouse.

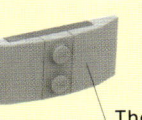

Downward
You can add pieces underneath your models, such as legs on animals or wheels on cars.

Sideways
Slightly trickier models use a lot of pieces with knobs on their sides for building sideways.

These pieces attach sideways

All around
Build moving parts into your models using pieces that can move together, such as clips and bars.

Each arm clips onto a bar

WAYS TO SCALE

Minifigure size
If you want minifigures to get inside your models, think about how wide or tall your minifigures are and how much space they will need inside.

Minus the minifigures
If you're happy to imagine who or what's inside your model, you could create builds of any size without leaving space for minifigures or interior details.

Microscale
Create tiny models by building in microscale. Microscale builds like this car use small pieces in inventive ways.

Houses

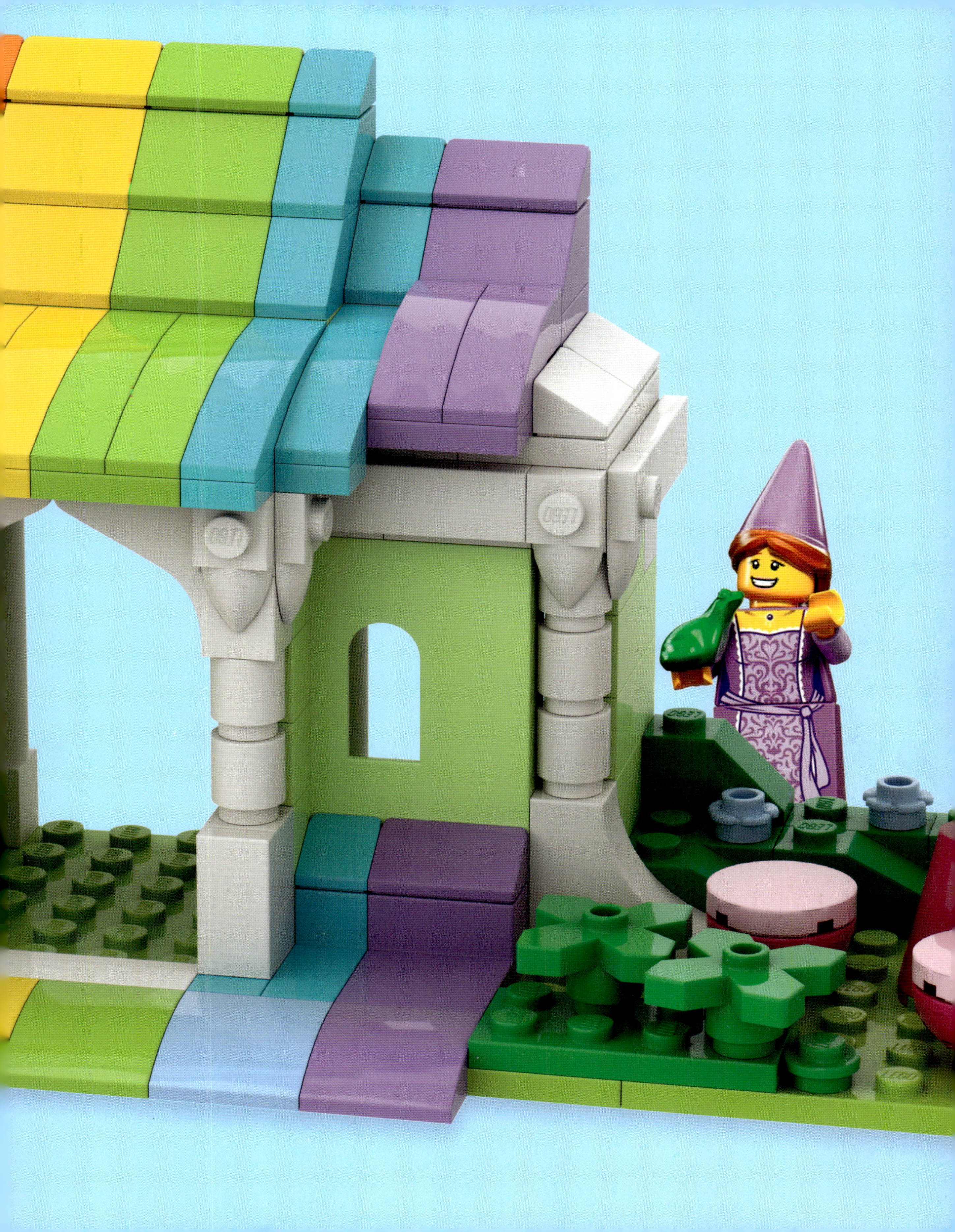

SIMPLE HOUSE

This little dwelling is for minifigures and builders who enjoy the simple things in life. Its perfectly square proportions make this a good one to start with if you're new to house building. The house and yard are all built on one 16x16 base plate.

If you live simply, you have more time for fun hobbies.

Like kendo. Let's go, Dad!

- 1x1 round brick chimney pot
- The roof has three layers of slope bricks
- Simple cool yellow brick walls
- 2x2x4 prickly bush
- The green base plate is the lawn
- Tile and jumper plate sidewalk

SKILL LEVEL

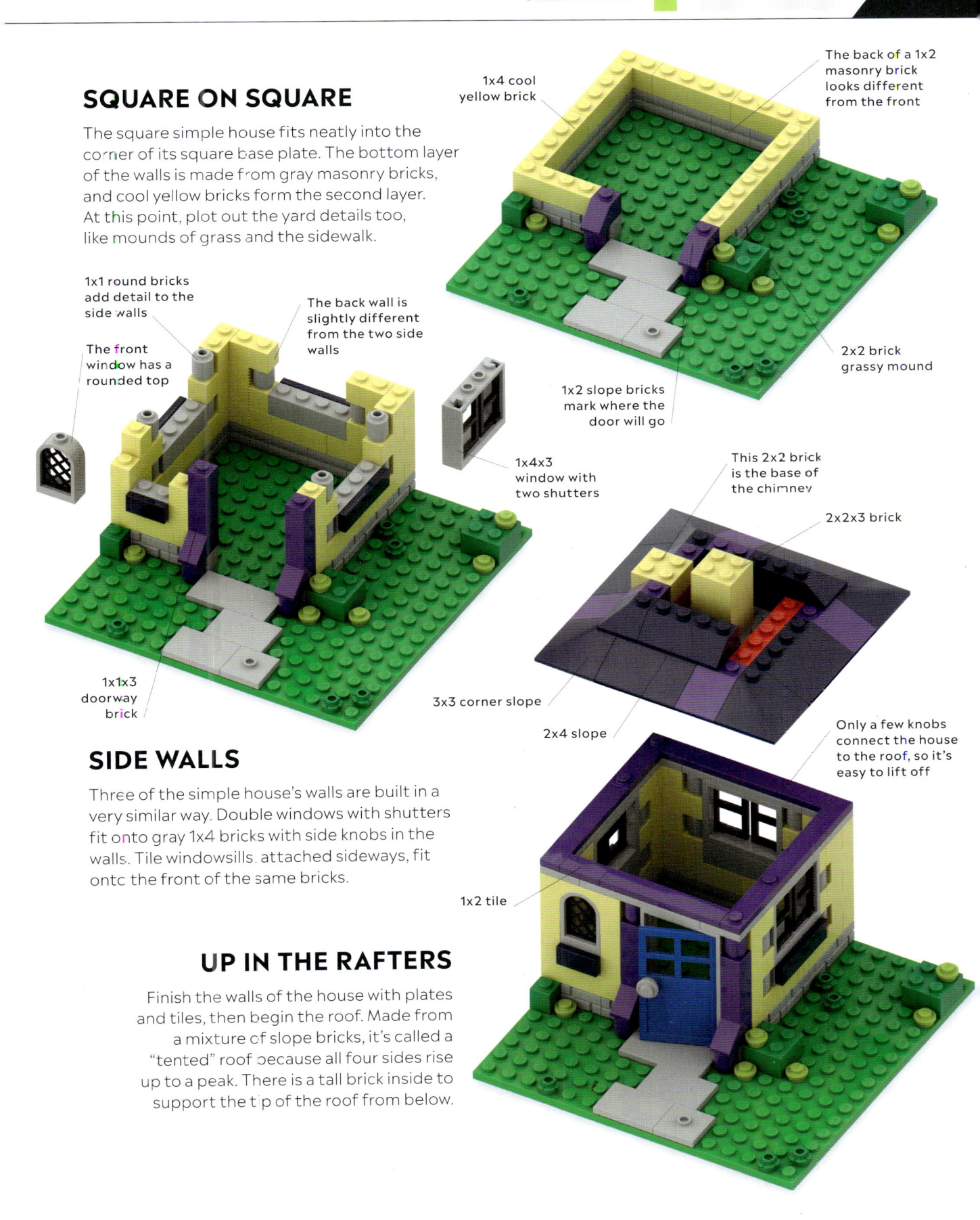

SQUARE ON SQUARE

The square simple house fits neatly into the corner of its square base plate. The bottom layer of the walls is made from gray masonry bricks, and cool yellow bricks form the second layer. At this point, plot out the yard details too, like mounds of grass and the sidewalk.

- The back of a 1x2 masonry brick looks different from the front
- 1x4 cool yellow brick
- 2x2 brick grassy mound
- 1x2 slope bricks mark where the door will go
- 1x1 round bricks add detail to the side walls
- The front window has a rounded top
- The back wall is slightly different from the two side walls
- 1x4x3 window with two shutters
- 1x1x3 doorway brick
- This 2x2 brick is the base of the chimney
- 2x2x3 brick
- 3x3 corner slope
- 2x4 slope

SIDE WALLS

Three of the simple house's walls are built in a very similar way. Double windows with shutters fit onto gray 1x4 bricks with side knobs in the walls. Tile windowsills, attached sideways, fit onto the front of the same bricks.

- 1x2 tile
- Only a few knobs connect the house to the roof, so it's easy to lift off

UP IN THE RAFTERS

Finish the walls of the house with plates and tiles, then begin the roof. Made from a mixture of slope bricks, it's called a "tented" roof because all four sides rise up to a peak. There is a tall brick inside to support the tip of the roof from below.

LOG CABIN

Deep in the LEGO® woods lies this traditional log cabin. It has walls made from horizontal logs that interlock at the corners. Minifigures come here to get away from it all and relax by the tranquil waters of the lake ... but the local wildlife seems to have other ideas!

Tiles at alternating heights create a corrugated metal roof

If you don't have tree pieces, you could build your own trees

Wooden support post made from 1x1 round bricks

That wasn't in the brochure!

Gray pieces look like rocks

A double curved slope and two eyes are the submerged body of a lake monster!

Open fire is a flame piece on a 1x1 round plate

SKILL LEVEL

13

Two 1x4 tiles form the step

8x8 plate cabin floor

ON THE LAKE

Green and blue plates form the pretty lakeside location of the log cabin. The bottom of the cabin is made from gray bricks that look like rocks on the lake shore.

6x10 plate "lake"

1x2x3 window frame

1x1x5 brick

1x2 plate with rail windowsill matches the window

Door with four panes

FENCE AND WALLS

Next, build the porch fence and the wooden walls of the cabin, leaving space for the door and window. The walls are made from a mixture of plain and ridged bricks to create a natural-looking texture.

1x6 roof tile

6x8 plate roof base

1x6 plates under some of the roof tiles make them higher

If you don't have these printed 1x1 tiles, plain tiles would work too

I love the scent of freshly cut wood.

The roof rests on this 1x10 plate

INTERLOCKING LOGS

Fill the gaps at the sides of the cabin walls with stacked headlight bricks placed at right angles to each other. Add round plates and printed log tiles to create the interlocking log effect. Complete your cabin by building porch posts and a red roof.

You could add a lake monster or keep the water calm!

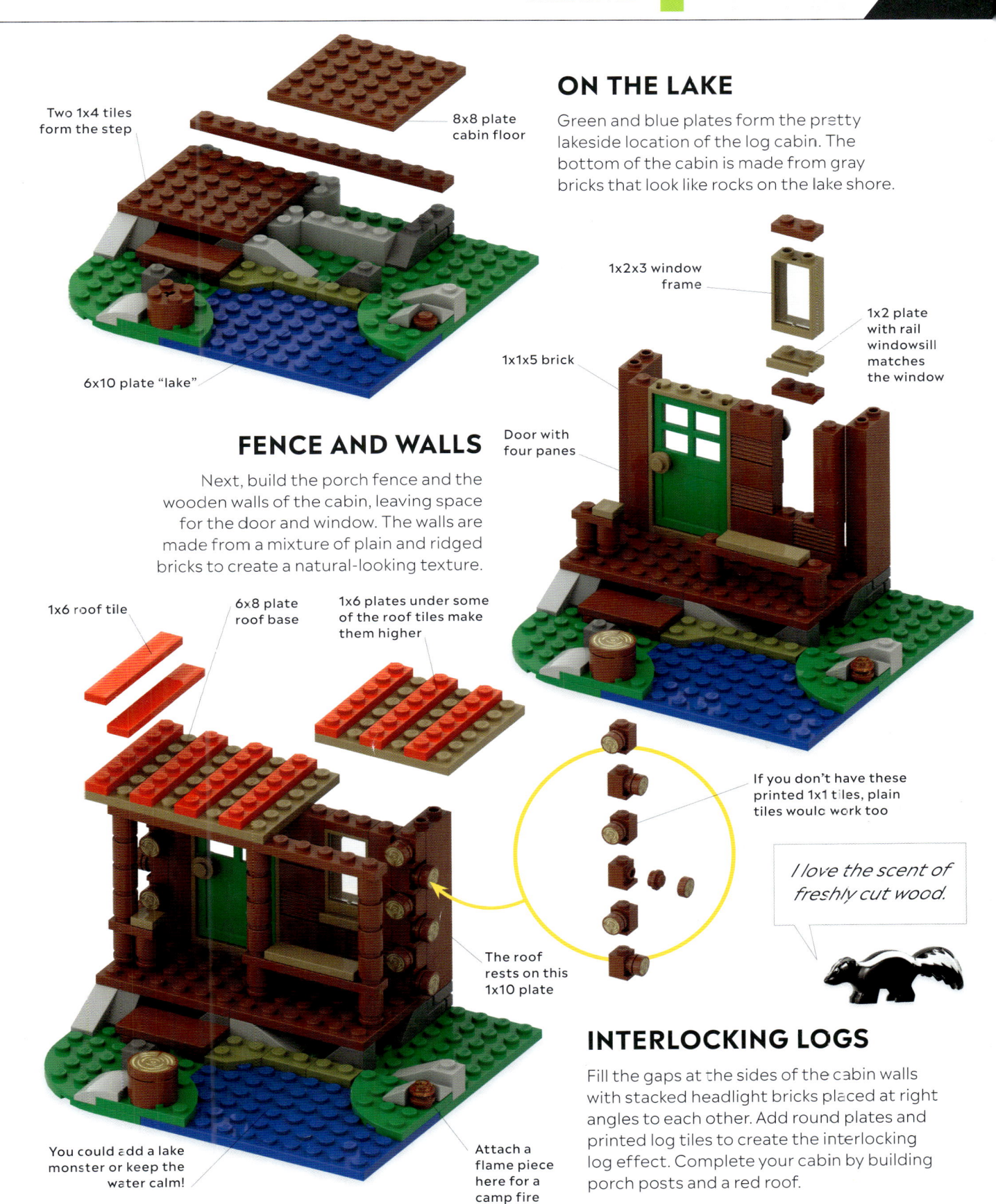

Attach a flame piece here for a camp fire

COLORS AND TEXTURES

Plain LEGO walls look great and serve their purpose perfectly, but building in more texture and color can take your building to the next level. Try out some of these wall techniques to give your homes a more natural or unusual look.

SIMPLE CHANGES

Adding just one other color to a plain wall can make a big difference to the look of it. This country cottage has lots of "palisade" or log bricks in a color (nougat) that works well with yellow. The curved edges of the palisade bricks create texture too.

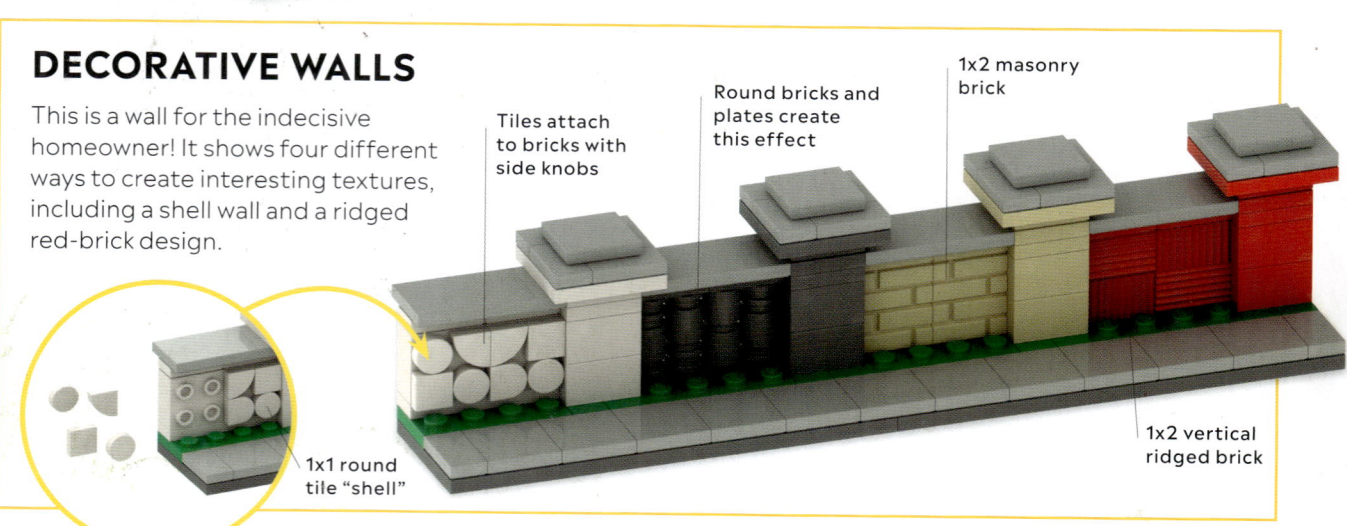

The roof slopes are two shades of red

1x2 palisade brick

Place the different colors at irregular intervals

DECORATIVE WALLS

This is a wall for the indecisive homeowner! It shows four different ways to create interesting textures, including a shell wall and a ridged red-brick design.

1x1 round tile "shell"

Tiles attach to bricks with side knobs

Round bricks and plates create this effect

1x2 masonry brick

1x2 vertical ridged brick

IDEAS GALLERY

Pink bricks give the same wall a totally different look

OFFSET BRICKS

Most walls aren't perfectly even, especially walls on older buildings. This offsetting technique makes it look like some bricks in the wall stick out more than others.

This brown flower stem looks like dried-up flowers!

It's all about texture.

This "brick" is actually a 1x2 tile attached sideways

OLD WALLS

The thatched cottage's well-worn walls have lots of character. They are constructed from bricks in a variety of shapes and sizes. Some of the bricks have clips and side knobs so other pieces can attach sideways to them.

Find out more about how to make this cottage on pages 52–53

2x2 round tile attaches to side knobs in the wall

I say it's all about color.

Add round bricks next to square ones for texture

BEACH HOUSE

Narrow tiles make a flat rooftop

Looks like he's having a swell time!

Stacked round bricks are porch fence posts

Darker tan plates are wet sand

Transparent 1x1 slopes look like breaking waves

Perched right on the seashore, this small wooden beach hut is the perfect place for a minifigure to relax on a sunny day. It's raised off the sand on wooden stilts so it won't be flooded when high tide comes.

SKILL LEVEL

STILTED START

Begin the beach hut with sandy- and sea-colored 8x16 plates, then plot out the shape of your hut by adding the stilts and stair pieces. A 2x4 brick will provide extra stability for the beach hut's floor.

1x2 plate and brick stairs

Rounded plates create an uneven shoreline

Blue 8x16 plate is the sea

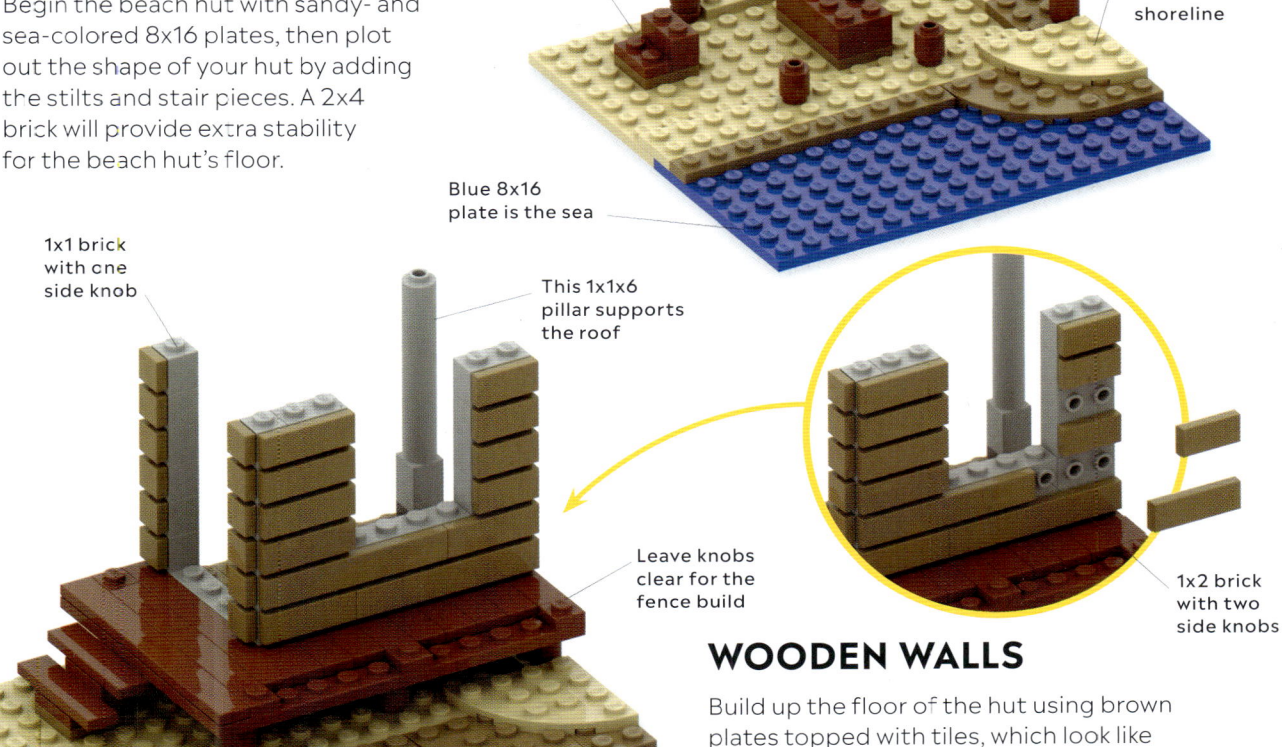

1x1 brick with one side knob

This 1x1x6 pillar supports the roof

Leave knobs clear for the fence build

1x2 brick with two side knobs

WOODEN WALLS

Build up the floor of the hut using brown plates topped with tiles, which look like smooth planks of wood. The walls come next—they're made from horizontal tiles attached to bricks with side knobs.

3x4 slope brick

1x4 tile step

Life buoy hangs from a 1x1 plate with bar

This 2x4 brick supports the second layer of the roof

SHORELINE SHELTER

Finally, add a window, door, and roof to give your minifigures shelter from the sea breeze. The roof has sides made from slope bricks. To finish off, add a second layer of slope bricks and top them off with tiles.

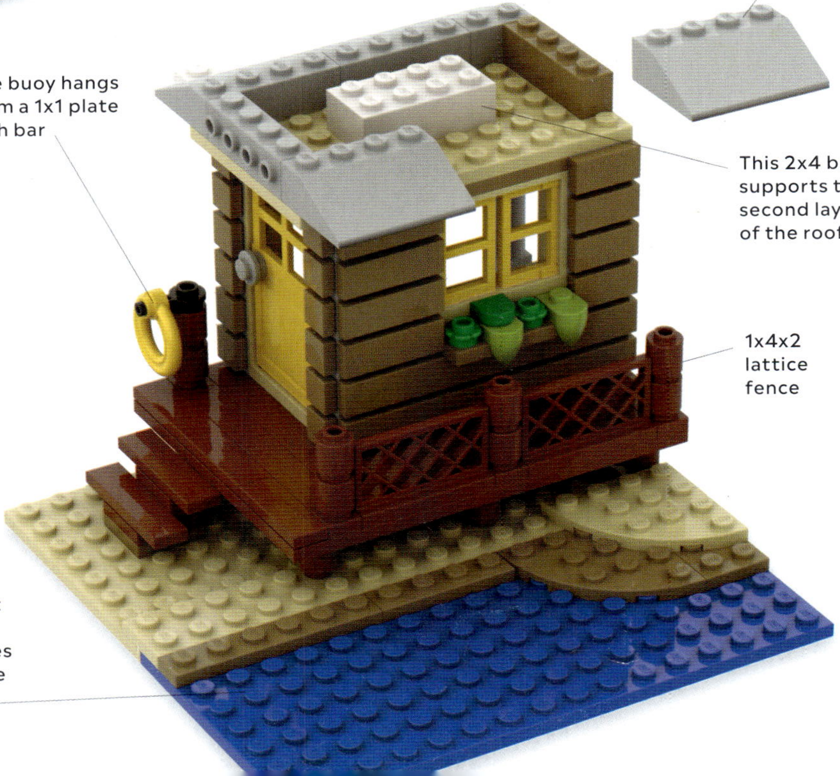

1x4x2 lattice fence

Nip nip!

Add transparent "wave" pieces and sea creatures here to bring the scene to life!

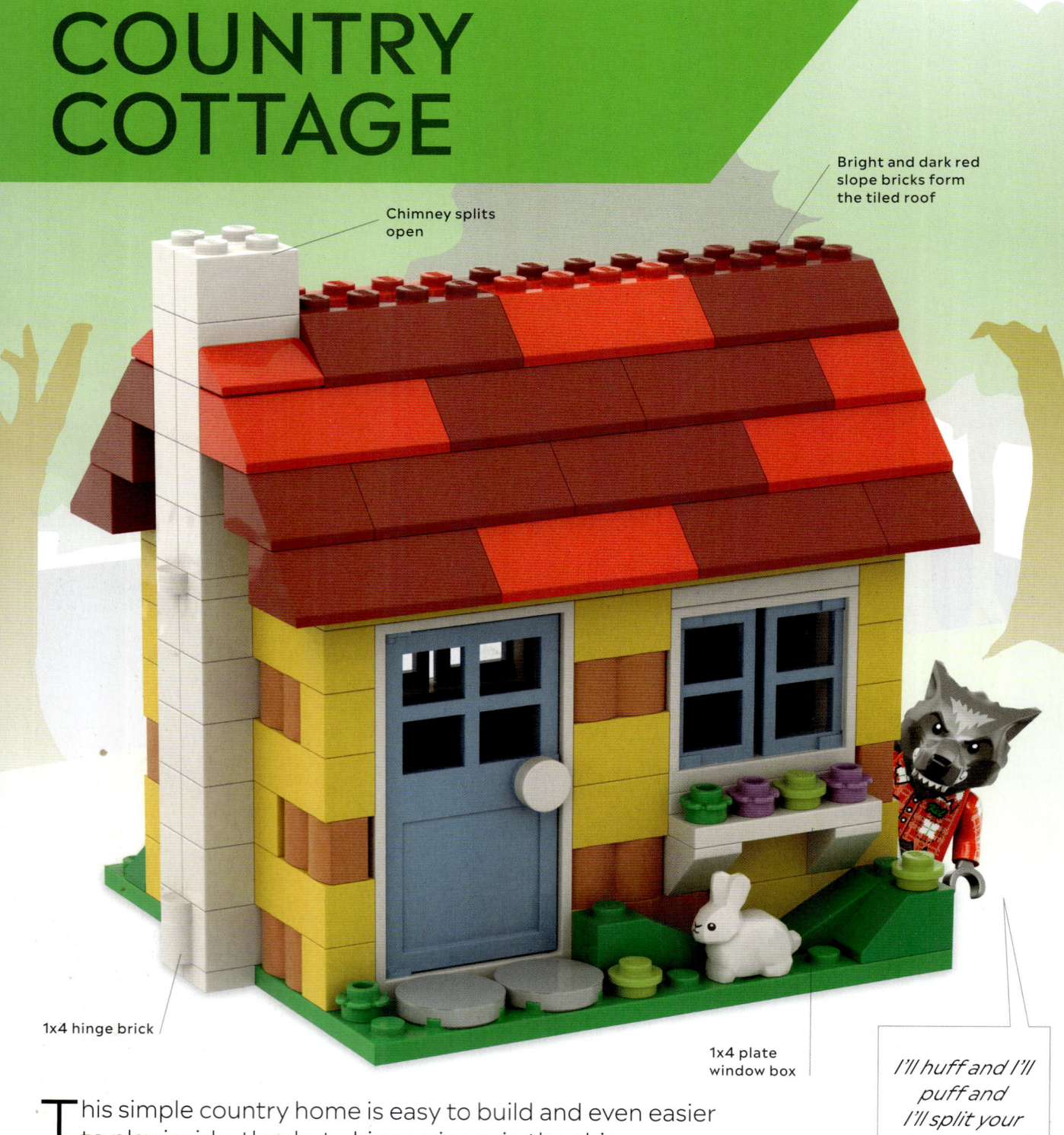

COUNTRY COTTAGE

Bright and dark red slope bricks form the tiled roof

Chimney splits open

1x4 hinge brick

1x4 plate window box

I'll huff and I'll puff and I'll split your house open!

This simple country home is easy to build and even easier to play inside, thanks to hinge pieces in the chimney. They allow the front and back walls of the house to open up. The back part of the house has a matching blue window and another window box filled with flowers.

SKILL LEVEL | 19

TWO SIDES

The country cottage is built on two identical 6x12 plates. Plan out the shape of the home by attaching the lowest bricks of the walls to each plate. Also make a start on the splitting chimney.

1x4 hinge brick swivel top

1x4 hinge brick swivel base

2x2 round tile doorstep

Small slopes are long grass in the yard

1x1 plate sits under half of the hinge brick

1x4 brick

1x4x3 window with shutters

2x4 plate windowsill

Stacked 1x1 bricks

MATCHING WINDOWS

The yellow and nougat brick walls of the cottage are taking shape, and it's time for the identical windows to be fitted into them. Build up the chimney at the same time as the walls.

Inverted 1x2 bricks support the windowsills

Each step of the side wall is one knob wide

2x4 slope brick

1x2 brick

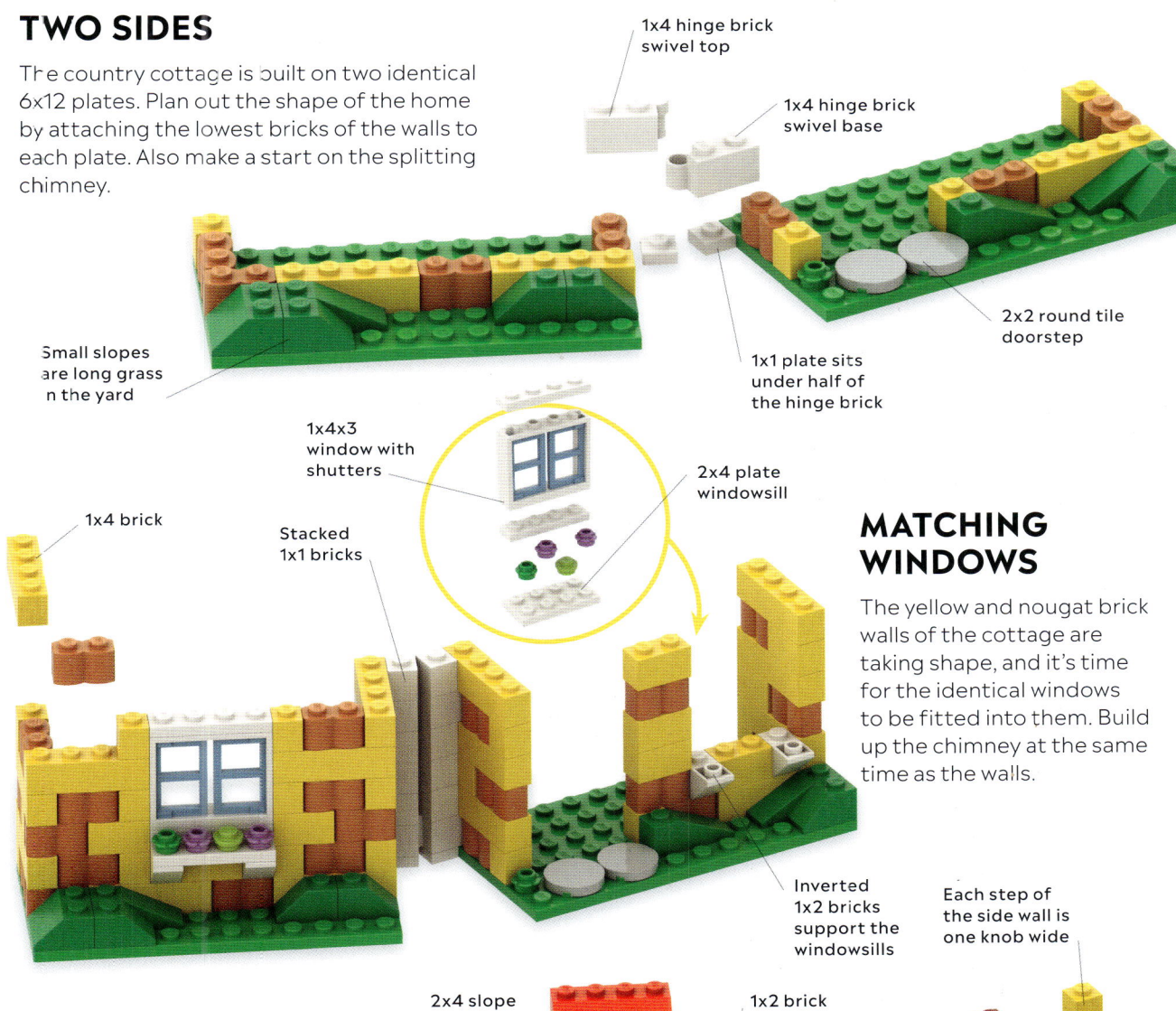

STEPPED ROOF

Once the windows and front door are in place, begin the roof section. This simple roof is mostly made from 2x4 slope bricks. They are supported by small, stepped bricks on either side of both parts of the cottage.

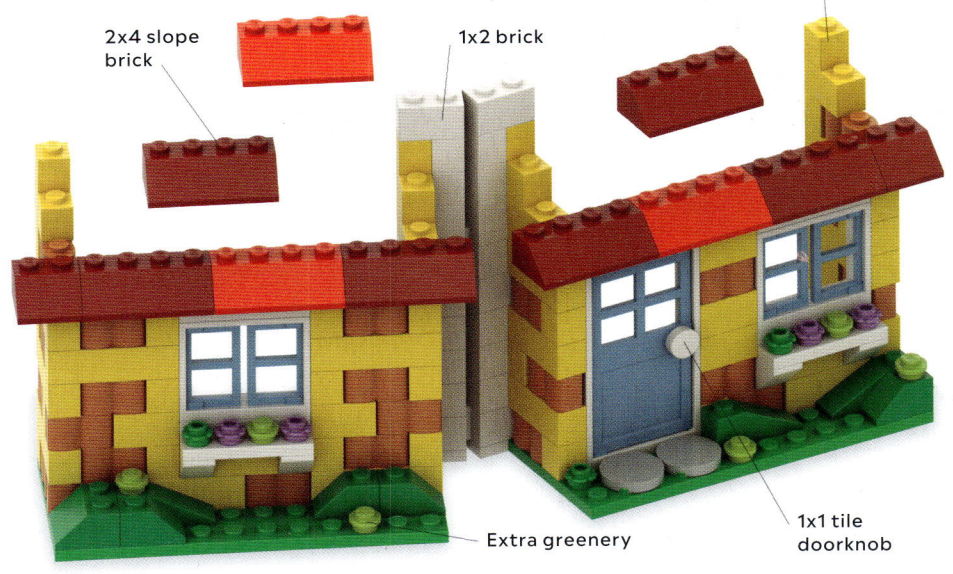

Extra greenery

1x1 tile doorknob

BUNGALOW

This charming microscale bungalow is built using small bricks, tiles, and plates in inventive ways. It's small in size but big on details, with a wooden front door, a perfectly paved driveway, and a colorful yard with manicured hedges.

2x2 double slope bricks and other sloped pieces form the top layer of the gable roof

Flower blooms made from 1x1 round plates with petals

1x1 round tiles form the winding path

Tan bricks and plates create a natural stone effect

The driveway is made from four 1x2 grille tiles

Hedges are 1x2 tiles with vertical teeth

This house is the perfect size for my rubber duck!

SKILL LEVEL

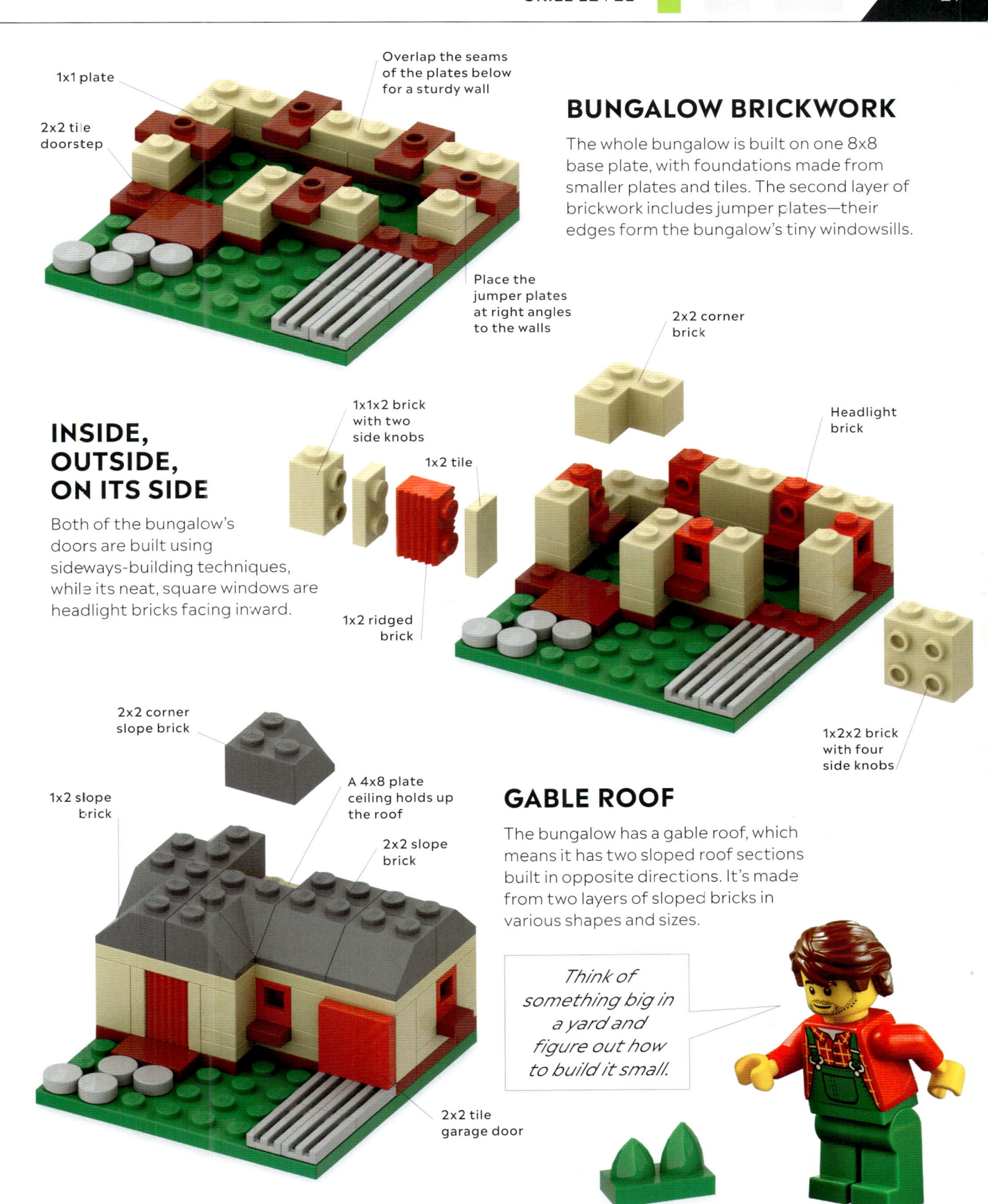

- 1x1 plate
- 2x2 tile doorstep
- Overlap the seams of the plates below for a sturdy wall

BUNGALOW BRICKWORK

The whole bungalow is built on one 8x8 base plate, with foundations made from smaller plates and tiles. The second layer of brickwork includes jumper plates—their edges form the bungalow's tiny windowsills.

- Place the jumper plates at right angles to the walls
- 1x1x2 brick with two side knobs
- 1x2 tile
- 2x2 corner brick
- Headlight brick
- 1x2 ridged brick
- 1x2x2 brick with four side knobs

INSIDE, OUTSIDE, ON ITS SIDE

Both of the bungalow's doors are built using sideways-building techniques, while its neat, square windows are headlight bricks facing inward.

- 2x2 corner slope brick
- 1x2 slope brick
- A 4x8 plate ceiling holds up the roof
- 2x2 slope brick
- 2x2 tile garage door

GABLE ROOF

The bungalow has a gable roof, which means it has two sloped roof sections built in opposite directions. It's made from two layers of sloped bricks in various shapes and sizes.

Think of something big in a yard and figure out how to build it small.

LIGHTHOUSE

Ships watch for this lighthouse's bright, flashing beacon of light at night. It tells them where land is so they can steer clear of hazardous rocks. The lighthouse's red-and-white striped tower has a lantern room for its powerful lamp at the top and an adjoining lighthouse keeper's cottage at the bottom.

Lantern room balcony

Red and white stripes are easy for ships to spot

Hey, I've just washed that roof! Pesky gulls.

Gray slopes are craggy rocks

Small transparent pieces look like swirling water

SKILL LEVEL

ROCKY BASE

The lighthouse is built on a base that's half-land, half-sea thanks to large dark gray and blue plates. The sharp, uneven-looking rocks on the coastline are made from slope pieces in different sizes, placed at right angles.

1x4 bricks build up the base for the lighthouse

1x3 slope brick

1x2x5 bricks are the glass panes of the lantern room

6x16 "sea" plate

Stacked 2x2 round bricks are the lamp

LAND AHOY!

After adding a layer of gray plates onto the rocky base, add grass and a paved path on the land. Then begin building the bottom of the lighthouse and the walls of the keeper's cottage.

This layer of plates supports the roof

Simple walls made from blue bricks

Each layer of the lighthouse is built in a similar way

2x2 macaroni bricks form the curved edges of the tower

2x2 double concave slope

Black bricks at the bottom match the lighthouse

Overhanging curved bricks make the coastline look more natural

BEACON BUILD

Finish off the keeper's cottage with a black roof, then build up the lighthouse tower. Once it's tall enough to see for miles out to sea, build the light-filled lantern room at the top.

FAIRY HOUSE

Never step on a toadstool while out walking in the woods—it might be a fairy's home, like this one! The mushroom's recognizable red cap with white spots is its tiny roof, with a fairy-built chimney poking out. There's a large door built into the mushroom's stem for welcoming tiny visitors.

3x3 radar dish is the chimney top

"Cheese" slopes and curved slopes form the top layer of the roof

Round-top door with latticed window

So this next house is one of our most unique properties.

It's nice, but there's not mush-room.

Green base plates are the forest floor

1x1 plates with petals are tiny plants or moss

SKILL LEVEL

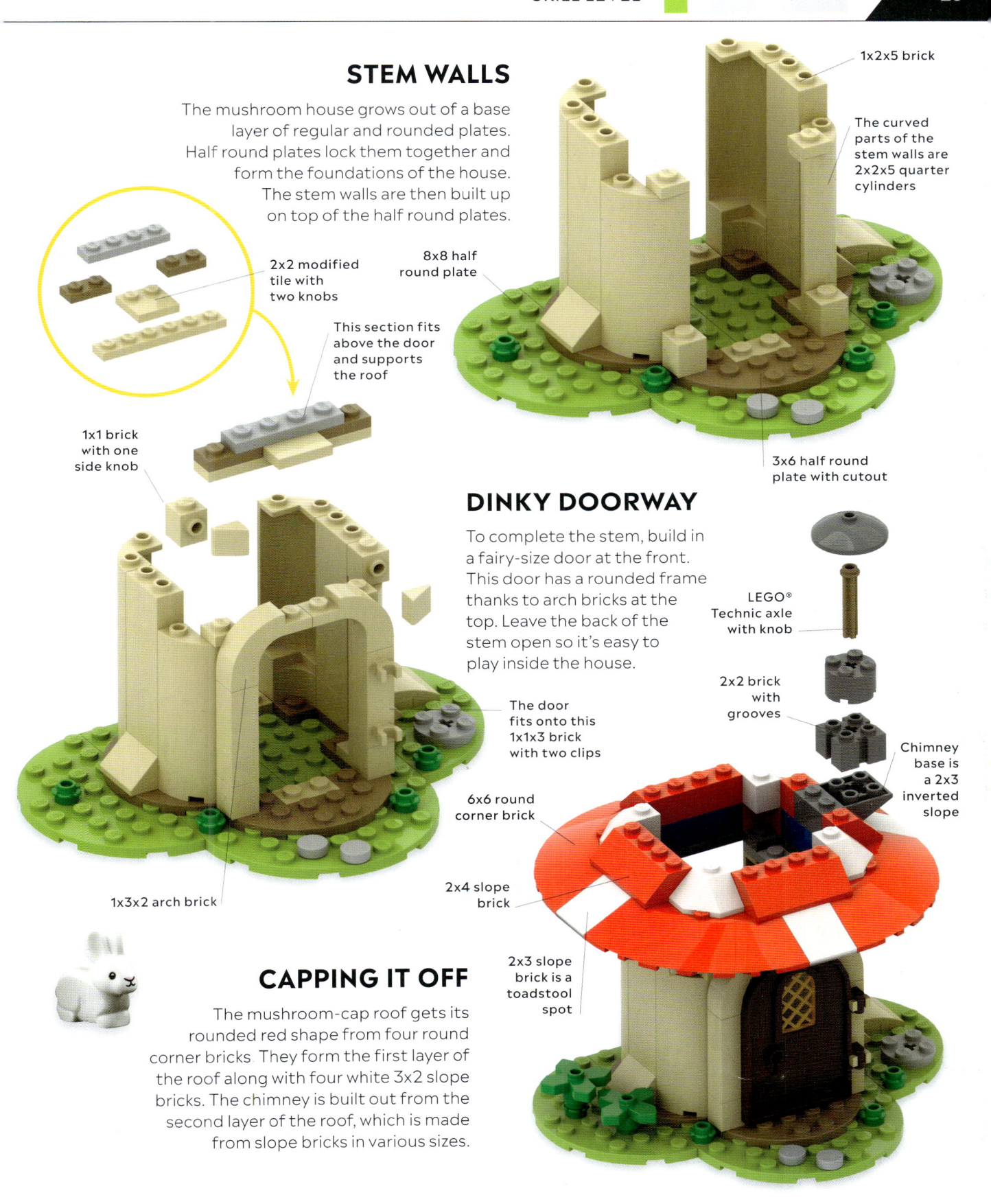

25

STEM WALLS

The mushroom house grows out of a base layer of regular and rounded plates. Half round plates lock them together and form the foundations of the house. The stem walls are then built up on top of the half round plates.

1x2x5 brick

The curved parts of the stem walls are 2x2x5 quarter cylinders

2x2 modified tile with two knobs

8x8 half round plate

This section fits above the door and supports the roof

1x1 brick with one side knob

3x6 half round plate with cutout

DINKY DOORWAY

To complete the stem, build in a fairy-size door at the front. This door has a rounded frame thanks to arch bricks at the top. Leave the back of the stem open so it's easy to play inside the house.

LEGO® Technic axle with knob

2x2 brick with grooves

The door fits onto this 1x1x3 brick with two clips

Chimney base is a 2x3 inverted slope

6x6 round corner brick

2x4 slope brick

1x3x2 arch brick

2x3 slope brick is a toadstool spot

CAPPING IT OFF

The mushroom-cap roof gets its rounded red shape from four round corner bricks. They form the first layer of the roof along with four white 3x2 slope bricks. The chimney is built out from the second layer of the roof, which is made from slope bricks in various sizes.

FESTIVE HOUSE

This cozy home is a welcome sight in the cold and dark days and nights of winter. Illuminated with colorful lights, a freshly built snowman, and frosted foliage, it's beautifully decorated inside and out for the festive season.

White pieces in the roof look like snow

Colorful string lights are light bulb pieces

External chimney promises a warming fire inside

Outdoor Christmas tree

1x1 round tiles create snow-covered plants

Have you seen the broom?

SKILL LEVEL

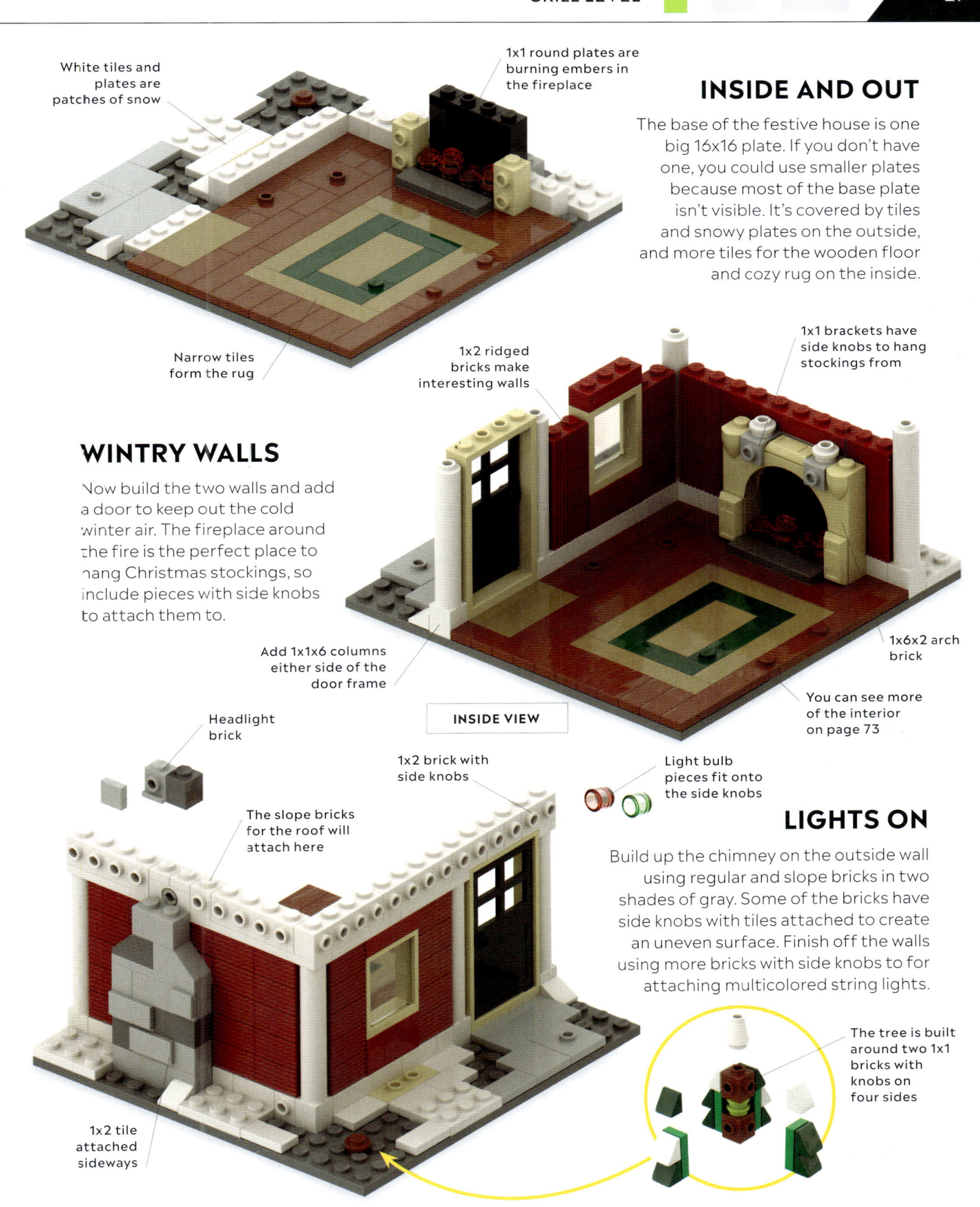

INSIDE AND OUT

The base of the festive house is one big 16x16 plate. If you don't have one, you could use smaller plates because most of the base plate isn't visible. It's covered by tiles and snowy plates on the outside, and more tiles for the wooden floor and cozy rug on the inside.

- White tiles and plates are patches of snow
- 1x1 round plates are burning embers in the fireplace
- Narrow tiles form the rug

WINTRY WALLS

Now build the two walls and add a door to keep out the cold winter air. The fireplace around the fire is the perfect place to hang Christmas stockings, so include pieces with side knobs to attach them to.

- 1x2 ridged bricks make interesting walls
- 1x1 brackets have side knobs to hang stockings from
- Add 1x1x6 columns either side of the door frame
- 1x6x2 arch brick
- You can see more of the interior on page 73

INSIDE VIEW

LIGHTS ON

Build up the chimney on the outside wall using regular and slope bricks in two shades of gray. Some of the bricks have side knobs with tiles attached to create an uneven surface. Finish off the walls using more bricks with side knobs to for attaching multicolored string lights.

- Headlight brick
- 1x2 brick with side knobs
- Light bulb pieces fit onto the side knobs
- The slope bricks for the roof will attach here
- 1x2 tile attached sideways
- The tree is built around two 1x1 bricks with knobs on four sides

VIKING LONGHOUSE

Minifigures of the Viking Age lived in longhouses. Many Viking families lived, slept, and ate together in one big living space alongside their farm animals. This longhouse has a narrow wooden frame and a turf roof that keeps it warm in freezing temperatures.

Layers of olive and lime green plates form the turf roof

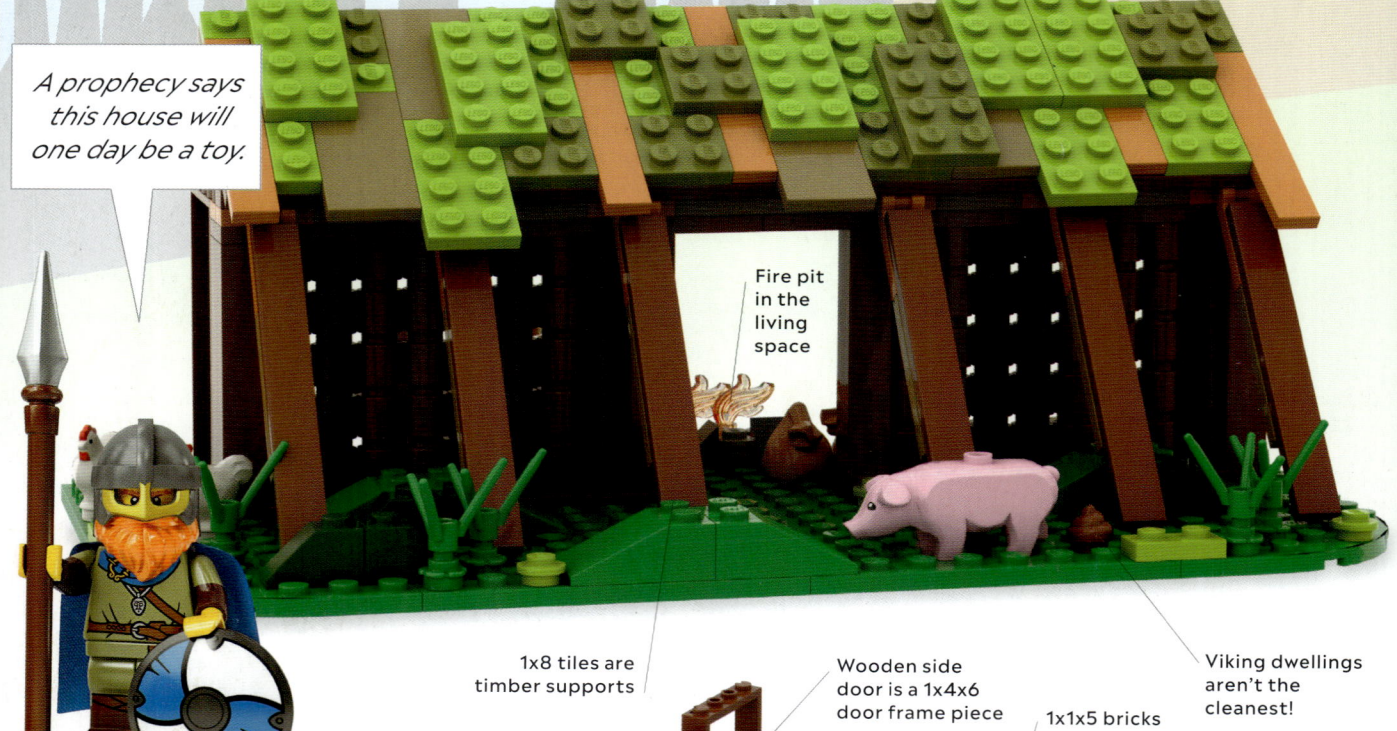

A prophecy says this house will one day be a toy.

Fire pit in the living space

1x8 tiles are timber supports

Wooden side door is a 1x4x6 door frame piece

1x1x5 bricks will be part of the front doorway

Viking dwellings aren't the cleanest!

1x1x3 brick

LONG FRAME

A longhouse needs a long base. This one is made by combining four plate pieces. Mounds of turf cover the front half of the base, and the timber frame of the house covers the back.

Two 2x2 slope bricks make a turf mound

6x6 quarter round plate

SKILL LEVEL

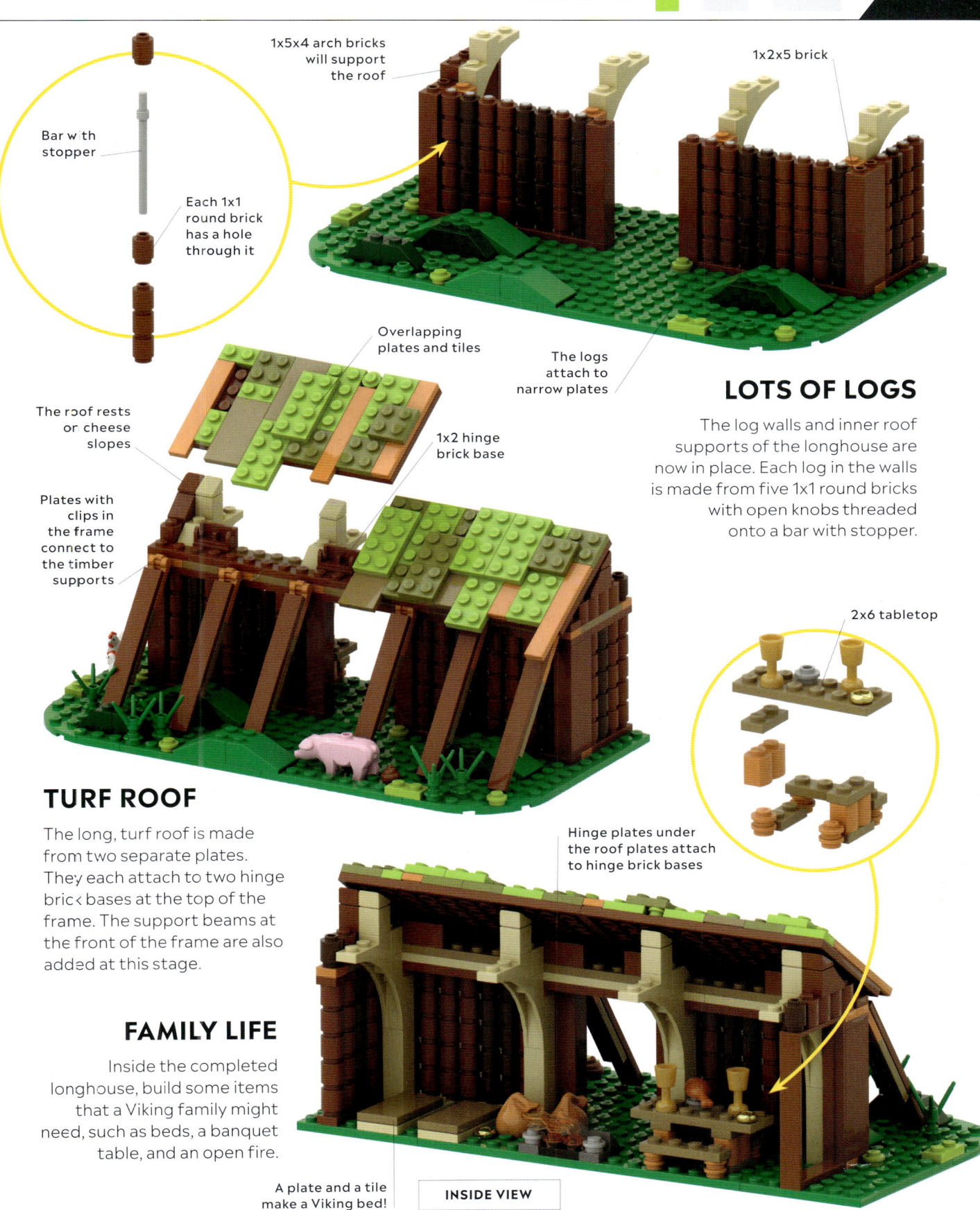

Bar with stopper

Each 1x1 round brick has a hole through it

1x5x4 arch bricks will support the roof

1x2x5 brick

Overlapping plates and tiles

The logs attach to narrow plates

LOTS OF LOGS

The log walls and inner roof supports of the longhouse are now in place. Each log in the walls is made from five 1x1 round bricks with open knobs threaded onto a bar with stopper.

The roof rests on cheese slopes

1x2 hinge brick base

Plates with clips in the frame connect to the timber supports

2x6 tabletop

TURF ROOF

The long, turf roof is made from two separate plates. They each attach to two hinge brick bases at the top of the frame. The support beams at the front of the frame are also added at this stage.

Hinge plates under the roof plates attach to hinge brick bases

FAMILY LIFE

Inside the completed longhouse, build some items that a Viking family might need, such as beds, a banquet table, and an open fire.

A plate and a tile make a Viking bed!

INSIDE VIEW

SKI CHALET

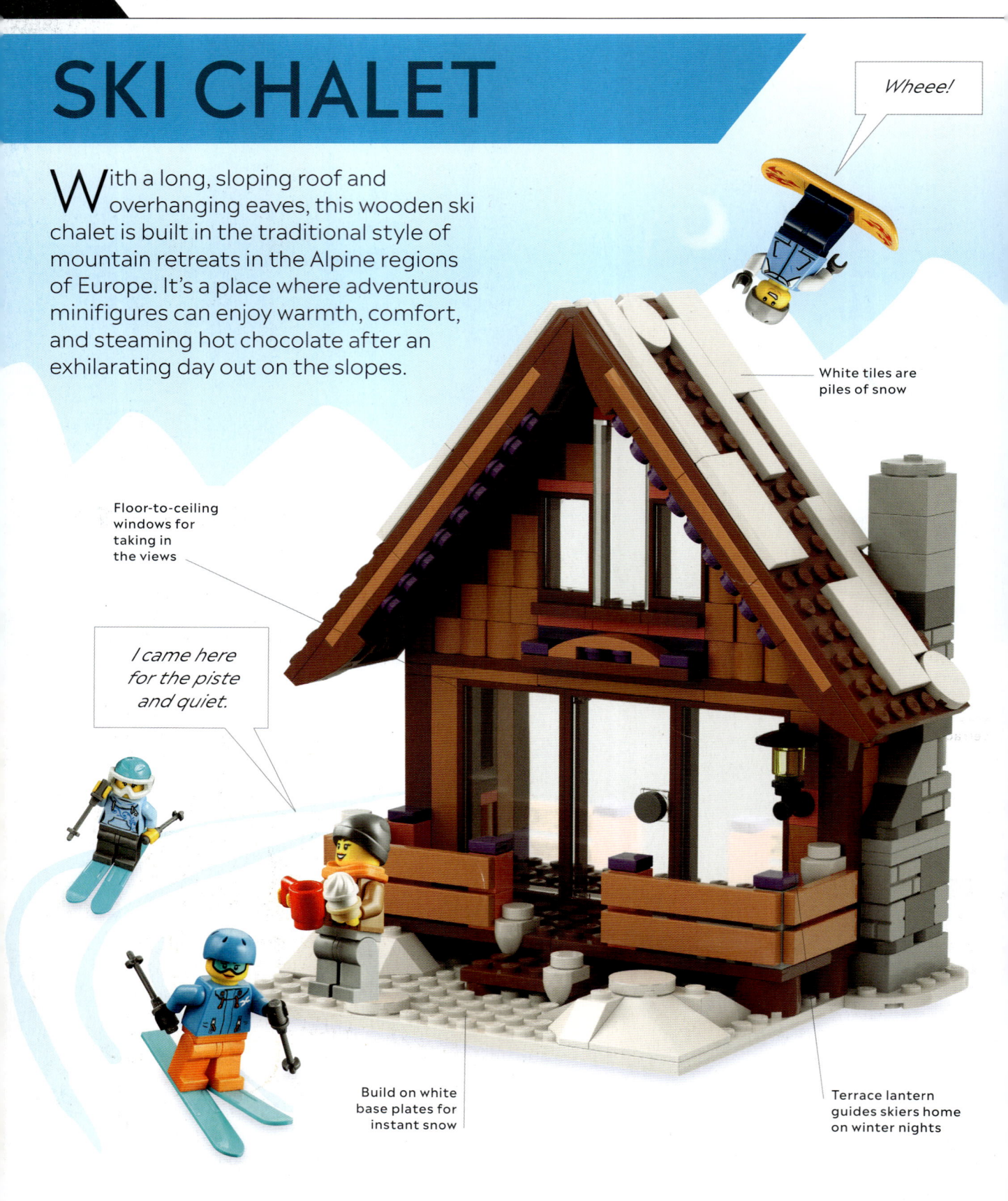

With a long, sloping roof and overhanging eaves, this wooden ski chalet is built in the traditional style of mountain retreats in the Alpine regions of Europe. It's a place where adventurous minifigures can enjoy warmth, comfort, and steaming hot chocolate after an exhilarating day out on the slopes.

Wheee!

White tiles are piles of snow

Floor-to-ceiling windows for taking in the views

I came here for the piste and quiet.

Build on white base plates for instant snow

Terrace lantern guides skiers home on winter nights

SKILL LEVEL 31

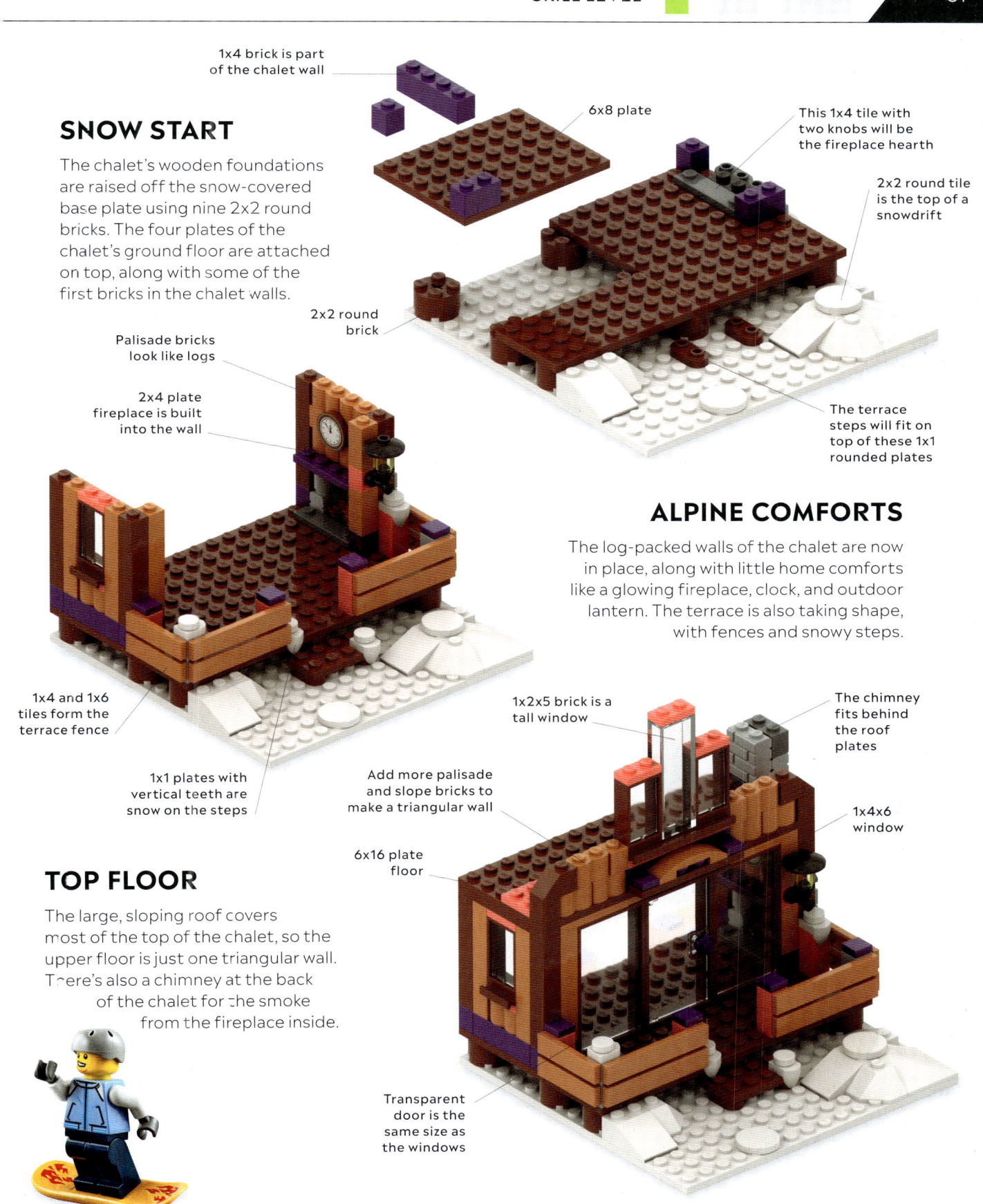

SNOW START

The chalet's wooden foundations are raised off the snow-covered base plate using nine 2x2 round bricks. The four plates of the chalet's ground floor are attached on top, along with some of the first bricks in the chalet walls.

1x4 brick is part of the chalet wall

6x8 plate

This 1x4 tile with two knobs will be the fireplace hearth

2x2 round tile is the top of a snowdrift

2x2 round brick

The terrace steps will fit on top of these 1x1 rounded plates

Palisade bricks look like logs

2x4 plate fireplace is built into the wall

ALPINE COMFORTS

The log-packed walls of the chalet are now in place, along with little home comforts like a glowing fireplace, clock, and outdoor lantern. The terrace is also taking shape, with fences and snowy steps.

1x4 and 1x6 tiles form the terrace fence

1x1 plates with vertical teeth are snow on the steps

1x2x5 brick is a tall window

The chimney fits behind the roof plates

Add more palisade and slope bricks to make a triangular wall

1x4x6 window

6x16 plate floor

TOP FLOOR

The large, sloping roof covers most of the top of the chalet, so the upper floor is just one triangular wall. There's also a chimney at the back of the chalet for the smoke from the fireplace inside.

Transparent door is the same size as the windows

MODERN HOUSE

It's no wonder this homeowner looks perfectly content in her two-story modern home. It has solar panels that provide power and heat, sweet-scented flower boxes, and a rooftop terrace for watching the world go by.

I'm sure my coffee tastes better up here.

Side windows create light-filled spaces inside

Using different shades of green gives the plants a natural look

Vertical tooth plates make lovely hanging plants!

White column pieces support the terrace and door canopy

Palisade bricks make tidy flower boxes

Build onto a green base plate to create a neatly mowed lawn

The door canopy colors match the roof

SKILL LEVEL   33

TEXTURED WALLS

The modern house's walls are made from bricks in different colors and textures to give it a natural look. The white-framed windows on the front and side walls sit at the perfect height for a minifigure to peer from.

These 1x2 bricks have grooves to look like house bricks

These columns support the first-floor ceiling

Windowsills are 1x2 plates with rails

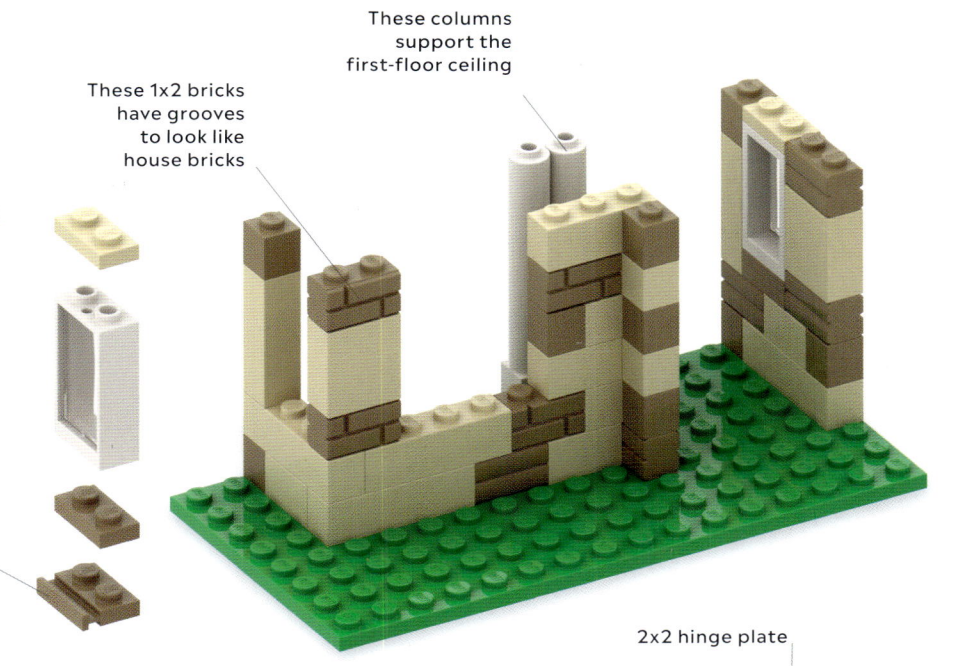

SECOND FLOOR

Two plate pieces form the first-floor ceiling of the modern house. The walls of the second floor and the fences of the terrace are built on top of them.

1x4x2 arch brick forms the top of this doorway

Swiveling terrace chair sits on a 2x2 turntable plate

This 1x1x3 brick supports the roof

2x2 hinge plate

1x2 hinge brick base on top of a 1x2 brick

Solar panels are two printed 1x4 tiles

1x3 slope brick

3x4 slope brick

Terrace trellises made from 1x4x2 fence pieces

SOLAR-PANEL ROOF

The stepped roof can be built separately from the main home, on a 4x6 plate. The solar panels rest at an angle on the roof thanks to a hinge brick and plate connection.

SUBURBAN STREET

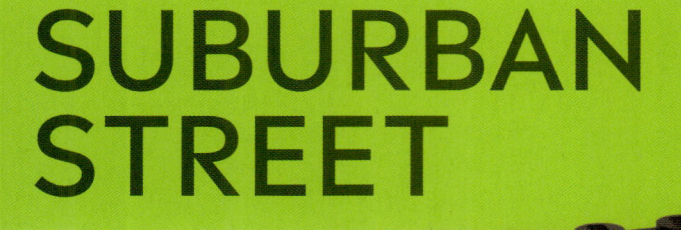

Treetops are 2x2 flower pieces attached upside down

The road is made from dark gray tiles

This neat microscale street has a row of six matching houses in complementary colors. At this size, ice-cream-scoop pieces become bushes and panel pieces are boundary walls. Each of the four-knob-wide houses is built separately so they can be arranged in different ways. What would your street look like in microscale?

STREET LEVEL

The base of the street is made from two long, narrow plates. They're locked together in the middle with tile pieces that will become a crossroad on the final model.

Keep your street clean!

2x4 tile fits over the seam of the plates

6x16 plate

SKILL LEVEL

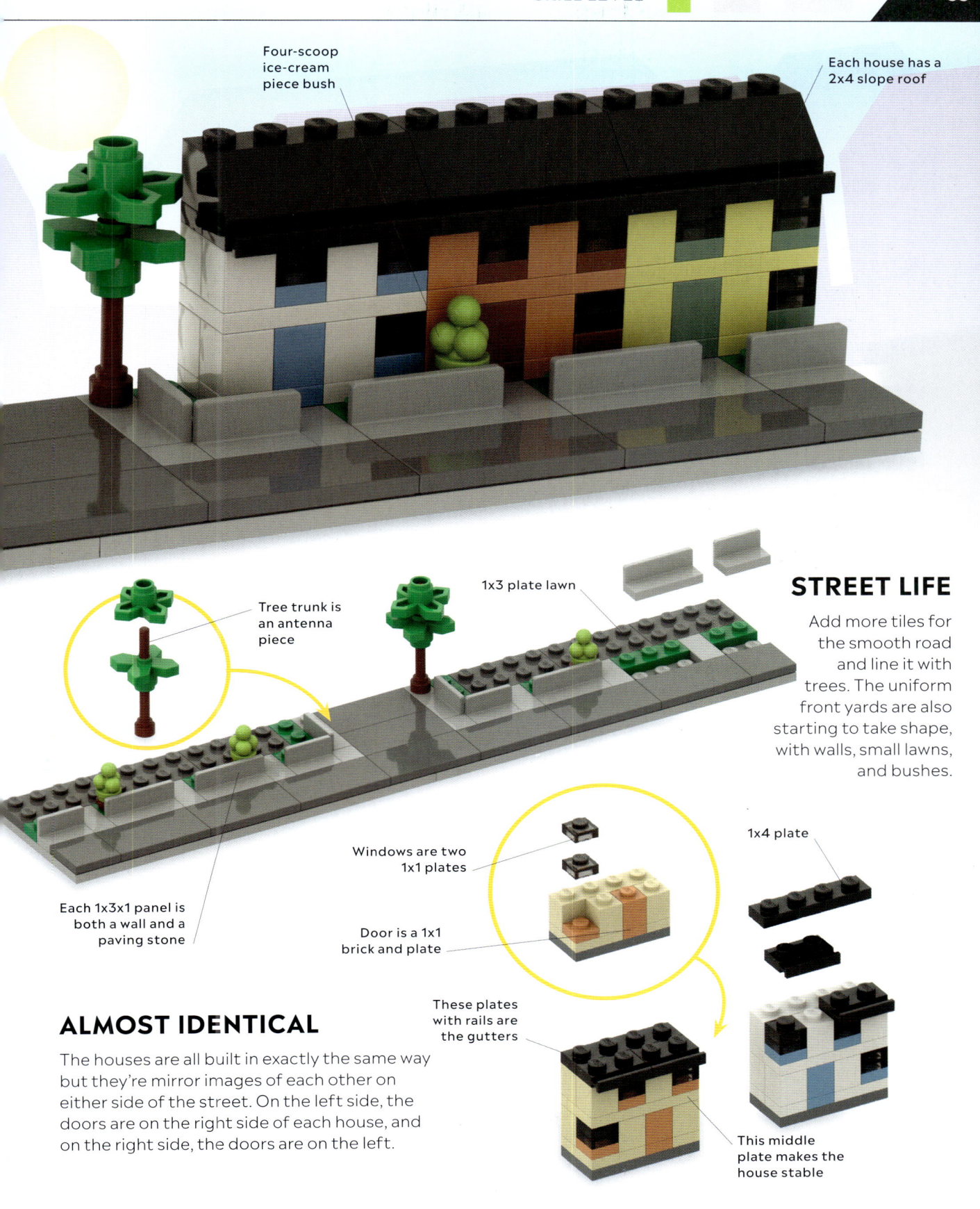

35

Four-scoop ice-cream piece bush

Each house has a 2x4 slope roof

1x3 plate lawn

STREET LIFE

Add more tiles for the smooth road and line it with trees. The uniform front yards are also starting to take shape, with walls, small lawns, and bushes.

Tree trunk is an antenna piece

Each 1x3x1 panel is both a wall and a paving stone

Windows are two 1x1 plates

Door is a 1x1 brick and plate

1x4 plate

These plates with rails are the gutters

This middle plate makes the house stable

ALMOST IDENTICAL

The houses are all built in exactly the same way but they're mirror images of each other on either side of the street. On the left side, the doors are on the right side of each house, and on the right side, the doors are on the left.

RAINBOW HOUSE

When the sun is shining and it's raining somewhere, a rainbow appears for a little while. It looks like one such arc of color has left its mark on this house, turning its roof and front yard red and yellow and purple and green ...

Small slopes form the tapered tip of the roof

2x2 slopes form the second layer of the roof

Parasol shields minifigures from the bright colors

2x2 curved slopes continue the rainbow's arc

This part of the front wall is one 1x4x5 wall panel

Make an outdoor seat from just two pieces

I feel instantly at home here.

SKILL LEVEL

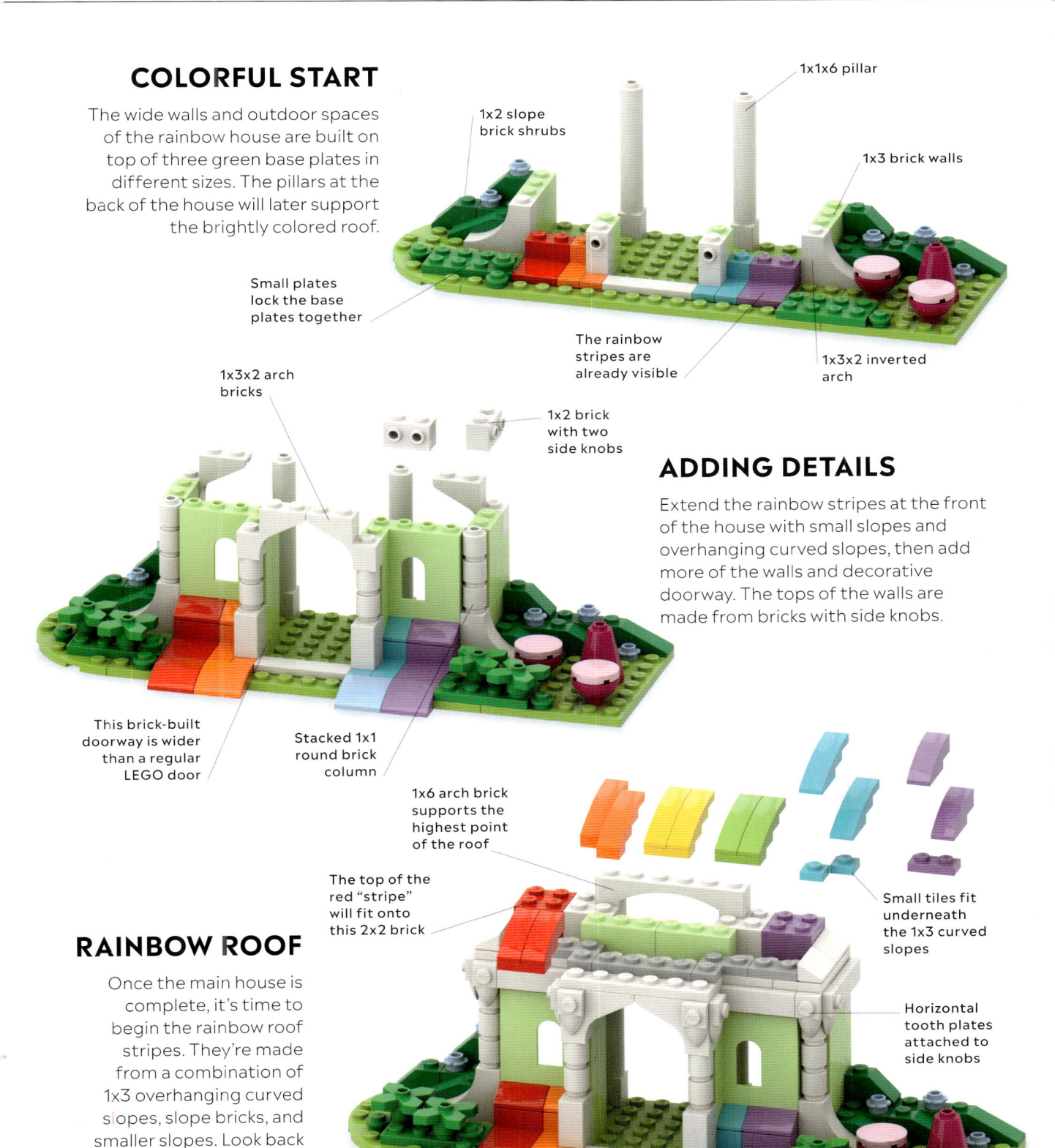

COLORFUL START

The wide walls and outdoor spaces of the rainbow house are built on top of three green base plates in different sizes. The pillars at the back of the house will later support the brightly colored roof.

1x1x6 pillar

1x2 slope brick shrubs

1x3 brick walls

Small plates lock the base plates together

The rainbow stripes are already visible

1x3x2 inverted arch

1x3x2 arch bricks

1x2 brick with two side knobs

ADDING DETAILS

Extend the rainbow stripes at the front of the house with small slopes and overhanging curved slopes, then add more of the walls and decorative doorway. The tops of the walls are made from bricks with side knobs.

This brick-built doorway is wider than a regular LEGO door

Stacked 1x1 round brick column

1x6 arch brick supports the highest point of the roof

The top of the red "stripe" will fit onto this 2x2 brick

Small tiles fit underneath the 1x3 curved slopes

RAINBOW ROOF

Once the main house is complete, it's time to begin the rainbow roof stripes. They're made from a combination of 1x3 overhanging curved slopes, slope bricks, and smaller slopes. Look back at the main picture to see the final roof build.

Horizontal tooth plates attached to side knobs

37

TOWNHOUSE

Blooming 1x4 plate window boxes

This two-story townhouse is a home found in many towns and cities around the world. It's made from two shades of red bricks, with a wide bay window on the first floor that lets in lots of light and provides the perfect place to relax. This home stands alone, but it could also form part of a row of similar townhouses.

Have a good day, Colin. Can you remember to pick up a cabbage for dinner?

Mailbox built into the wall

Empty milk bottles, ready for the milk delivery service

Part of the bay window will rest on this tile

1x1 brick with side knob fits onto the front wall

HOUSE BRICKS

The townhouse covers the entire width of its 8x16 plate base. The walls are a mixture of dark red and bright red bricks—a combination that creates a natural red-brick effect. The doorstep is in place already, as is an inverted slope that will hold up the bay window.

Smooth 1x4 tile doorstep

1x2 dark red masonry brick

SKILL LEVEL

DOWNSTAIRS

The first floor of the townhouse now has a door, a mailbox, and a bay window. The large window is made from three separate panels, each made from two window pieces. The three window panels are held in place with hinge plates.

This 2x4 plate built into the wall will support a door canopy

Two 1x2 plates fit above this window

This 6x16 plate is the downstairs ceiling and the upstairs floor

1x2x2 window

This part of the hinge plate attaches to the wall

1x4 hinge plate

1x6 tile

This base plate is 16 knobs wide, like the house

Slope bricks protrude at the front

Two 1x4 plates fit above the window

2x4 plate windowsill

Bay window canopy hides gaps above the window

2x2 curved slope canopy

UPSTAIRS

The top floor of the townhouse is now complete. It has two matching windows. Perhaps the three-panel one is for the primary bedroom and the two-panel one is for the second bedroom. A simple roof made from a 6x16 plate, regular and slope bricks, and tiles tops off the townhouse.

Porch light attaches to a knob in the wall

TOWNHOUSE
PAGES 38-39

Grandma, there's a lion and an Allosaurus chasing the ice-cream truck down the street!

That's nice, dear.

ICE-CREAM TRUCK
PAGES 206-207

MODEL MASH-UP

ALLOSAURUS
PAGES 252–253

LION
PAGES 134–135

SPORTS CAR
PAGES 186–187

They just ran that red light!

GINGERBREAD HOUSE

With its winding cookie path, candy-cane doorway, and frosted icing roof, this little home looks good enough to eat. But don't try that, or you might need a trip to the dentist—and it won't be because you ate too much sugar!

Vertical tooth plates make icing icicles

Candy-cane columns made from round bricks and plates

Tempting trees topped with ice-cream-scoop pieces

We just brought our baby home from the bakery!

Plates with rails work well as small windowsills

"Caramel" 1x1 round plate

This caramel-latticed door pane matches the windows

Leave space for the windows

Pink-iced fence posts are 1x1 round plates with swirled tops

1x1 tile with cookie pattern

2x2 corner slope

GINGERBREAD BEGINNINGS

Place cookie-patterned 1x1 tiles on a 16x16 base plate to form the sidewalk. Then create the iced outline of the house and start building the gingerbread and caramel walls.

SKILL LEVEL   43

- 1x4 arch brick
- 1x6x2 arch brick
- 1x1 round brick

SWEET DETAILS

The walls of a real gingerbread house would be joined at the corners with icing. Create that look by adding white corner columns made from stacked 1x1 round bricks. Then add the door frame arches and windows.

- The roof will rest on these 1x1 slopes
- 1x2 hinge brick
- 1x2x2 window with rounded top
- 1x1 round brick
- 1x1 plate with vertical tooth

DELICIOUS DOORWAY

Use a 4x6 plate to make the doorway canopy and add icing icicles and round tiles in the bright colors of hard candy. There are now smaller plates and slopes on top of the walls—these will support the house's upper wall and roof.

- Bar with stopper
- 1x2/1x2 bracket
- Small slope pieces give this wall a tapered shape
- 1x1 plate with vertical tooth
- 1x1 round tile with swirl pattern

ICED ROOF

The gingerbread roof is made in two parts that rest on the tapered top wall of the house. Both roof pieces are made from brown 6x10 plates trimmed with white brackets and vertical tooth plates to look like perfectly piped icing.

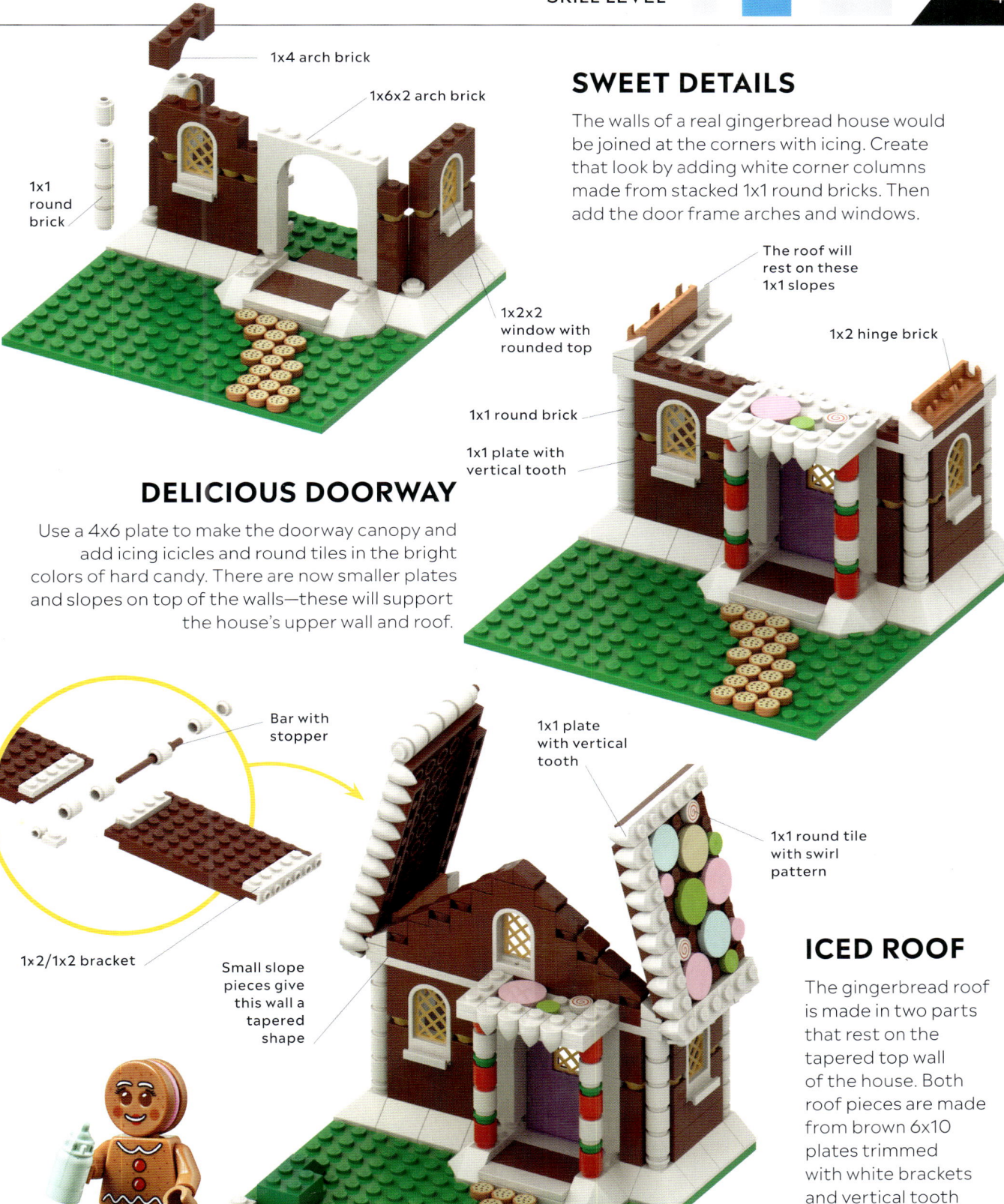

MODULAR APARTMENTS

In bustling towns and cities, many people live in apartment buildings like this one. The two separate, open-plan apartments in this building are stacked on top of each other, but they can also be joined at the side—this makes them modular builds. You could build even more apartments to make a tower!

I think I'll build my apartment right on top.

Details like this air-conditioning unit bring the building to life

The top floor has a roof terrace

Modern balcony railing made from window panels on their sides

Oh no, I can't ride it. It just looks cool!

Apartment mailboxes on the first floor

Neat floor made from tiles

Other apartment buildings could be attached here using LEGO Technic pins

SKILL LEVEL　　　45

BOTTOM UP

Each modular apartment is built on a square 16x16 plate, with walls made from a mixture of plain and textured masonry bricks. The first-floor apartment has a tiled front entrance with mailboxes built into the wall.

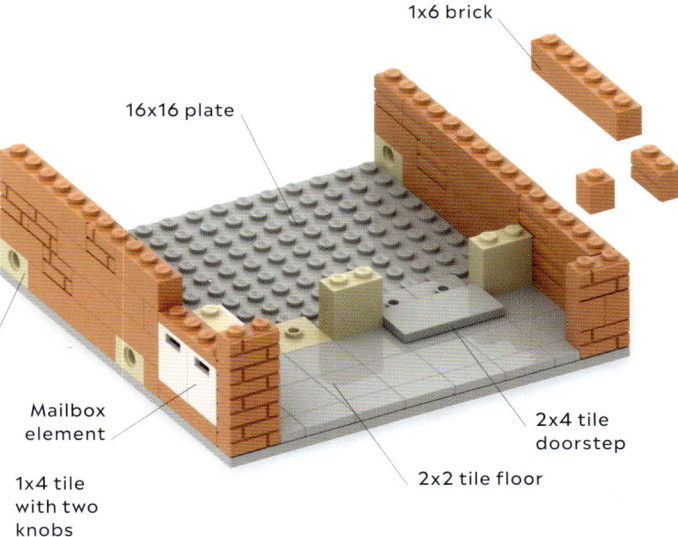

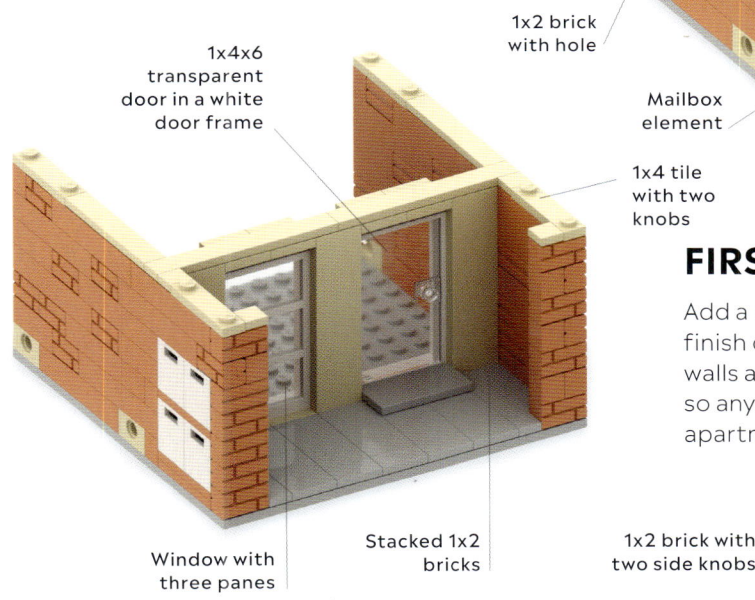

- 1x6 brick
- 16x16 plate
- 1x2 brick with hole
- Mailbox element
- 1x4 tile with two knobs
- 2x4 tile doorstep
- 2x2 tile floor
- 1x4x6 transparent door in a white door frame
- Window with three panes
- Stacked 1x2 bricks

FIRST FLOOR

Add a modern door and window and finish off the walls. The tops of the walls are lined with mostly smooth tiles, so any floors that are added above the apartment can easily be lifted off.

- 2x2 printed tile
- 1x4 brick side walls
- 1x2 brick with two side knobs
- 1x2 grille tile
- 1x4 brick with side knobs
- 1x1 round elbow brick
- 1x8 plate with rail
- 1x8 tile

ROOF

This square roof could be added to any number of modular apartments. Like the apartments, it's built on a 16x16 base plate. Add low walls around its sides to make it a roof terrace, and add practical details like pipes and air-conditioning units.

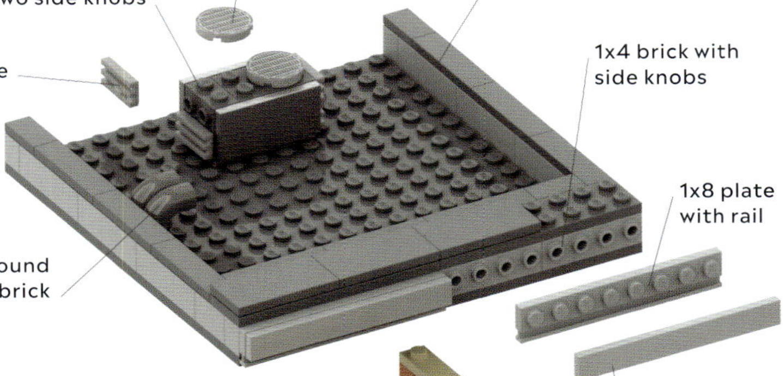

UPPER FLOOR

The second floor of the modular apartment block is built in the same way as the first-floor apartment, except it has a balcony with railings, and a potted plant in place of mailboxes.

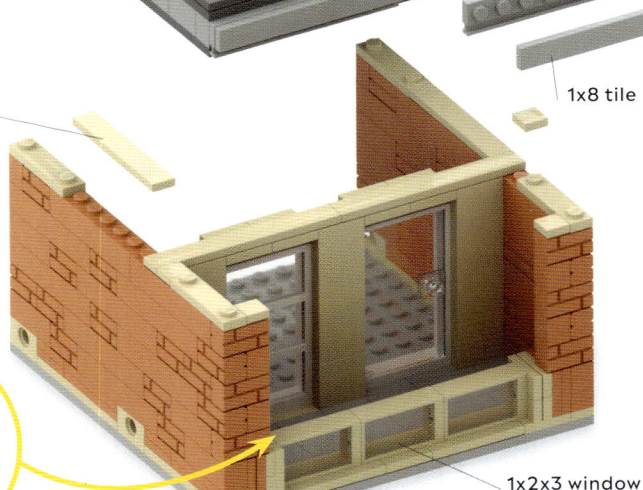

- 1x6 tile
- Stacked 1x1 round plates with three leaves
- 2x2 inverted dome pot
- 1x2x3 window

HILLSIDE VILLAGE

This should be easy enough to invade.

When building in microscale, you can literally have a whole world in your hands. This entire hilltop village—with leafy greenery, buildings of all shapes and sizes, and a gently winding road running through it—fits onto one 8x16 base plate.

Church made from 1x1 bricks and slopes

Trees made from 1x1 cones and bar pieces

1x1 double curved slope roof

1x4 palisade bricks add interesting texture

Bricks on the edge will be seen, so they're gray

GOING UPHILL

To begin the gently sloping hill, add bricks at the back, where the highest part of the hill will be, and plates at the front. Since most of the pieces on the bottom layer won't be seen, they can be any color you like.

This 1x8 plate won't be seen later

8x16 base plate

2x2 round plate

SKILL LEVEL

47

STEEP CLIMB
Now there are more bricks and plates on the hillside, in natural-looking grays, browns, and greens. The dark gray winding road is also starting to take shape. Two 1x10 curved slopes form the steepest parts of the road.

1x2 rounded plate

Bricks at the back add height

This is the highest point of the hill

These 1x1 cones will become a cluster of bushes

2x2 macaroni tile

LUSH LANDSCAPE
Complete the winding road with curved tiles, then add more greenery to the hillside using cones for bushes and round plates for tree-trunk bases. The gray round plates are rocky parts of the hillside.

This 1x4 tile is the only flat, straight part of the road

1x1 round plate rock

Build houses at different heights by using a mixture of plates and bricks

The bar piece trunk fits into this cone

You build it, I knock it down.

Stacked 1x1 bricks are the church spire

VILLAGE LIFE
Last come the buildings that bring the hillside village to life. Use some of the tiniest bricks you have, such as 1x1 bricks and cheese slopes, to create all kinds of homes and functional buildings.

1x1 double curved slopes make rounded roofs

PLAYHOUSE

This is one *fun* house! No one lives in this colorful wooden structure, but plenty of children visit it every day to play. There's a climbing wall, a tire swing, a spiral slide to whiz down, and plenty of cozy corners to play house or hide-and-seek in.

Ladder leads to an attic hidey-hole

I can't believe they haven't found me yet...

Spiral slide is all one piece

A tire on a chain makes a tire swing!

Want to swap your phone for my drink?

Whee!

These quarter tiles are climbing wall footholds

LET'S PLAY

The playhouse has a wooden base of nougat brown plates topped with tiles around the edges. The blue panels on the wall are 1x2 tiles attached sideways to knobs in the wall. The house fits neatly in the corner of a 16x16 plate.

1x2/2x2 bracket piece

16x16 plate

1x1 plate

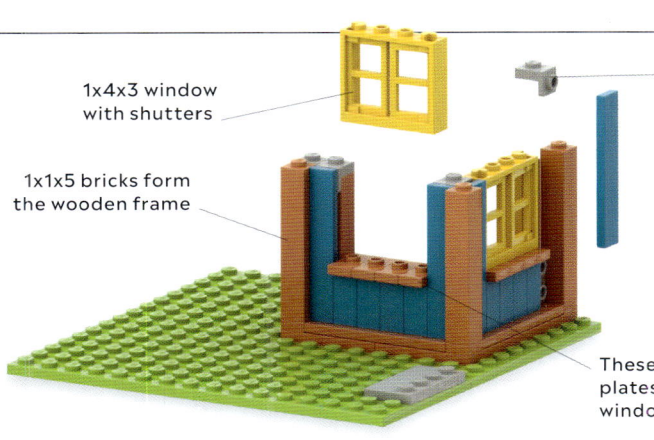

SKILL LEVEL

49

1x4x3 window with shutters

1x1/1x1 bracket holds up the longer 1x6 tile

1x1x5 bricks form the wooden frame

PEEKABOO

Build up the wall by adding more blue panels and a wooden frame. Then add wide windows with shutters for children to peek out from.

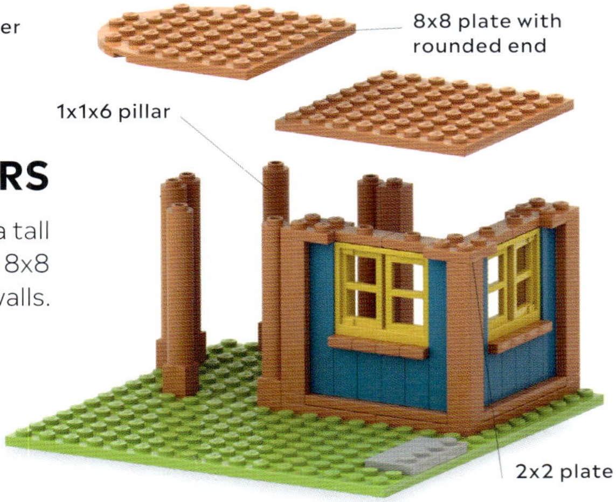

These four jumper plates are the windowsill

8x8 plate with rounded end

1x1x6 pillar

SUPPORT PILLARS

Once the main house is finished, add pillars and a tall brick to support a wide second floor. A square 8x8 plate fits onto knobs on top of the house's walls.

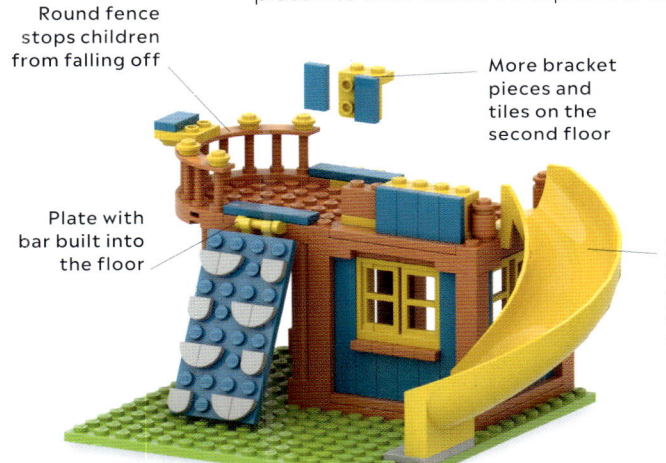

2x2 plate

Round fence stops children from falling off

More bracket pieces and tiles on the second floor

Plate with bar built into the floor

If you don't have a slide piece, you could build your own

1x2 hinge plate connects the roof to the house

6x8 plate roof slope

1x2 hinge brick base

SLIDE AND CLIMB

Now this house is starting to look much more fun, with its slide and climbing wall in place. The slide attaches to the 8x8 plate on the first floor, while the wall hangs from a clip and bar connection.

Climb the ladder to get in

4x4 macaroni tiles top off the fence

ATTIC ROOM

Finish off the playhouse by building a third floor. First, build a second wooden frame and place another 8x8 plate on top. Next, attach sloping roof plates to this floor to make a snug attic room with a ladder.

SNAIL HOUSE

There are no limits to the kinds of houses you can build with LEGO pieces—as this imaginative home proves! A shell is a snail's house, so why can't it be a minifigure's too? Its bright pink shell has latticed windows and a mailbox for any snail mail.

Ocular tentacles are bar pieces with stop rings

Lever pieces are the snail's lower tentacles

Snails love to eat plants!

Radar dishes give the top of the shell its rounded shape

Four round corner bricks form the front of the shell

Sometimes I have a hard time coming out of my shell.

Mailbox piece built into the shell

The head will be built up from here

6x8 plate

2x4 slope brick

BEST FOOT FORWARD

The foundations of the snail house look just like the bottom part of a snail's body, which is called a foot. Slope bricks in different sizes form the low, flat, slithering shape. The 6x8 plate that fits on top of the foot section will become the floor of the house.

SKILL LEVEL

51

Bricks adds more height to the front

1x4 slope brick

2x4 curved slope

The shell pieces will attach to these side knobs

HOUSE WALLS

Complete the foot of the snail by adding curved slopes to finish the back and more bricks at the front. At this stage, you can also start constructing the gray inner walls of the shell house.

1x2x3 latticed window pane

2x4 plate windowsill

Stacked 1x2 bricks

SNAIL MAIL

Now the walls of the shell house are taking shape and the two latticed windows are ready to go in. The front wall of the house includes a box to mail letters in—if you don't mind your letters taking a while to arrive!

LEGO Technic eyeball fits onto the tentacle

LEGO Technic axle with knob

Bar with stopper

2x2 plate with axle hole

8x8 radar dish

Mailbox is a 2x2x2 container box

2x4 curved slope adds more curves to the front of the snail's foot

Bar with tow ball

Curved slopes shape the head

The axle (above) fits into this hole

5x5x1 round corner brick

A tiny lever tentacle fits onto this headlight brick

SHELL SHAPES

The snail house looks almost ready to slither away (very slowly!). It now has a head with tentacles and a shell-shaped wall thanks to four round corner bricks attached sideways. Add a large radar dish to the roof to continue the rounded shape of the shell.

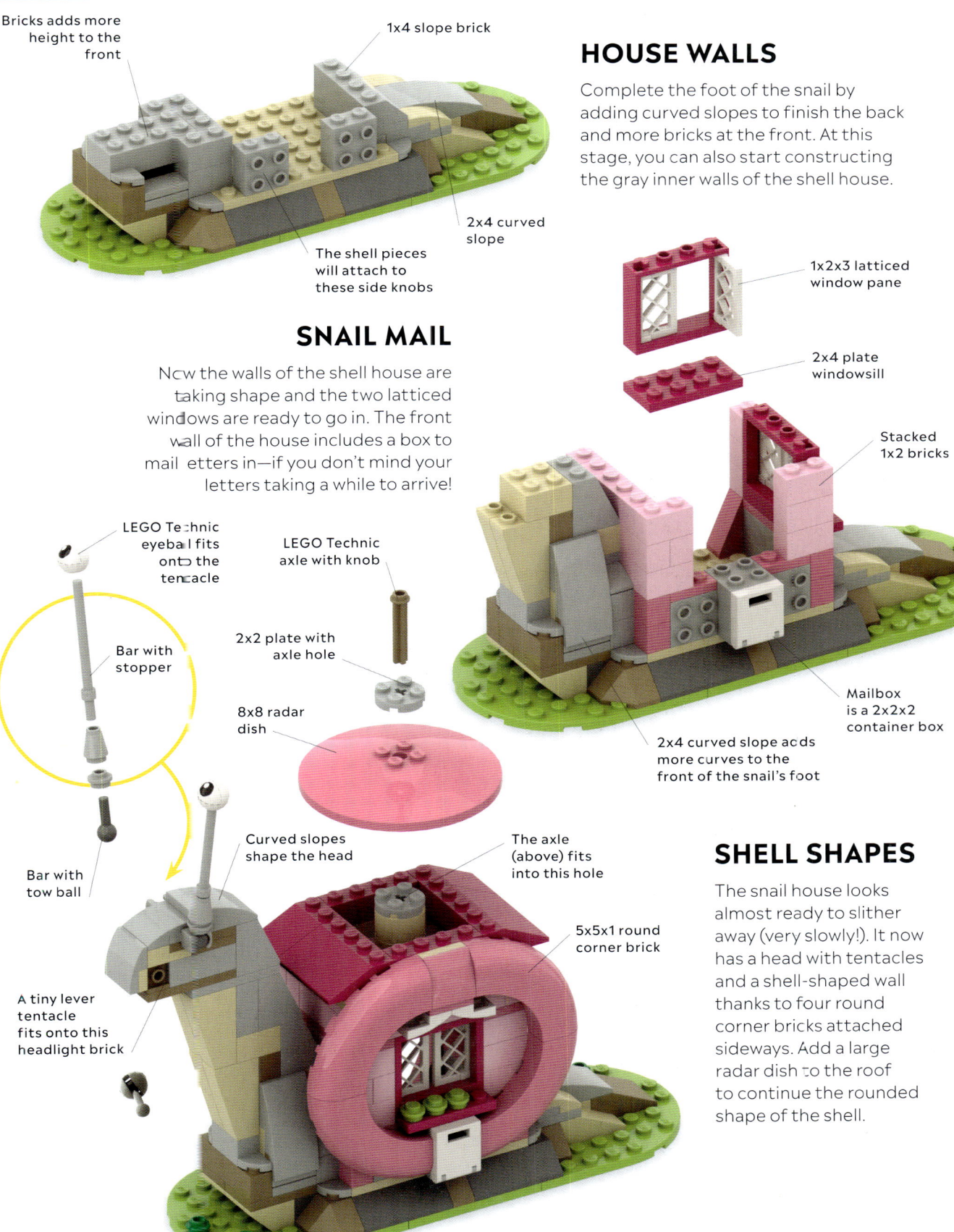

THATCHED COTTAGE

This cozy countryside cottage looks like it could have been home to minifigures for hundreds of years. It has a thatched roof, which means it is made from layers of straw or other dried plants, and uneven gray walls that have stood the test of time.

1x2 grille plates look like stalks of straw

Round tiles attach to side knobs in the walls

Uneven path made from round tiles in different sizes

These green round plates with petals could be small plants or moss

This rabbit likes the wild country yard

I hope grandma will be home.

2x4 tile doorstep

Colorful 1x1 round plate flowers in the window box

Two 1x1 round plates make a fire

1x2 palisade bricks add texture

HUMBLE BEGINNINGS

The walls of the thatched cottage are made from a variety of gray bricks. Using pieces in different shades, shapes, and sizes gives the walls a weathered look. An open fire on the cottage floor heats the home.

1x2 slopes look like overgrown grass

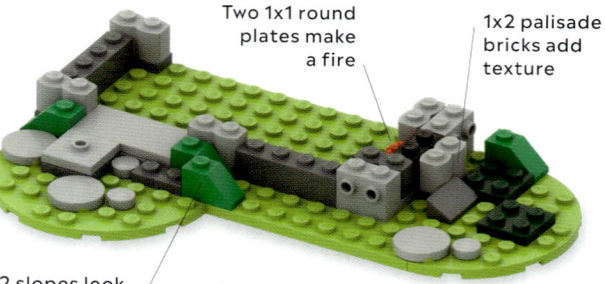

SKILL LEVEL 53

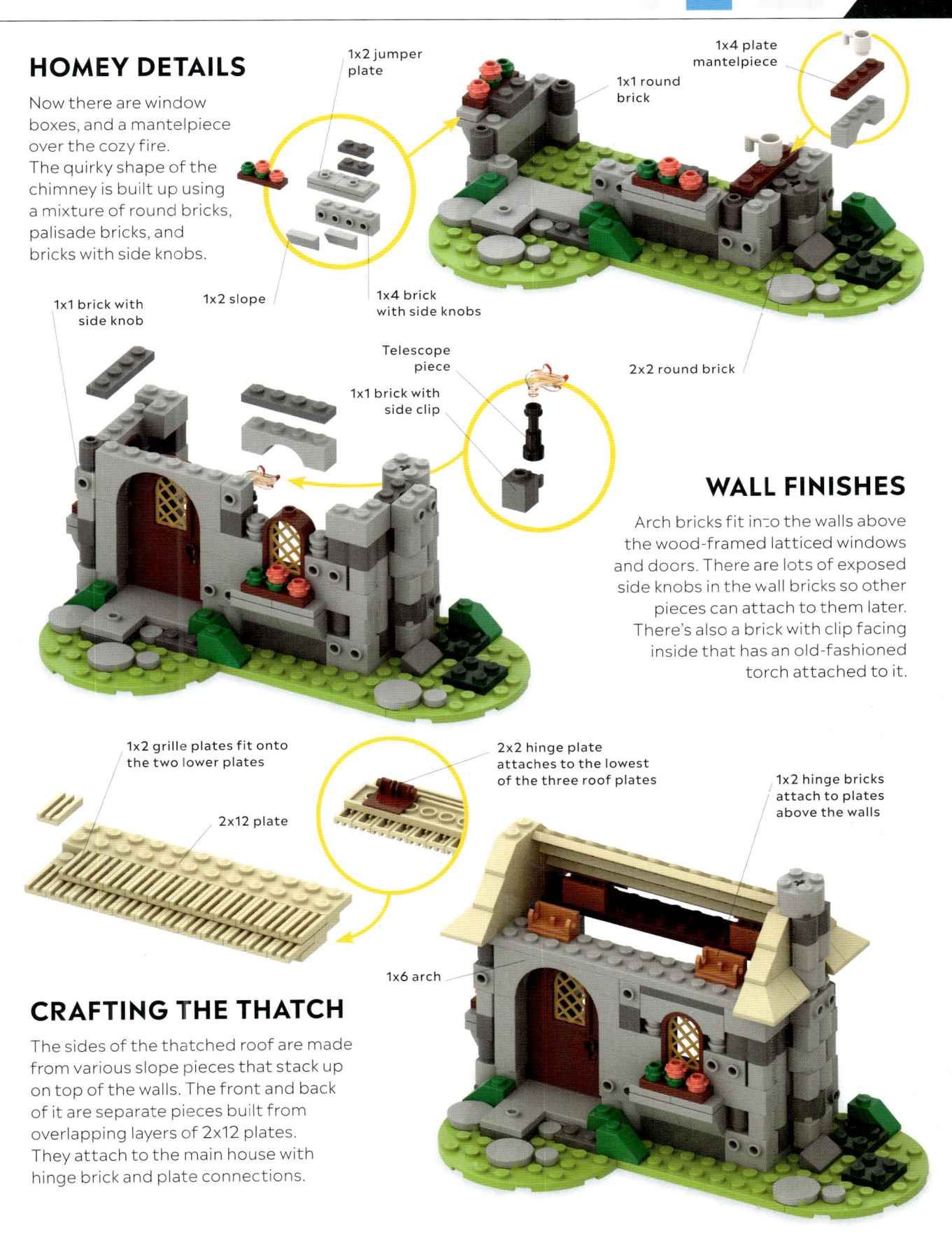

HOMEY DETAILS

Now there are window boxes, and a mantelpiece over the cozy fire. The quirky shape of the chimney is built up using a mixture of round bricks, palisade bricks, and bricks with side knobs.

- 1x2 jumper plate
- 1x4 plate mantelpiece
- 1x1 round brick
- 1x1 brick with side knob
- 1x2 slope
- 1x4 brick with side knobs
- 2x2 round brick
- Telescope piece
- 1x1 brick with side clip

WALL FINISHES

Arch bricks fit into the walls above the wood-framed latticed windows and doors. There are lots of exposed side knobs in the wall bricks so other pieces can attach to them later. There's also a brick with clip facing inside that has an old-fashioned torch attached to it.

- 1x2 grille plates fit onto the two lower plates
- 2x12 plate
- 2x2 hinge plate attaches to the lowest of the three roof plates
- 1x2 hinge bricks attach to plates above the walls
- 1x6 arch

CRAFTING THE THATCH

The sides of the thatched roof are made from various slope pieces that stack up on top of the walls. The front and back of it are separate pieces built from overlapping layers of 2x12 plates. They attach to the main house with hinge brick and plate connections.

HAMBURGER HOUSE

The minifigure owner of this house relishes the thought of coming home every day ... because it's shaped like his favorite fast food! Beneath the bun roof are layers of lettuce, cheese, tomatoes, and a burger patty. Inside, there's a well-equipped kitchen with plenty of room for cooking—you guessed it—more burgers!

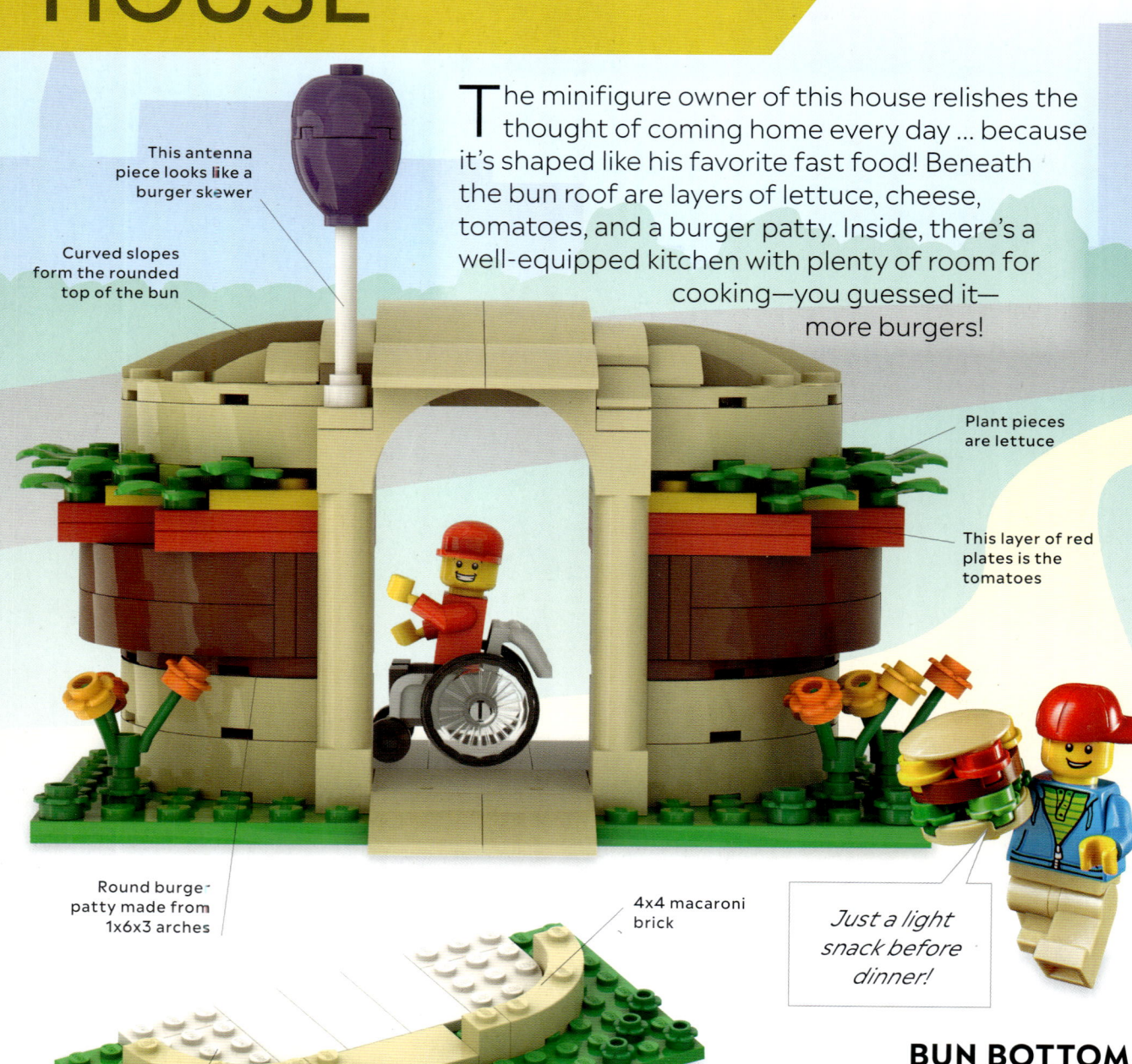

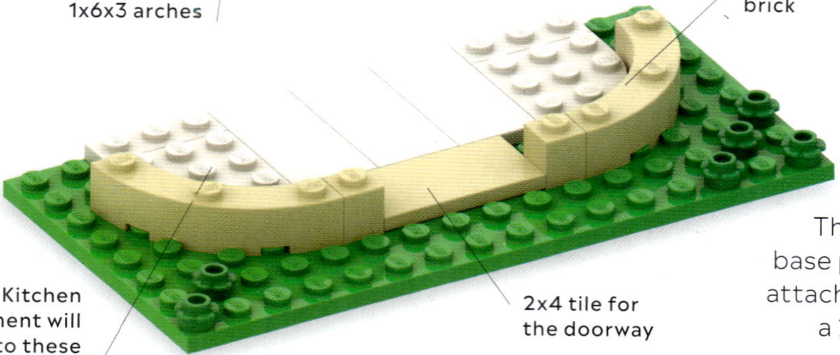

This antenna piece looks like a burger skewer

Curved slopes form the rounded top of the bun

Plant pieces are lettuce

This layer of red plates is the tomatoes

Round burger patty made from 1x6x3 arches

4x4 macaroni brick

Kitchen equipment will fit onto these knobs later

2x4 tile for the doorway

Just a light snack before dinner!

BUN BOTTOM

The hamburger house is served up on an 8x16 base plate. To begin its bun-shaped foundations, attach large macaroni bricks for the house's walls, a 2x4 tile for the doorway opening, and white plates and tiles for the kitchen floor.

SKILL LEVEL   55

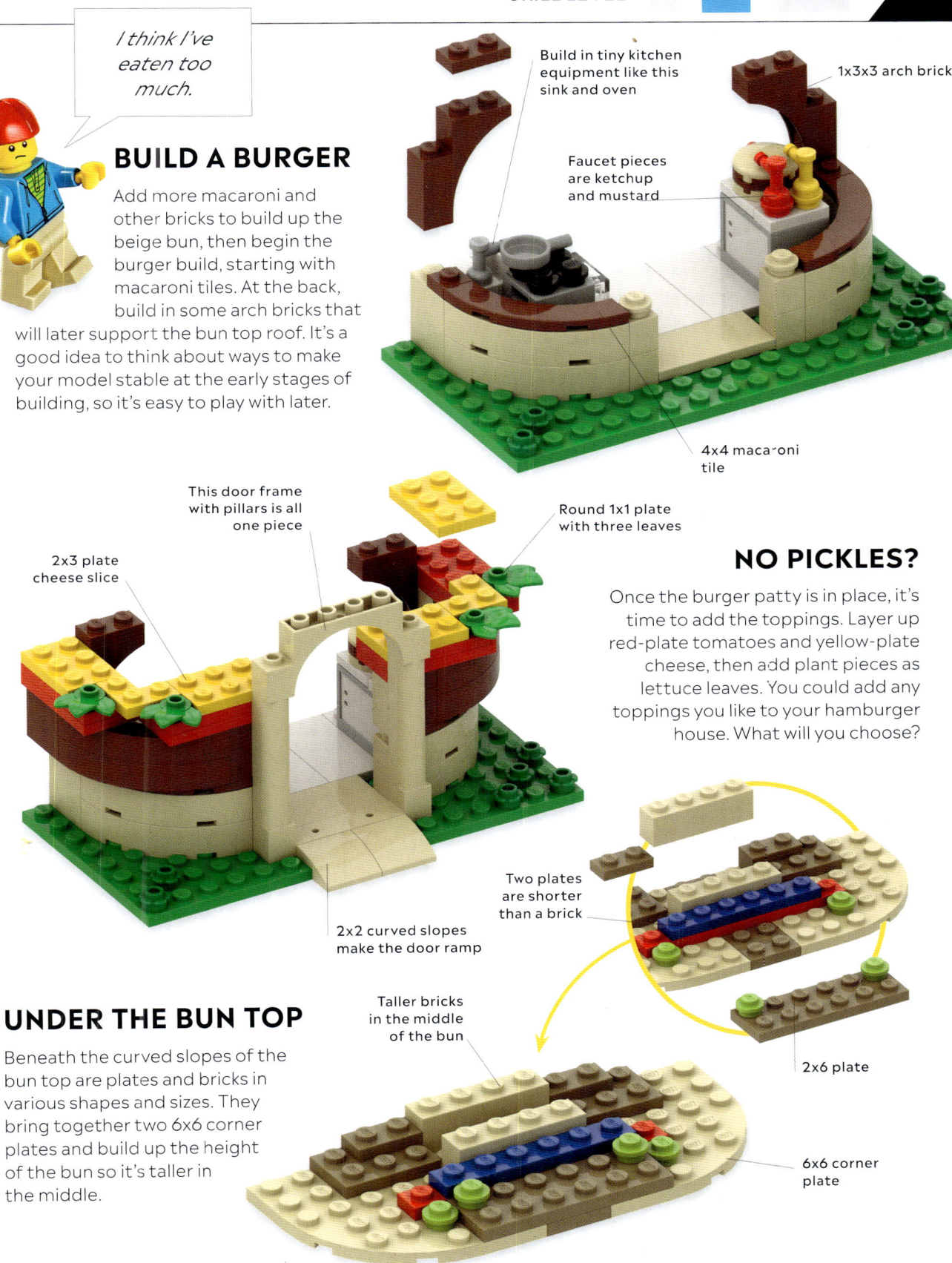

I think I've eaten too much.

BUILD A BURGER

Add more macaroni and other bricks to build up the beige bun, then begin the burger build, starting with macaroni tiles. At the back, build in some arch bricks that will later support the bun top roof. It's a good idea to think about ways to make your model stable at the early stages of building, so it's easy to play with later.

Build in tiny kitchen equipment like this sink and oven

1x3x3 arch brick

Faucet pieces are ketchup and mustard

4x4 macaroni tile

2x3 plate cheese slice

This door frame with pillars is all one piece

Round 1x1 plate with three leaves

NO PICKLES?

Once the burger patty is in place, it's time to add the toppings. Layer up red-plate tomatoes and yellow-plate cheese, then add plant pieces as lettuce leaves. You could add any toppings you like to your hamburger house. What will you choose?

2x2 curved slopes make the door ramp

Two plates are shorter than a brick

2x6 plate

UNDER THE BUN TOP

Beneath the curved slopes of the bun top are plates and bricks in various shapes and sizes. They bring together two 6x6 corner plates and build up the height of the bun so it's taller in the middle.

Taller bricks in the middle of the bun

6x6 corner plate

EXTENSIONS AND EXTRAS

Once you've built a house, you might want to th nk about how you could extend it or add extra features. These features could be attached to the house or in the yard. Here are some ideas to extend your imagination!

I'd like a greenhouse, or a pool, or a garage.

GREENHOUSE

This lean-to greenhouse is the perfect place to grow sun-loving plants because it gives them lots of light and protects them from the cold. Made from rows of window pieces, it fits snugly against the wall of a house.

- Add thriving potted plants inside
- Use tiles in different shapes to make an interesting path
- The top of the roof leans against the wall
- Small transparent pieces fill this gap
- 1x2 hinge brick and plates angle the roof
- 1x2x3 window pieces form the walls and roof
- "Glass" door adds even more light

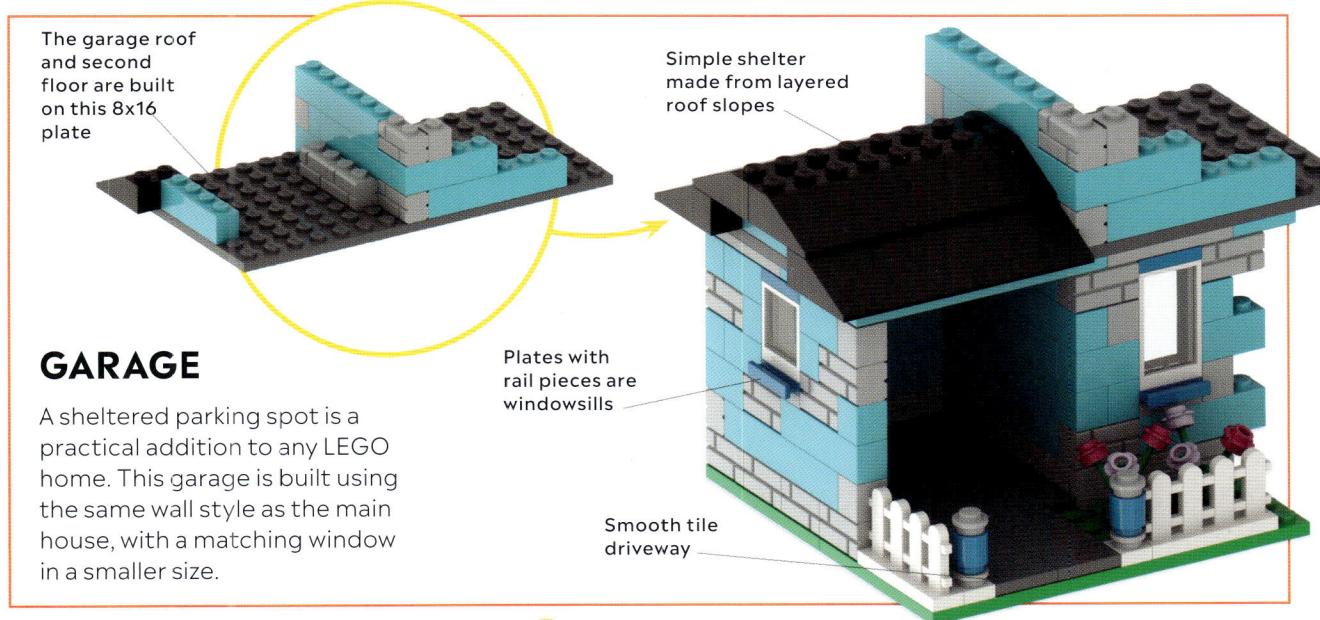

IDEAS GALLERY 57

The garage roof and second floor are built on this 8x16 plate

Simple shelter made from layered roof slopes

GARAGE

A sheltered parking spot is a practical addition to any LEGO home. This garage is built using the same wall style as the main house, with a matching window in a smaller size.

Plates with rail pieces are windowsills

Smooth tile driveway

French doors made from a large window and transparent door

This 2x4 tile is a paved step

Build a patio table for morning coffee!

PATIO

Perhaps your minifigures would like to relax on a patio on a sunny day! Add square tiles onto a base layer of plates to create a paved area in the yard.

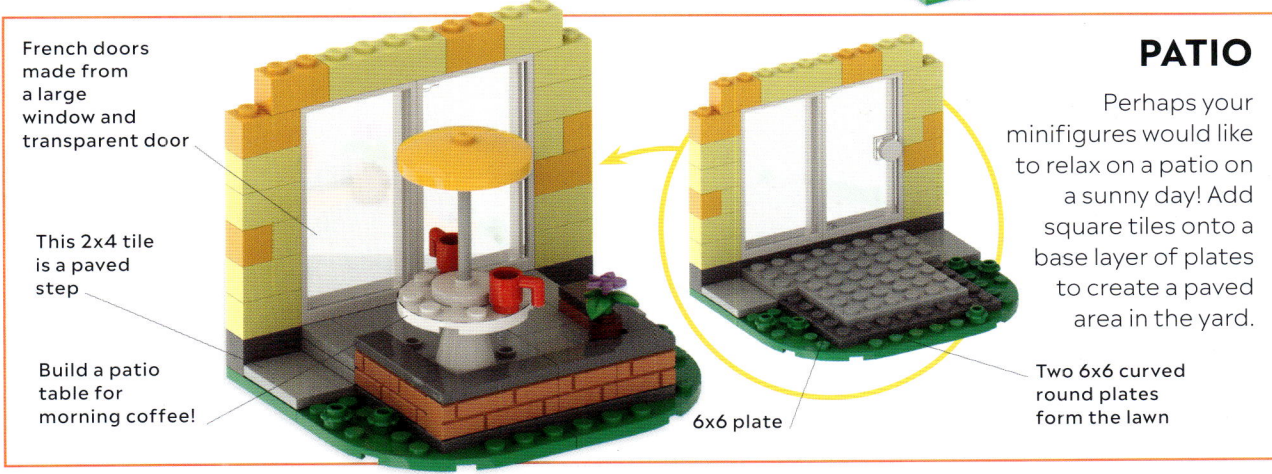

Two 6x6 curved round plates form the lawn

6x6 plate

SWIMMING POOL

Pool party! Let your minifigures live a life of luxury by building them a swimming pool. This one has glistening transparent blue water with bubbles caused by a surprise swimmer!

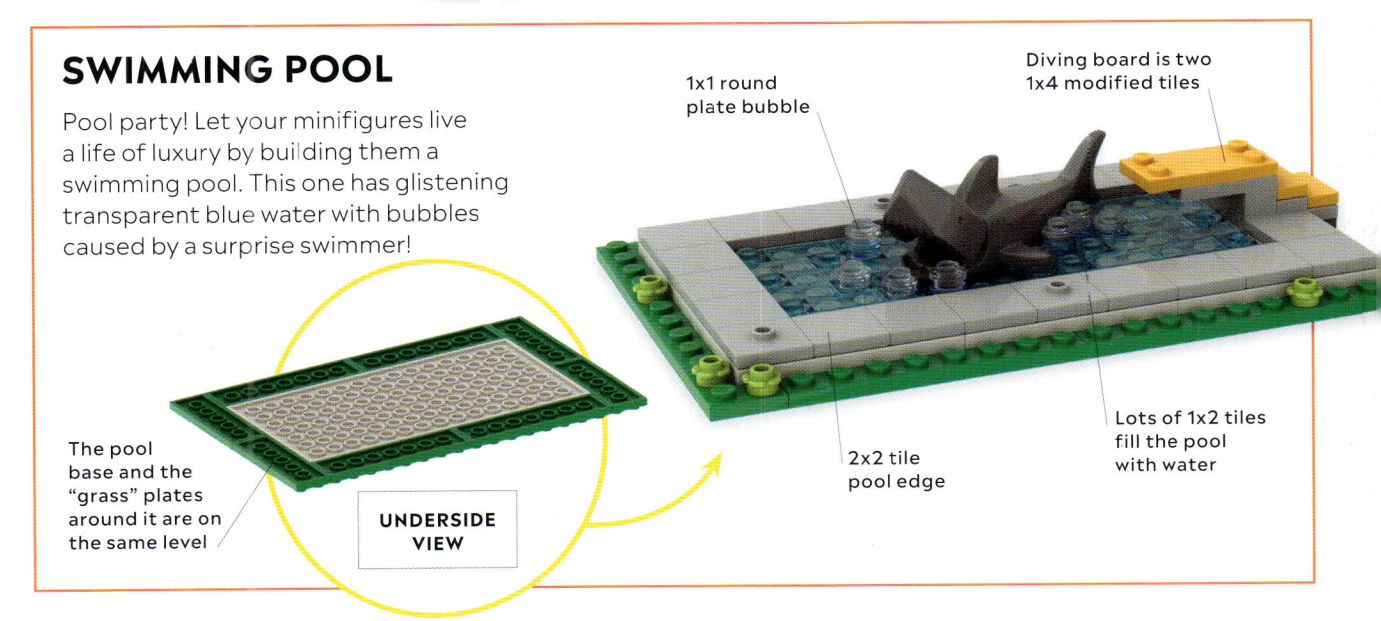

1x1 round plate bubble

Diving board is two 1x4 modified tiles

2x2 tile pool edge

Lots of 1x2 tiles fill the pool with water

The pool base and the "grass" plates around it are on the same level

UNDERSIDE VIEW

UNDERWATER HOUSE

This aquatic home is located in the dark depths of the ocean, on the seabed. All kinds of marine life whiz past its three bubble windows, wondering what on earth it is. Half-submarine, half-home, the underwater house has a rudder and propeller so it can relocate in a flash if any predators approach.

The home's residents can peer out of this periscope

These pieces are more often used on LEGO cars

Small propeller spins on a tile with pin

The front door opens from the bottom

Watertight bubble window attaches to a 4x4 round plate with hole

1x1x6 pillars support the roof but leave the back of the house open

1x4 brick wall

1x2 tile doorstep

Layered round tiles create a ripple effect

SANDY START

The curvy, uneven base of the underwater house looks like ripples of sand on the seabed. It's made from one rectangular 8x16 plate surrounded by round plates in different sizes. Plates form the lowest parts of the house's walls, with a layer of bricks on top.

SKILL LEVEL

SUBMARINE SIDES

The walls of the underwater house are deceiving—they look round but they're actually square! Each square wall contains a brick with four side knobs so separate rounded walls can be attached sideways.

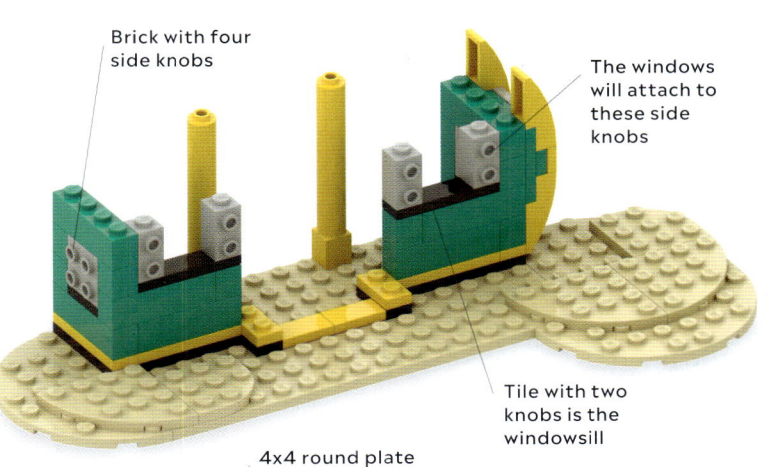

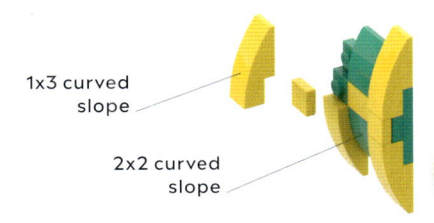

- 1x3 curved slope
- 2x2 curved slope
- Brick with four side knobs
- The windows will attach to these side knobs
- Tile with two knobs is the windowsill
- The roof will rest on this 1x1 brick
- 4x4 round plate with 2x2 hole
- 4x4 cylinder

BUBBLE WINDOWS

A house underwater needs big bubble windows! These are made from cylinders attached to round plates with holes. They fit onto bricks with side knobs on either side of the window openings in the wall.

- Stacked 1x2 bricks create an extra-thick door frame
- Add tiles to make the door look heavier and more watertight
- 1x2 plate with two clips

WATERTIGHT DOOR

The heavy, submarine-style door has a matching bubble window. It opens from the handle at the bottom and hangs from a plate with bar built into the roof.

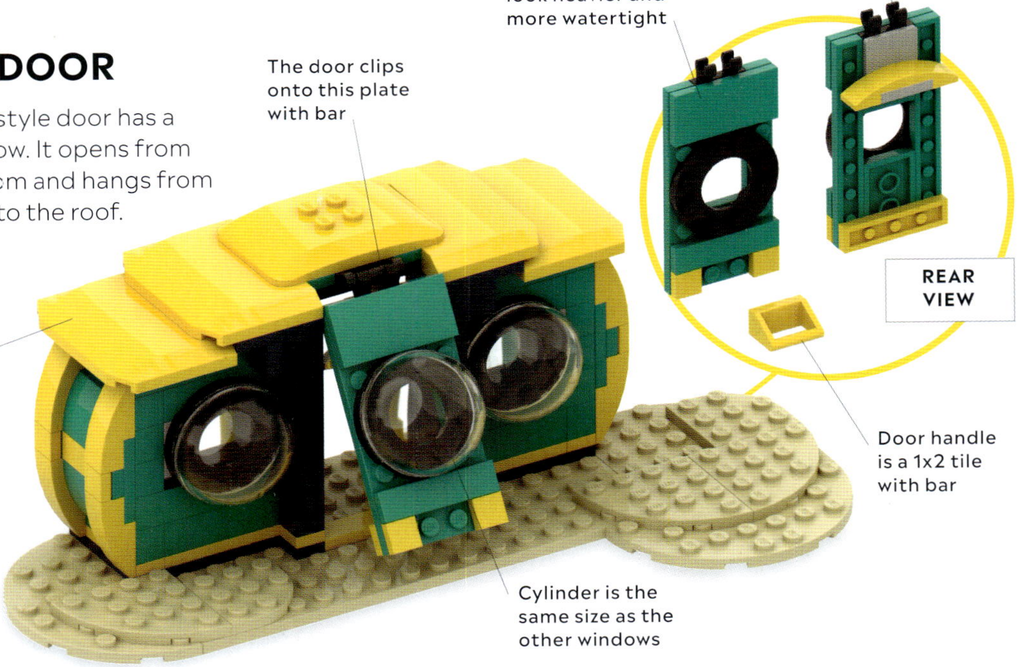

- 4x6 car hood wedge plates overlap to form the roof
- The door clips onto this plate with bar
- REAR VIEW
- Door handle is a 1x2 tile with bar
- Cylinder is the same size as the other windows

FLYING HOUSE

And they say I'm make-believe!

Who wants to live in a subdivision when you can live high up in the clouds? This bird-shaped house with brightly colored plumage is not only a delightful imaginary home—it's also a form of transportation. It can take off from its cloud perch and soar through the skies.

Arch bricks are pink tail feathers

Large, flapping wings

Curved-slope beak

Curved bricks in different sizes make fluffy clouds

Horn pieces make sharp talons

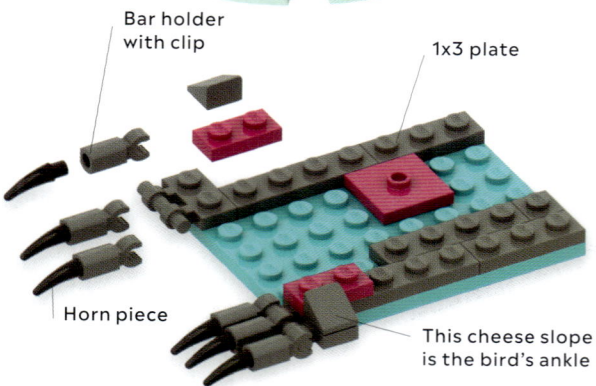

Bar holder with clip

1x3 plate

Horn piece

This cheese slope is the bird's ankle

FEET FIRST

Begin this birdlike build with a 6x8 plate, adding narrow plates for the bases of the side walls. Next, make the bird's feet and claws. Start with a 1x2 plate with bar handle and clip three bar holders onto it. Add horn pieces to create the talons.

SKILL LEVEL

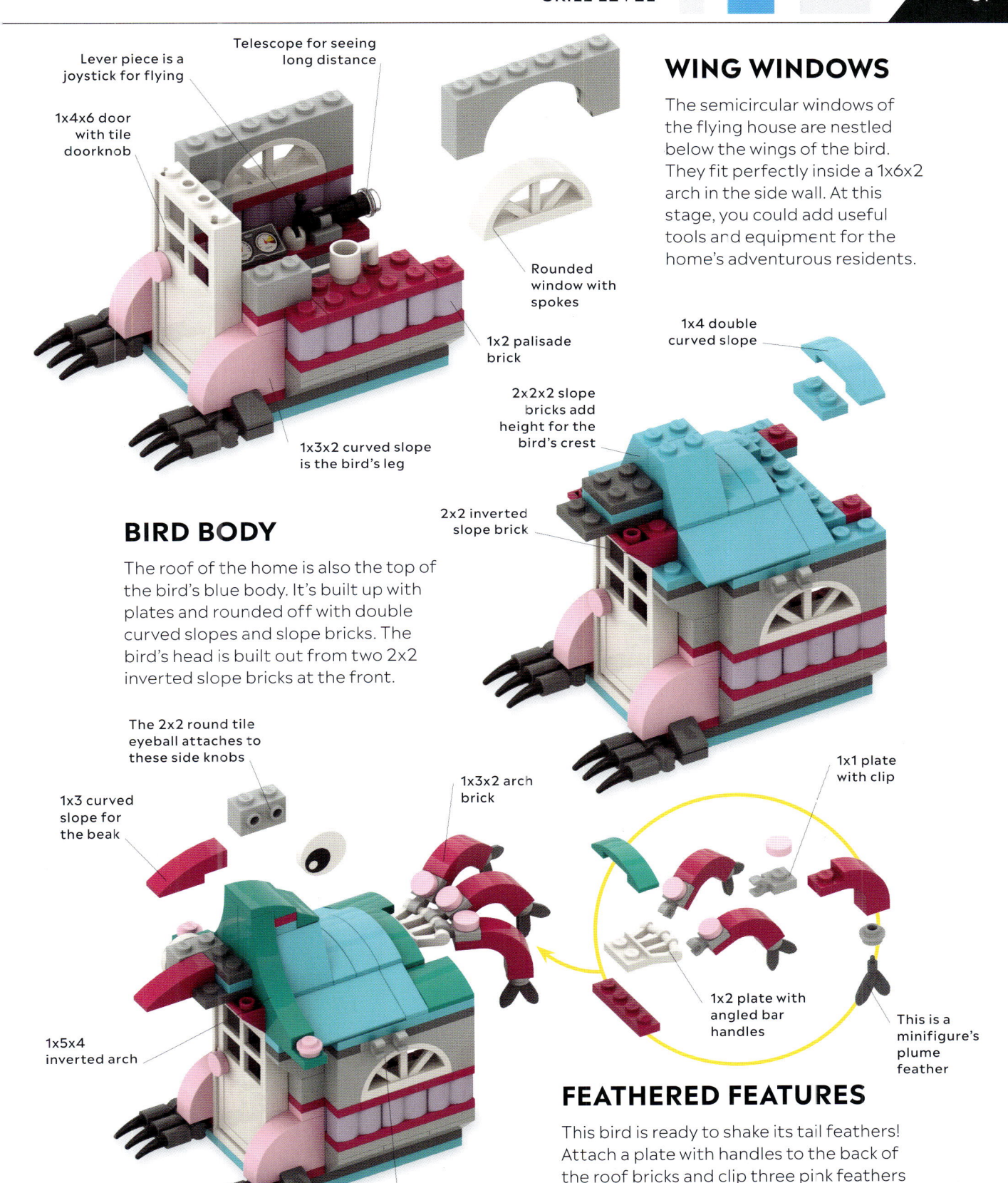

WING WINDOWS

The semicircular windows of the flying house are nestled below the wings of the bird. They fit perfectly inside a 1x6x2 arch in the side wall. At this stage, you could add useful tools and equipment for the home's adventurous residents.

- Lever piece is a joystick for flying
- Telescope for seeing long distance
- 1x4x6 door with tile doorknob
- Rounded window with spokes
- 1x2 palisade brick
- 1x4 double curved slope
- 2x2x2 slope bricks add height for the bird's crest
- 2x2 inverted slope brick
- 1x3x2 curved slope is the bird's leg

BIRD BODY

The roof of the home is also the top of the bird's blue body. It's built up with plates and rounded off with double curved slopes and slope bricks. The bird's head is built out from two 2x2 inverted slope bricks at the front.

- The 2x2 round tile eyeball attaches to these side knobs
- 1x3 curved slope for the beak
- 1x3x2 arch brick
- 1x1 plate with clip
- 1x2 plate with angled bar handles
- This is a minifigure's plume feather
- 1x5x4 inverted arch
- The wings fit onto these plates with clips

FEATHERED FEATURES

This bird is ready to shake its tail feathers! Attach a plate with handles to the back of the roof bricks and clip three pink feathers to it. To complete the bird's head, add a round tile eyeball and a curved slope beak.

OLD BOOT HOUSE

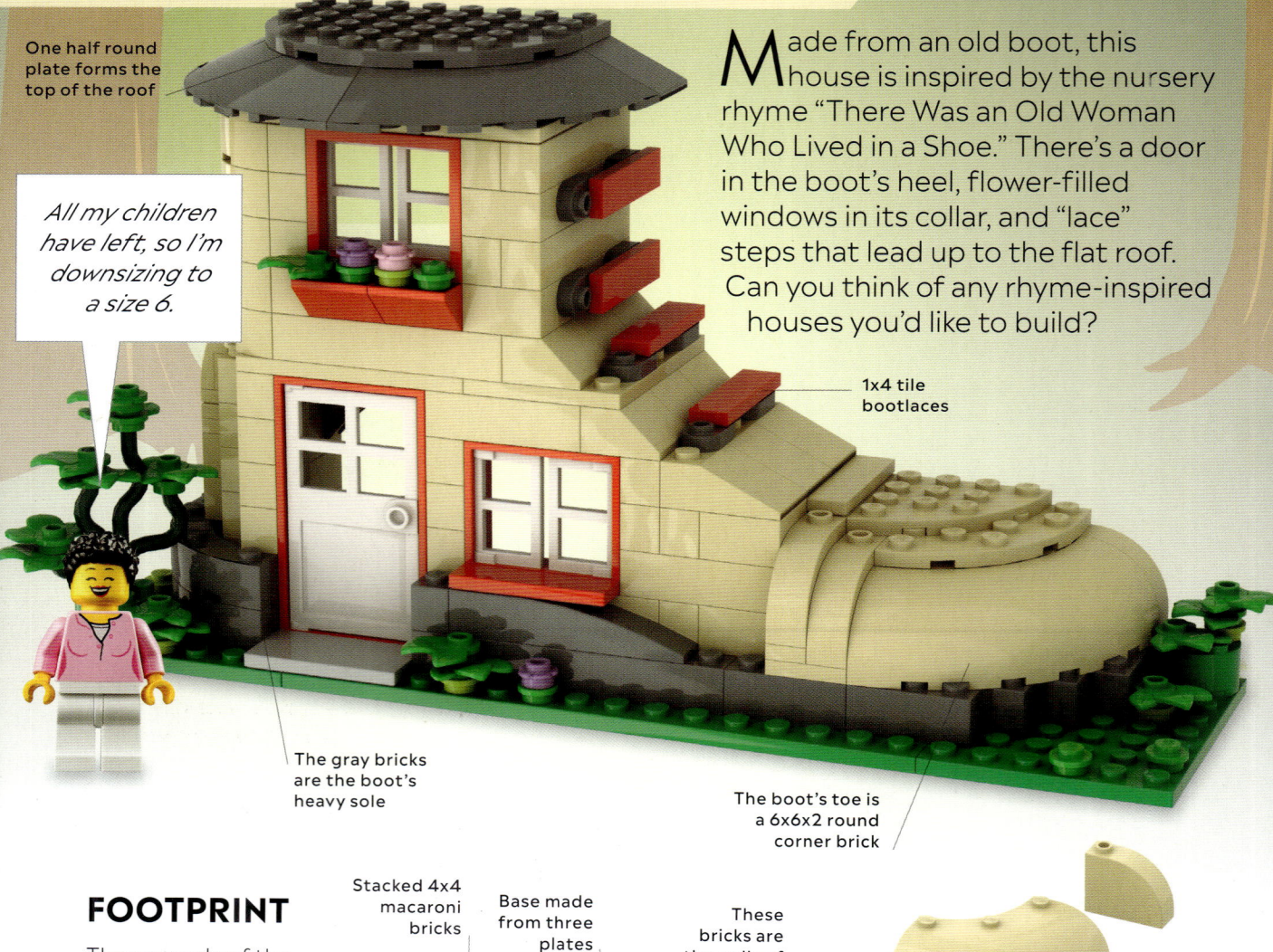

One half round plate forms the top of the roof

Made from an old boot, this house is inspired by the nursery rhyme "There Was an Old Woman Who Lived in a Shoe." There's a door in the boot's heel, flower-filled windows in its collar, and "lace" steps that lead up to the flat roof. Can you think of any rhyme-inspired houses you'd like to build?

All my children have left, so I'm downsizing to a size 6.

1x4 tile bootlaces

The gray bricks are the boot's heavy sole

The boot's toe is a 6x6x2 round corner brick

FOOTPRINT

The gray sole of the old boot is, fittingly, the foundations of the house. The heel gets its rounded shape from 4x4 macaroni bricks, while the shape of the toe is made by offsetting 1x3 bricks. Leave space in the sole for the doorway.

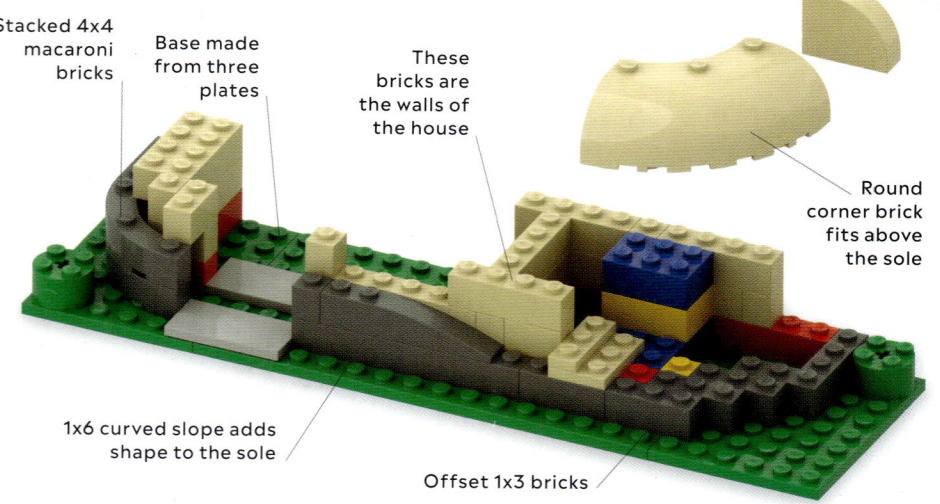

Stacked 4x4 macaroni bricks

Base made from three plates

These bricks are the walls of the house

Round corner brick fits above the sole

1x6 curved slope adds shape to the sole

Offset 1x3 bricks

SKILL LEVEL ■ ■ ■ 63

BOOT ROOM

The bottom of the boot is the lower room of the house. Add a window and door and a large arch at the back to keep it open for play. Also build up the toe with round plates and slope pieces.

- 1x12x3 arch
- These stepped slope bricks are the laced-up part of the boot
- 1x3x2 curved slope
- 2x2 inverted slope
- 4x6 plate floor
- Pieces for the laces will attach here

COLLAR FLOOR

The upstairs room of the boot house is inside the collar. Its red windowsills jut out of the walls thanks to inverted slope bricks.

- Stacked macaroni bricks make this curved wall
- The bottom two tiles lie flat
- Leave the outside knob exposed
- 6x6 round corner brick with sloping sides

LACE IT UP

Once the top floor is complete with windows and flower boxes, the old boot house is ready for some laces! Two 1x2 rounded plates underneath each tile look like the holes the shoelaces thread through.

- 2x3 slope brick
- Add more greenery in the yard as a finishing touch

ROOF COVER

This boot is built to be lived in, not worn, so it needs a roof at the top of the collar. The round roof has a first layer made from two sloped round corner bricks and two slope bricks.

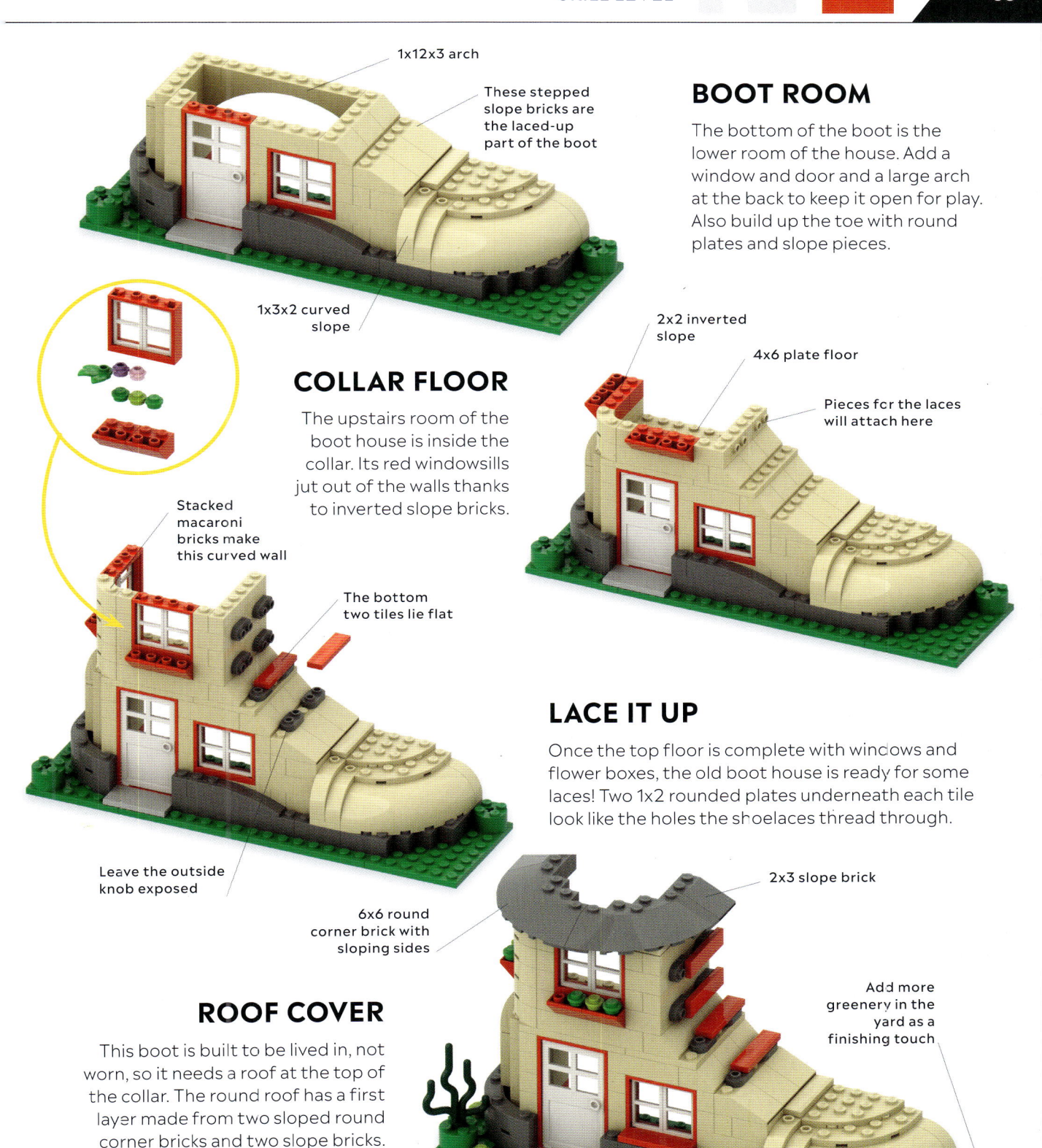

JUNGLE HUT

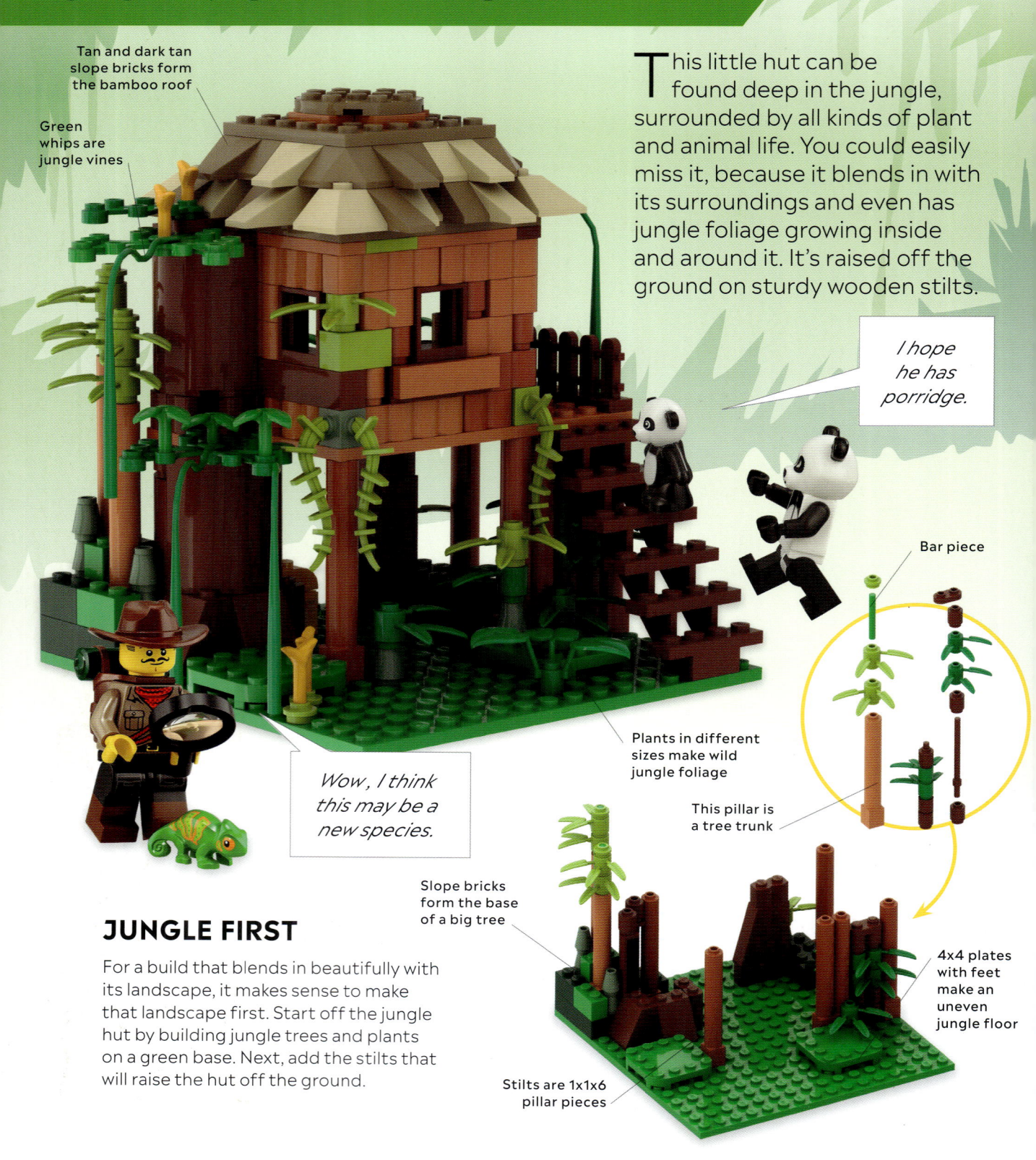

Tan and dark tan slope bricks form the bamboo roof

Green whips are jungle vines

This little hut can be found deep in the jungle, surrounded by all kinds of plant and animal life. You could easily miss it, because it blends in with its surroundings and even has jungle foliage growing inside and around it. It's raised off the ground on sturdy wooden stilts.

I hope he has porridge.

Bar piece

Plants in different sizes make wild jungle foliage

This pillar is a tree trunk

Wow, I think this may be a new species.

Slope bricks form the base of a big tree

4x4 plates with feet make an uneven jungle floor

Stilts are 1x1x6 pillar pieces

JUNGLE FIRST

For a build that blends in beautifully with its landscape, it makes sense to make that landscape first. Start off the jungle hut by building jungle trees and plants on a green base. Next, add the stilts that will raise the hut off the ground.

SKILL LEVEL 65

STEP START

This jungle is growing rapidly! Now there are even more trees, plants, and vines. There may not be a hut yet, but the steps are there already. They fit onto a sturdy section of stilts and jungle plants.

Plant leaves and stems

Stairs are all one piece

Carrot tops make interesting jungle plants

This tree trunk is growing with the hut

2x2 round brick log

Balcony railing

1x2 palisade brick

HUT HIDEAWAY

The plate floor of the hut is now in place and the walls are taking shape. They're made from a combination of curvy palisade bricks and round bricks to create a log effect.

Ceiling plate is the same dimensions as the hut

All kinds of plant pieces can be used for a jungle

1x8 plate is the top of the door frame

These vines hang from bricks with side knobs

2x2 wedge slope

1x3 slope

CEILING

Once the walls are finished and topped with plates, it's time to attach the hut's square 8x8 plate ceiling. While the walls were growing, the jungle foliage was too!

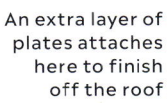

An extra layer of plates attaches here to finish off the roof

BAMBOO ROOF

Bamboo is a jungle plant with a very hard stem that is often used to make roofs in hot countries. The two layers of small slope pieces in the jungle hut's roof overlap each other so they look like thatched bamboo.

FANTASY CASTLE

Castles don't need to be big to be spectacular! The pristine white towers and golden spires of this microscale fantasy castle are as awe-inspiring as any regular-size royal residence. It is perched high on a rocky mountaintop so its minuscule residents can watch for minifigure-scale invaders.

This castle will be mine, even if I can't fit inside!

- Spear tip pieces form the tops of the spires
- Stacked round brick tower
- Windows are the backs of headlight bricks
- Tiny 1x1 slope rooftop
- Craggy surface formed from slope pieces in various sizes
- Modified 8x8 plate
- Plates in all shapes and sizes fit around the 8x8 plate

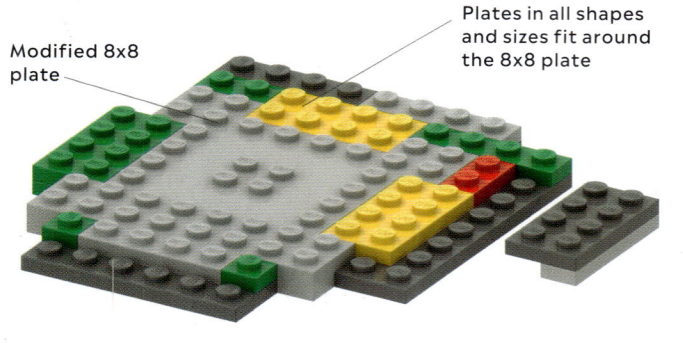

ROCKY START

Before building the fantasy castle, make the mountaintop it stands proudly upon. The rocky terrain has a characteristically uneven base, which is built around an 8x8 plate with cutout parts.

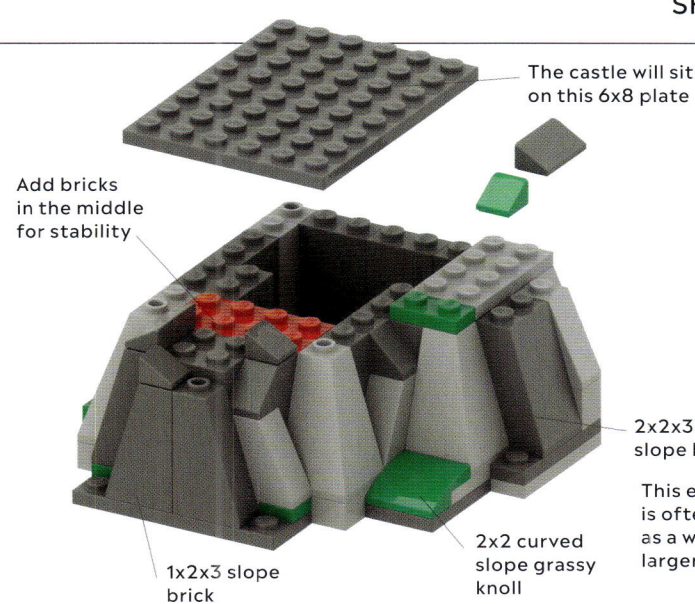

SKILL LEVEL

67

The castle will sit on this 6x8 plate

Add bricks in the middle for stability

MAJESTIC MOUNTAIN

Carve the jagged surface of the mountain out of slope bricks in two shades of gray. Add small green pieces among the gray bricks to make small sections of grass and moss on the rocks.

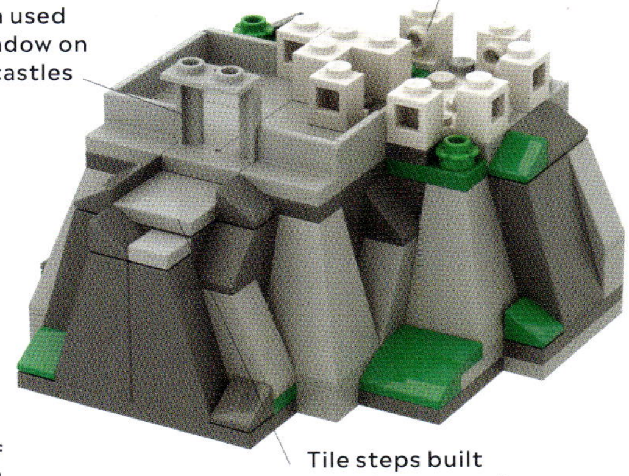

2x2x3 slope brick

This entrance is often used as a window on larger castles

1x2x3 slope brick

2x2 curved slope grassy knoll

Each square window is the back of a headlight brick

CASTLE WALLS

The castle is now emerging from the rocks. It has solid square walls with a large entrance. Inside the walls, there's a smooth courtyard and the first layer of the castle buildings.

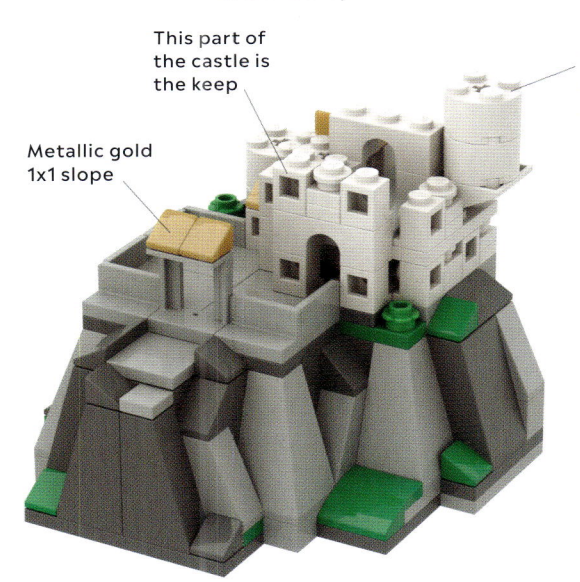

Tile steps built into the mountain

This part of the castle is the keep

The start of a 2x2 round brick tower

Metallic gold 1x1 slope

GRAND DETAILS

The castle keep and other buildings are taking shape. Metallic gold cheese slopes above the entrance and on some of the castle rooftops make this castle look even more grand.

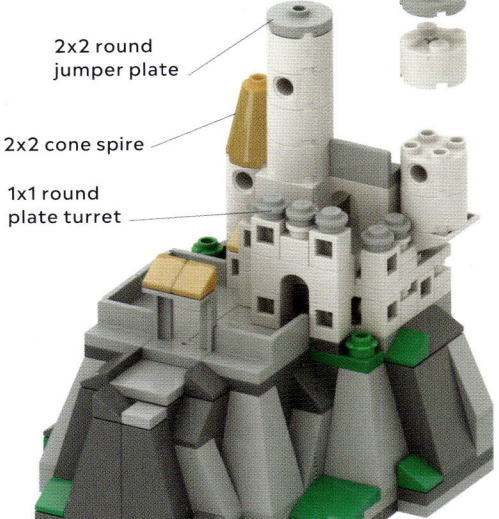

The larger towers have radar dish roofs

2x2 round jumper plate

2x2 cone spire

1x1 round plate turret

NOT-SO-TALL TOWERS

An entire castle tower can be made from three or four round bricks at microscale. The fantasy castle has three gold-topped towers with windows made from holes in some of the round bricks.

MEDIEVAL INN

Inns or taverns like this one were places where minifigures from that time and beyond would gather to rest and feast. This inn is a traditional half-timbered building, which means it has a wooden frame that's filled with brick or stone. It has an overhanging upper story to hurl garbage or insults from!

I've heard better...

Black pieces for the stained wood frame

Outdoor storage for casks of ale

A round and a square tile form the well-worn doorstep

Wooden window surrounds made from cones, round bricks, and plates

The bottom of the building is made from bricks

This wall will only be seen from the inside

Low wall for the outdoor storage

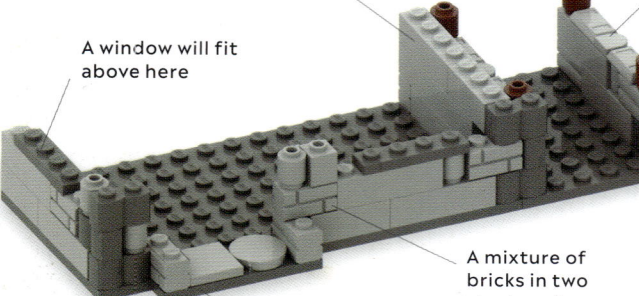

A window will fit above here

A mixture of bricks in two shades of gray looks like natural stone

The doorstep is built on a 1x6 plate

STONE BASE

While the top half of the medieval inn has a wooden frame, the bottom half is made from stone. The whole building covers its base plate, with no extra space around the edges, because this medieval building would be right next to other houses on a street.

SKILL LEVEL

69

Gold latticed window pane

Timber supports for second floor

1x1 round plate timber post

TIMBER DETAILS

Now the windows and door are fitted inside the walls along with more timber details. There are carved wood panels on either side of the windows and a matching frame around the door.

Windowsill is two plates with rails

Use any small, round pieces you have to make carved wood

REAR VIEW

Flaming torch attached to the wall

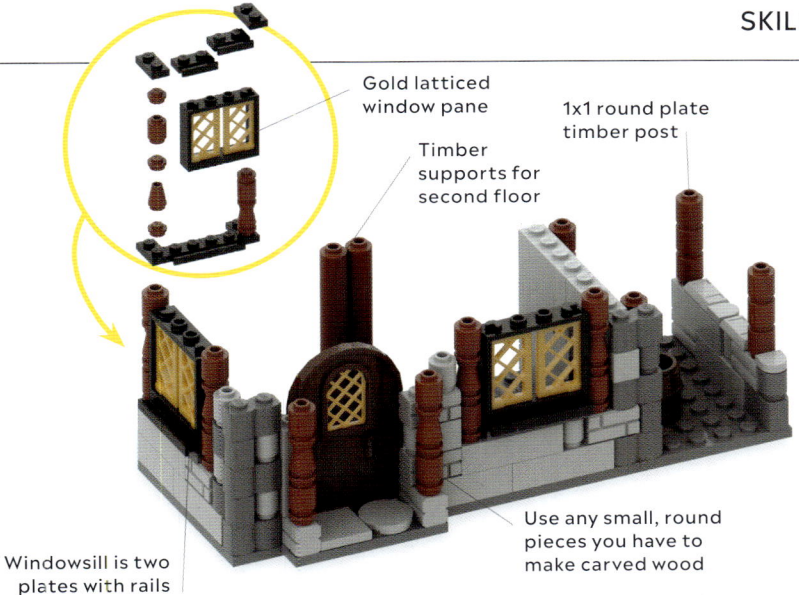

INSIDE THE INN

Once you have an inn with solid walls, you could add some interior details, such as a desk, cabinets, benches, tables, and lighting, to make it look inviting.

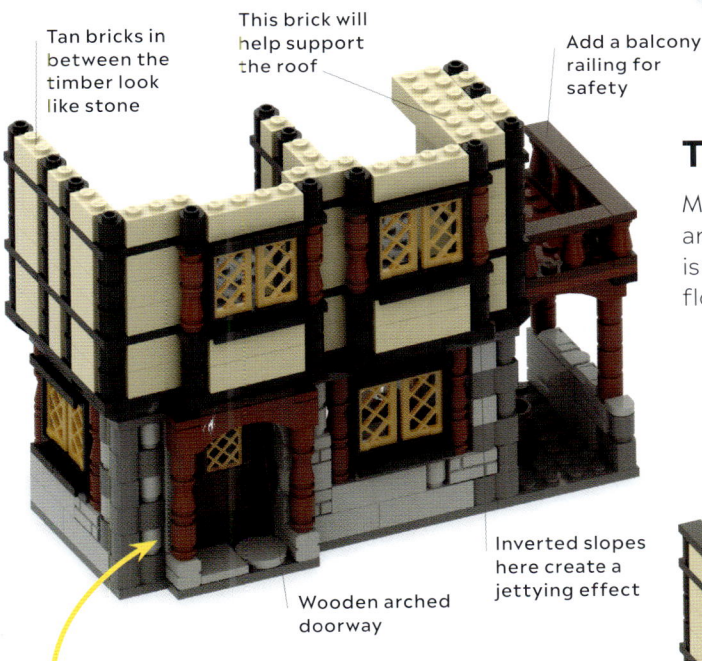

Jumper plate desk drawer

Table made from bricks and a round jumper plate

Tan bricks in between the timber look like stone

This brick will help support the roof

Add a balcony railing for safety

TIMBERED TOP FLOOR

Many medieval buildings have upper floors that are bigger than the floor below, like the inn. This is called jettying. The inn's timber-framed top floor also has a balcony overlooking the street.

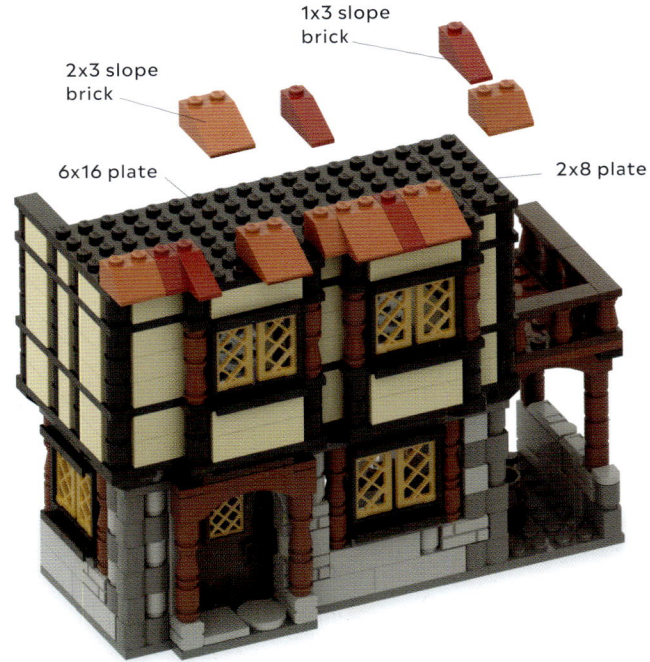

1x3 slope brick

2x3 slope brick

6x16 plate

2x8 plate

Inverted slopes here create a jettying effect

Wooden arched doorway

TILED ROOF

Complete the medieval inn with a clay tile roof. The slope-brick tiles above the windows are angled slightly differently from the rest to protect the windows from rain and snow.

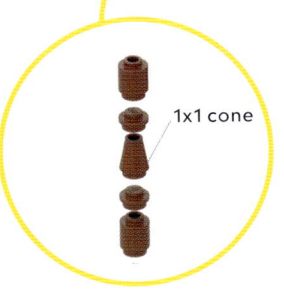

1x1 cone

TREE HOUSE

Up on the top perch of this tree house, high among the canopy of the surrounding trees, your minifigures can escape the modern world and return to nature. Perhaps too much so! It has two platforms that are built out of the thick tree trunk and a rope for climbing up and down.

Round corner brick roof

I'd better go now, Chimpy.

Kevin, your boss is on the phone. She's wondering why you aren't at work.

Minifigures can hold onto this hanging rope

Branches are plant leaf pieces

Slope brick tree root

Stacked 2x2 round bricks

1x2x2 slope brick

Layered round plates make an uneven forest floor

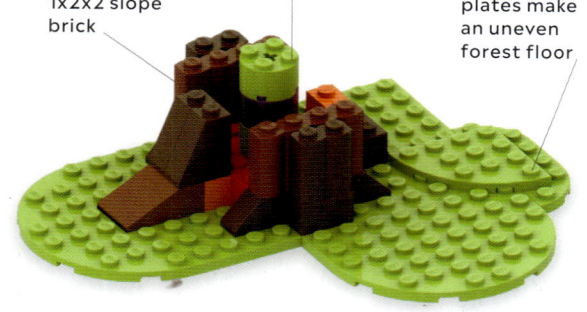

PUTTING DOWN ROOTS

Build gnarly tree roots from any slope bricks and rounded bricks you have in your collection. The bricks in the middle of the trunk won't be seen on the final model, so they can be any color you like.

SKILL LEVEL 71

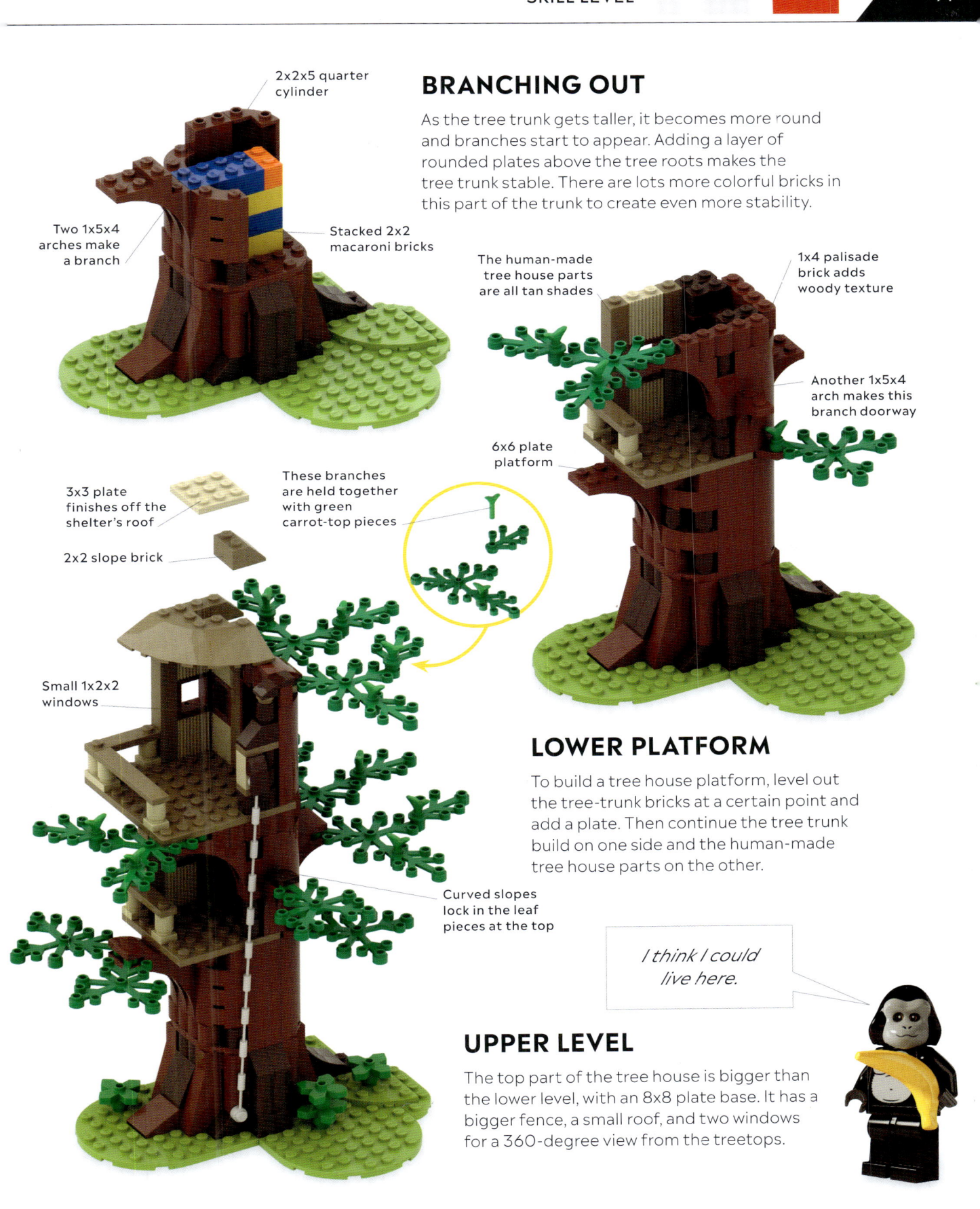

BRANCHING OUT

As the tree trunk gets taller, it becomes more round and branches start to appear. Adding a layer of rounded plates above the tree roots makes the tree trunk stable. There are lots more colorful bricks in this part of the trunk to create even more stability.

2x2x5 quarter cylinder

Two 1x5x4 arches make a branch

Stacked 2x2 macaroni bricks

The human-made tree house parts are all tan shades

1x4 palisade brick adds woody texture

Another 1x5x4 arch makes this branch doorway

6x6 plate platform

3x3 plate finishes off the shelter's roof

2x2 slope brick

These branches are held together with green carrot-top pieces

Small 1x2x2 windows

LOWER PLATFORM

To build a tree house platform, level out the tree-trunk bricks at a certain point and add a plate. Then continue the tree trunk build on one side and the human-made tree house parts on the other.

Curved slopes lock in the leaf pieces at the top

I think I could live here.

UPPER LEVEL

The top part of the tree house is bigger than the lower level, with an 8x8 plate base. It has a bigger fence, a small roof, and two windows for a 360-degree view from the treetops.

FURNITURE

Even minifigures like to flop down on a soft-looking sofa after a hard day's work! Adding movable items like chairs, tables, beds, and cabinets will bring comfort and character to the interiors of your LEGO homes.

Don't forget to make your bed!

BED

- 1x4x1 bow window
- 4x6 plate bed base
- 1x1 cheese slopes form the edges of the duvet
- 1x1 plate bedpost

Give your minifigures a warm bed to come home to. This single one has a smooth tile and cheese slope duvet, and a headboard made from a bow window piece.

SIDEBOARD

A sideboard is often the place where the best glasses, plates, and cutlery in the house are stored. This sideboard is particularly grand looking, with carved wood and gold drawers and details.

- Two headlight bricks hold each drawer in place
- 1x4 curved slopes make it rounded at the top
- 1x1 plate with swirl
- Jumper plate drawer
- 1x1 round bricks and plates look like carved wood

SOFA

- 1x3 curved slope
- 1x2 curved slope sofa arm
- 1x1 brick with side knob
- 1x1 round plates are wooden feet
- 1x8 plate

This brightly colored LEGO sofa may not be soft, but it's perfectly suited to a minifigure's needs. It has tile cushions attached to sideways-facing knobs built into the back of the sofa.

IDEAS GALLERY

HOME OFFICE

Hardworking minifigures may need a home office like this one, complete with a lamp, desk, and swivel chair. On the desk are neatly organized books, a desktop computer, and a cell phone.

- Books are 2x2 plates and a tile, turned sideways
- If you don't have specially printed tiles, plain ones would also work
- 1x1 cheese slope bookends
- 2x2 round jumper plate chair base

COFFEE TABLE

This tiny table has two jumper plates on the top to attach a mug of coffee and a plate of cookies to!

- It looks like someone has taken one already!
- 1x2 rounded plate table legs
- 2x4 plate

ARMCHAIR

This snug armchair has just the right dimensions for a minifigure's bottom! Its curved back and arms are made from the same curved slope pieces.

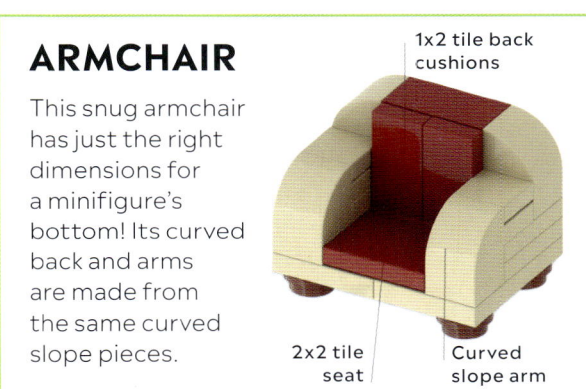

- 1x2 tile back cushions
- 2x2 tile seat
- Curved slope arm

COZY ROOM

Once you've got the hang of LEGO furniture building, you could fill a room with your new creations, as in this cozy scene in the festive house from pages 26-27.

Of course I won't do the cream pie trick again this year!

- Dome-topped floor lamp
- Corner tile stockings by the fire
- Tile floor rug

MEDIEVAL CASTLE

Real-life medieval castles were built to stand in place for centuries, but a LEGO castle can be changed around regularly to suit its royal residents' needs! This imposing castle has two turrets, red and gold banners, and a catapult to scare away anyone who wants to capture it.

The winner will marry my pet frog!

Modified 3x3 bricks shape the castle turrets

Banners hang from a bar piece attached to clips

Erm, I'm not sure about this...

Olive pieces give the walls an aged look

Leave gaps for arrowslit windows

Palisade bricks add a stony texture to the walls

1x1x6 pillar

Flaming torches clipped to the inside walls

CASTLE PLOT

Begin the castle by plotting out where the gate, turrets, and walls will go on a base plate. At this point, you could also add in smaller details like a well-worn path and flaming torches inside.

8x16 base plate

Different-size tiles look like worn-down stones

SKILL LEVEL 75

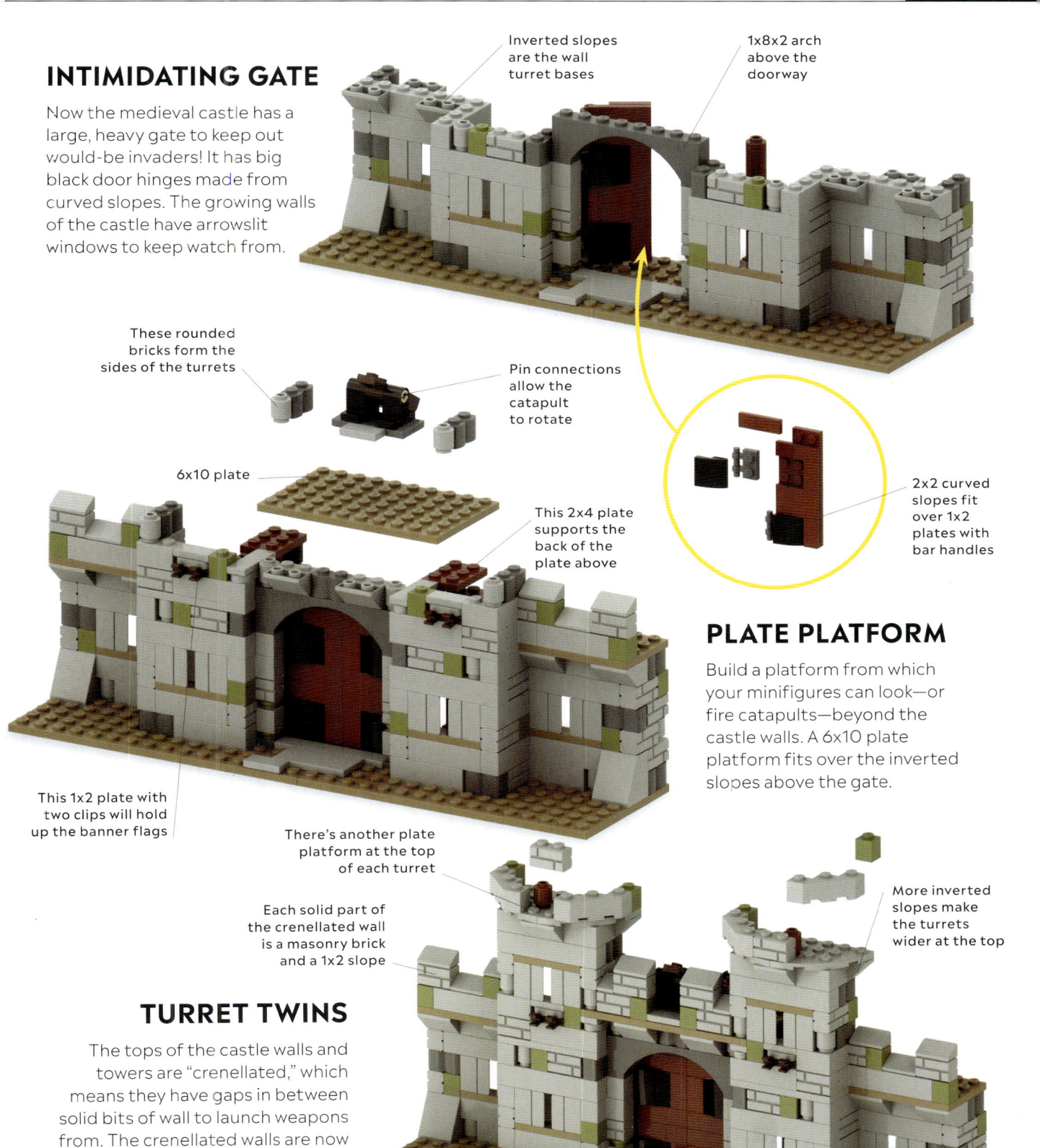

INTIMIDATING GATE

Now the medieval castle has a large, heavy gate to keep out would-be invaders! It has big black door hinges made from curved slopes. The growing walls of the castle have arrowslit windows to keep watch from.

Inverted slopes are the wall turret bases

1x8x2 arch above the doorway

These rounded bricks form the sides of the turrets

Pin connections allow the catapult to rotate

6x10 plate

This 2x4 plate supports the back of the plate above

2x2 curved slopes fit over 1x2 plates with bar handles

PLATE PLATFORM

Build a platform from which your minifigures can look—or fire catapults—beyond the castle walls. A 6x10 plate platform fits over the inverted slopes above the gate.

This 1x2 plate with two clips will hold up the banner flags

There's another plate platform at the top of each turret

Each solid part of the crenellated wall is a masonry brick and a 1x2 slope

More inverted slopes make the turrets wider at the top

TURRET TWINS

The tops of the castle walls and towers are "crenellated," which means they have gaps in between solid bits of wall to launch weapons from. The crenellated walls are now complete—now add crenellated tops to the turrets to finish off the castle.

HAUNTED HOUSE

Take care if you step beyond the tall iron gates of this home—it's the kind of place where things go bump in the night. With a creepy color scheme, a shadowy basement, and a creaky old tree in the garden, this haunted house is sure to be filled with hidden horrors.

No cheese, please—it gives me nightmares.

This 1x4x1 fence adds detail to the top roof section

Railings made from skeleton legs clipped to a bar

1x1 plates with vertical teeth create an unwelcoming doorway

Bare tree made from arch bricks in different sizes

Slope pieces make tree roots for the autumnal tree

Attach skittering critters to 2x2 jumper plates in the basement

This 1x4 brick forms the base of the second step

FRIGHTFUL FOUNDATIONS

The whole haunted house is built on a 16x16 base plate. Begin by building the outlines of the basement floor and the front steps, then attach tiles to the front of the base plate to create a jaggedly winding sidewalk.

SKILL LEVEL

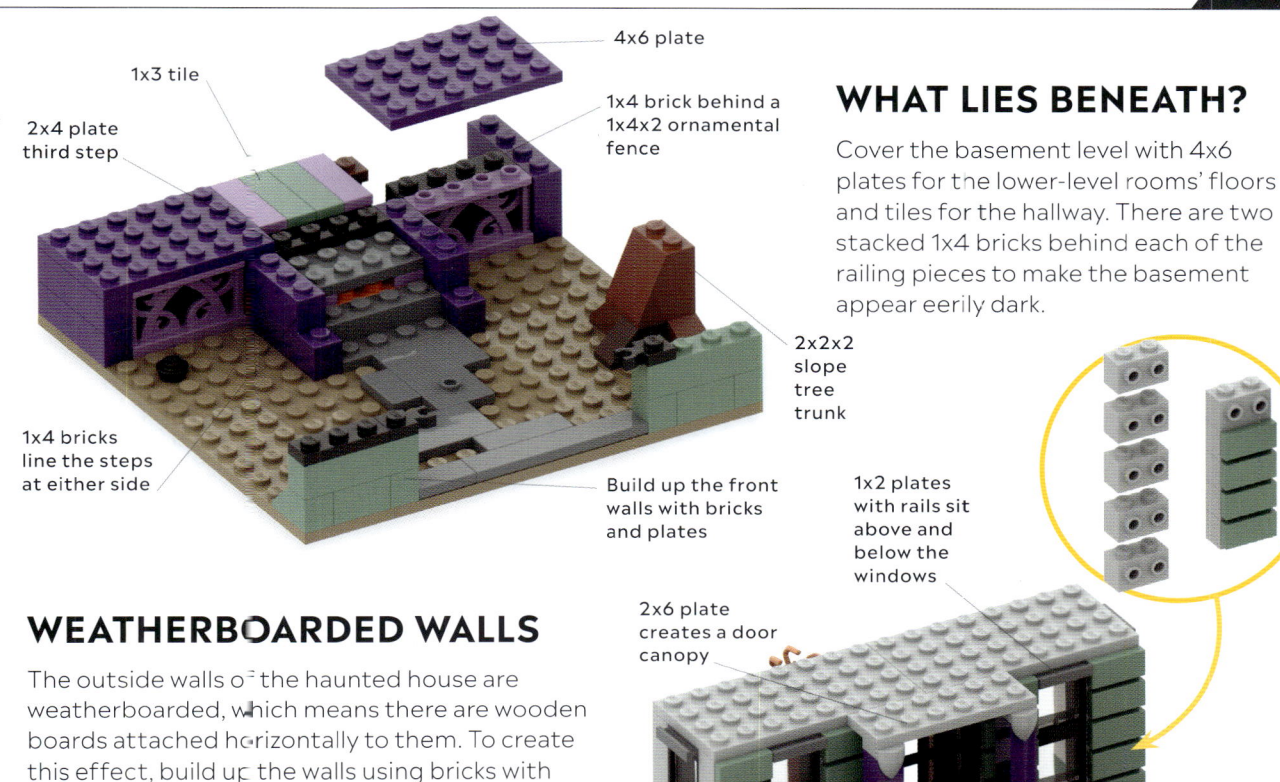

WHAT LIES BENEATH?

Cover the basement level with 4x6 plates for the lower-level rooms' floors and tiles for the hallway. There are two stacked 1x4 bricks behind each of the railing pieces to make the basement appear eerily dark.

- 2x4 plate third step
- 1x3 tile
- 4x6 plate
- 1x4 brick behind a 1x4x2 ornamental fence
- 2x2x2 slope tree trunk
- 1x4 bricks line the steps at either side
- Build up the front walls with bricks and plates
- 1x2 plates with rails sit above and below the windows
- 2x6 plate creates a door canopy
- 1x1x5 doorway column
- 1x4x2 spindled fence

WEATHERBOARDED WALLS

The outside walls of the haunted house are weatherboarded, which means there are wooden boards attached horizontally to them. To create this effect, build up the walls using bricks with side knobs, then attach tiles across them.

- 1x4 tile
- 1x2x3 slope
- Arch bricks inside look like wooden beams
- Modified 4x4 tile with knobs on the edge
- Tall windows made from stacked 1x2x2 window pieces
- Spider piece attached to a sideways-facing knob
- Bat hangs from a bar in the wall

CREEPY NOOKS

Build up the second floor and the three parts of the mansard roof. There are plenty of dark spaces on the top floor where you can add creepy details such as skeleton bones and bats.

REAR VIEW

Animals

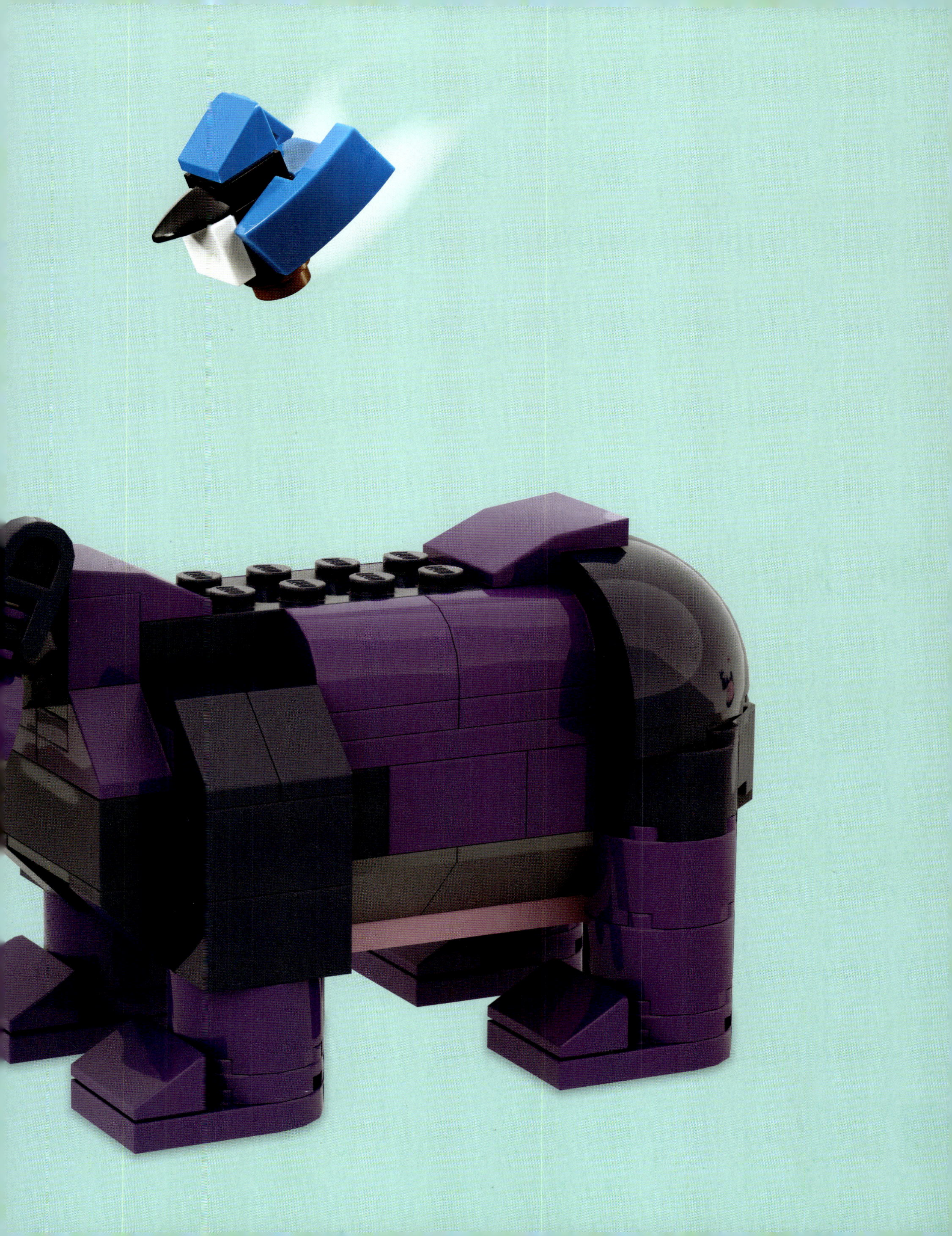

GIRAFFE

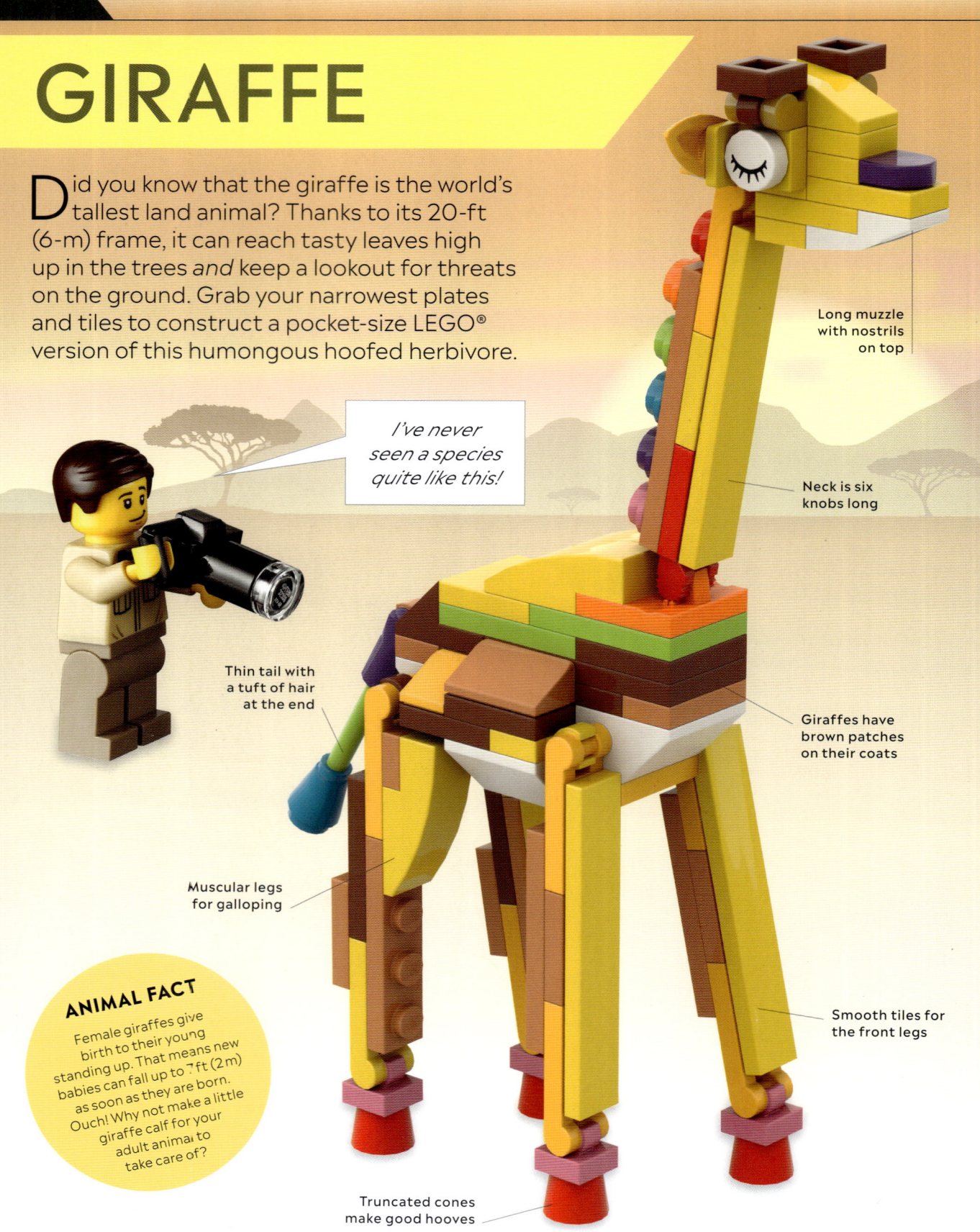

Did you know that the giraffe is the world's tallest land animal? Thanks to its 20-ft (6-m) frame, it can reach tasty leaves high up in the trees *and* keep a lookout for threats on the ground. Grab your narrowest plates and tiles to construct a pocket-size LEGO® version of this humongous hoofed herbivore.

I've never seen a species quite like this!

Long muzzle with nostrils on top

Neck is six knobs long

Giraffes have brown patches on their coats

Thin tail with a tuft of hair at the end

Muscular legs for galloping

Smooth tiles for the front legs

Truncated cones make good hooves

ANIMAL FACT
Female giraffes give birth to their young standing up. That means new babies can fall up to 7 ft (2 m) as soon as they are born. Ouch! Why not make a little giraffe calf for your adult animal to take care of?

SKILL LEVEL

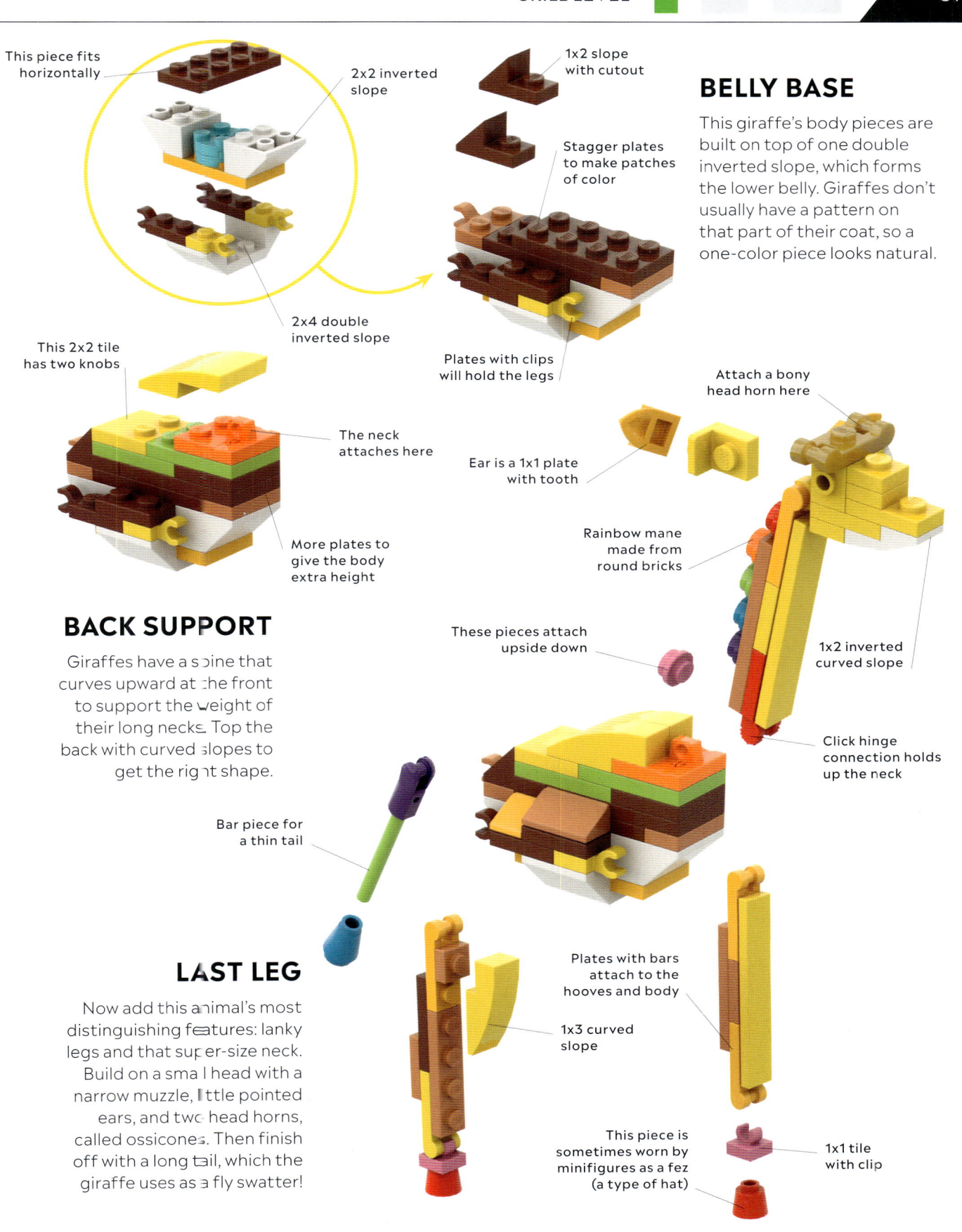

BELLY BASE

This giraffe's body pieces are built on top of one double inverted slope, which forms the lower belly. Giraffes don't usually have a pattern on that part of their coat, so a one-color piece looks natural.

- This piece fits horizontally
- 2x2 inverted slope
- 1x2 slope with cutout
- Stagger plates to make patches of color
- 2x4 double inverted slope
- Plates with clips will hold the legs

BACK SUPPORT

Giraffes have a spine that curves upward at the front to support the weight of their long necks. Top the back with curved slopes to get the right shape.

- This 2x2 tile has two knobs
- The neck attaches here
- More plates to give the body extra height
- Attach a bony head horn here
- Ear is a 1x1 plate with tooth
- Rainbow mane made from round bricks
- These pieces attach upside down
- 1x2 inverted curved slope
- Click hinge connection holds up the neck

LAST LEG

Now add this animal's most distinguishing features: lanky legs and that super-size neck. Build on a small head with a narrow muzzle, little pointed ears, and two head horns, called ossicones. Then finish off with a long tail, which the giraffe uses as a fly swatter!

- Bar piece for a thin tail
- Plates with bars attach to the hooves and body
- 1x3 curved slope
- This piece is sometimes worn by minifigures as a fez (a type of hat)
- 1x1 tile with clip

81

PIGLET

This little hog looks ready for a hug! Everything about this baby-faced farmyard animal is adorable, from the corners of its big pink ears to the tip of its curly tail. And the best thing? It's "sow" easy to make! Design your LEGO piglet plopped down in a sitting position, like this one, or build yours standing up, ready to trot to the food trough.

This little piggy... ate all my lunch!

These knobs look like tufts of hair

The underside of a wedge plate makes a good ear

Outstretched front leg

Tiny 1x1 round plate snout

Trotters are two pyramid slopes

Plump tummy

ANIMAL FACT
Pigs have a reputation for being dirty and smelly because they like to roll in mud, but they're actually quite clean by nature. They only roll in mud to keep cool and to protect their skin from the sun.

SKILL LEVEL

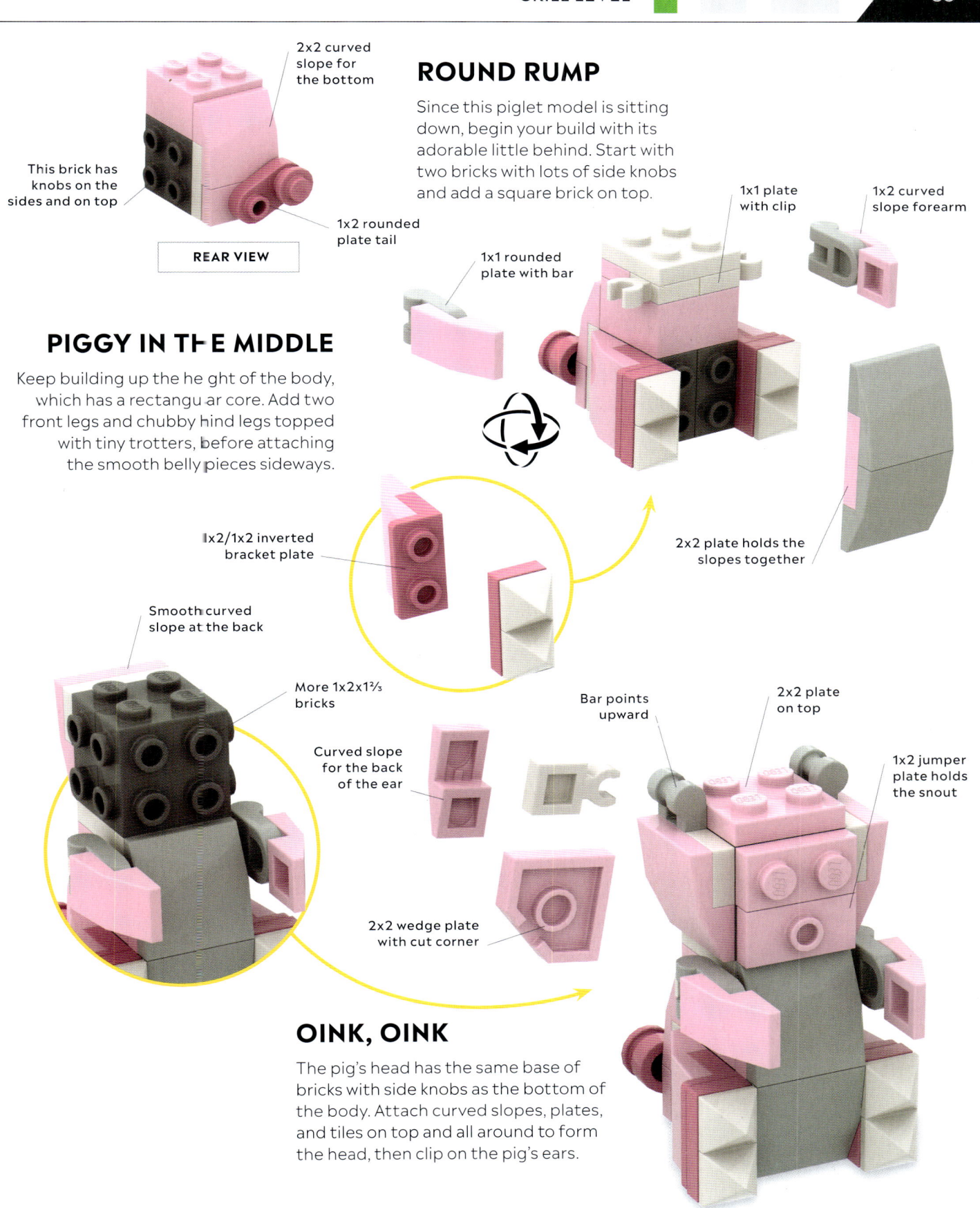

83

ROUND RUMP

Since this piglet model is sitting down, begin your build with its adorable little behind. Start with two bricks with lots of side knobs and add a square brick on top.

- 2x2 curved slope for the bottom
- This brick has knobs on the sides and on top
- 1x2 rounded plate tail

REAR VIEW

- 1x1 rounded plate with bar
- 1x1 plate with clip
- 1x2 curved slope forearm
- 2x2 plate holds the slopes together

PIGGY IN THE MIDDLE

Keep building up the height of the body, which has a rectangular core. Add two front legs and chubby hind legs topped with tiny trotters, before attaching the smooth belly pieces sideways.

- 1x2/1x2 inverted bracket plate
- Smooth curved slope at the back
- More 1x2x1²⁄₃ bricks
- Curved slope for the back of the ear
- 2x2 wedge plate with cut corner
- Bar points upward
- 2x2 plate on top
- 1x2 jumper plate holds the snout

OINK, OINK

The pig's head has the same base of bricks with side knobs as the bottom of the body. Attach curved slopes, plates, and tiles on top and all around to form the head, then clip on the pig's ears.

ANGELFISH

A serene-looking swimmer, the angelfish gets its name from its long, flowing fins. This one has a silvery-yellow body with black stripes, but there are many varieties of angelfish in different colors, so use whatever color pieces you have. They all have narrow bodies for weaving in and out of underwater plants and coral.

What a beauty!

Upside-down tile for the tapering dorsal fin

Caudal fin propels the body forward

Wide body shaped like an arrowhead

Pointed mouth is a 1x2 double slope

Transparent pieces look like shimmering scales

Ventral fin helps the fish steer through water

ANIMAL FACT
The colors of an angelfish reflect its mood. The colors can appear very bright when the fish is happy and well-nourished, and much paler when its not so content. How is your angelfish feeling?

SKILL LEVEL 85

SIDE STROKE

The angelfish's body is mostly made from lots of plates and bricks that face sideways. Begin your build with one long plate, like the yellow 1x8 plate shown here, and add more plates either side of it. Include plates with clips to attach two of the fish's fins to later.

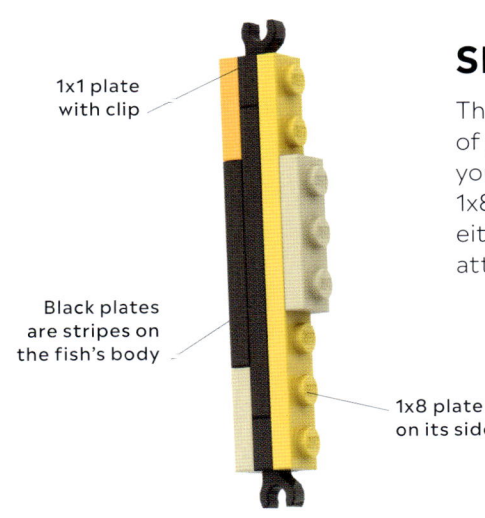

- 1x1 plate with clip
- Black plates are stripes on the fish's body
- 1x8 plate on its side

FISHY FEATURES

Keep building out the body in both sideways directions. Two slope bricks at the front of the fish's body create its recognizable arrowhead shape, while two smaller slopes at the back form the caudal fin.

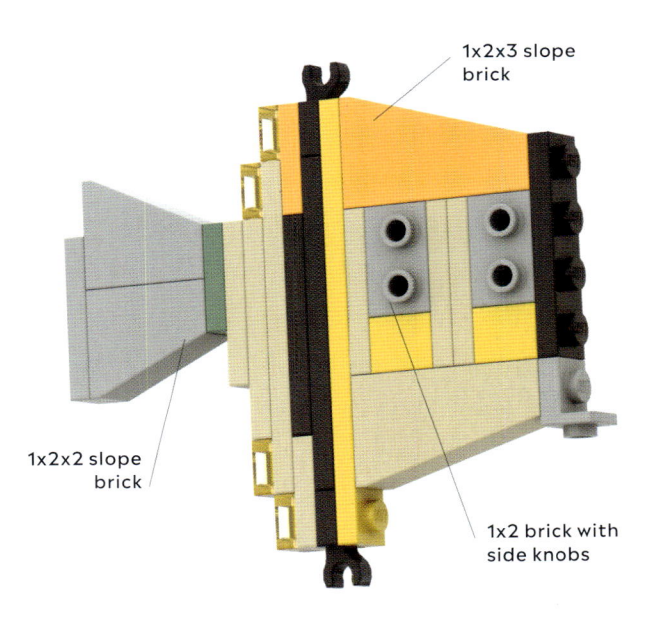

- 1x2x3 slope brick
- 1x2x2 slope brick
- 1x2 brick with side knobs

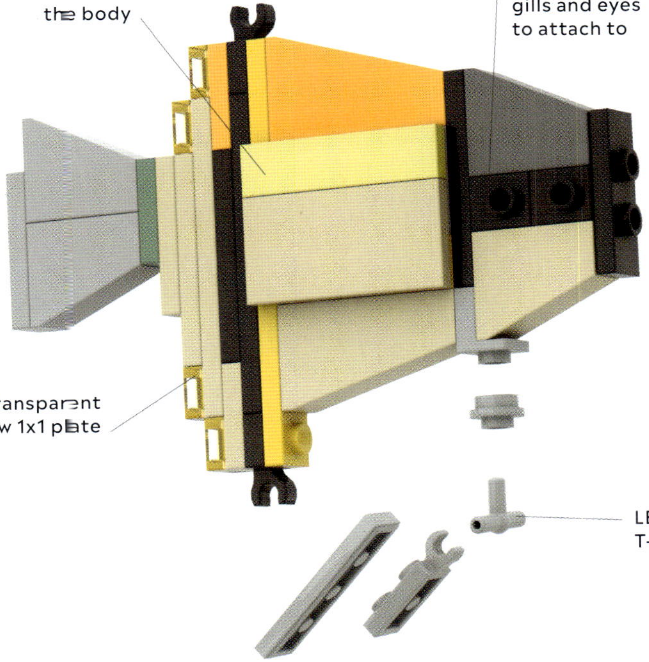

FIN-ISHING OFF

The fish's facial features and more fins are the last pieces to add. Some of the fins attach to the fish's body with clip and bar connections, but the ventral fin's connection is a little different. The two parts of the fin hang from a T-piece that fits to the underside of the body.

- Tiles widen the middle of the body
- Include side knobs for the gills and eyes to attach to
- Transparent yellow 1x1 plate
- LEGO® Technic T-piece

MOUSE

Squeak, squeak! This fast-moving little creature is just as quick to build. Give your mouse model big ears, a long tail, and a tiny twitching nose to make it instantly recognizable. Mice are known to grow in number very quickly, so you could build a whole "mischief" of them to live in your home! Just don't let them near your kitchen cabinets ...

Pointed snout

Extra-large ears with pink centers

Small, furry body

Long, hairless tail made from dinosaur tail elements

Clips make tiny clawed feet

Fancy a game of hide and squeak?

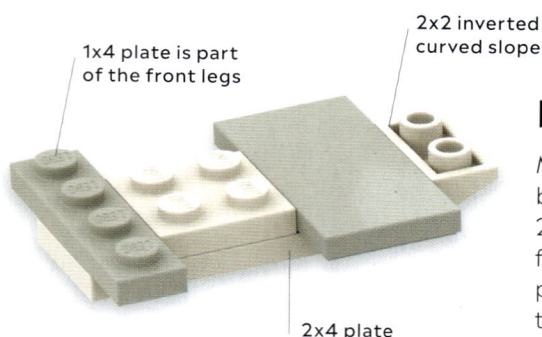

ANIMAL FACT
Mice belong to a group of small mammals called rodents. Rats, squirrels, hamsters, porcupines, and beavers are other examples of rodents. What other LEGO rodents could you build?

1x4 plate is part of the front legs

2x2 inverted curved slope

2x4 plate

BELLY BASE

Mice have soft, light-colored fur on their bellies, so begin your build with a white 2x4 plate. Add an inverted curved slope for the rounded rear, then place a gray plate and tile on top—these will become the upper parts of the mouse's legs.

SKILL LEVEL

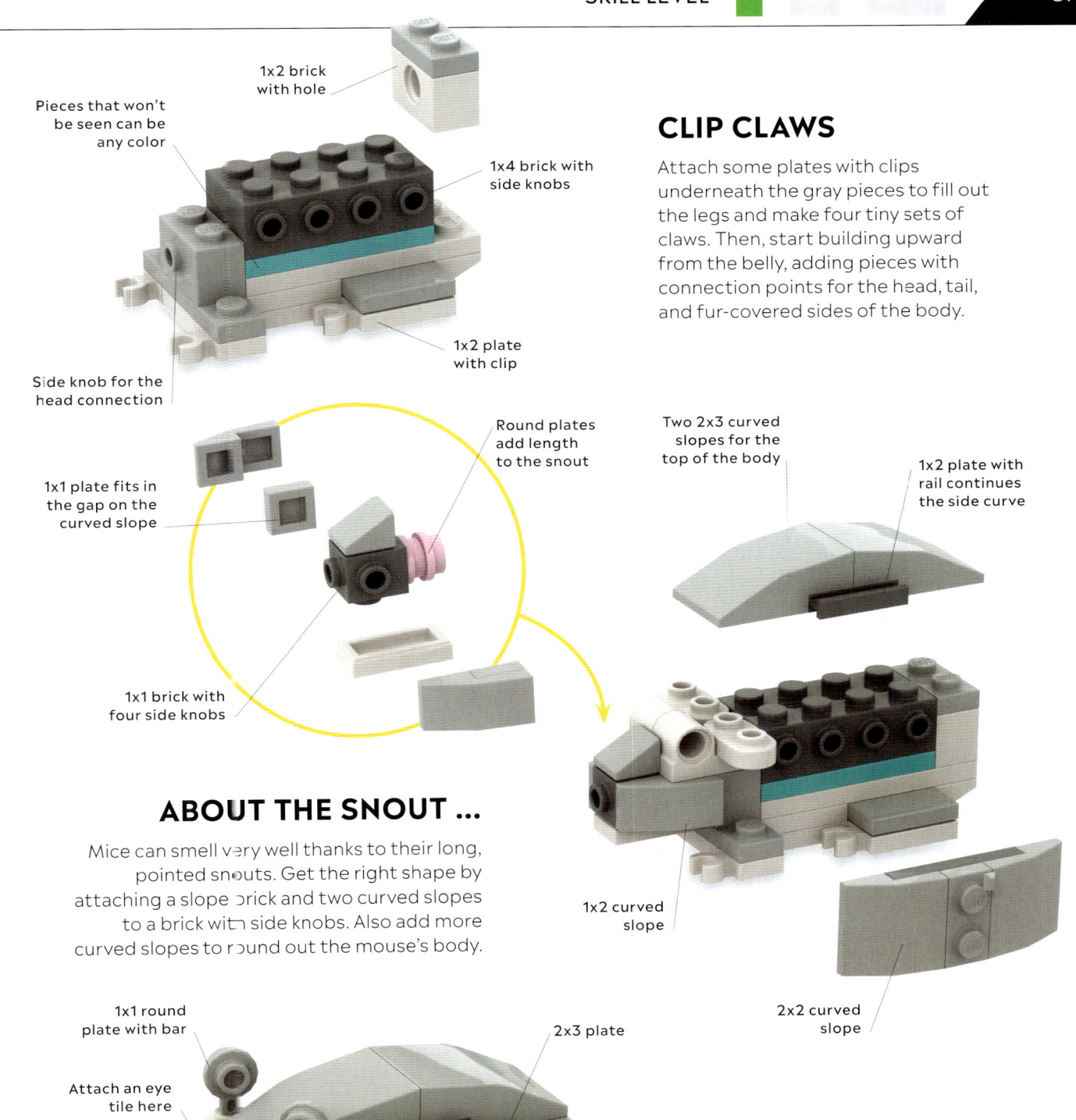

87

1x2 brick with hole

Pieces that won't be seen can be any color

1x4 brick with side knobs

CLIP CLAWS

Attach some plates with clips underneath the gray pieces to fill out the legs and make four tiny sets of claws. Then, start building upward from the belly, adding pieces with connection points for the head, tail, and fur-covered sides of the body.

1x2 plate with clip

Side knob for the head connection

Round plates add length to the snout

Two 2x3 curved slopes for the top of the body

1x2 plate with rail continues the side curve

1x1 plate fits in the gap on the curved slope

1x1 brick with four side knobs

ABOUT THE SNOUT ...

Mice can smell very well thanks to their long, pointed snouts. Get the right shape by attaching a slope brick and two curved slopes to a brick with side knobs. Also add more curved slopes to round out the mouse's body.

1x2 curved slope

2x2 curved slope

1x1 round plate with bar

Attach an eye tile here

2x3 plate

N-EAR-LY THERE

Every mouse needs big ears for listening for predators. To build some just like this model's, plug round plates with bars vertically into open knobs. Then attach a radar dish to the back of each round plate. Add eyes, a nose, and a tail, and your mouse is ready to make mischief.

1x2 rounded plate

FUR AND FEATHERS

The animal kingdom is incredibly diverse. Experiment with your LEGO collection to find just the right combination of parts to create your animals' unique and varied furs, fleeces, coats, feathers, or other fuzzy features. Here are some ideas to get you thinking about how to approach all types of fur and feathers. Flick through the pages of this chapter for further inspiration.

1x1 round plate with three leaves

Plant pieces look like the center and barbs of a feather

FEATHER FAN

There are so many ways to make LEGO feathers. Small, leaflike pieces are particularly useful, as shown here on the peacock's fantastic fan (learn how to build it on pages 128–129). Heart and quarter circle tiles also create interesting effects.

CURLY COAT

Some animals have unusual textures on their coats and fleeces, giving you an opportunity to play with your LEGO pieces to get the look just right. This sheep (see more of it on pages 98–99) has a thick, curly fleece formed by layered round plates, with drooping locks around its legs and on its chest.

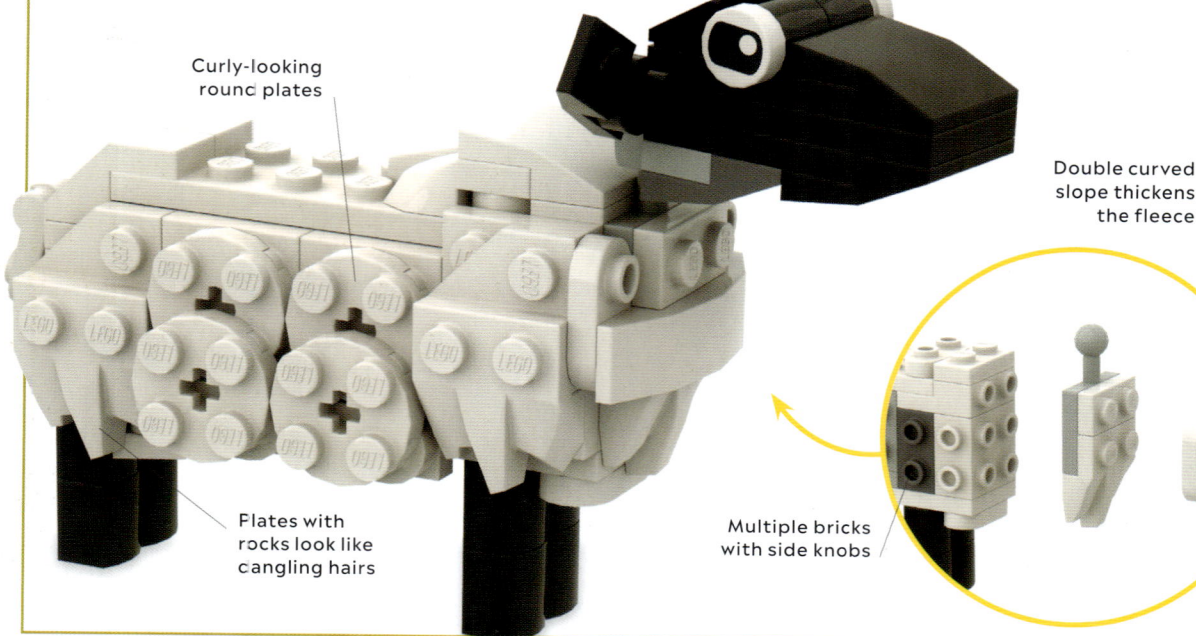

Curly-looking round plates

Double curved slope thickens the fleece

Plates with rocks look like dangling hairs

Multiple bricks with side knobs

IDEAS GALLERY   89

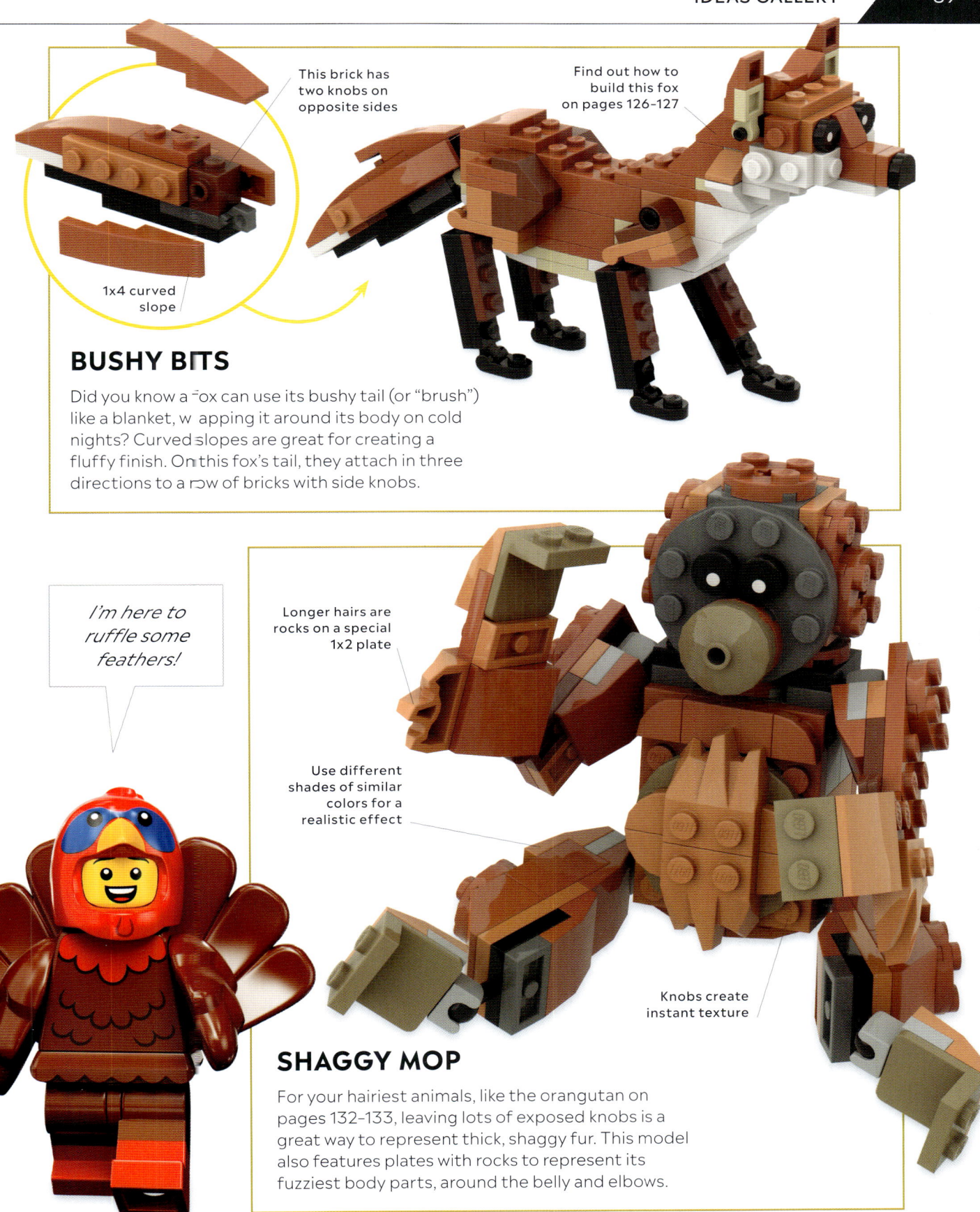

This brick has two knobs on opposite sides

Find out how to build this fox on pages 126–127

1x4 curved slope

BUSHY BITS

Did you know a fox can use its bushy tail (or "brush") like a blanket, wrapping it around its body on cold nights? Curved slopes are great for creating a fluffy finish. On this fox's tail, they attach in three directions to a row of bricks with side knobs.

I'm here to ruffle some feathers!

Longer hairs are rocks on a special 1x2 plate

Use different shades of similar colors for a realistic effect

Knobs create instant texture

SHAGGY MOP

For your hairiest animals, like the orangutan on pages 132–133, leaving lots of exposed knobs is a great way to represent thick, shaggy fur. This model also features plates with rocks to represent its fuzziest body parts, around the belly and elbows.

PARROT

You don't need any "pieces of eight" for this build—just a handful of colorful LEGO bricks and plates! Parrots live in the forests of warm, tropical countries. Give your parrot beautiful plumage, large wings, a strong hooked beak, and gripping claws. The ability to talk is optional!

Many parrots are vividly colored

Hooked beak is one curved slope piece

Wide, flapping wings

Clawed toes are good for climbing and grasping things

Long tail feathers

I'm going to need a bigger shoulder!

ANIMAL FACT
Many parrots are monogamous, which means they only have one mate for as long as they can. Parrot partners live near each other and look for food together. Build your LEGO parrot a loyal mate!

SKILL LEVEL

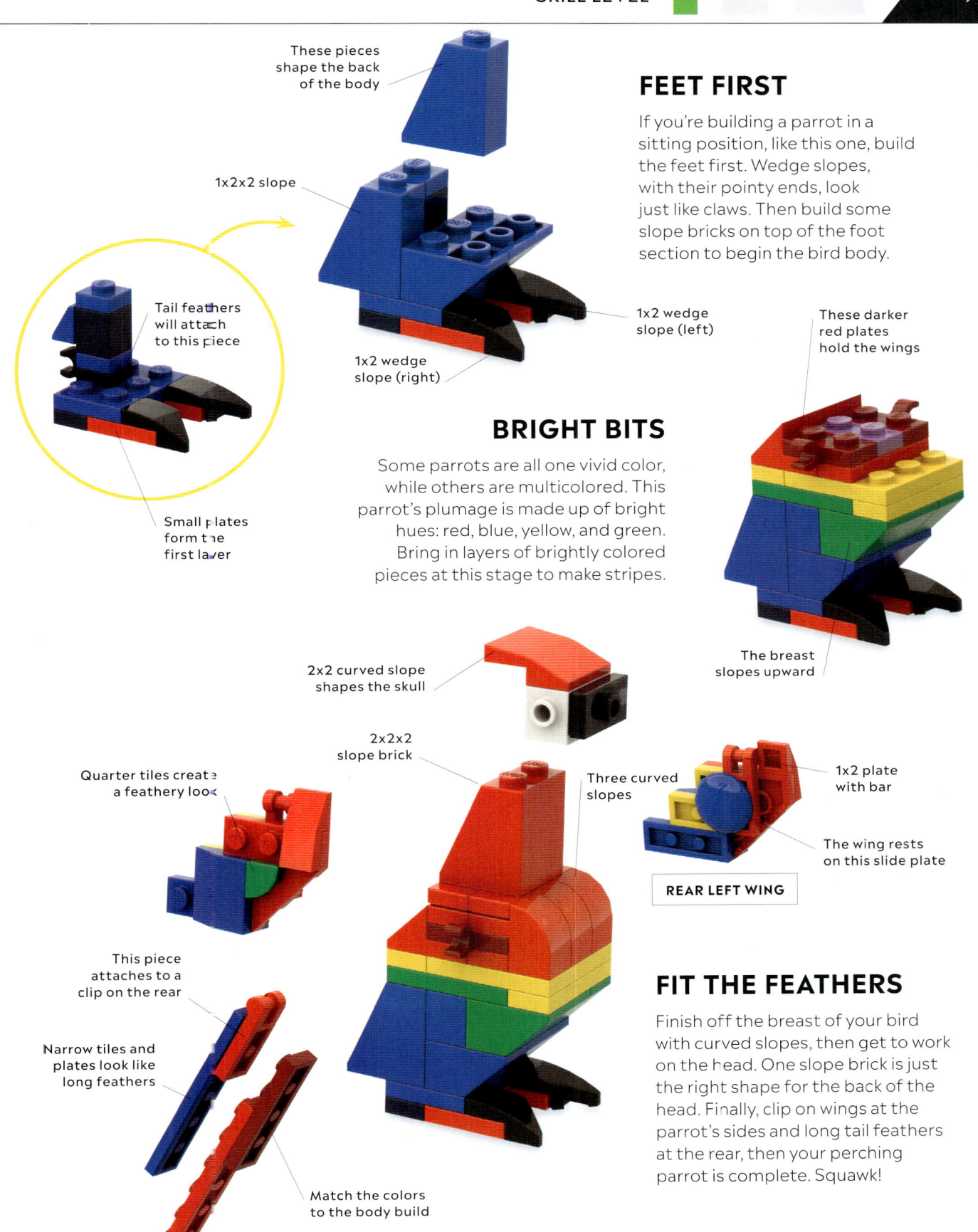

91

FEET FIRST

If you're building a parrot in a sitting position, like this one, build the feet first. Wedge slopes, with their pointy ends, look just like claws. Then build some slope bricks on top of the foot section to begin the bird body.

These pieces shape the back of the body

1x2x2 slope

1x2 wedge slope (left)

1x2 wedge slope (right)

Tail feathers will attach to this piece

Small plates form the first layer

These darker red plates hold the wings

BRIGHT BITS

Some parrots are all one vivid color, while others are multicolored. This parrot's plumage is made up of bright hues: red, blue, yellow, and green. Bring in layers of brightly colored pieces at this stage to make stripes.

The breast slopes upward

2x2 curved slope shapes the skull

2x2x2 slope brick

Three curved slopes

1x2 plate with bar

The wing rests on this slide plate

REAR LEFT WING

Quarter tiles create a feathery look

This piece attaches to a clip on the rear

Narrow tiles and plates look like long feathers

Match the colors to the body build

FIT THE FEATHERS

Finish off the breast of your bird with curved slopes, then get to work on the head. One slope brick is just the right shape for the back of the head. Finally, clip on wings at the parrot's sides and long tail feathers at the rear, then your perching parrot is complete. Squawk!

CAMEL

This amazing animal is perfectly adapted for living and working in hot, dry terrains. It can trek across deserts with no food or water thanks to its helpful hump, which is filled with fat that can nourish it for days. Build your own camel with a long neck, strong limbs, and a coat of sandy fur. How many humps will yours have?

Add a saddle for a rider

Curved neck made from click hinge plates

Smooth hump

Only 1,000 miles of desert to go!

Long, lean legs

Angle a foot for a walking posture

ANIMAL FACT
Camels have been used for transporting humans and heavy loads across deserts for thousands of years. Can you put your camel to work by building it some LEGO cargo to carry?

SKILL LEVEL

SADDLE UP

Begin your camel with the bulging belly section, then build a colorful cloth saddle right on top of it. The saddle fits neatly over the long sides of the belly bricks thanks to two bracket plates, which are smooth, like tiles, on the inside.

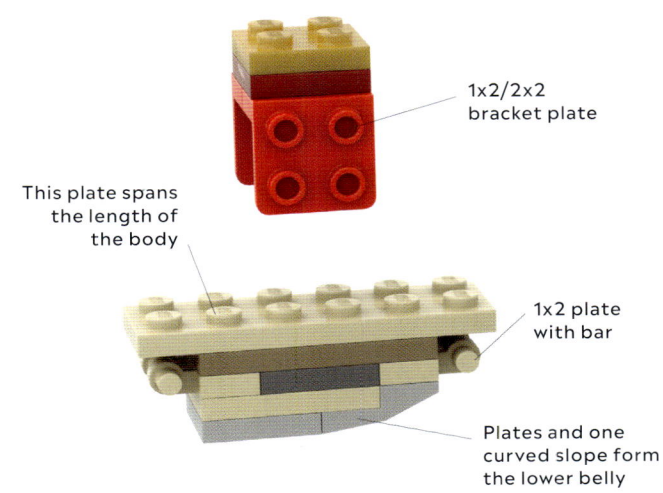

1x2/2x2 bracket plate

This plate spans the length of the body

1x2 plate with bar

Plates and one curved slope form the lower belly

1x2x1 curved slope padding

1x2 click hinge plate

Hump is two quarter tiles

GOT THE HUMP

Now build up the top sections of the camel's body and the saddle. Give the top of the back a downward curve shape and at least one hump. At this middle stage, don't forget to build in connecting pieces for the neck and tail.

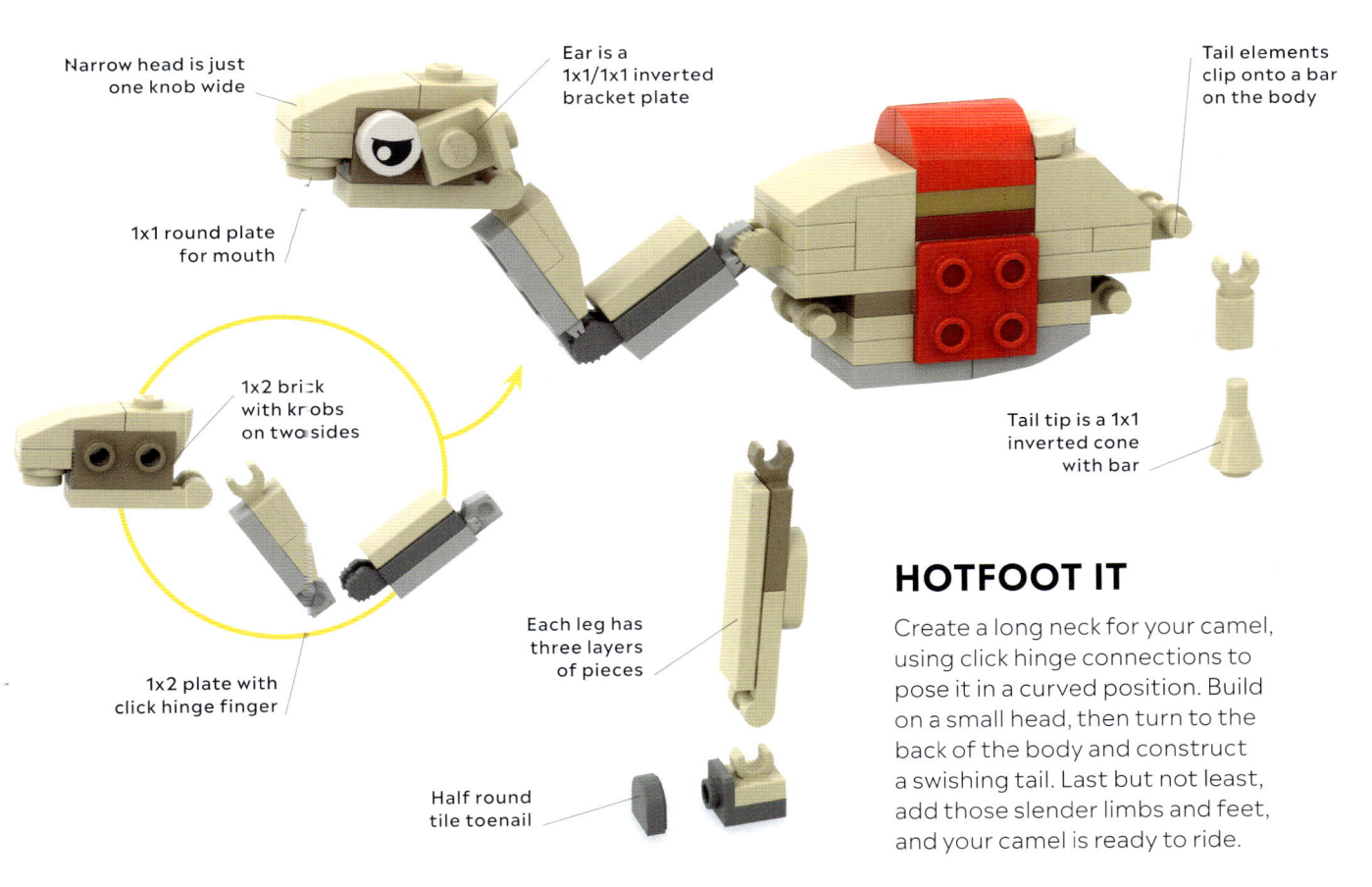

Some camels have two humps

Narrow head is just one knob wide

Ear is a 1x1/1x1 inverted bracket plate

Tail elements clip onto a bar on the body

1x1 round plate for mouth

1x2 brick with knobs on two sides

Tail tip is a 1x1 inverted cone with bar

1x2 plate with click hinge finger

Each leg has three layers of pieces

HOTFOOT IT

Create a long neck for your camel, using click hinge connections to pose it in a curved position. Build on a small head, then turn to the back of the body and construct a swishing tail. Last but not least, add those slender limbs and feet, and your camel is ready to ride.

Half round tile toenail

OCTOPUS

This legendary creature of the deep is famous for its eight long, powerful arms. They dangle behind the octopus's rounded body as it swims through the darkest depths of the ocean. Its two bulging eyes have excellent eyesight for spotting prey, such as crabs, clams, and lobsters, and sometimes larger creatures like sharks.

ANIMAL FACT
Octopuses can change color to communicate with others of their species or to blend in with their surroundings. Perhaps you could build yours to match a room in your home!

Rotund, muscular "mantle"

Large eye at the top of the head

Grasping arms can move in all directions

Each limb is wider at the top

The end of this piece looks like a sucker

Tapering limb is a dinosaur tail tip

When we first met, we shook hands eight times!

SKILL LEVEL   95

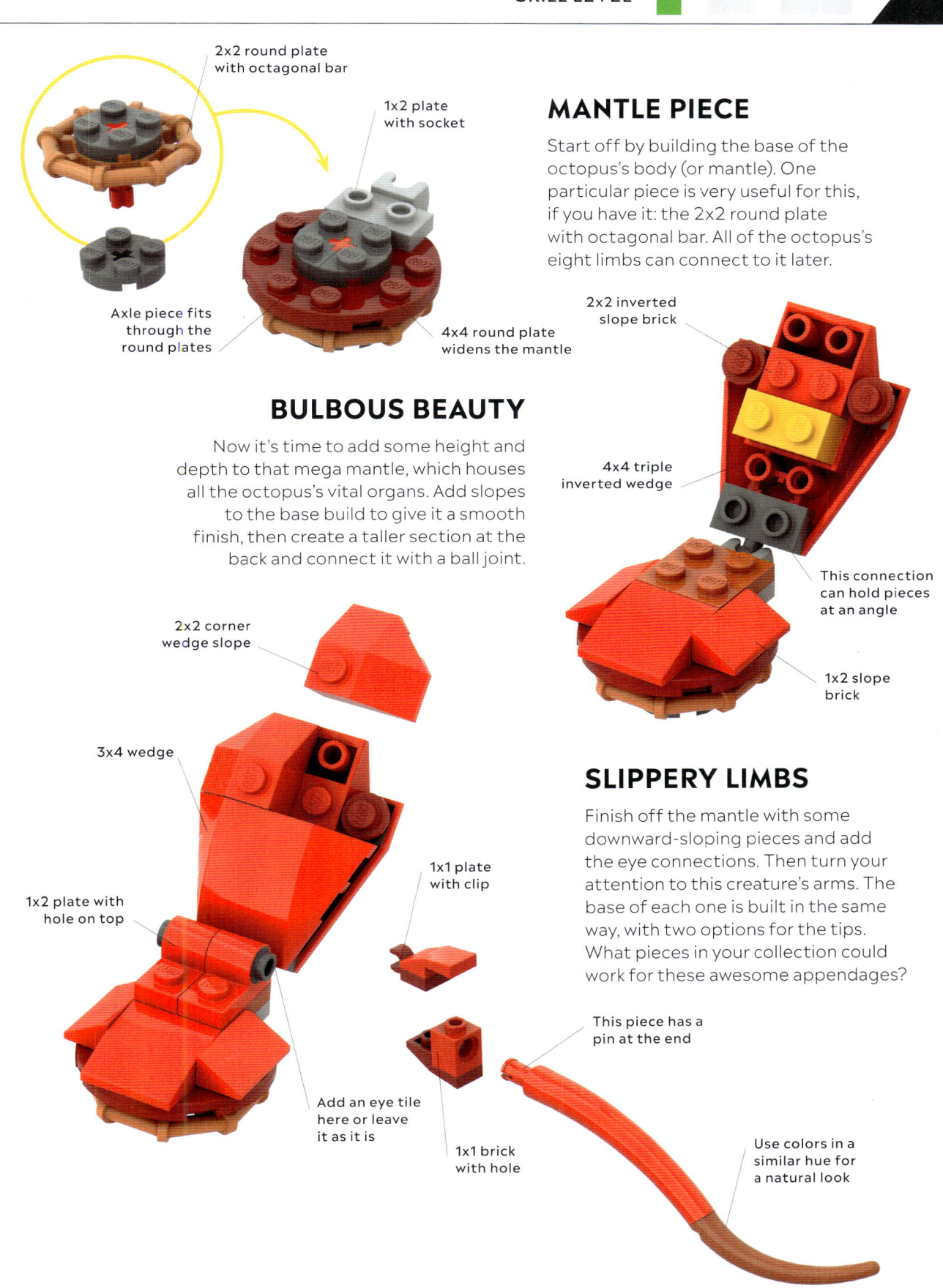

MANTLE PIECE

Start off by building the base of the octopus's body (or mantle). One particular piece is very useful for this, if you have it: the 2x2 round plate with octagonal bar. All of the octopus's eight limbs can connect to it later.

- 2x2 round plate with octagonal bar
- 1x2 plate with socket
- Axle piece fits through the round plates
- 4x4 round plate widens the mantle
- 2x2 inverted slope brick
- 4x4 triple inverted wedge
- This connection can hold pieces at an angle
- 1x2 slope brick

BULBOUS BEAUTY

Now it's time to add some height and depth to that mega mantle, which houses all the octopus's vital organs. Add slopes to the base build to give it a smooth finish, then create a taller section at the back and connect it with a ball joint.

- 2x2 corner wedge slope
- 3x4 wedge
- 1x2 plate with hole on top
- 1x1 plate with clip
- Add an eye tile here or leave it as it is
- 1x1 brick with hole

SLIPPERY LIMBS

Finish off the mantle with some downward-sloping pieces and add the eye connections. Then turn your attention to this creature's arms. The base of each one is built in the same way, with two options for the tips. What pieces in your collection could work for these awesome appendages?

- This piece has a pin at the end
- Use colors in a similar hue for a natural look

PLESIOSAURUS
PAGES 264–265

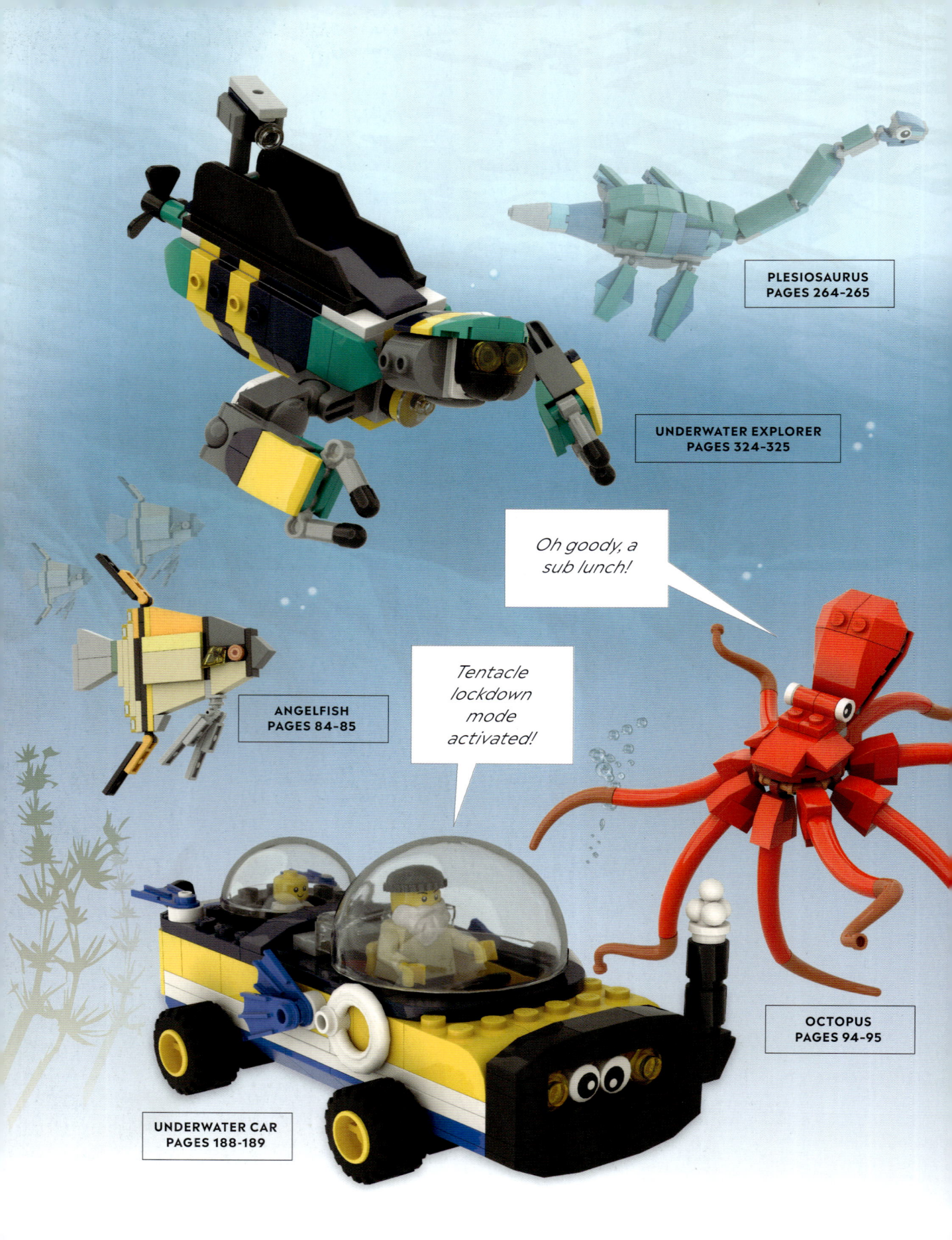

UNDERWATER EXPLORER
PAGES 324–325

Oh goody, a sub lunch!

Tentacle lockdown mode activated!

ANGELFISH
PAGES 84–85

OCTOPUS
PAGES 94–95

UNDERWATER CAR
PAGES 188–189

MODEL MASH-UP

ICHTHYOSAURUS
PAGES 262-263

I don't mean to be orc-ward, but could you please not refer to me as a killer whale?

ORCA
PAGES 122-123

UNDERWATER HOUSE
PAGES 58-59

SHEEP

This woolly wonder is a regular sight in fields and farmyards. Grab your black, white, and gray pieces and see if you can construct something similar, with a shaggy white fleece and a small black face and legs. Sheep like to live in large groups, called flocks, to protect themselves—why not build a whole flock?

ANIMAL FACT
Sheep have woolly bodies to keep them cozy in the winter. In the summer, their fleece is sheared off. Perhaps you could shear your LEGO sheep, building more closely cropped curls for warm weather!

Time for a cut. Just the ewe-sual?

Some sheep have white or gray faces

Protruding black ears

Gray neck fuzz

1x2 plate with rocks looks like longer wool

Layered round plates create a curly fleece

Narrow legs with hooves at the end

SKILL LEVEL

BUILDING BLOCK

This LEGO sheep starts off life as mostly a big block of bricks with side knobs. Those knobs will be used to attach the woolly fleece later. Add a piece with a clip or other connecting element to attach the tiny tail in the next steps.

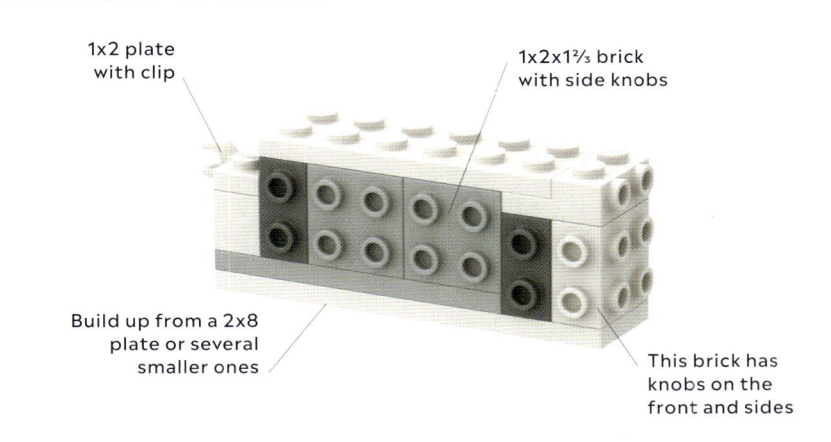

1x2 plate with clip

1x2x1²⁄₃ brick with side knobs

Build up from a 2x8 plate or several smaller ones

This brick has knobs on the front and sides

1x2 plate with ball for the head connection

LEGS AND HOOVES

Just two stacked-up bricks at the front and back of the body make four slender legs for the sheep to stand up on. Sheep have straight hooves at the ends of their legs, so there's no need to build feet! At the front, create a fleecy chest and add shape to the sheep's back.

1x2 log bricks for the legs

1x4 double curved slope

1x1 round tile with bar

Half pin is the eye connector

1x1 rounded plate with bar for a tail

Layer up pieces in different shapes

Plate with clip attaches to the ear pieces

This piece has a socket

1x2 rounded plate

FLUFFY FINISH

Find your woolliest-looking white pieces to fashion a fabulous fleece. Using lots of rounded and jagged-edged pieces gives the fleece a naturally curly, shaggy look. To finish, build on a little black head and a short woolly tail.

2x2 curved slope for the lower muzzle

99

TURTLE

When building LEGO animals, you don't have to be scientifically accurate. Have fun trying out unusual colors or features to come up with your own unique species. This rainbow reptile would brighten up any ocean with its multicolored shell!

Big, friendly eyes

Bony shell with bright "scutes"

Beak shaped from slope bricks

Fore flipper for swimming

Short, sturdy hind flipper

Wow! I turtley need a shell-fie.

ANIMAL FACT
A turtle's shell is made from tough bone. The top is called the carapace, and the lower part is the plastron. Most turtles can tuck their head, legs, and tail inside the shell if they're in danger.

1x4 inverted curved slope

2x6 brick

2x6 double round corner plate

4x6 plate

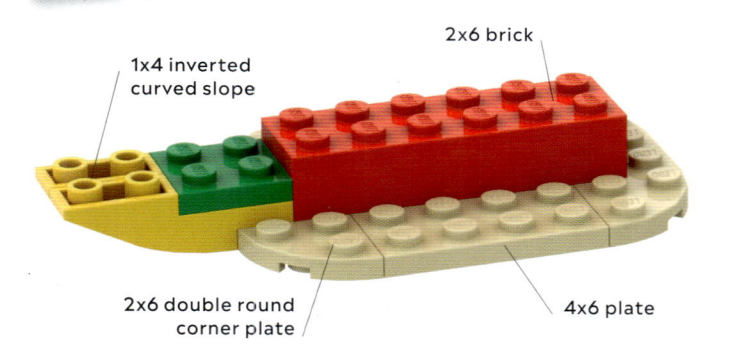

SHELL WE BEGIN?

A turtle's shell wraps right around its ribs. Start off your build by creating that part. Grab some plates in the right sort of shape and add a brick on top to begin the turtle's body. Then place inverted curved slopes at the front to form the upturned neck.

SKILL LEVEL

101

FLOATING FLIPPERS

Now the turtle's body is really taking shape. Build four identical flippers, placing the fore flippers facing toward the front and the hind flippers toward the back. Add a small wide tail at the rear and start building up the head and shell.

Include connection points for face details

Two 1x2 wedge slopes for the tail top

2x2 slide plate

1x2x1⅓ curved slope

Fill out the shell with square and curved bricks

UNDERNEATH VIEW

Little tiles give the skin a scaly look

1x2 brick connects the leg to the body

1x2 slope with cutout

This brick will soon be hidden

Extra plate layer

BUILD A BEAK

Finish off the turtle's head by topping it with two slope bricks. Their slanted angles combined with the inverted slope brick used for the lower part of the turtle's head make a pointed beak. Once the body is almost complete, start building up the height of the shell.

2x3 inverted slope

1x1 round plates are little claws

1x2 curved slope

2x2 round plates fit under the top layer

2x2 slope brick

COLORFUL CRITTER

Gather your brightest bricks and use them for the loud and proud upper shell. A first layer of sloped bricks in different shapes and sizes gives it a gentle curve. Create a final layer of curved slopes and round tiles to smooth off the top, then let your turtle take to the waves!

2x2 corner wedge slope

CHAMELEON

This colorful character is one of the coolest creatures around. Chameleons can change color to match their environment, to regulate their body temperature, or to communicate. What color will your chameleon be? They also have long tails for balancing on tree branches and two large, bulging eyes for looking around in all directions.

ANIMAL FACT
Chameleons have extra-long tongues for catching insects. Can you build a tongue into your model so it can grab some dinner?

Smooth back made from curved slopes

Small, bendy legs

Curved tail can grab or wrap around things

Rounded, radar dish eyelid

I feel like I'm being watched.

Match the environment to your chameleon's colors

Large cylinders form a tree-trunk perch

SKILL LEVEL

103

START LOW

Begin your build with the chameleon's flat lower body. Include an inverted curved slope at the front so it slopes upward toward the head. At the back, include a connecting piece to attach the tail to, like a plate with bar.

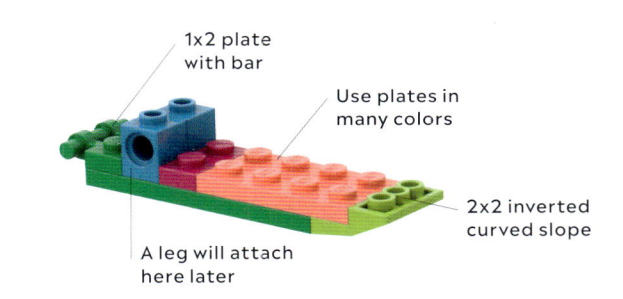

1x2 plate with bar

Use plates in many colors

2x2 inverted curved slope

A leg will attach here later

1x3 curved slope

1x4 curved slope

Pins are the leg connectors

The head will attach here

COLORFUL CURVES

Add more bricks and plates in lots of colors to build up the chameleon's bright body. The curve of the dorsal crest at the top of the body is made with curved slopes in two sizes. Slope bricks continue the curve at the back.

1x2 curved slope with cutout

1x2x1 curved slope is the head crest

Bar piece fits through two 1x1 bricks

BENDY BITS

Give your chameleon bendable legs and a curling tail by building lots of clip and bar connections into them. Finally, add the head and those distinctive eyes, which both slot onto a single bar piece.

Curved slopes widen this end of the tail

1x2 plate with two bars

Clip attaches to the body

This hole fits onto a pin on the body

This piece is also used in the tail

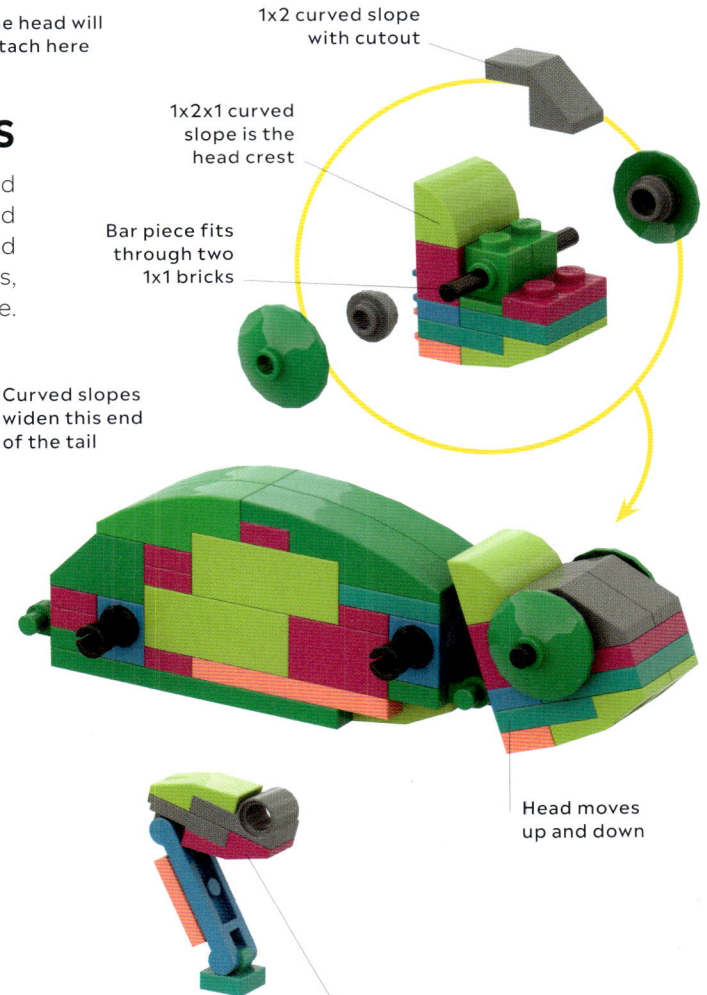

Foot is a 1x1 tile with clip

Head moves up and down

1x2 curved slope

REINDEER

Reindeer are a familiar sight at Christmas, when the most famous of their kind, red-nosed Rudolph, gets set to pull Santa's sleigh! Regular reindeer like this one can survive in some of the coldest regions on Earth. They live in large herds, moving around a lot in search of food. Both sexes have large antlers, and they have lots of waterproof fur to keep them warm and dry.

ANIMAL FACT
Did you know reindeer are strong swimmers? As they migrate across freezing landscapes, they sometimes have no choice but to plunge into icy rivers on their route. Their broad feet help them swim.

Are you available to pull a sleigh this December?

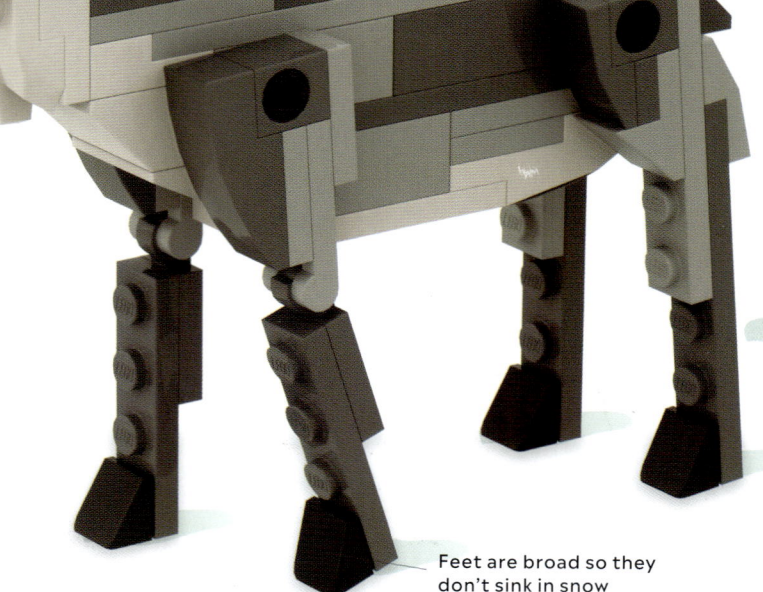

- Tough, twisted antlers
- Small, upright ears
- Thick, gray winter coat
- Long snout for finding food
- Feet are broad so they don't sink in snow

SKILL LEVEL 105

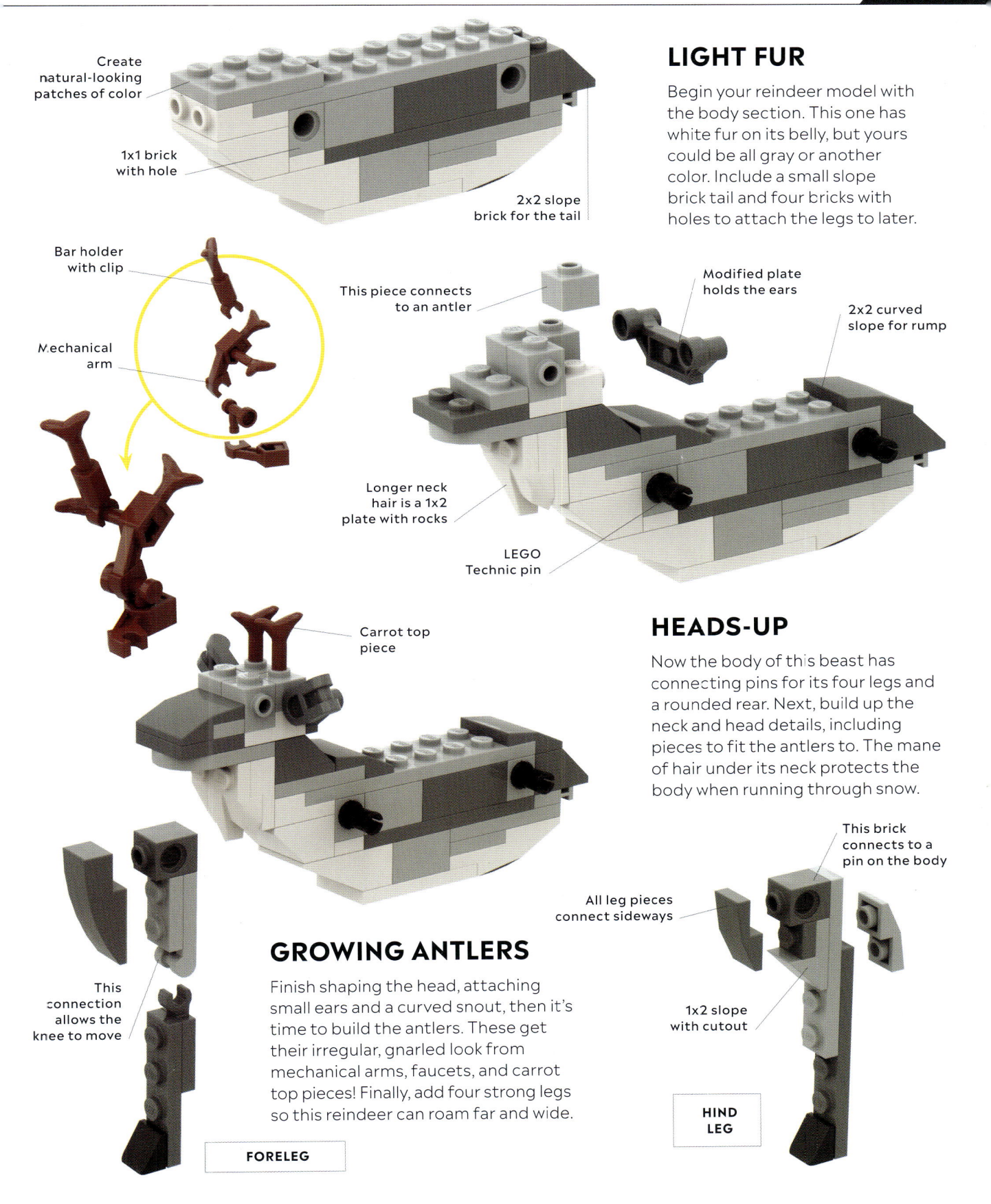

LIGHT FUR

Begin your reindeer model with the body section. This one has white fur on its belly, but yours could be all gray or another color. Include a small slope brick tail and four bricks with holes to attach the legs to later.

- Create natural-looking patches of color
- 1x1 brick with hole
- 2x2 slope brick for the tail

- Bar holder with clip
- Mechanical arm
- This piece connects to an antler
- Modified plate holds the ears
- 2x2 curved slope for rump
- Longer neck hair is a 1x2 plate with rocks
- LEGO Technic pin

HEADS-UP

Now the body of this beast has connecting pins for its four legs and a rounded rear. Next, build up the neck and head details, including pieces to fit the antlers to. The mane of hair under its neck protects the body when running through snow.

- Carrot top piece

- This connection allows the knee to move

- This brick connects to a pin on the body
- All leg pieces connect sideways
- 1x2 slope with cutout

GROWING ANTLERS

Finish shaping the head, attaching small ears and a curved snout, then it's time to build the antlers. These get their irregular, gnarled look from mechanical arms, faucets, and carrot top pieces! Finally, add four strong legs so this reindeer can roam far and wide.

FORELEG

HIND LEG

KOI CARP

Sit back and relax as you gaze at this beautiful fish. Often found in tranquil ornamental ponds, colorful koi carp create a calming environment in gardens and other beautiful places. They are symbols of love, courage, and good luck in Japan, where they have been admired for centuries. Perhaps you could build a koi as a gift for someone you love.

Konnichiwa, koi.

ANIMAL FACT
Koi are very tough and hardy fish. Japanese people fly koinobori—streamers shaped like koi—on an annual holiday called Children's Day because they represent growing up strong and brave.

- Long, silver dorsal fin
- This connection lets the tail swish
- Transparent pieces make the koi look luminous
- Bright scales in patches of color
- Little whiskers called barbels

SKILL LEVEL   107

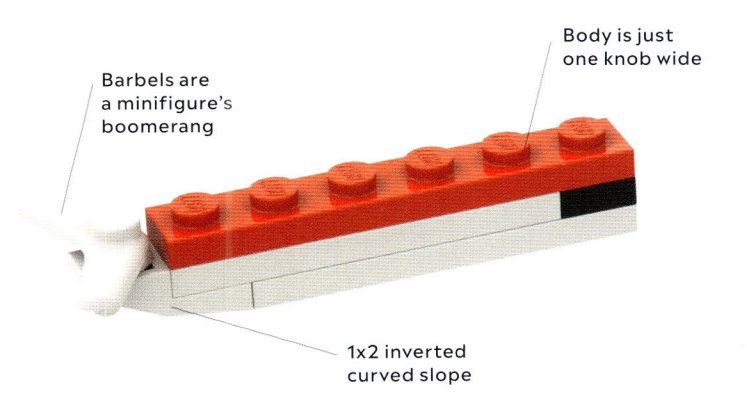

FOOD FIRST

Start off the fish's scaly body with long plates, placing a small curved slope at the front to give shape to the mouth. Koi have little whiskers near their mouths, called barbels, which they use to taste when looking for food in water. Add those at this stage too.

- Barbels are a minifigure's boomerang
- Body is just one knob wide
- 1x2 inverted curved slope

SCALE IT UP

This koi model is built in three parts. Finish off the largest section of the body first, including bright bricks for splashes of color. Use some bricks with side knobs for attaching more fishy features at the next stage, and a brick with bar for connecting the next section.

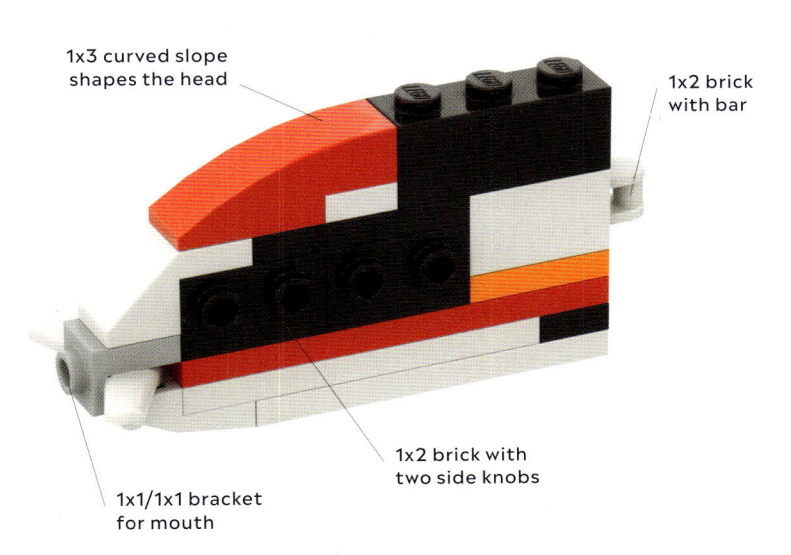

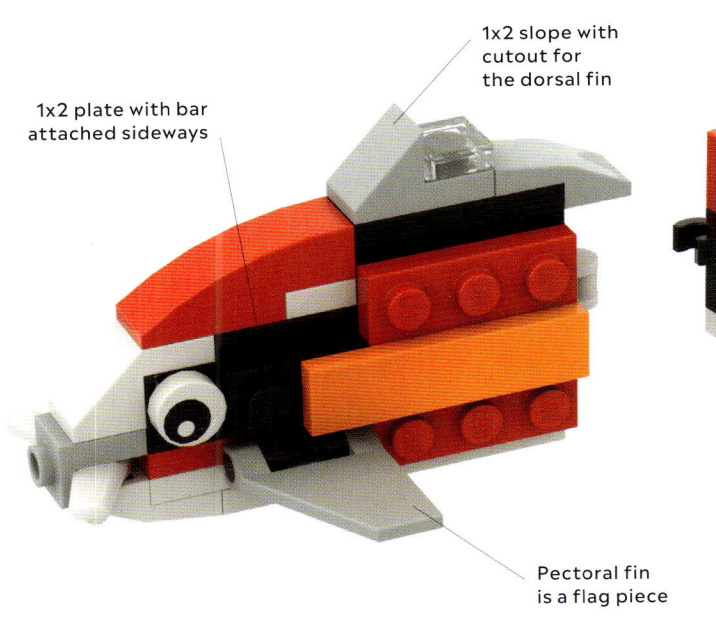

- 1x3 curved slope shapes the head
- 1x2 brick with bar
- 1x2 brick with two side knobs
- 1x1/1x1 bracket for mouth

- 1x2 plate with bar attached sideways
- 1x2 slope with cutout for the dorsal fin
- 1x3 slope brick tops the tapering tail
- 1x1 plate with clip
- Pectoral fin is a flag piece

MAKE A SPLASH

Attach fins and more bright scales to the chunkiest part of the carp, then construct the tail end. Using pieces that can move together to connect those parts means the koi can splash and swim in a realistic way.

OWL

This animal is a hoot to build! A beautiful bird of prey, an owl is nocturnal, which means it hunts for prey at night and snoozes during the day, usually high up in trees. Did you know owls can rotate their heads almost all the way around? Well, this LEGO version can rotate the full 360 degrees. Talk about eyes in the back of your head!

ANIMAL FACT
Owls' eyes are fixed in place, meaning they can't move them from side to side like humans can. This is why owls rotate their heads so far back—they need to in order to see around them.

It'll never spot me in this acorn hat!

Tufts of feathers called plumicorns

Hooked bill

Paler breast feathers

Dark-colored tail and wing feathers

Sharp talons for grasping prey

FEET FIRST

An owl's two powerful feet are an important part of its anatomy. They're used for catching and holding onto prey. Two curved plates are a good starting point for this section. They have holes at one end to attach two rear-facing talons (claws) to.

Upside-down plate with bar

1x4 plate holds the feet together

2x3 curved plate with hole

SKILL LEVEL   109

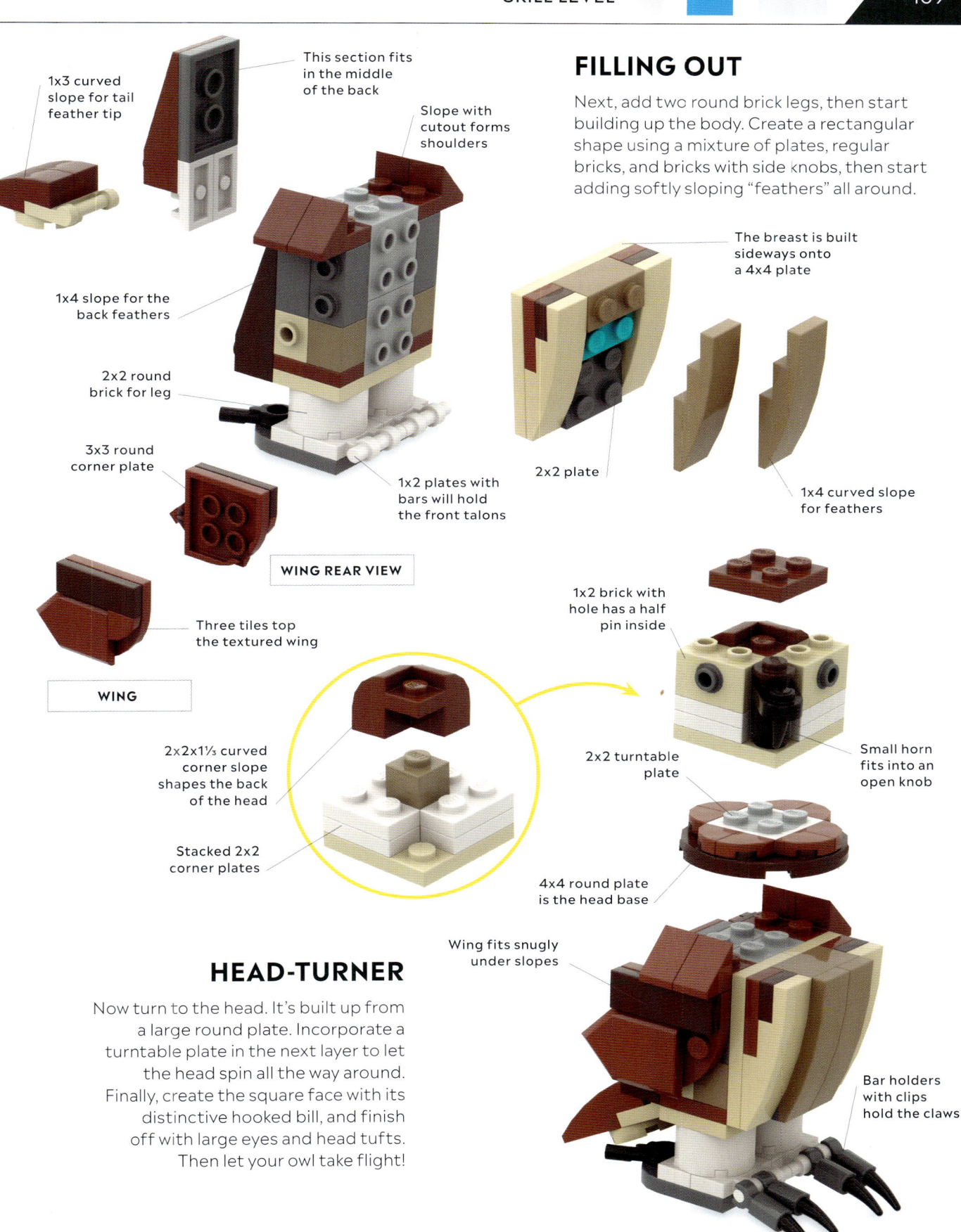

- 1x3 curved slope for tail feather tip
- This section fits in the middle of the back
- Slope with cutout forms shoulders
- 1x4 slope for the back feathers
- 2x2 round brick for leg
- 3x3 round corner plate
- 1x2 plates with bars will hold the front talons

FILLING OUT

Next, add two round brick legs, then start building up the body. Create a rectangular shape using a mixture of plates, regular bricks, and bricks with side knobs, then start adding softly sloping "feathers" all around.

- The breast is built sideways onto a 4x4 plate
- 2x2 plate
- 1x4 curved slope for feathers

WING REAR VIEW

- Three tiles top the textured wing

WING

- 2x2x1⅓ curved corner slope shapes the back of the head
- Stacked 2x2 corner plates

- 1x2 brick with hole has a half pin inside
- 2x2 turntable plate
- Small horn fits into an open knob
- 4x4 round plate is the head base
- Wing fits snugly under slopes
- Bar holders with clips hold the claws

HEAD-TURNER

Now turn to the head. It's built up from a large round plate. Incorporate a turntable plate in the next layer to let the head spin all the way around. Finally, create the square face with its distinctive hooked bill, and finish off with large eyes and head tufts. Then let your owl take flight!

CAT

This furry feline is the purrrrfect building project. Its multicolored, striped fur identifies it as a tabby cat, but you could make yours any color or pattern. Be sure to add classic kitty features like long whiskers, a little pink nose, a curling tail, and four soft paws before you declare your model complete.

I'm sure I just filled this bottle with fresh milk.

- Pointed, upright ears
- Whiskers are a minifigure hand claw
- Half round tile nose
- Tabbies have flecks, spots, or stripes on their coats
- Soft body curves
- Elephant trunk for a tail

ANIMAL FACT
Ancient Egyptians believed that cats were like gods, so they made beautiful statues of them. Perhaps you could turn your cat model into an ancient-looking statue as an extra building challenge.

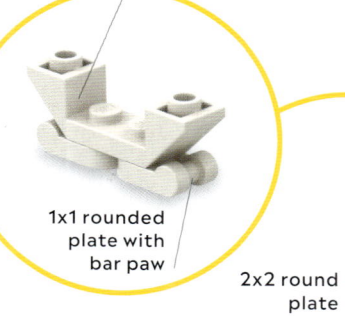

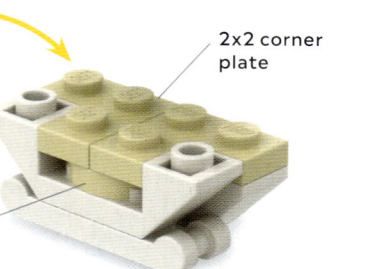

- 1x4 double inverted slope
- 1x1 rounded plate with bar paw
- 2x2 round plate
- 2x2 corner plate

START FROM SCRATCH

This cat is sitting down, so the lowest parts of its body are its rump and hind legs. Begin with those as a broad base, building upward from two double inverted slope bricks, which fit onto this section at different heights.

SKILL LEVEL   111

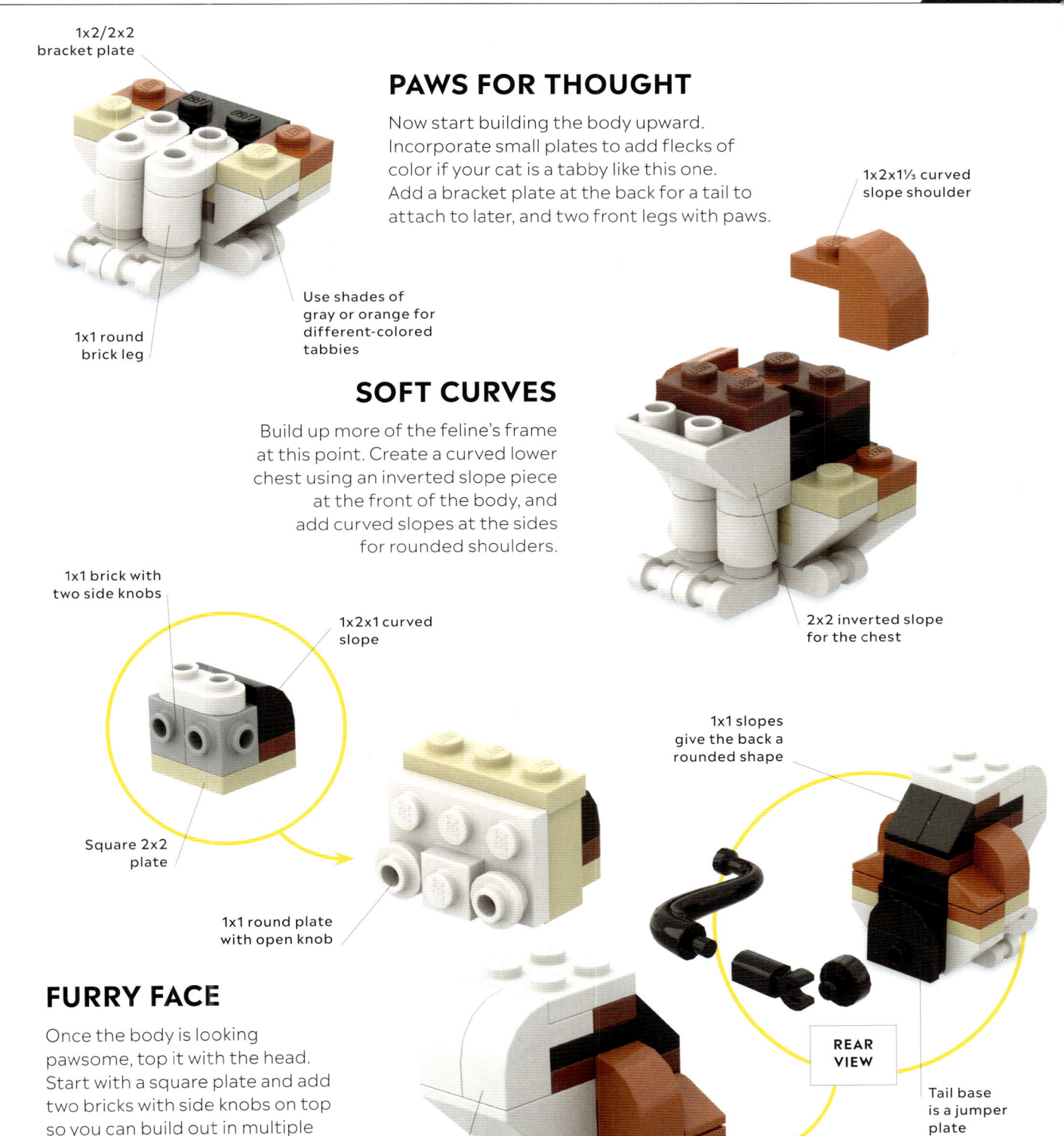

1x2/2x2 bracket plate

PAWS FOR THOUGHT

Now start building the body upward. Incorporate small plates to add flecks of color if your cat is a tabby like this one. Add a bracket plate at the back for a tail to attach to later, and two front legs with paws.

1x2x1⅓ curved slope shoulder

Use shades of gray or orange for different-colored tabbies

1x1 round brick leg

SOFT CURVES

Build up more of the feline's frame at this point. Create a curved lower chest using an inverted slope piece at the front of the body, and add curved slopes at the sides for rounded shoulders.

2x2 inverted slope for the chest

1x1 brick with two side knobs

1x2x1 curved slope

1x1 slopes give the back a rounded shape

Square 2x2 plate

1x1 round plate with open knob

FURRY FACE

Once the body is looking pawsome, top it with the head. Start with a square plate and add two bricks with side knobs on top so you can build out in multiple directions. Attach more plates at the top and front for the ears and face details. Don't forget to add the cat's long tail at the back of the body as a final furry flourish.

Curved slopes round out the chest

REAR VIEW

Tail base is a jumper plate

KOALA

This marvelous marsupial is native only to Australia. Herbivorous koalas spend their days high up in eucalyptus trees, eating leaves and sleeping for up to 20 hours a day. Create your own LEGO koala, with a round head; large, black nose; and stout, cuddly body. Don't forget some curved claws to help it hang onto trees.

ANIMAL FACT
Koalas are related to their fellow marsupials, kangaroos. Like female kangaroos, female koalas have pouches on their bellies that are used to raise their young. Can you create a koala with a baby joey in its pouch?

I have a lot of koalafications for this job.

- Circular ears
- Gray, fur-covered head and body
- Long, leathery black nose
- Large, padded rump for resting on for hours
- Feet and claws built for clinging
- Soft, white belly fur

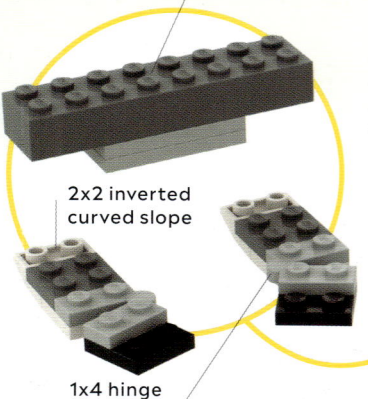

- 2x8 brick
- 2x2 inverted curved slope
- 1x4 hinge plate
- Build more plates and bricks on top of the legs
- 1x2x1⅓ curved slope
- 1x2 plate with three teeth

BUM BRICK

Koalas rest *a lot*, so they need a comfortable place to sit. Start off your koala build with a wide, flat rump made from one long brick. Attach two flexible legs below the brick rump, using hinge plates to let the legs move inward.

SKILL LEVEL

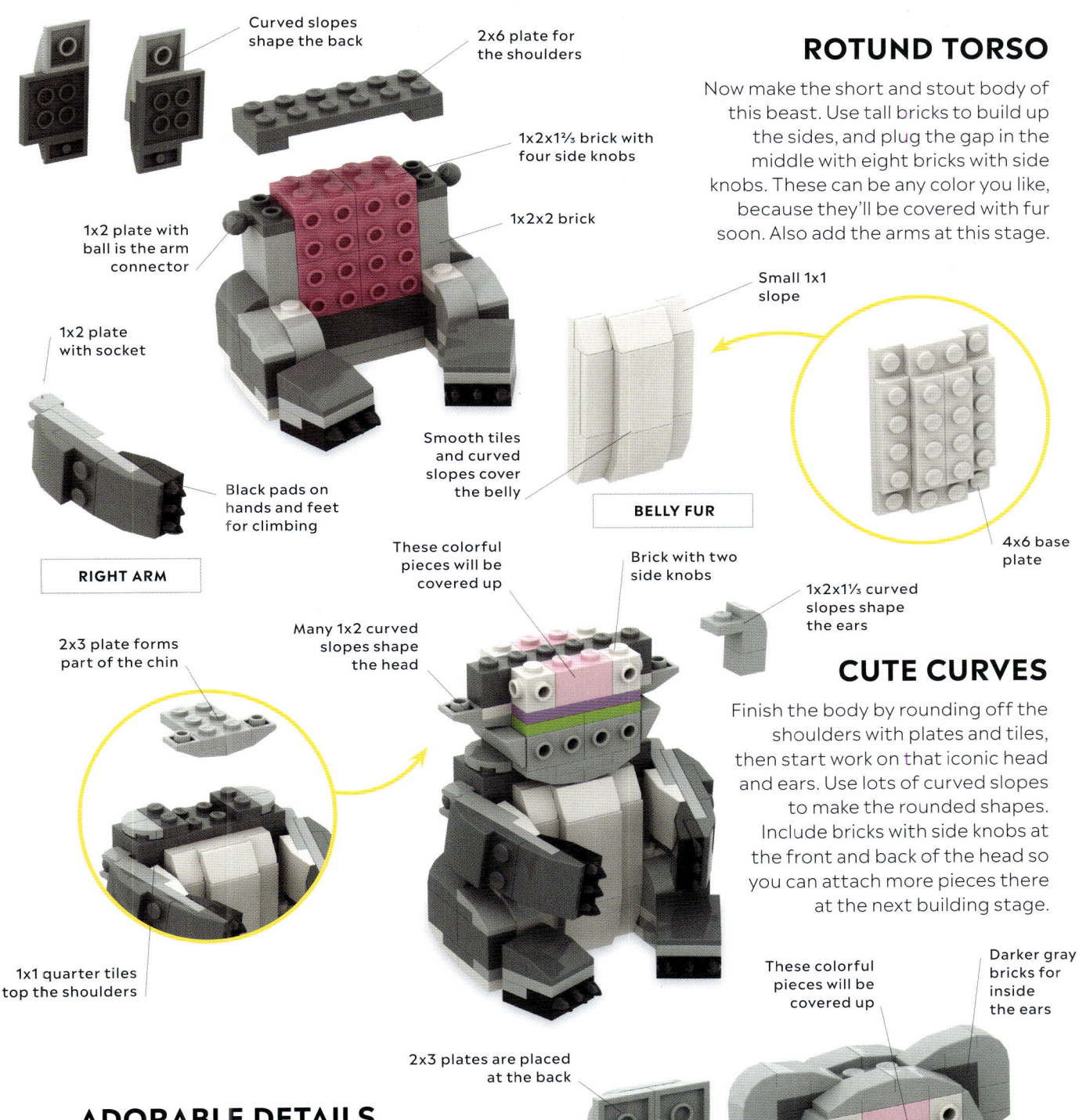

ROTUND TORSO

Now make the short and stout body of this beast. Use tall bricks to build up the sides, and plug the gap in the middle with eight bricks with side knobs. These can be any color you like, because they'll be covered with fur soon. Also add the arms at this stage.

- Curved slopes shape the back
- 2x6 plate for the shoulders
- 1x2x1⅔ brick with four side knobs
- 1x2x2 brick
- 1x2 plate with ball is the arm connector
- 1x2 plate with socket
- Black pads on hands and feet for climbing
- Small 1x1 slope
- Smooth tiles and curved slopes cover the belly
- 4x6 base plate

BELLY FUR

RIGHT ARM

- 2x3 plate forms part of the chin
- These colorful pieces will be covered up
- Many 1x2 curved slopes shape the head
- Brick with two side knobs
- 1x2x1⅓ curved slopes shape the ears

CUTE CURVES

Finish the body by rounding off the shoulders with plates and tiles, then start work on that iconic head and ears. Use lots of curved slopes to make the rounded shapes. Include bricks with side knobs at the front and back of the head so you can attach more pieces there at the next building stage.

- 1x1 quarter tiles top the shoulders
- These colorful pieces will be covered up
- Darker gray bricks for inside the ears
- 2x3 plates are placed at the back
- 1x2 brick with two side knobs

ADORABLE DETAILS

Finally, round out the back of the head and build a sweet face using sideways building techniques. Two 2x3 plates attach in both directions, creating a rounded 3D shape. Add jumper plates and small tiles to those plates at the front to give your koala pal lots of personality.

113

PUPPY

Woof, woof! Make a playful puppy using your LEGO pieces and you'll have a loyal friend for life—without all the feeding, walking, and cleaning up that comes with owning a real pet dog. With those big floppy ears, cuddly looking fur, and wagging tail, who could resist building this little fella?

ANIMAL FACT
Dogs come in all different shapes, sizes, and colors. There are around 200 official dog breeds in the world, with more being created all the time. Challenge yourself to build a variety of canine companions!

My pup loves to come to work with me!

- Long, spaniel-like ears
- Multicolored fur is shades of black and brown
- Panting tongue
- Layered pieces are long leg fur
- Half circle tile paws
- This piece will soon be hidden
- Top layer of plates
- 2x4 double inverted slope brick
- 2x3 inverted slope is the neck

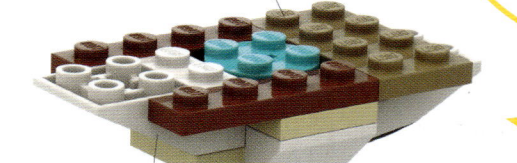

BELLY RUB
Dogs love to have their bellies scratched, so start off your build with that sweet spot. The lowest point of this section is an inverted slope brick. Build several layers of plates and bricks at a right angle on top.

SKILL LEVEL   115

BUSY LEGS

Build four legs onto your belly build for your pooch to pad around on. One arch brick forms the two forelegs, while small bricks and bracket plates create the hind legs. Include slope bricks above all four legs for the thighs.

- One 1x4 arch makes two legs
- 1x2 inverted slope brick for thigh
- 1x2 curved slope for brow
- 1x1 brick with two side knobs
- 1x1/1x2 inverted bracket plate
- Plates and tiles shape the back of the head
- 2x2 curved slope rear
- Row of 1x2 jumper plates
- An eye tile will connect here
- Muzzle is a 1x2 brick with side knobs
- 3x3 cross plate
- 1x2 plate for paw pad

HOUND'S HEAD

Finish off your furball's body with more plates and slopes, then turn your attention to its head. The head is built upward from a cross-shaped plate base. Include bracket plates with side knobs for the ears to attach to at the next step.

- 2x3 wedge plate ear
- 2x2 wedge plate jowl
- Small slopes are the tops of the ears
- Dinosaur tail tip piece

TAIL END

Attach the last of the furry features to the face—two floppy ears and a pair of jiggling jowls—and this pup is pawsitively perfect! Finish off by adding a long tail for your dog to wag when it's feeling happy.

- 1x2 brick with one side knob
- 1x2 jumper plate chin

REAR VIEW

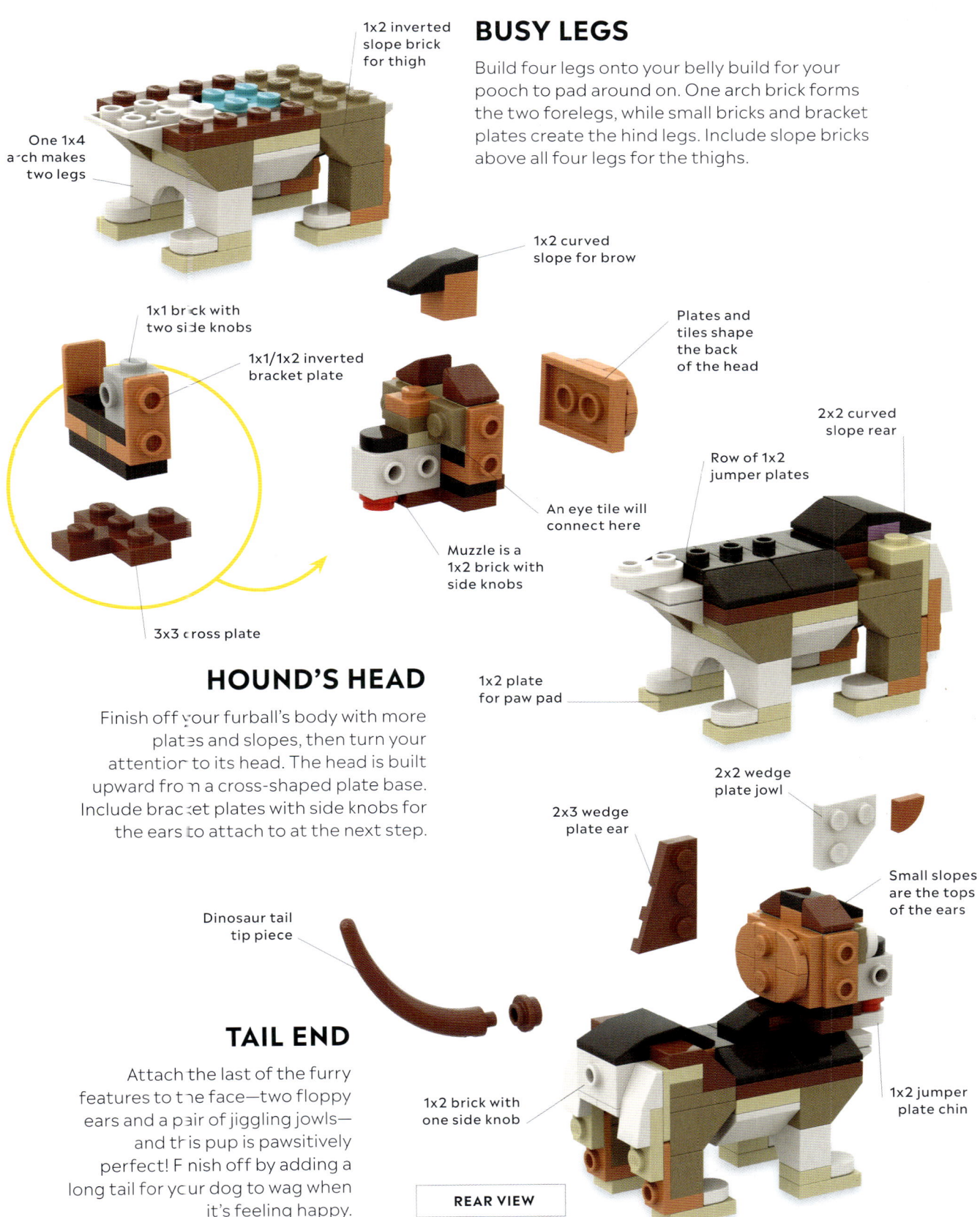

IMAGINARY CRITTERS

When it comes to LEGO animal building, there's no rule that says you have to build creatures from the real world. Build fantastical beasts from books, movies, computer games, or straight out of your imagination. If you like, base your creations on everyday animals, then go wild with colors, body shapes, and facial features.

1x2 brick with side knobs holds the wing

1x2 curved slopes for the comb

1x2 wedge slope wing

Is this chicken wearing shoes?

FLOATING FELLOW

How will your imaginary critter get around? Many animals walk, hop, fly, or slither, but what about an animal that floats? It'll need a belly filled with air for buoyancy. This animal's rounded snout makes it look like its cheeks could be full of air too.

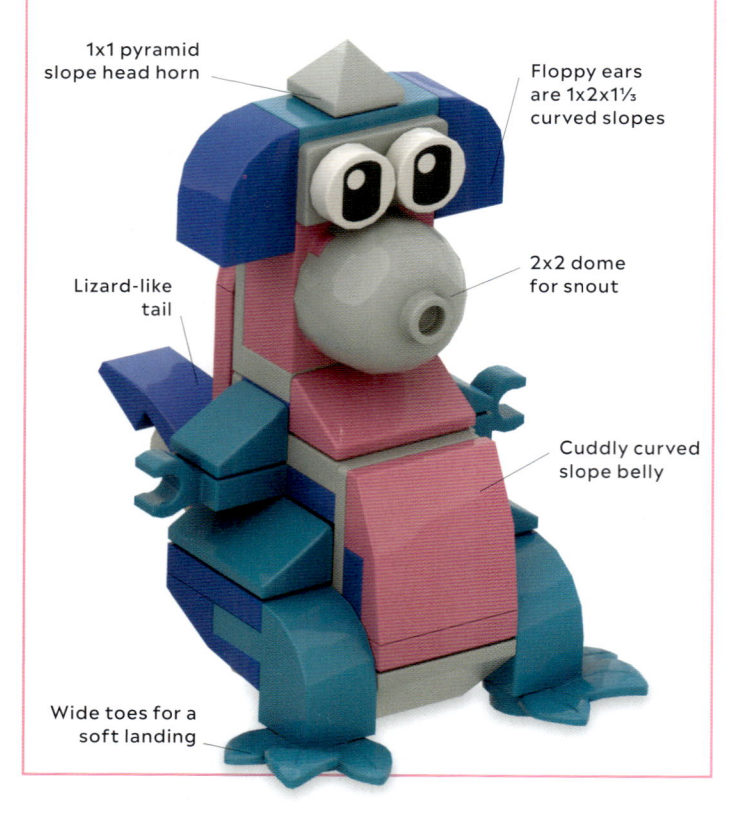

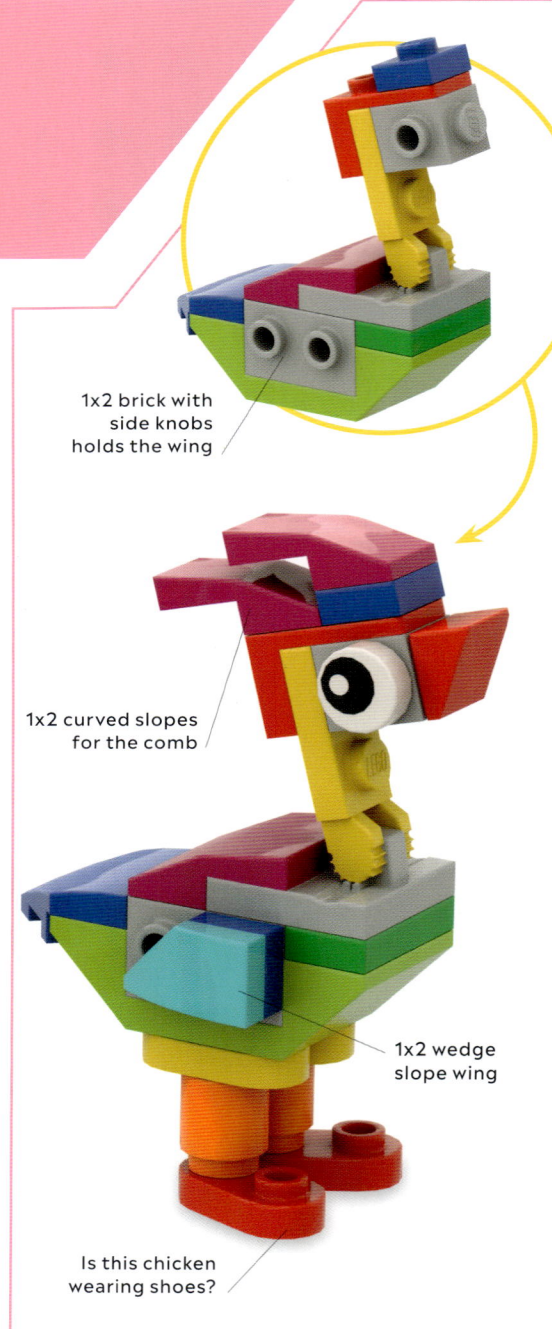

1x1 pyramid slope head horn

Floppy ears are 1x2x1⅓ curved slopes

Lizard-like tail

2x2 dome for snout

Cuddly curved slope belly

Wide toes for a soft landing

FUNKY CHICKEN

With familiar features like wings, a beak, and a head plume (or comb), this creature's body shape looks a bit like a chicken's, but the bright breed can't be found anywhere on Earth. Clashing primary colors can turn an everyday animal into a funky, feathered phenomenon!

IDEAS GALLERY 117

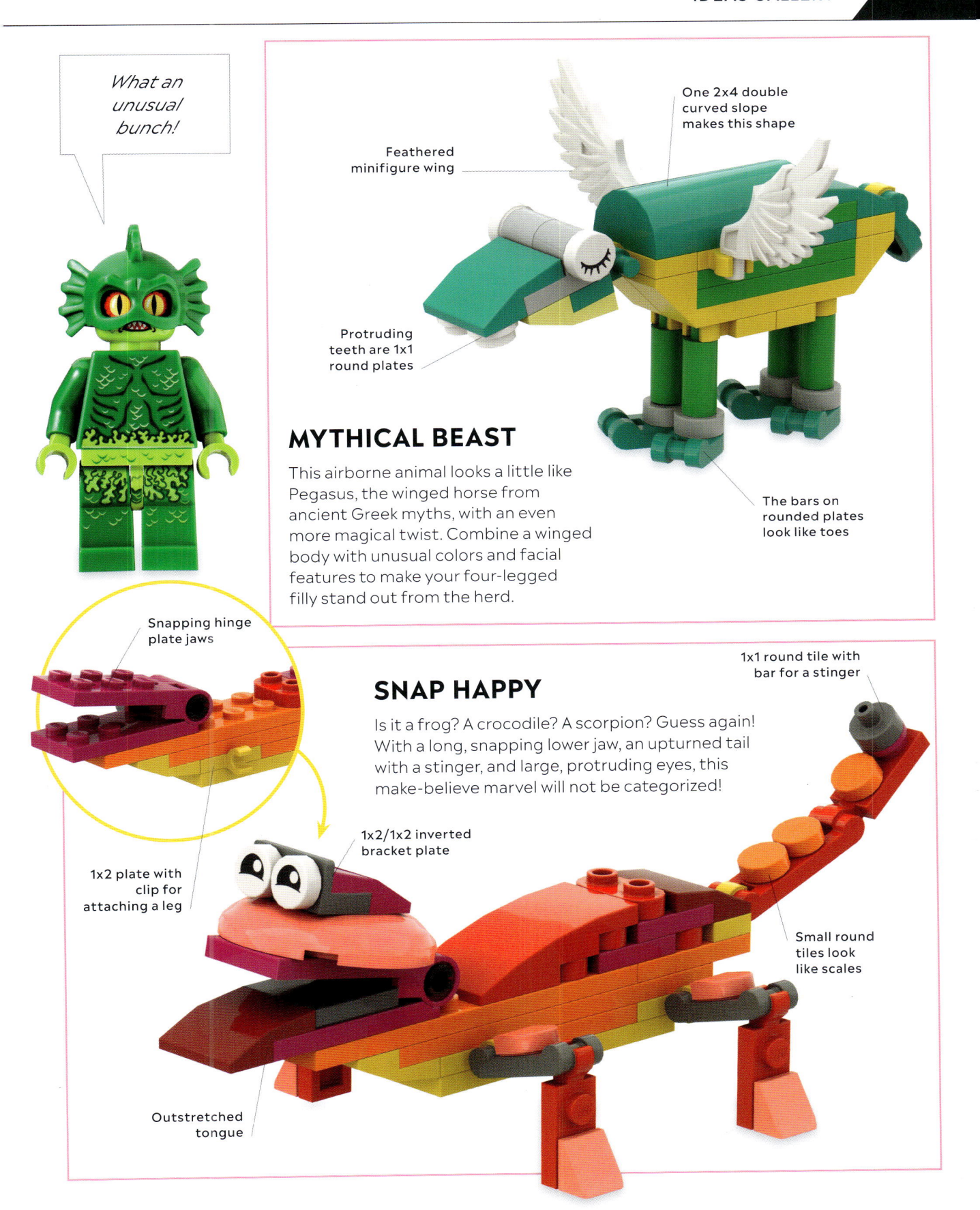

What an unusual bunch!

One 2x4 double curved slope makes this shape

Feathered minifigure wing

Protruding teeth are 1x1 round plates

MYTHICAL BEAST

This airborne animal looks a little like Pegasus, the winged horse from ancient Greek myths, with an even more magical twist. Combine a winged body with unusual colors and facial features to make your four-legged filly stand out from the herd.

The bars on rounded plates look like toes

Snapping hinge plate jaws

SNAP HAPPY

Is it a frog? A crocodile? A scorpion? Guess again! With a long, snapping lower jaw, an upturned tail with a stinger, and large, protruding eyes, this make-believe marvel will not be categorized!

1x1 round tile with bar for a stinger

1x2/1x2 inverted bracket plate

1x2 plate with clip for attaching a leg

Small round tiles look like scales

Outstretched tongue

SLOTH

This slow-moving mammal loves nothing more than to snooze the day away in a rainforest treetop, waiting for (hopefully) nothing to happen. It has a rounded head and body covered in shaggy fur, and long limbs for all that hanging around. Thanks to a very slow metabolism, sloths can sleep for up to 20 hours a day, so take your time with this model …

ANIMAL FACT
Sloths can't move around well on the ground, but they are good swimmers. They sometimes drop down from the trees to take a dip in water. Perhaps you could build a river for this model to splash into.

- Small, circular head
- Big, drowsy eyes
- Grasping claws on hands and feet
- Limbs ready to cling to a tree branch

I've been here for days just waiting for it to move.

- 2x2 round brick
- 2x2 inverted slope brick bottom

REAR VIEW

First of all, give your sloth a nice comfortable rump for it to flop down on. Start with a layer of plates, then add a brick layer on top. Include slopes and round bricks to build up the curved body shape.

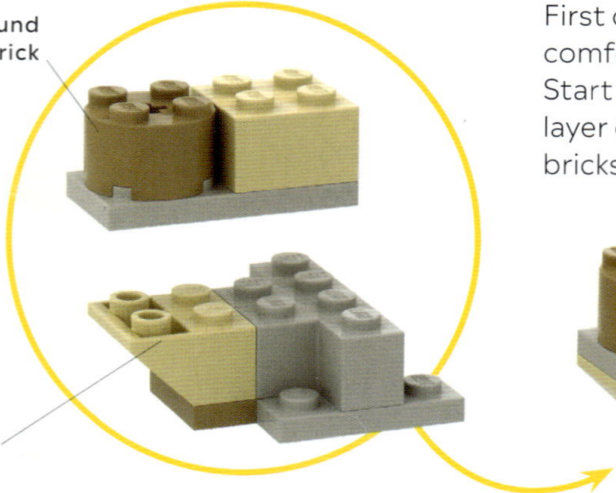

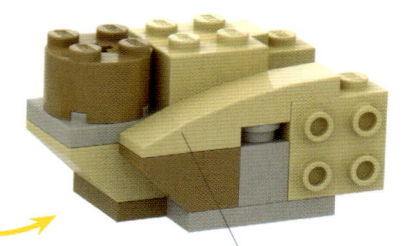

- Curved slope is the edge of the lower limb

SKILL LEVEL

119

FRONT LIMBS

Now build two upper limbs onto the base layers, plus more bricks and plates for the upper body. Include bricks with side knobs so you can attach soft-looking fur to the limbs later.

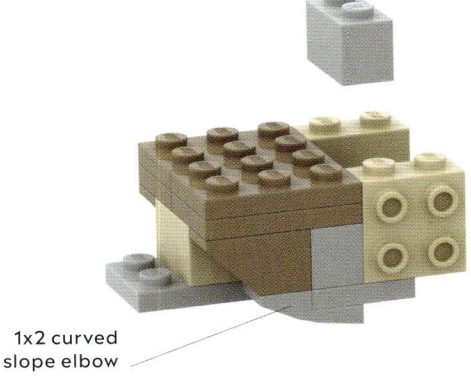

1x2 curved slope elbow

Use dark and light tan pieces for natural-looking fur

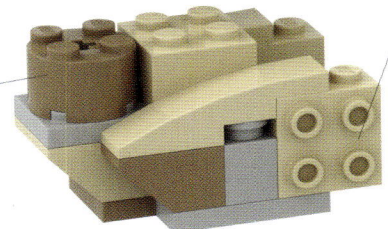

1x2x1⅔ brick with four side knobs

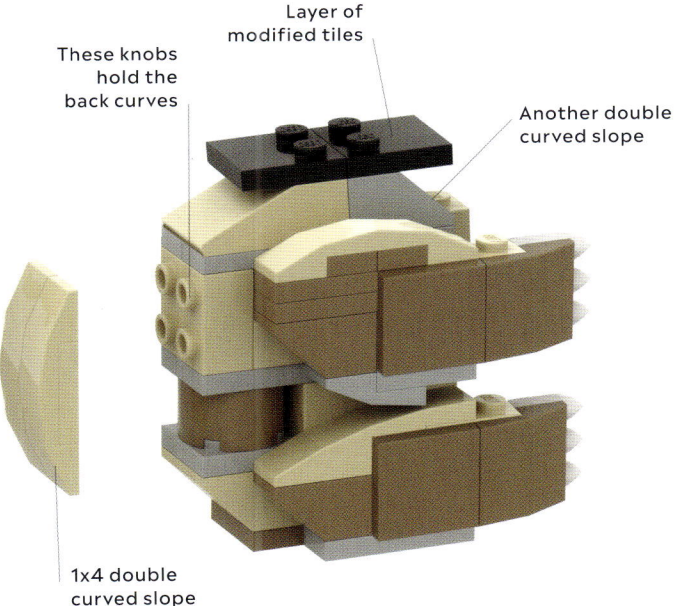

Layer of modified tiles

These knobs hold the back curves

Another double curved slope

ROUND IT OUT

Finish off the sloth's cuddly curves by adding two double curved slopes for its rounded back, and two slope bricks to form the neck and shoulders. Then cover the knobs on the limbs with smooth tiles and more curved slopes.

1x4 double curved slope

4x4 round plate

2x2 curved slope

These knobs look like ruffled fur

The face plate attaches here

SLEEPY HEAD

Build a smaller rounded shape on top of the body for the sloth's head. Start with modified tiles and add bricks with side knobs on top for building out in three directions. Finally, add an adorable flat face with a round plate and tiles, then celebrate sloth-style with a snooze!

RABBIT

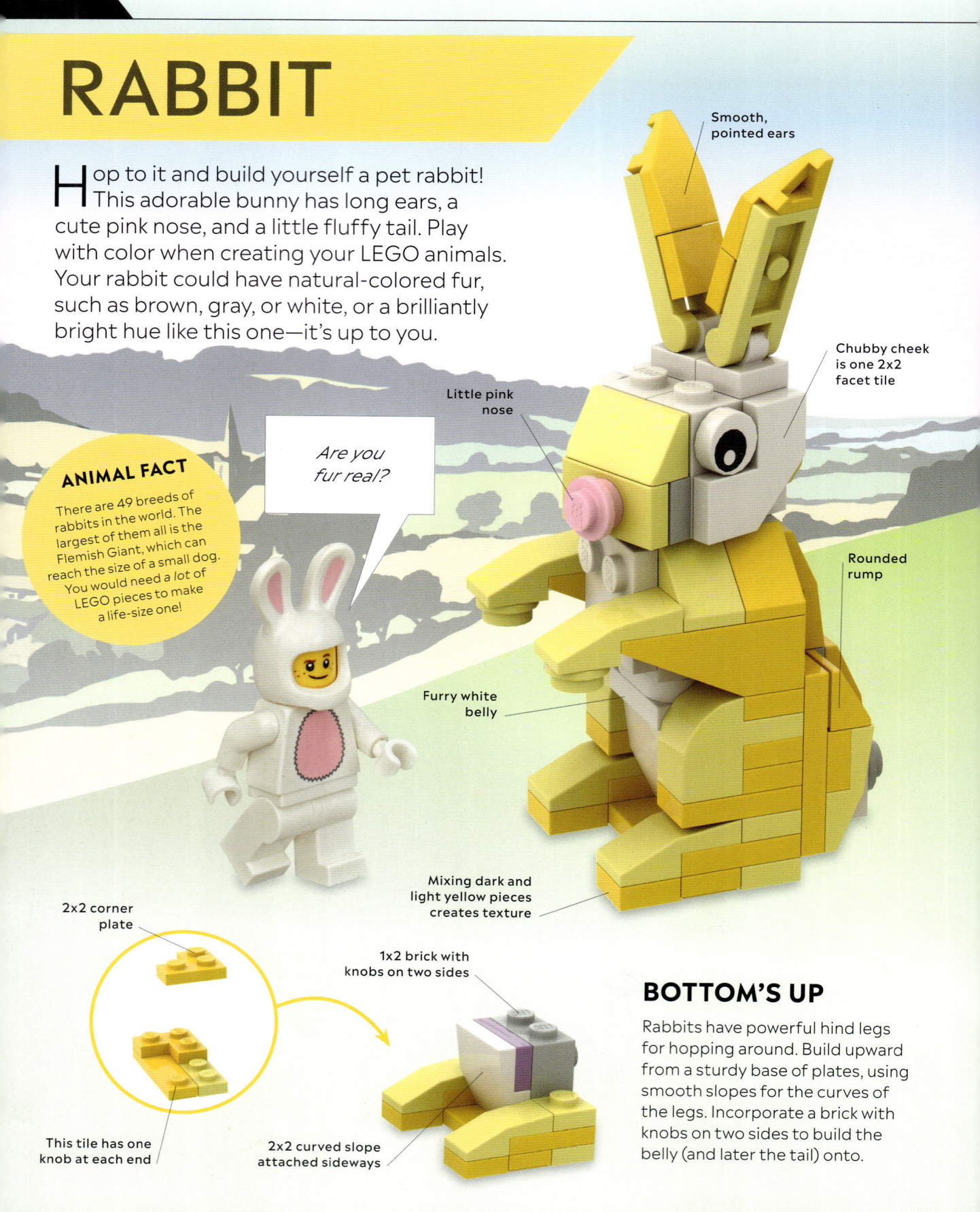

Hop to it and build yourself a pet rabbit! This adorable bunny has long ears, a cute pink nose, and a little fluffy tail. Play with color when creating your LEGO animals. Your rabbit could have natural-colored fur, such as brown, gray, or white, or a brilliantly bright hue like this one—it's up to you.

ANIMAL FACT

There are 49 breeds of rabbits in the world. The largest of them all is the Flemish Giant, which can reach the size of a small dog. You would need a *lot* of LEGO pieces to make a life-size one!

Are you fur real?

- Smooth, pointed ears
- Chubby cheek is one 2x2 facet tile
- Little pink nose
- Rounded rump
- Furry white belly
- Mixing dark and light yellow pieces creates texture
- 2x2 corner plate
- This tile has one knob at each end
- 1x2 brick with knobs on two sides
- 2x2 curved slope attached sideways

BOTTOM'S UP

Rabbits have powerful hind legs for hopping around. Build upward from a sturdy base of plates, using smooth slopes for the curves of the legs. Incorporate a brick with knobs on two sides to build the belly (and later the tail) onto.

SKILL LEVEL

121

BUNNY BODY

The belly and chest of the rabbit get gradually wider toward the top. There's a middle section of white plates, built upward, with two bricks with side knobs on top. At this stage, also add rounded thigh sections on top of the hind legs.

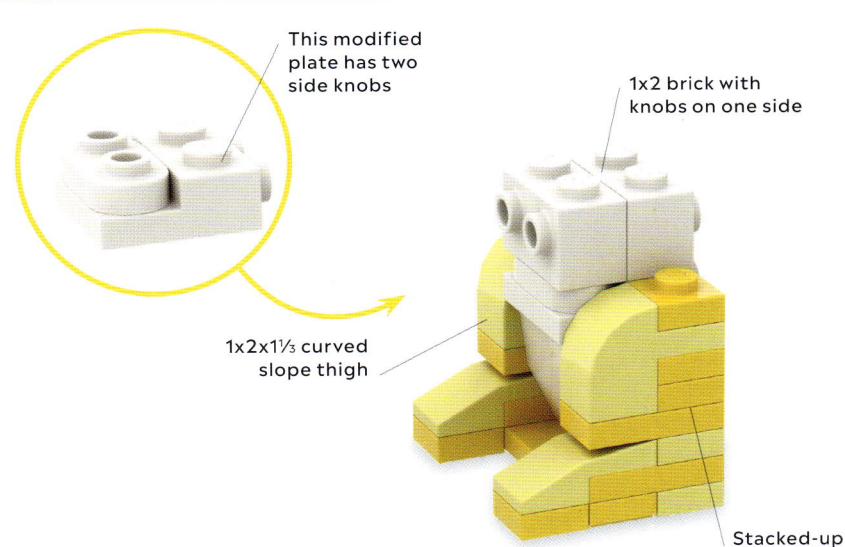

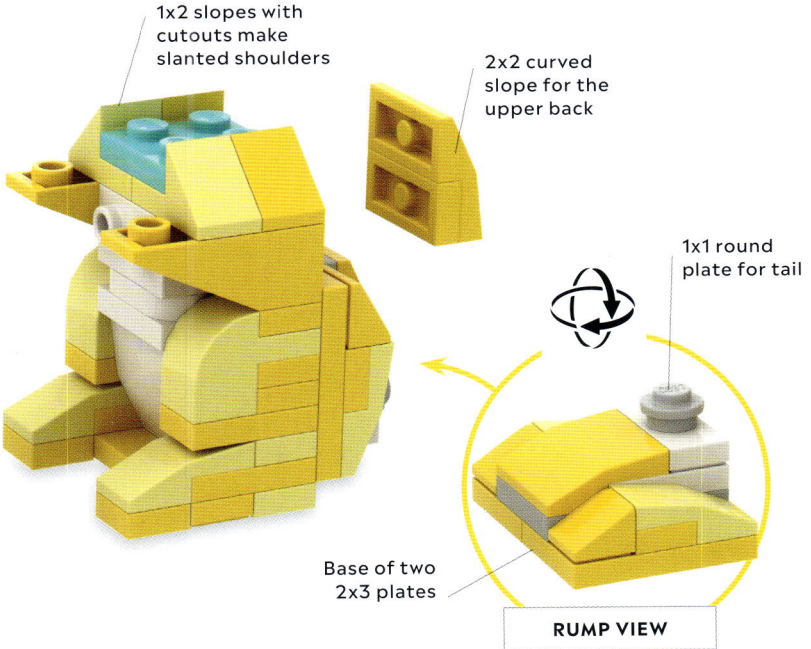

This modified plate has two side knobs

1x2x1⅓ curved slope thigh

1x2 brick with knobs on one side

Stacked-up small plates

1x2 slopes with cutouts make slanted shoulders

2x2 curved slope for the upper back

1x1 round plate for tail

Base of two 2x3 plates

RUMP VIEW

CREATING CURVES

Build on the beginnings of the rabbit's forelegs and sloping shoulders, then start building some curves at the back. The rounded rump, with its little tuft of a tail, is built up in one section then attached sideways to the main body.

1x1 tiles with clips hold up the ears

1x2 jumper plate connects to the nose

Cheek and eye attach here

FURRY FLOURISHES

It's time to build your bunny a rounded head. Bricks with side knobs create a central cube shape, which more pieces can fit onto in five different directions. Finally, add eyes, a nose, and two upright ears to finish off your furry friend.

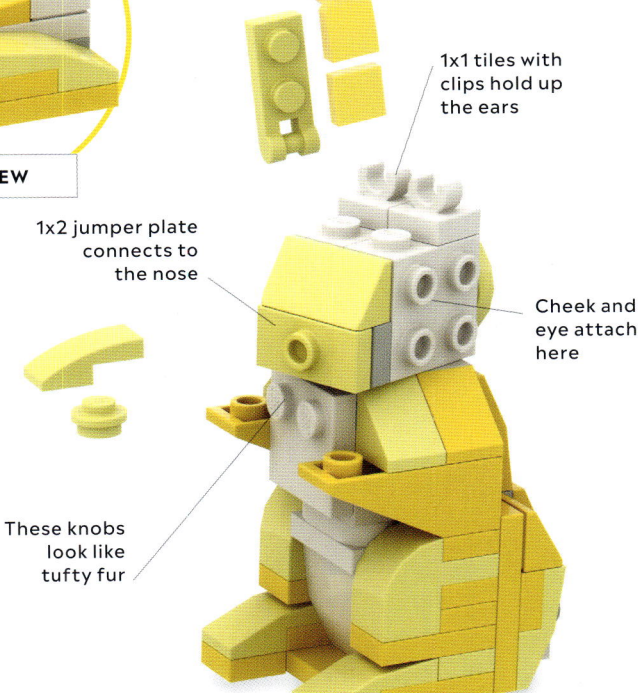

These knobs look like tufty fur

ORCA

Have a splashing time creating this marine model! With its distinctive black-and-white coloring, this is unmistakably an orca, which is the largest member of the dolphin family. You could use the same shape in other colors to create different species of dolphins or whales. Build yours with a thrashing tail and moving flippers so it can power through the ocean.

ANIMAL FACT
Orcas swim the seas hunting for fish, birds, and seals in large groups called pods, which can number up to 50 members. They can communicate and work together to capture prey.

This part of the head is called the melon

Large dorsal fin

The tail can move in two places

Caudal fin flips up and down

Flippers steer as it swims

White skin underneath

2x3 wedge plate (left)

This yellow plate will eventually be covered

What an orc-some sight!

BLUBBER BUILD

Using two black 4x4 plates as a flat base, start building up from the middle of the marine mammal's body. The tapered front of the first section, made from two wedge plates, will form the top of its mouth.

The eye will attach here

SKILL LEVEL

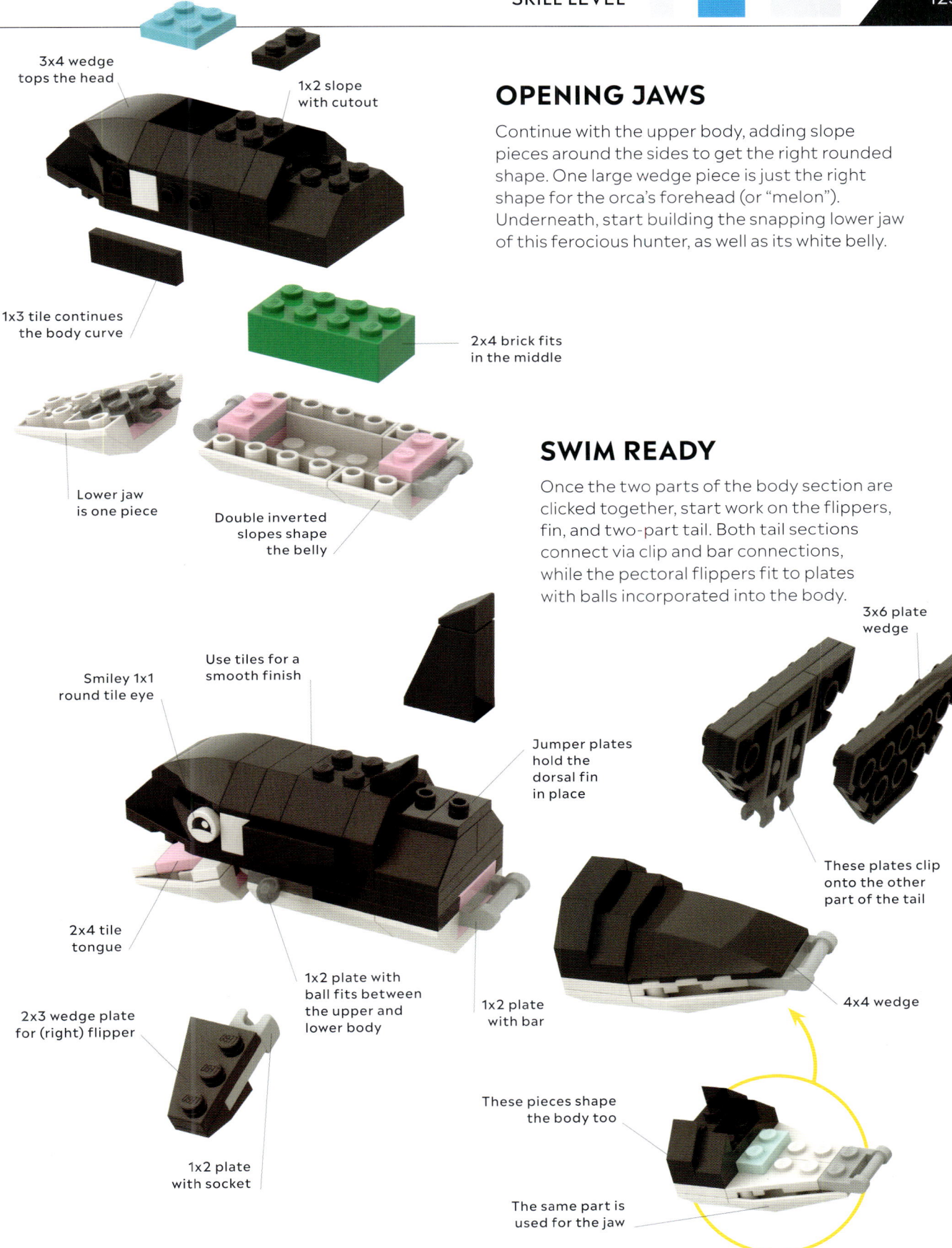

123

OPENING JAWS

Continue with the upper body, adding slope pieces around the sides to get the right rounded shape. One large wedge piece is just the right shape for the orca's forehead (or "melon"). Underneath, start building the snapping lower jaw of this ferocious hunter, as well as its white belly.

3x4 wedge tops the head

1x2 slope with cutout

1x3 tile continues the body curve

2x4 brick fits in the middle

Lower jaw is one piece

Double inverted slopes shape the belly

SWIM READY

Once the two parts of the body section are clicked together, start work on the flippers, fin, and two-part tail. Both tail sections connect via clip and bar connections, while the pectoral flippers fit to plates with balls incorporated into the body.

Smiley 1x1 round tile eye

Use tiles for a smooth finish

Jumper plates hold the dorsal fin in place

3x6 plate wedge

These plates clip onto the other part of the tail

2x4 tile tongue

1x2 plate with ball fits between the upper and lower body

1x2 plate with bar

4x4 wedge

2x3 wedge plate for (right) flipper

1x2 plate with socket

These pieces shape the body too

The same part is used for the jaw

ROOSTER

Get "clucking" your bricks together to make this colorful rooster! Male chickens like this one strut their stuff in farmyards, crowing at dawn to let everyone know a new day has begun. You can tell roosters apart from female hens because they have bigger tail feathers and longer crests (combs) on their heads.

Brightly colored comb

Large, pointy tail feathers

ANIMAL FACT
Roosters are omnivores, which means they will eat just about anything, from fruit, grains, and seeds to insects and worms. Find your smallest pieces and build your rooster a snack!

Does this guy ever sleep in?

Wide wings flap up and down

Puffed-out breast

Curved feet with toe claws

2x6 plate between the bricks

Breast feathers will attach here

EARLY BIRD
Start off in the middle of your rooster's body. Use bricks and plates to build a shape that's curved at the front, where the breast is, and staggered at the back, where the tail feathers will sit.

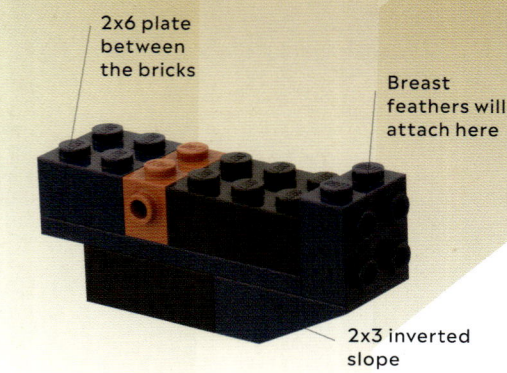

2x3 inverted slope

SKILL LEVEL

POULTRY PLANNING

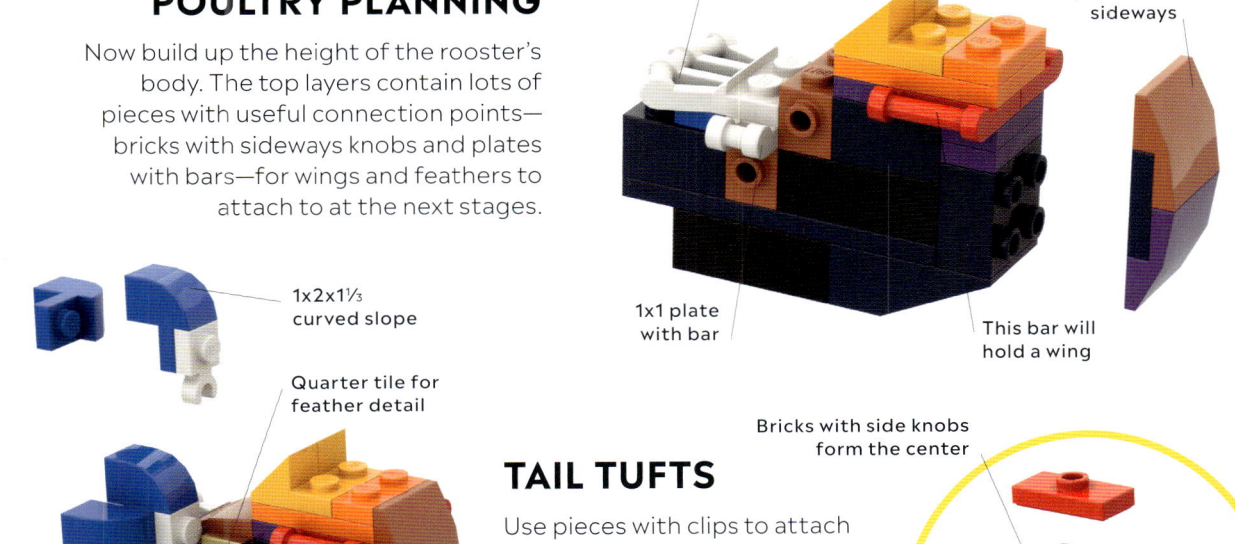

Now build up the height of the rooster's body. The top layers contain lots of pieces with useful connection points—bricks with sideways knobs and plates with bars—for wings and feathers to attach to at the next stages.

1x2 plate with three bars

These curved slopes attach sideways

1x1 plate with bar

This bar will hold a wing

1x2x1⅓ curved slope

Quarter tile for feather detail

TAIL TUFTS

Use pieces with clips to attach the first feathers to the bars at the back of the rooster's body. The blue feathers are made from curved slopes attached at right angles.

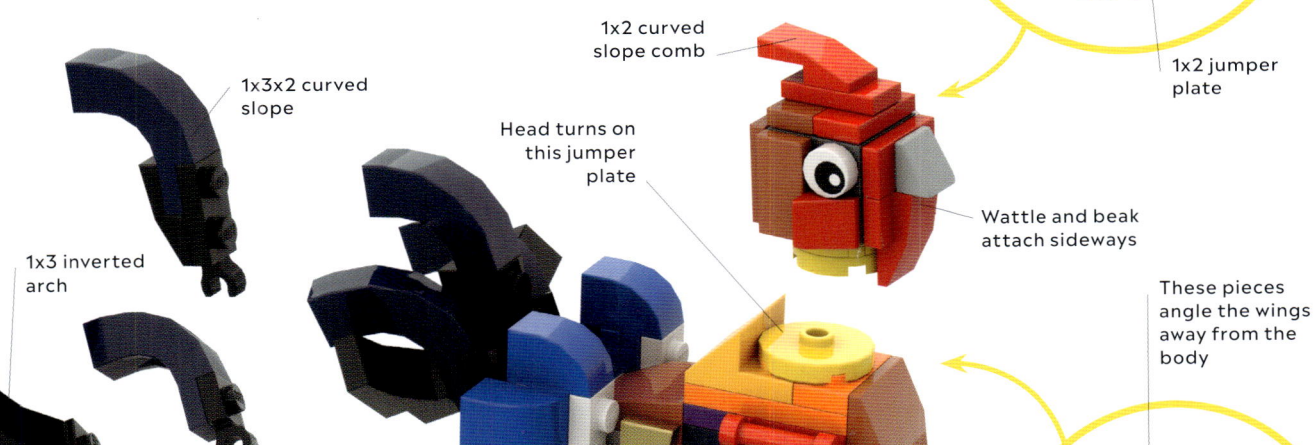

Bricks with side knobs form the center

1x2 jumper plate

1x3x2 curved slope

1x2 curved slope comb

Head turns on this jumper plate

Wattle and beak attach sideways

1x3 inverted arch

These pieces angle the wings away from the body

READY TO ROOST

Complete the tail with more proud plumage made from curved and inverted arch pieces, then get to work on the head, wings, and feet. The wings attach with clips, while the head uses lots of sideways building to incorporate classic rooster features. Cock-a-doodle-doo!

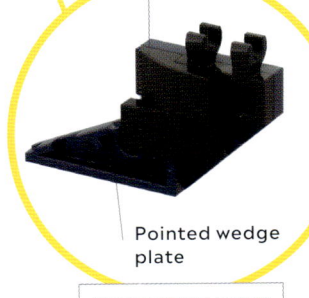

Inverted slope for the thigh

1x2 wedge slope claw

Pointed wedge plate

WING REAR VIEW

125

FOX

This shy nocturnal scavenger can be found in many towns and cities and also in the countryside. Foxes belong to the canine family of animals, so their body shapes are a lot like a dog's. Give yours reddish-brown fur, alert ears, a pointed snout ... and don't forget that trademark bushy tail, known as a "brush."

ANIMAL FACT
Unlike wolves, who are also members of the canine animal family, foxes like to hunt alone rather than in packs. They eat almost anything, from berries to birds to the leftovers in your trash.

Upright, pointed ears

Ruffled white chest fur

Bushy tail points outward

Black legs and feet

Strong, muscular thighs for running fast

Why shouldn't I wear this outfit tonight? Oh ...

The reddish fur starts here

Light tan 2x4 plate

1x4 inverted curved slope

FILL THAT BELLY

Begin by building a belly for your wild canine to fill. The fur on the underside of a fox is usually lighter, so use white and light tan plates and slopes to get the right effect.

LEG UP

The rear legs are the next focal point of this build. After adding bricks to add height to the base of the body, build on inverted slope bricks at either side of the body to form the wide thighs.

1x2 inverted slope brick for thigh

Bricks build up the body quickly

The thigh pieces attach here

2x2 inverted slope brick is the chest

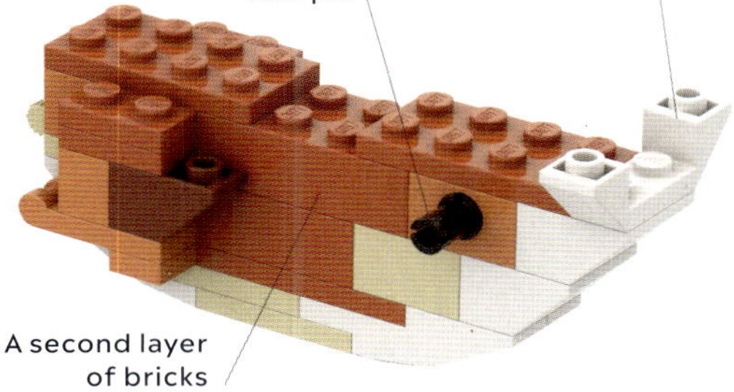

1x4 double inverted slope brick with cutout

LEGO Technic half pin

A second layer of bricks

FRONT FEATURES

Next, build on more bricks at the front of the fox's body, including a brick with hole. Plug half pins into that hole at either side of the body—the front legs will attach to these pins later.

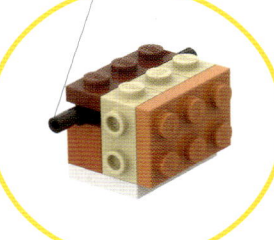

1x1 round plate with bar is the ear connector

BRUSH BUILD

This fox's brush is shaped around bricks with side knobs. Build long curved slopes above and on either side of the bricks to make it fantastically furry.

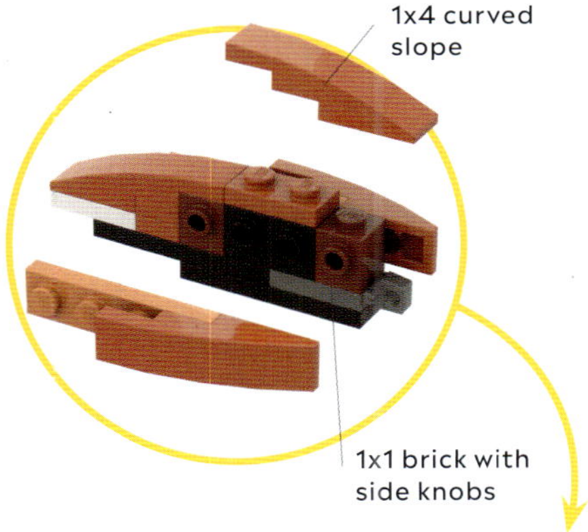

1x4 curved slope

1x1 brick with side knobs

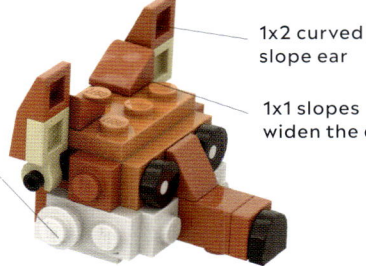

1x2 curved slope ear

1x1 slopes widen the ears

Add more white fur around the chin

Slope bricks for the neck

1x1 half circle tile nose

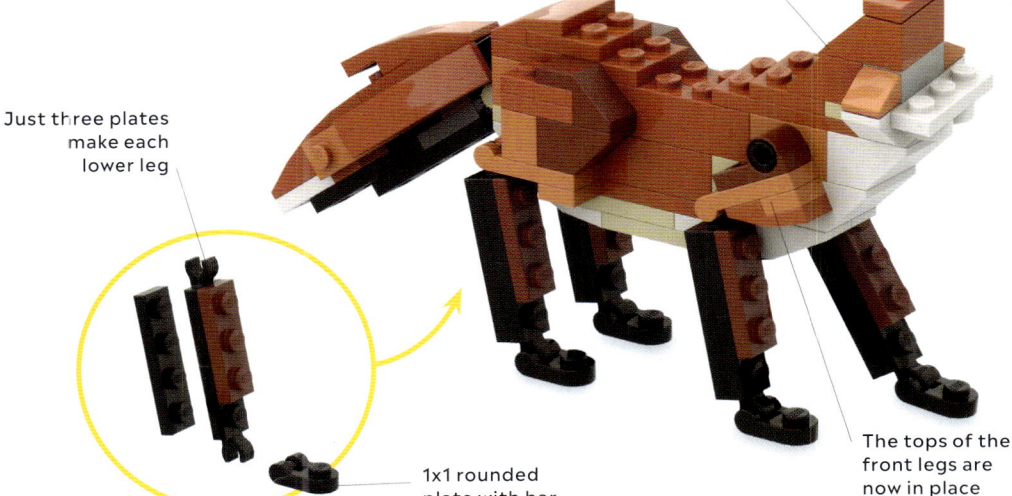

Just three plates make each lower leg

GO AHEAD

Add the head, with its pricked-up ears, long snout, and focused eyes, then finish off with four (identical) narrow black legs. Your fox is ready to prowl!

1x1 rounded plate with bar

The tops of the front legs are now in place

SKILL LEVEL

127

PEACOCK

Dazzle your friends by building this beautiful bird! A peacock is a male peafowl. It fans out its long, shimmering blue-and-green feathers, called a train, to attract female peahens. The more colorful, enormous, and downright fabulous the train is, the better! Find your prettiest pieces for a peacock in its prime.

ANIMAL FACT
Female peahens' bodies and trains are mostly a muted gray color. That's to ensure they don't attract the attention of predators, who could steal their eggs or baby peafowl, called peachicks.

I want to dedicate this next song to my favorite animal.

This peacock wears its heart on its train!

Curved beak is a horn piece

Long neck with a small head

Cleverly placed quarter tiles are wing feathers

Wide, clawed feet for balance

START THE TRAIN

This peacock's body is built upward from two modified tiles, which each have one knob at either end. The space they create allows you to start building up the train sideways from this first step.

1x1 round tile with bar is part of the train

1x4 tile with two knobs

1x2 rounded plate attaches under the tiles

1x2x2/3 brick with side knobs

SKILL LEVEL

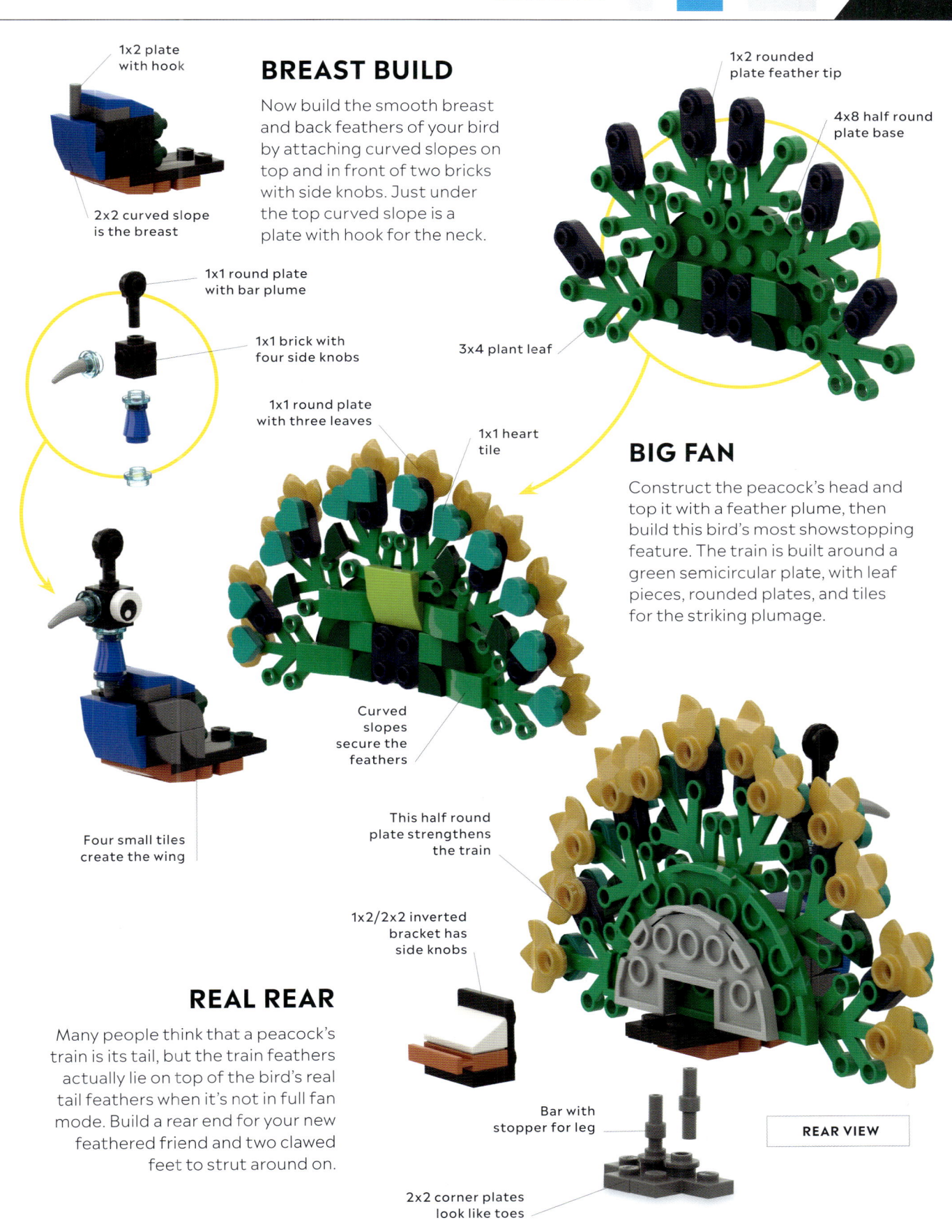

129

1x2 plate with hook

2x2 curved slope is the breast

BREAST BUILD

Now build the smooth breast and back feathers of your bird by attaching curved slopes on top and in front of two bricks with side knobs. Just under the top curved slope is a plate with hook for the neck.

1x1 round plate with bar plume

1x1 brick with four side knobs

1x1 round plate with three leaves

1x1 heart tile

1x2 rounded plate feather tip

4x8 half round plate base

3x4 plant leaf

BIG FAN

Construct the peacock's head and top it with a feather plume, then build this bird's most showstopping feature. The train is built around a green semicircular plate, with leaf pieces, rounded plates, and tiles for the striking plumage.

Four small tiles create the wing

Curved slopes secure the feathers

This half round plate strengthens the train

1x2/2x2 inverted bracket has side knobs

REAL REAR

Many people think that a peacock's train is its tail, but the train feathers actually lie on top of the bird's real tail feathers when it's not in full fan mode. Build a rear end for your new feathered friend and two clawed feet to strut around on.

Bar with stopper for leg

2x2 corner plates look like toes

REAR VIEW

HIPPOPOTAMUS

The hulking hippopotamus is one of the largest and heaviest land animals in the world. This lightweight LEGO version may be a lot easier to carry, but it has all the features of the real thing: thick, hairless skin; a barrel-shaped body; and large, tusk-like teeth that can be used as weapons if the hippo feels threatened. Let's hope that's a smile on its face!

ANIMAL FACT
Hippos love to loll around in water to keep their bodies cool. Their name translates from ancient Greek as "water horse," and they are more closely related to whales and dolphins than other land mammals.

Heavy, rounded rump

Nostrils stay out of water while bathing

Wide feet support the bulky body

Pink belly skin

Tusk is an upside-down 1x1 round plate

You can stop smiling now!

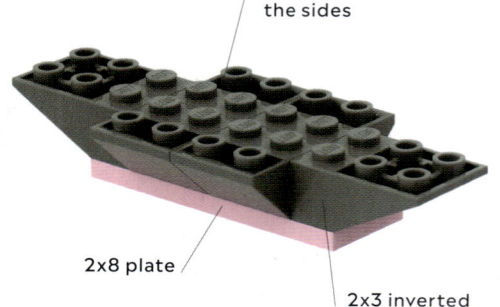

2x2 inverted slopes for the sides

2x8 plate

2x3 inverted slope

HEFTY START

Everything about a hippopotamus is massive, especially its belly, which it loves to fill with grass. Start off the belly with one base plate for the soft lower-body skin, then add slope bricks in two sizes to bulk up the belly right away.

SKILL LEVEL

COLOSSAL CURVES

Build up the body using more bricks and plates, then start adding some curves. Two round corner bricks combine at the rear to make the perfect hippopotamus rump! At this stage, also start thinking about where the legs will attach to the body.

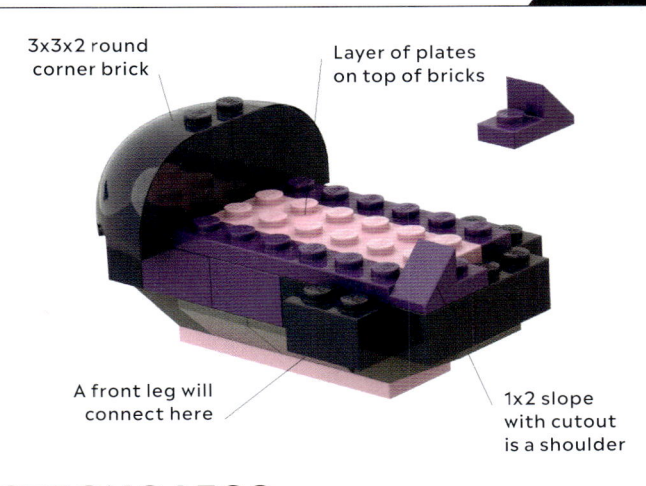

3x3x2 round corner brick

Layer of plates on top of bricks

A front leg will connect here

1x2 slope with cutout is a shoulder

Curved slopes create smooth folds of skin

Slope bricks for front thigh

STRONG LEGS

Finish off the body with more regular bricks, curved bricks, and slopes, then build some short, stumpy legs to support that chunky frame. The hind legs are made from stacked round bricks and plates, while the front ones are made from a mixture of round and square elements.

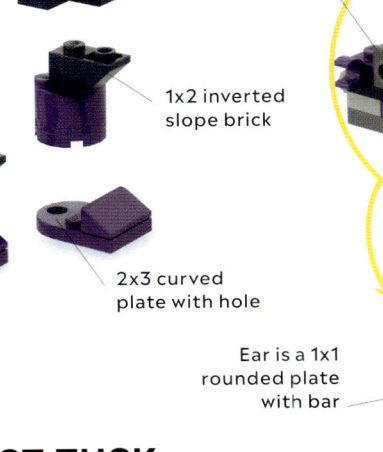

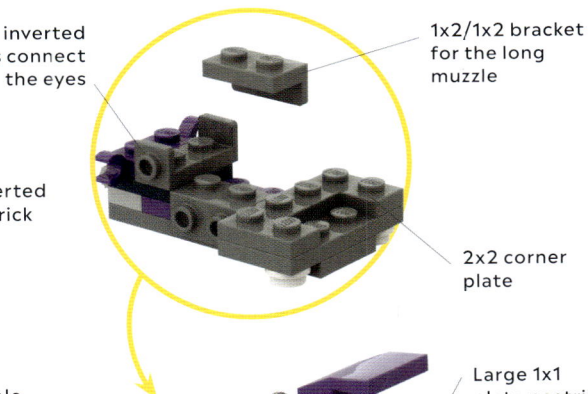

1x2x1 half cylinder

1x2 slope for toes

1x2 inverted slope brick

2x2 round plate

2x3 curved plate with hole

1x1/1x1 inverted brackets connect the eyes

1x2/1x2 bracket for the long muzzle

2x2 corner plate

Large 1x1 plate nostril

Ear is a 1x1 rounded plate with bar

LAST TUSK

Now all this hippo needs is a head, then it's ready to roll. Did you know that hippos have the biggest jaws of any land animal? And they can open them very wide—a whole 180 degrees! Build yours a movable lower jaw using a clip and bar connection. Include pink pieces to line the inside of the mouth, and don't forget the tusks!

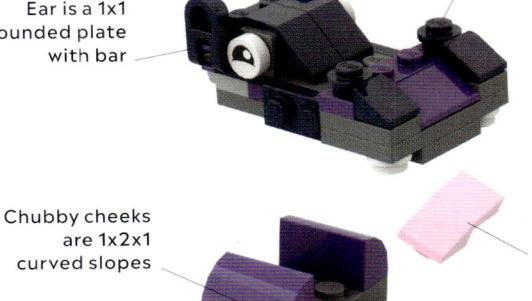

Chubby cheeks are 1x2x1 curved slopes

1x2 inverted curved slope for inside the mouth

Upside-down 1x2 plate with clip

Lower jaw is one wedge piece

ORANGUTAN

Look at this great ape swinging swiftly through the trees! These intelligent primates live in family groups in the rainforests of Southeast Asia. They have shaggy red fur and extra long arms that help them reach out and grasp hanging vines and branches.

ANIMAL FACT
Scientists think that orangutans are some of our closest living relatives. Humans and orangutans share almost 97 percent of their DNA!

- Rounded face
- Arms move in two places
- Longer arm fur
- Ankle joint is a ball and socket

That orangutan looks just like my brother!

BELLY BRICKS

The rounded belly of the orangutan starts off square. Place bricks with side knobs on top of regular 2x4 bricks. The side knobs will allow you to add more details to the mid section later.

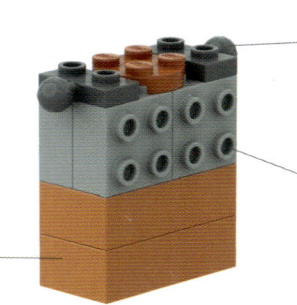

- An arm will attach here
- 1x2x1⅔ brick with four side knobs
- 2x4 brick

SKILL LEVEL   133

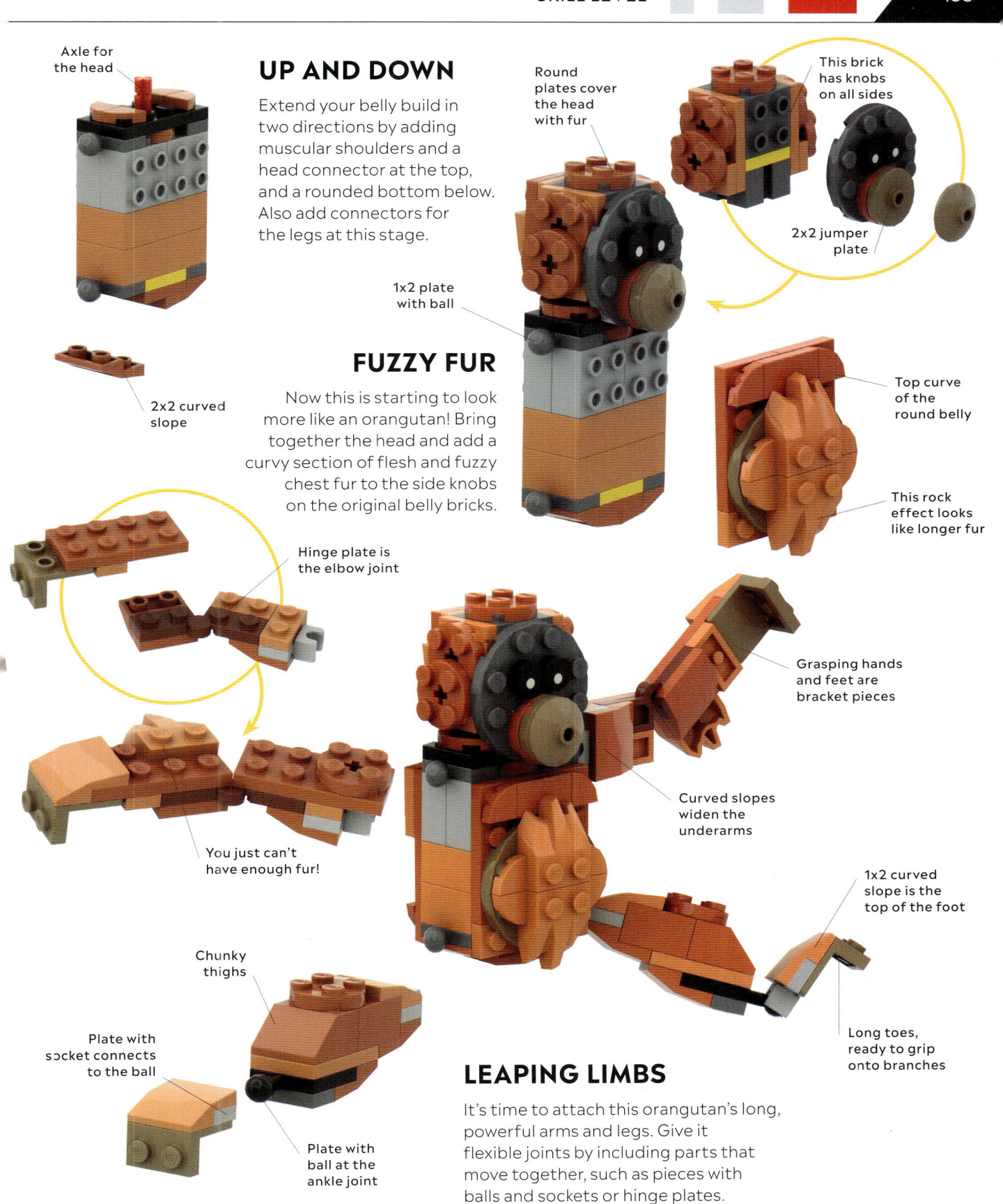

UP AND DOWN

Extend your belly build in two directions by adding muscular shoulders and a head connector at the top, and a rounded bottom below. Also add connectors for the legs at this stage.

Axle for the head

Round plates cover the head with fur

This brick has knobs on all sides

2x2 jumper plate

1x2 plate with ball

FUZZY FUR

Now this is starting to look more like an orangutan! Bring together the head and add a curvy section of flesh and fuzzy chest fur to the side knobs on the original belly bricks.

2x2 curved slope

Top curve of the round belly

This rock effect looks like longer fur

Hinge plate is the elbow joint

Grasping hands and feet are bracket pieces

You just can't have enough fur!

Curved slopes widen the underarms

1x2 curved slope is the top of the foot

Chunky thighs

Plate with socket connects to the ball

Plate with ball at the ankle joint

Long toes, ready to grip onto branches

LEAPING LIMBS

It's time to attach this orangutan's long, powerful arms and legs. Give it flexible joints by including parts that move together, such as pieces with balls and sockets or hinge plates.

LION

Roaaaaar! Known as the king of the beasts, this carnivorous big cat is a powerful hunter. Lions like this one live together in family groups called prides, mostly on the wide, open plains of Africa. The long, shaggy mane around this lion's head and neck shows that it is a male lion.

ANIMAL FACT
There are always more females in a lion pride than males. The females raise the cubs and do most of the hunting. The one or two males in a pride roar loudly and spray their scent to warn away other male lions.

Please don't eat me before I can check you over!

- Wild, uneven waves
- Hair-covered, muscular body
- Sharp teeth for chewing meat
- Tuft of hair on the tail tip
- Wide paws with long claws
- Fur will attach to these bricks with side knobs
- 2x2 inverted slope brick
- 1x4 inverted curved slope
- First plate layer

SOFT START

Lions have loose skin on their bellies that protects them if they are kicked by their prey, which tends to be animals with hard hooves! Begin your build in this soft spot, starting off with plates then adding curved slopes, bricks, and inverted slope bricks to shape it.

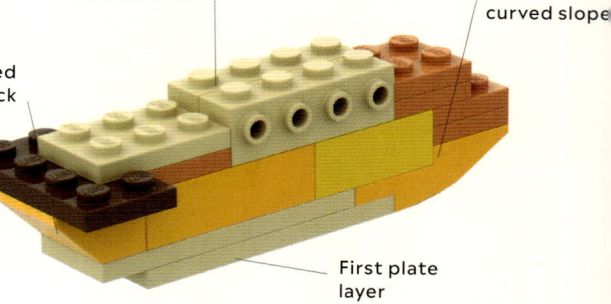

SKILL LEVEL 135

MANE EVENT
Continue building the body upward and outward by adding more plates, including four with balls to connect the legs to later. At the front of the body, start creating the lowest-hanging hairs of the mane.

- 1x2 plate with ball
- 1x2 plate with clip to hold the tail
- 1x2x1⅓ curved slopes start off the mane
- Curved slopes for the soft fur

ROARING JAW
Finish off the lion's body with some rolling curves formed by slopes, then continue with the mane and start the fierce head. There's a plate with bar built into the front of the body that the lion's lower jaw can attach to.

- This plate is the upper jaw
- 1x2 slope
- Top teeth are plates with clips
- This red plate doubles as the tongue
- 1x1x1⅔ brick with knobs on side
- 1x2 slope with cutout
- 2x2 inverted curved slope lower jaw
- 1x2 plate with bars at both ends
- Tail tip is a minifigure's fez
- Curved slopes are shaggy waves

READY TO LEAD
The bigger the mane, the more powerful a male lion is, so give yours a big cat bouffant! Then start shaping the face and legs, which all use lots of sideways building techniques. Don't forget to add a long tail with a black tip—lions use their tails to lead their pride in long grass.

- These jumper plates center the snout
- Ball-and-socket ankle joint
- Sockets connect to balls on the body
- Layers of sideways plates
- 1x2 plate with three claws
- Lions' huge paws help them hunt

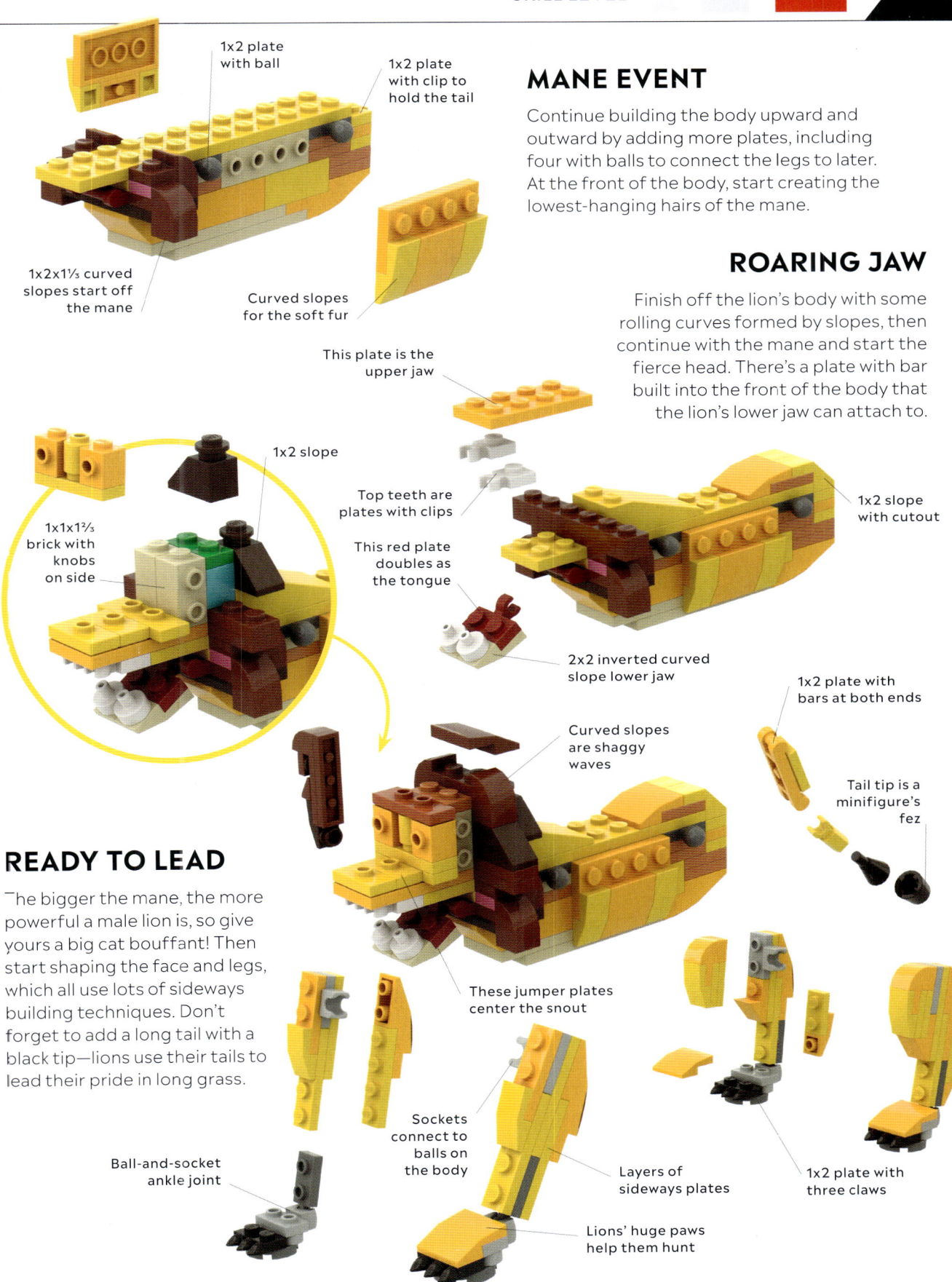

RACCOON

This creature of the night can often be found rummaging in trash cans for food while everyone's asleep! But not all raccoons live in towns and cities—some make homes in forests and on grasslands. Native to North America, raccoons are recognizable by their ringed tails and dark eye fur, which fittingly looks a bit like a bandit mask.

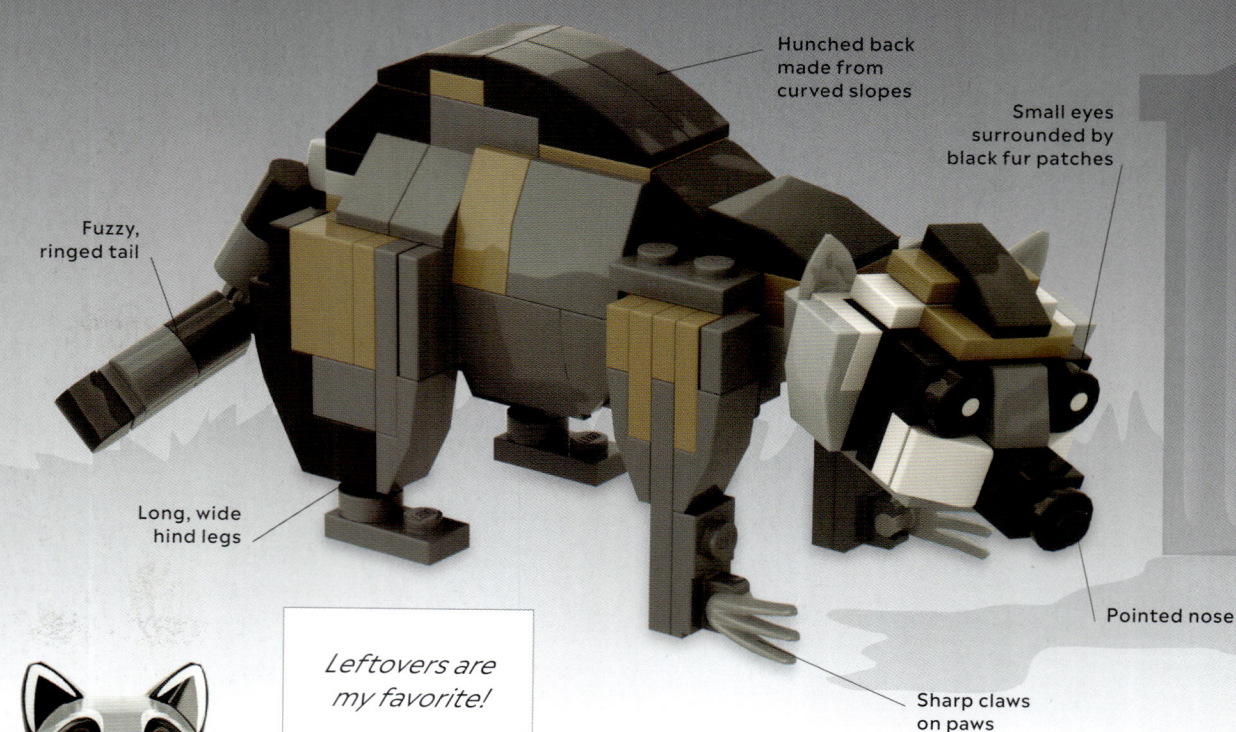

Hunched back made from curved slopes

Small eyes surrounded by black fur patches

Fuzzy, ringed tail

Long, wide hind legs

Pointed nose

Sharp claws on paws

Leftovers are my favorite!

This piece will hold the tail

Use any color pieces for fur-covered body parts

1x2/2x2 bracket plate

2x8 belly base plate

FIRST FUR

Raccoons' bodies are covered in thick gray fur that keeps them warm at night. Plan out how to give your raccoon a furry finish from the start by using pieces with side knobs to attach "fur" to later.

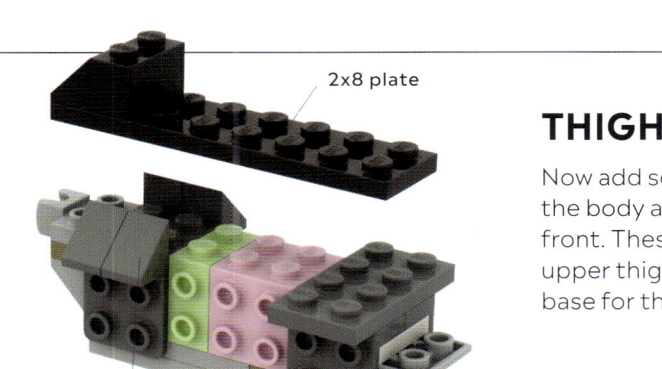

SKILL LEVEL   137

THIGH HIGH

Now add some sloping bricks at the back of the body and a plate at a right angle at the front. These pieces will form the raccoon's upper thighs. Next, add a long plate as a base for the arched back of the beast.

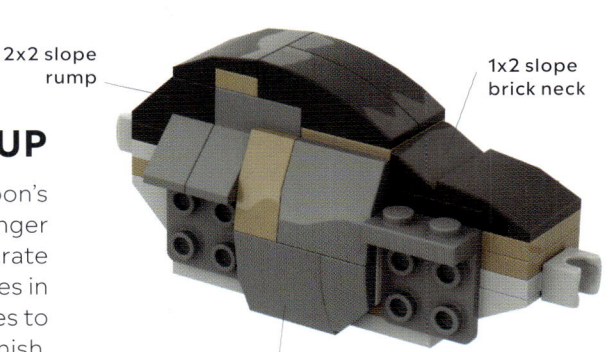

2x8 plate

2x1 slope with cutout

2x2 slope rump

1x2 slope brick neck

2x2 curved slope

SHAPE UP

There's a distinctive arch to a raccoon's back because the hind legs are longer than the front ones. Create an accurate shape using curved slopes and slopes in different sizes. Also add curved slopes to the sides of the body for a furry finish.

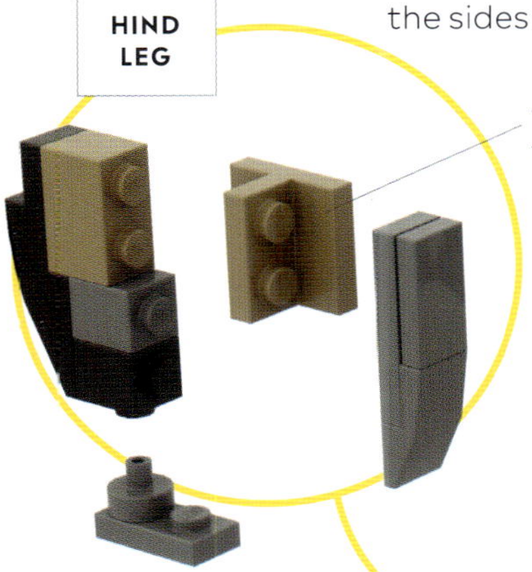

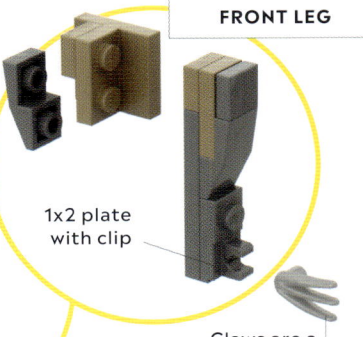

HIND LEG

This piece attaches to the body

The leg fits below this plate but doesn't attach

FRONT LEG

1x2 plate with clip

Claws are a minifigure weapon

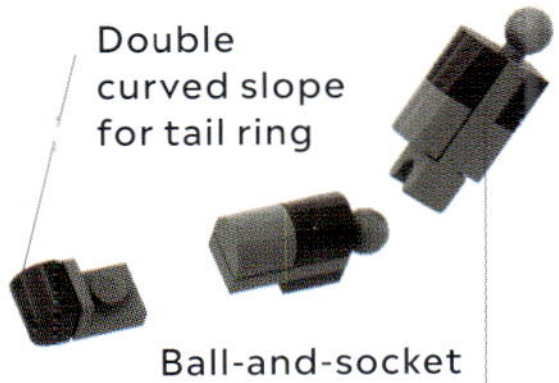

FUSSY FEATURES

This model is classed as hard because it uses lots of sideways building. Each leg is built sideways, then attached sideways (in a different direction) to the body. At this stage, also create the raccoon's ringed tail.

Double curved slope for tail ring

Ball-and-socket connections allow the tail to curve

1x4 plate

1x2 plate with ball connects to the body

1x1/1x2 bracket

1x1 plate with tooth for ear

1x2 curved slope for forehead

MASKED BANDIT

The back of the raccoon's head incorporates several bracket plates to create lots of sideways knobs. There are various small elements attached all around that section, creating the upright ears, furry cheeks, pointed nose, and those infamous masked eyes.

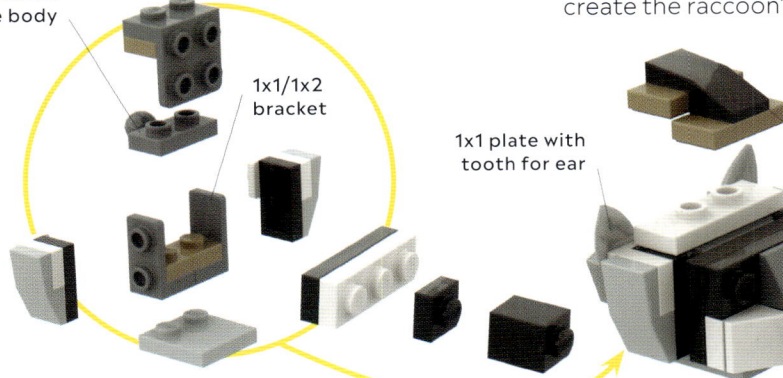

FEEDING TIME

All animals need to eat food for energy, but their diets are as varied as the animal kingdom itself. Some are mostly meat-eating carnivores while, at the other end of the spectrum, herbivores eat only plants, seeds, fruits, and vegetables. Build your LEGO creations a little of what they like to crunch, chomp, nibble, and gnaw.

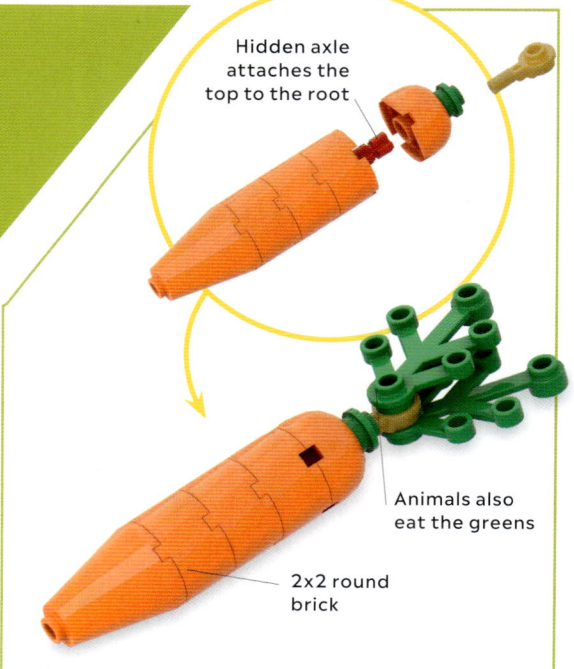

CRUNCHY CARROT

This carrot would be on the menu for rabbits, horses, squirrels, and many other vegetable-loving animals. Build a long orange root using round bricks and add a large cone for the tapering tip.

- Hidden axle attaches the top to the root
- Animals also eat the greens
- 2x2 round brick

RAW MEAT

Lions are hypercarnivores, which means at least 70 percent of their diet is meat. Build your LEGO lion a large hunk of meat on the bone to gnaw. Use bright red pieces for the fleshy part, light tan for fatty bits, and white and gray pieces for the bone.

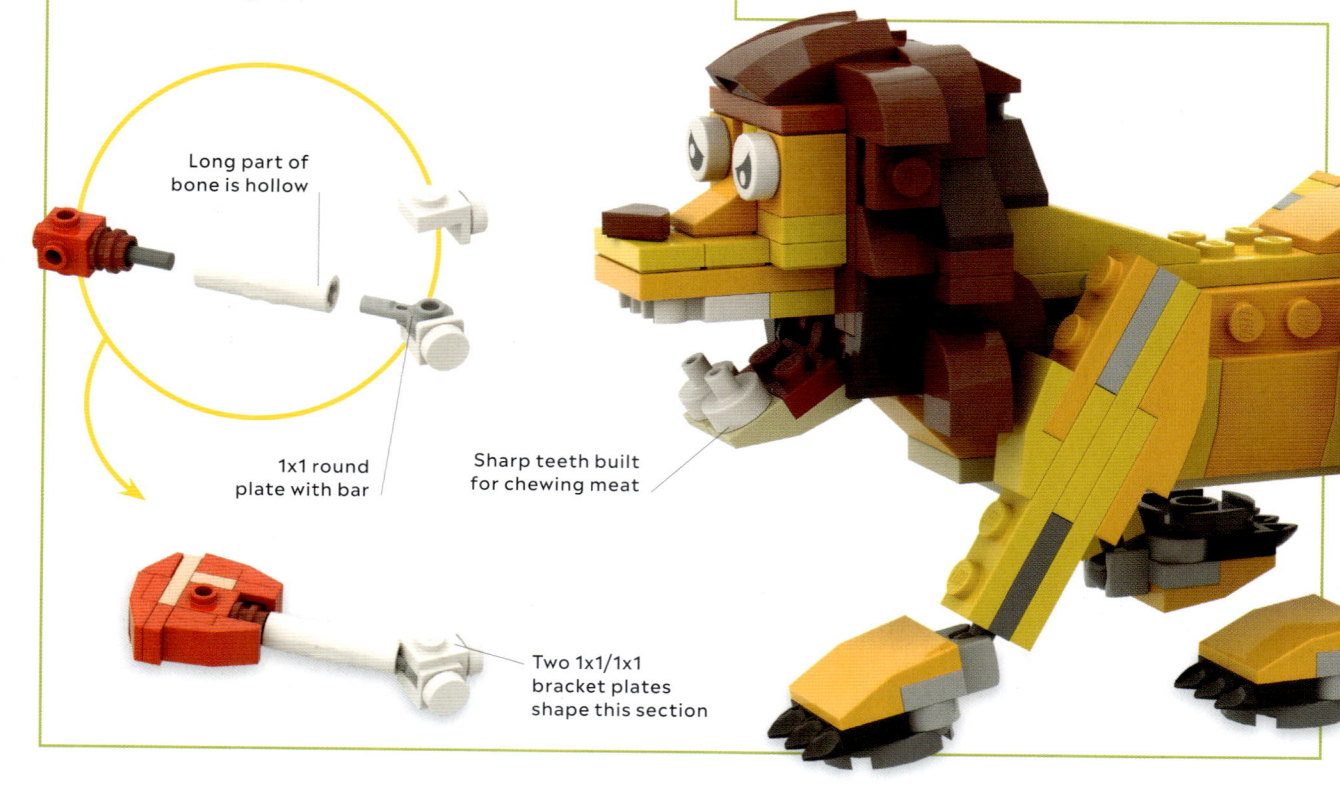

- Long part of bone is hollow
- 1x1 round plate with bar
- Sharp teeth built for chewing meat
- Two 1x1/1x1 bracket plates shape this section

IDEAS GALLERY 139

Surely she'll eat this gourmet breakfast kibble.

I'd rather have some tuna!

3x3x1 round corner brick

1x1 round plate

KITTY KIBBLE

Many pet owners serve their cats and dogs bowls of small, dry pellets of meat or grain called kibble. Build a food bowl like this one using round corner bricks, then fill it with lots of small round plates for instant pet food!

Pin attaches to the panda's paw

Stack up 1x1 round bricks with leaves

BAMBOO

Pandas have a diet that consists almost exclusively of bamboo. They have to munch through mountains of this woody plant every day in order to get the nutrients they need to survive. Build lots of tall bamboo shoots for this panda to dine on.

Roots of already-eaten shoots

POLAR BEAR

This powerful hunter lives in the Arctic Circle, thriving in a frozen land of snow, ice, and sea. Perfectly adapted for cold conditions, it has thick, shaggy fur with lots of blubber underneath to keep it warm and dry. Polar bears have white-looking fur that blends in with their landscape, but yours could weave in other colors, like this model, to give it a unique look.

ANIMAL FACT
A polar bear's fur is not actually white. Yes, you read that correctly! Each hair shaft is transparent, but when the hairs reflect the light of the Arctic sun, they usually appear white or yellow.

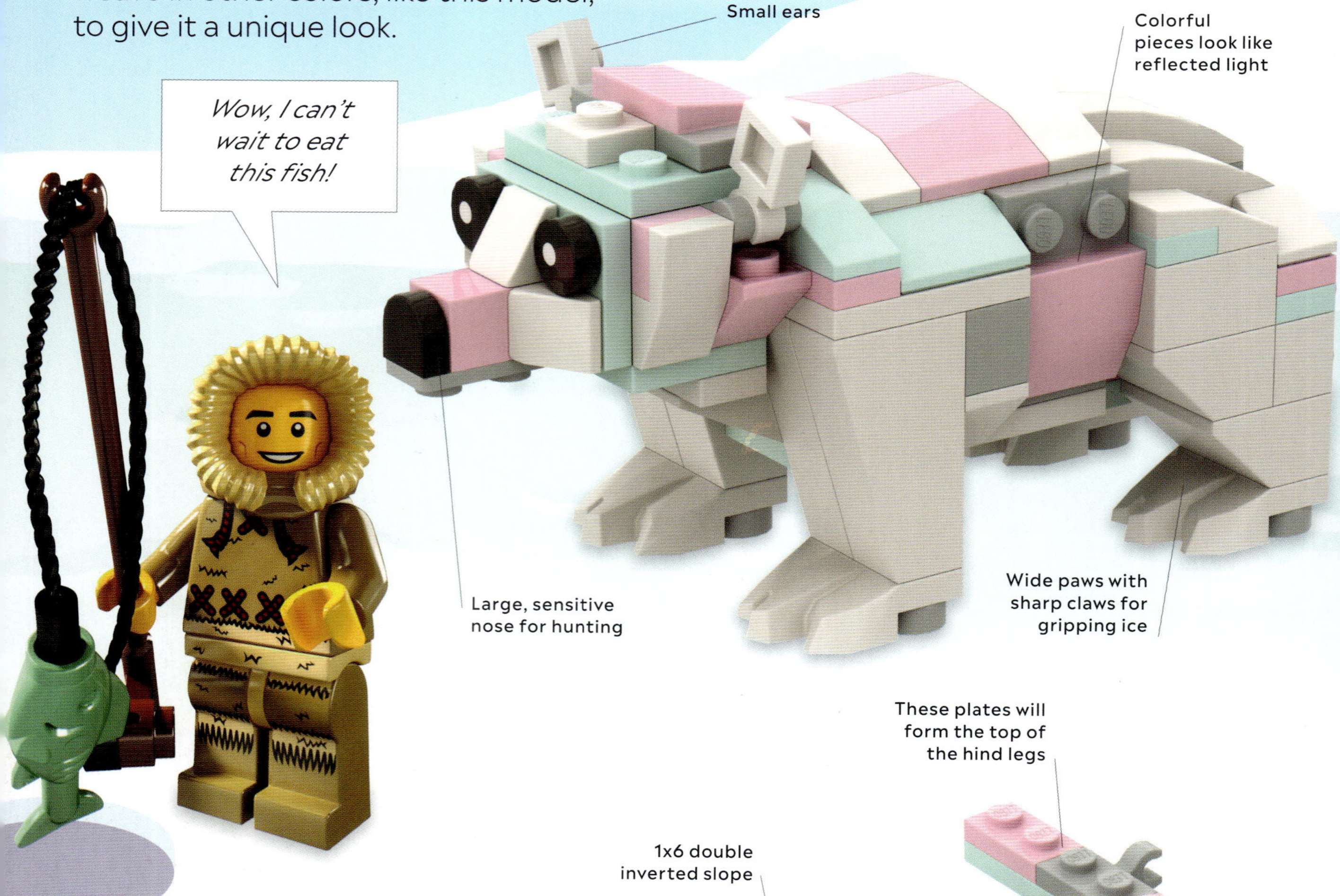

Wow, I can't wait to eat this fish!

- Small ears
- Colorful pieces look like reflected light
- Large, sensitive nose for hunting
- Wide paws with sharp claws for gripping ice

- These plates will form the top of the hind legs
- 1x6 double inverted slope

BIG BASE

Polar bears are the biggest and most dangerous bears on Earth, so build yours a large, muscular body. Begin with a blue belly base plate and build upward, gradually widening the body to give it an upward curve shape.

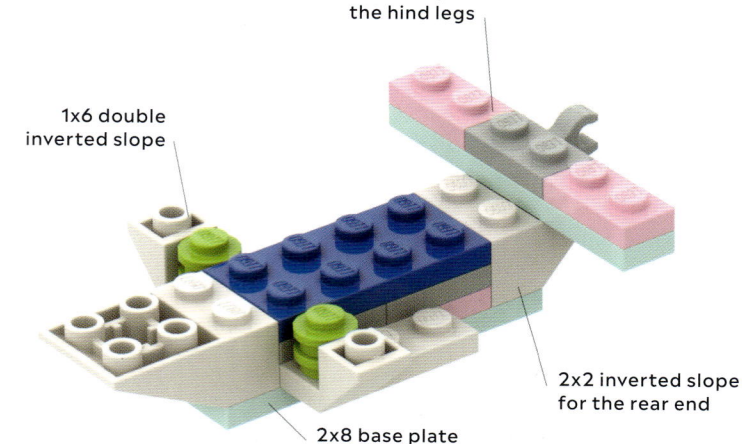

- 2x8 base plate
- 2x2 inverted slope for the rear end

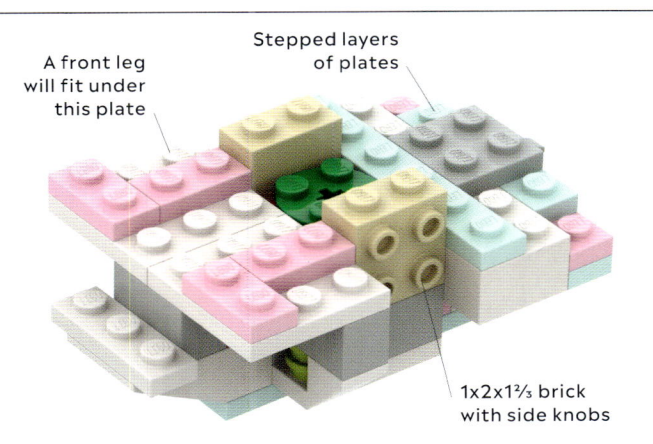

SKILL LEVEL 141

A front leg will fit under this plate

Stepped layers of plates

1x2x1⅔ brick with side knobs

POLAR PLANNING

The polar bear's mighty body has lots of shaggy fur and muscular curves. At this stage of the build, plan out where the outer curves and limbs will go by including bricks with side knobs and layers of plates. These elements will allow you to create interesting shapes at the next stage.

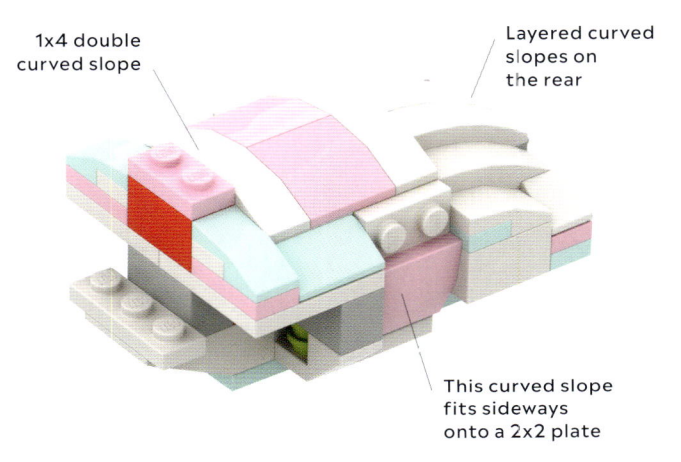

1x4 double curved slope

Layered curved slopes on the rear

SLIPPERY SLOPES

Add a brick in the center and build a few more plates around it before topping the body with curved slopes and slope bricks in various sizes for a furry finish. The curved slopes roll downward toward the rump.

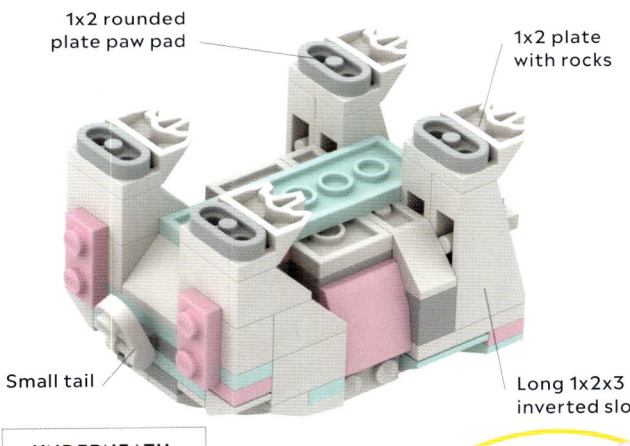

This curved slope fits sideways onto a 2x2 plate

1x2 rounded plate paw pad

1x2 plate with rocks

POWERFUL PAWS

Polar bears use their large, strong legs to run, walk, and swim. Your model will need wide paws like snowshoes, with big claws for clinging onto slippery terrains. This model's legs are made from inverted slope bricks, with paws of plates with rocks.

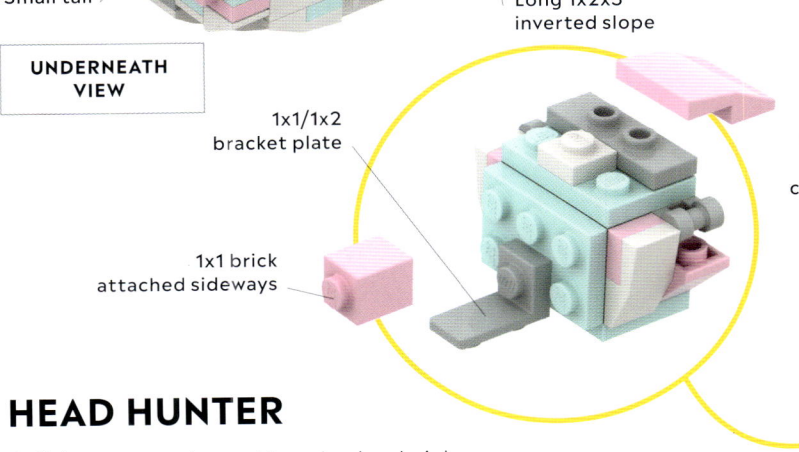

Small tail

UNDERNEATH VIEW

Long 1x2x3 inverted slope

1x1/1x2 bracket plate

1x1 brick attached sideways

1x1 plate with clip for ear

Furry cheeks are 1x2 curved slopes

HEAD HUNTER

Build a square-shaped head using bricks with side knobs so you can build out in four directions, then add a face with a long snout at the front. Lastly, give your Arctic animal its most powerful hunting tool: its nose. Polar bears can sniff out seals from many miles away.

Half circle tile for nostrils

WOLF

Wolves like this one can be found in parts of North America, Asia, and Europe. They live and work together in family packs led by an elder male and female, hunting and sharing their food and howling to warn other wolf packs away. They have thick, shaggy fur to keep them warm, expressive ears, and strong legs for chasing down prey.

ANIMAL FACT
A wolf's howl is 92 times more powerful than a human's voice, and it can be heard more than 7 miles (11km) away! This LEGO wolf's neck connection allows it to lift up its head to "howl" like the real thing.

I wonder if that dog knows where Grandma lives.

- Body like a domestic dog's
- Long, pointed muzzle
- Shaggy, black-tipped tail
- Claws for gripping the ground

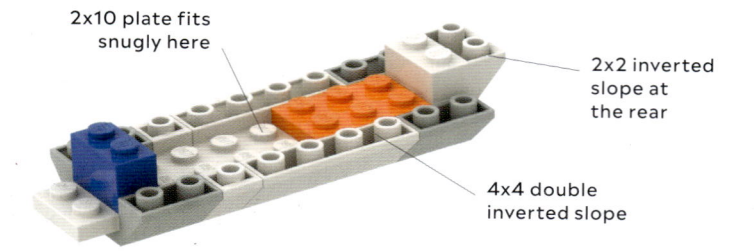

- 2x10 plate fits snugly here
- 2x2 inverted slope at the rear
- 4x4 double inverted slope

LIGHT PATCHES

This gray wolf has lighter patches of fur on its underbelly. Start building your model at this part of its body, using double inverted slope bricks to give it a rounded shape. There's a long plate inside the recessed parts of the slopes.

SKILL LEVEL 143

FURRY COAT

Add a layer of bricks at this next stage, using two shades of gray for the wolf's body fur. Also bring in some bricks with holes for the hind legs to attach to later, and a plate with a click hinge connection at the rear for the tail.

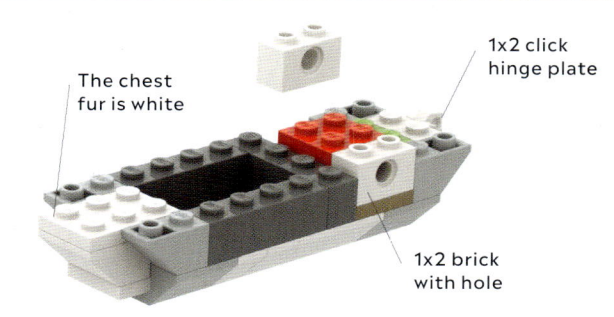

- The chest fur is white
- 1x2 click hinge plate
- 1x2 brick with hole

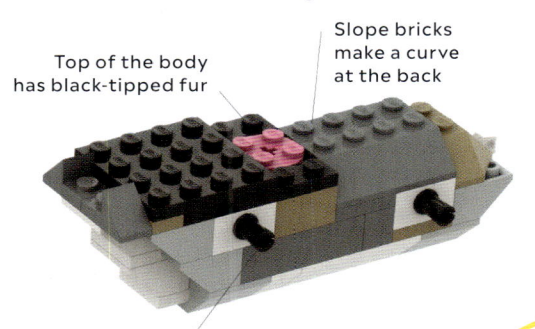

- Top of the body has black-tipped fur
- Slope bricks make a curve at the back
- Front legs attach higher than rear legs

BUILT FOR SPEED

A wolf's body has a streamlined shape—which means it gets narrower at the back—so it can run very fast. To get the right shape, make the front of the body a little higher with an extra layer of bricks and two layers of plates.

- Include lots of bricks with side knobs in the head
- Muzzle pieces attach sideways

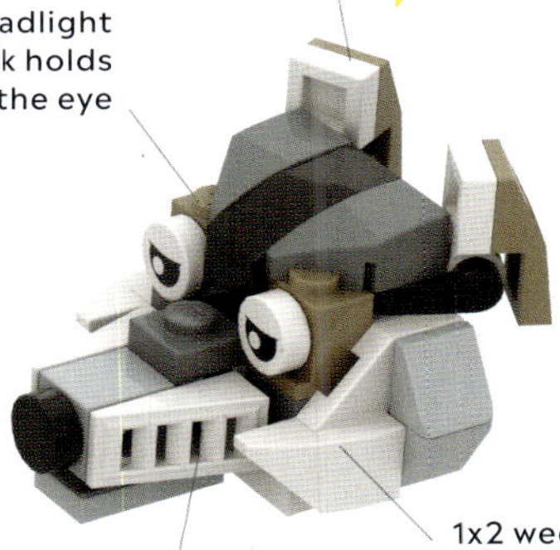

- Ears can move up and down
- 1x1 headlight brick holds the eye
- Grille slopes look like whiskers
- 1x2 wedge slope for cheek fur

- Many wolves have black tail tips
- 1x2 plate with two click hinge fingers
- Mix smooth pieces and plates for a shaggy look

WHAT BIG EYES YOU HAVE

"All the better to see you with!" Make sure you include at least two of the features of the famous fairy-tale wolf in Little Red Riding Hood. This model has big ears and big eyes, but it does not have big teeth! Will yours?

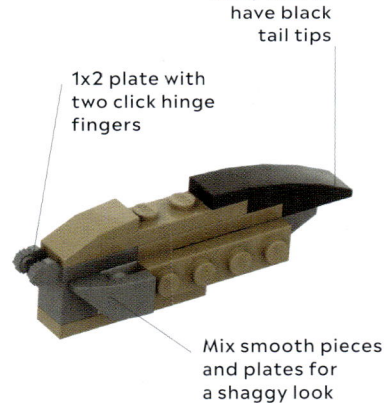

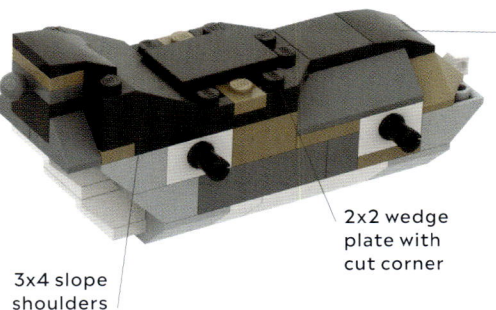

- This piece smooths off the top of the foreleg
- 1x4 curved slope thigh
- Click hinges connect the knee joint

- 2x2 curved slope rump
- 2x2 wedge plate with cut corner
- 3x4 slope shoulders

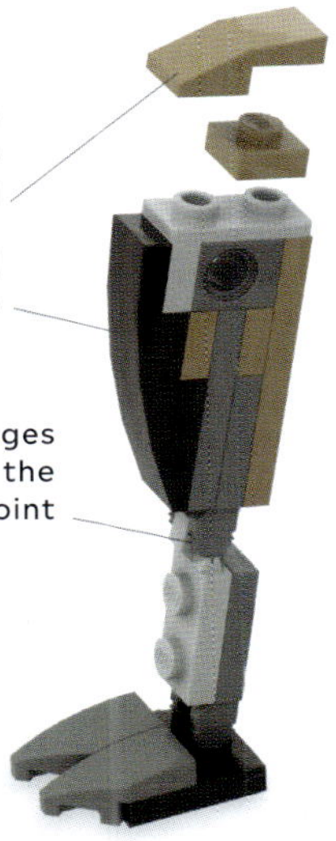

- 1x1 brick with hole fits onto the body pin
- 2x2 hinge plate with click hinge finger

ROAMING LEGS

Finish off the body build with muscular shoulders at the front and smooth tiles and curved slopes down its length. Then start work on the four long legs, which are mostly built sideways, before creating a big, bushy tail.

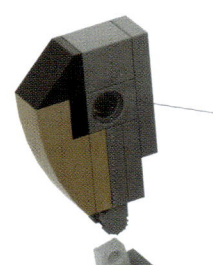

PANDA

This black-and-white bear only lives in the cool, wet forests of China. It spends most of its days sitting upright, eating a kind of tall grass called bamboo. Pandas eat so much of it that local people call them "bamboo bears"! Create your own with a thick, woolly coat to keep it warm, a bamboo-filled belly, and large paws for grasping its favorite food.

ANIMAL FACT
Pandas are a vulnerable species, which means there aren't many of them left in the wild. That's because lots of their habitat has been farmed or built on by humans. Now animal experts are trying to save the species.

Round black ears

Black patch of fur around the eyes

We need to build more bamboo!

Monochrome fur for camouflage in shady forests

Learn to build bamboo shoots on page 139

Hidden pieces make the middle stable

2x2 tile smooths out the back

1x2 plate with ball holds a leg

SOFT SEAT
Panda's sit down a lot because they spend up to 16 hours a day eating! Start your build by making it a soft rump to sit on from one double inverted slope. Build the bear's belly and back upward from there.

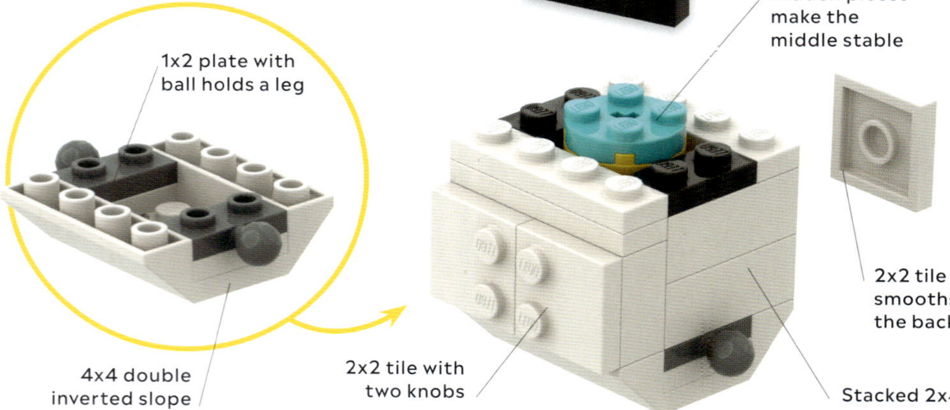

4x4 double inverted slope

2x2 tile with two knobs

Stacked 2x4 bricks

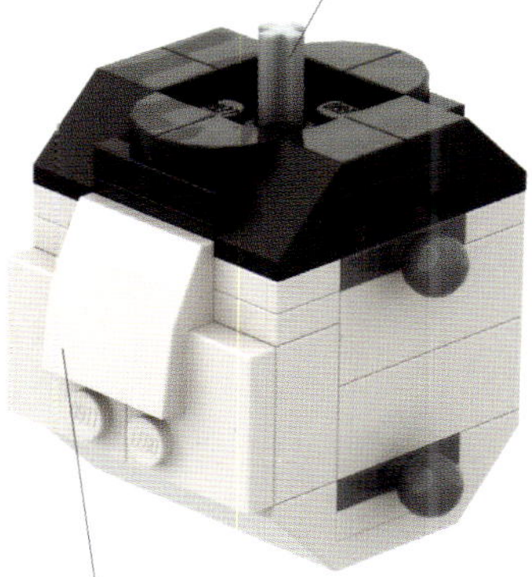

SKILL LEVEL 145

FURRY AND FULL

Now add more curves to the top of the body, bringing in some black bricks to make the fur monochrome. Include more plates with balls for the arm connections and slide an axle in the middle for the neck.

- Hidden axle supports the neck
- 2x2 curved slope rounds out the belly
- 1x3 plate
- This cross plate gives the back a rounder shape
- 2x2 cylinder neck
- 1x1 round tile with bar ear
- This brick has knobs on the front and sides

PANDA EYES

Next, build the head. At its core is a square of bricks with side knobs that has plates attached sideways all around. At the front, build a large face with black-ringed eyes and a wide snout.

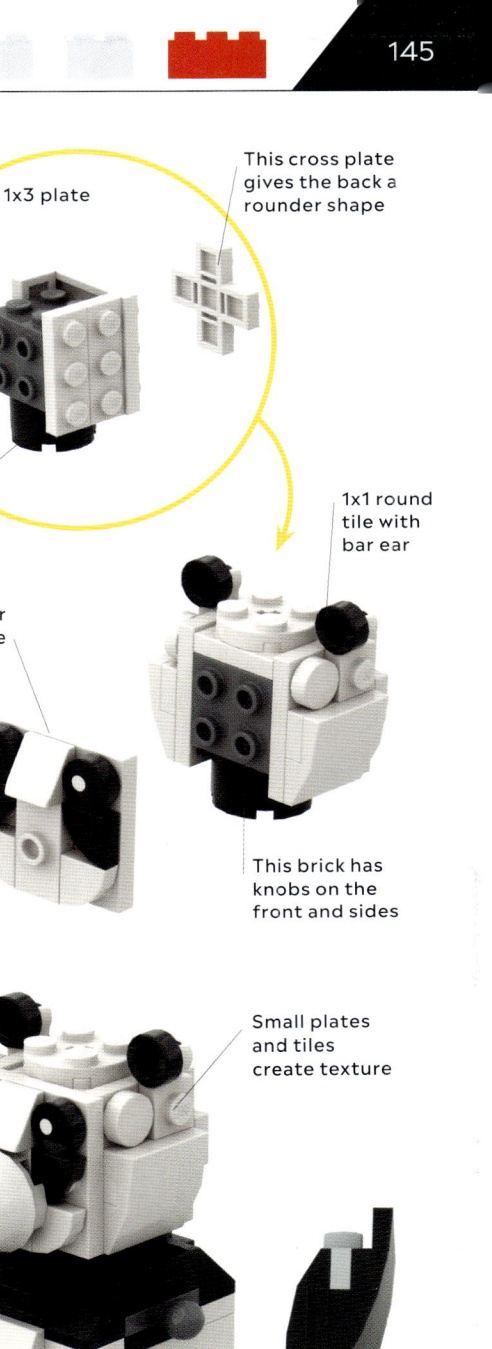

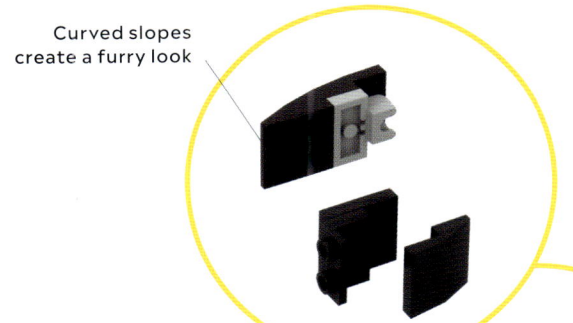

- 4x4 plate for base of face
- The snout is a truncated cone
- Curved slopes create a furry look

CHUNKY LIMBS

Finish off your panda bear with arms and grasping hands to hold bamboo shoots with, as well as plump legs to settle down on. The limbs have built-in sockets that fit onto balls on the body.

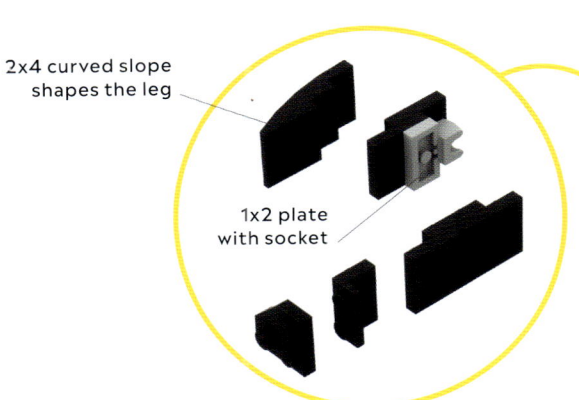

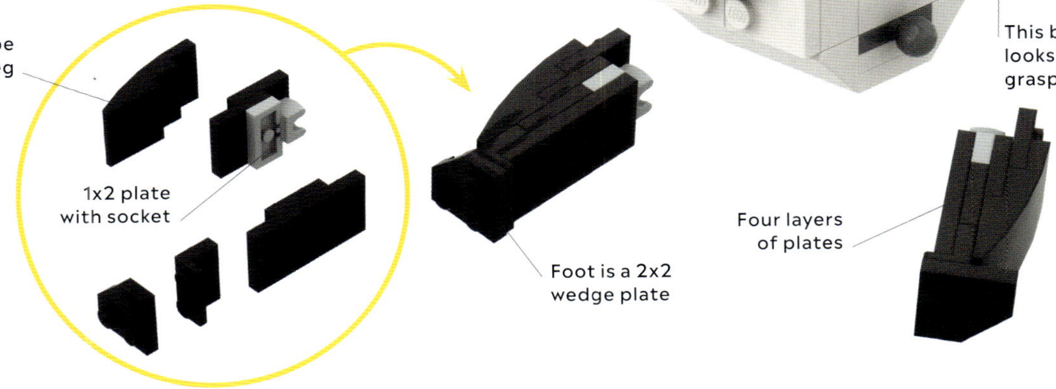

- 1x1 plate with socket
- Small plates and tiles create texture
- 2x4 curved slope shapes the leg
- 1x2 plate with socket
- Foot is a 2x2 wedge plate
- This bracket plate looks like a grasping hand
- Four layers of plates

Cars

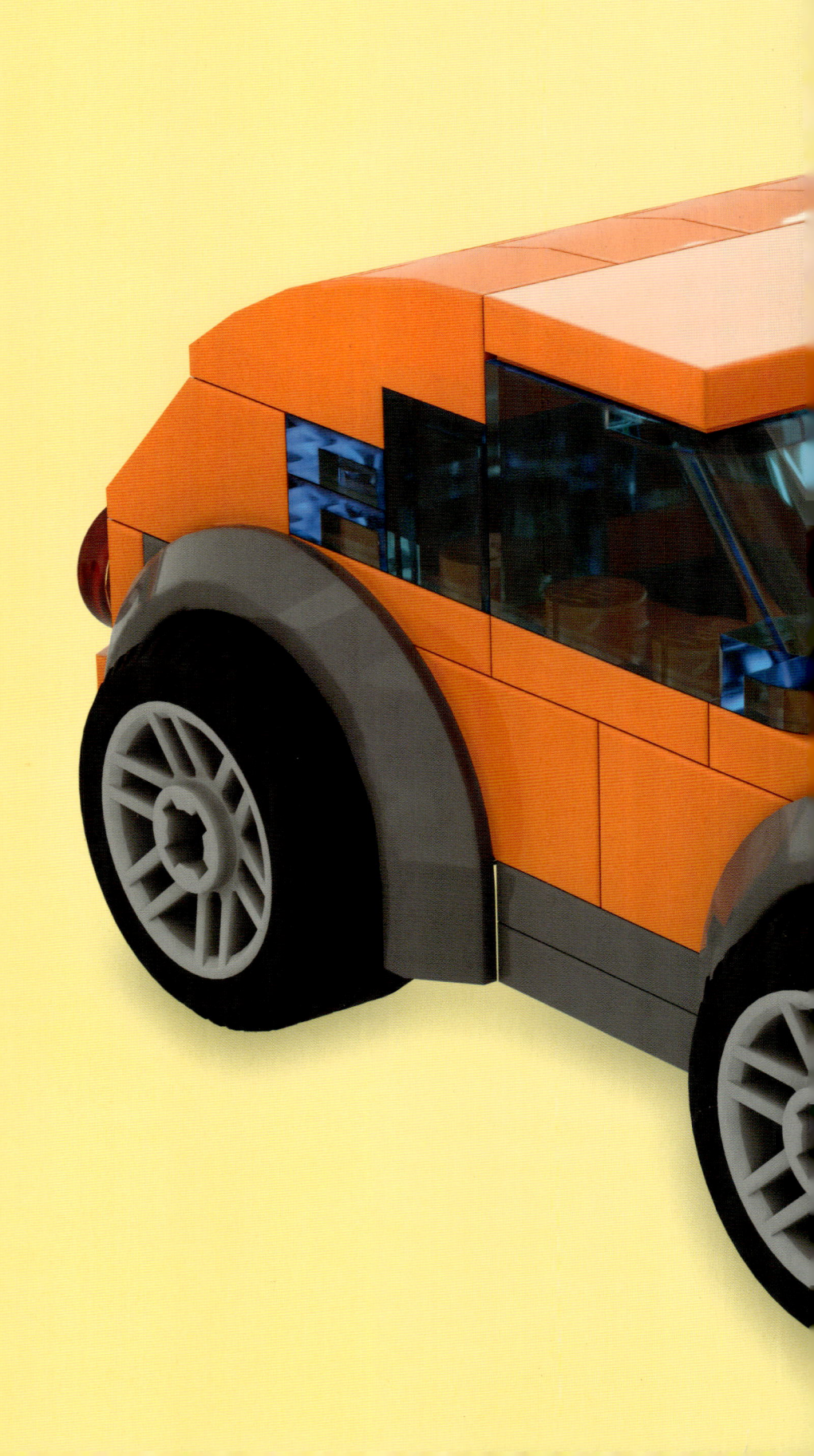

```
01001100
01000101
01000111
01001111
```

OFF-ROADER

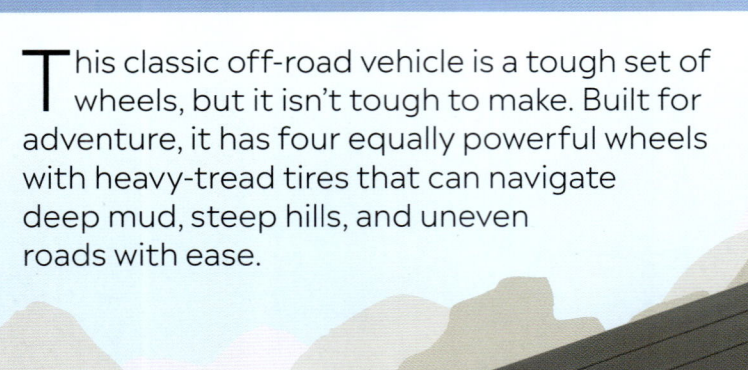

This classic off-road vehicle is a tough set of wheels, but it isn't tough to make. Built for adventure, it has four equally powerful wheels with heavy-tread tires that can navigate deep mud, steep hills, and uneven roads with ease.

1x4 tiles in a row create a flat, boxy roof

Gray plates for the metallic front grille

Wheel with offset treads

Tiles for smooth blue bodywork

Buckle up, Rolo! This may be a rough ride.

I love a woof ride!

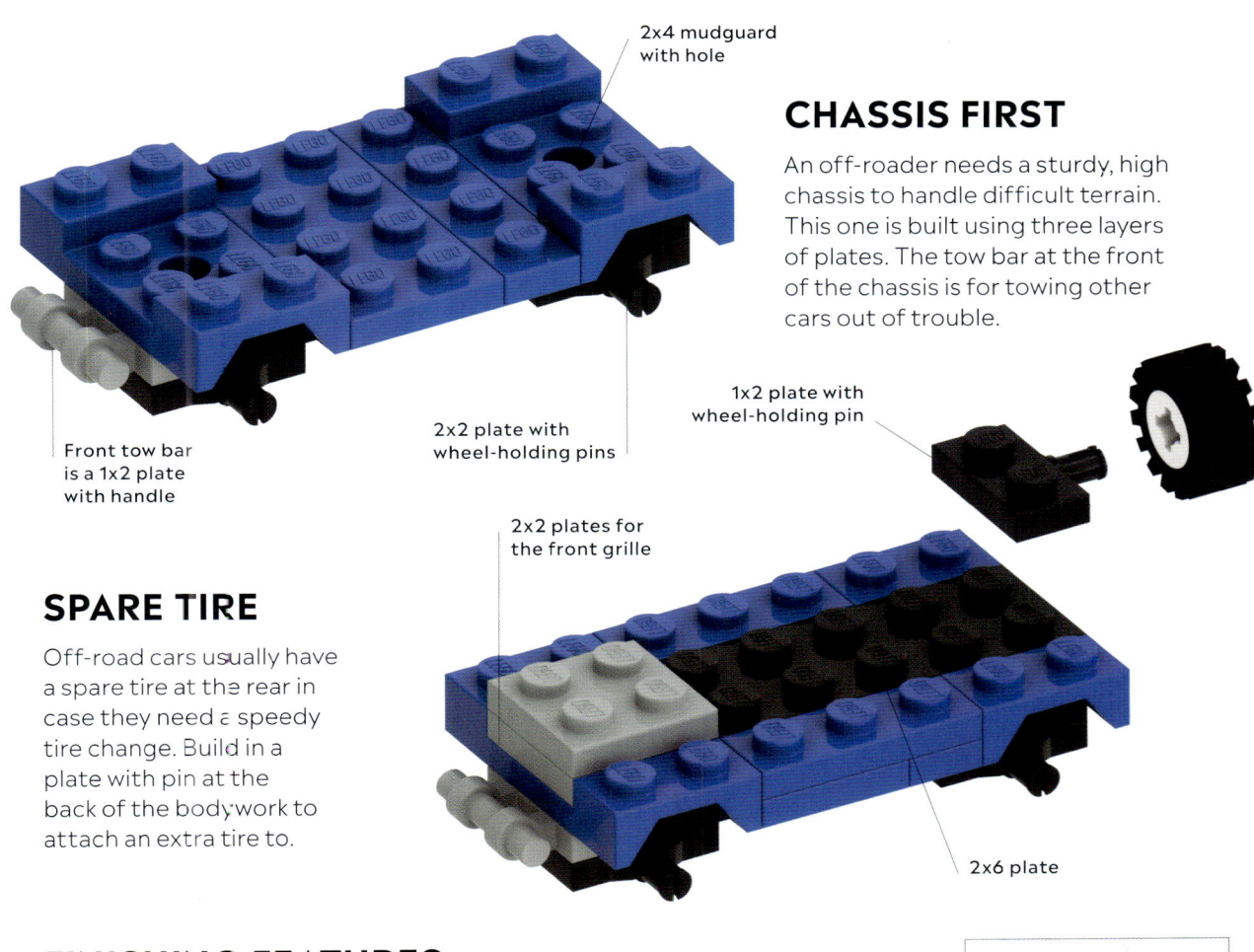

SKILL LEVEL 149

2x4 mudguard with hole

CHASSIS FIRST

An off-roader needs a sturdy, high chassis to handle difficult terrain. This one is built using three layers of plates. The tow bar at the front of the chassis is for towing other cars out of trouble.

Front tow bar is a 1x2 plate with handle

2x2 plate with wheel-holding pins

1x2 plate with wheel-holding pin

2x2 plates for the front grille

SPARE TIRE

Off-road cars usually have a spare tire at the rear in case they need a speedy tire change. Build in a plate with pin at the back of the bodywork to attach an extra tire to.

2x6 plate

FINISHING FEATURES

After creating the sharp-edged shell of the hood with tile pieces, build a wide windshield with two transparent slopes and assemble the soft-top roof.

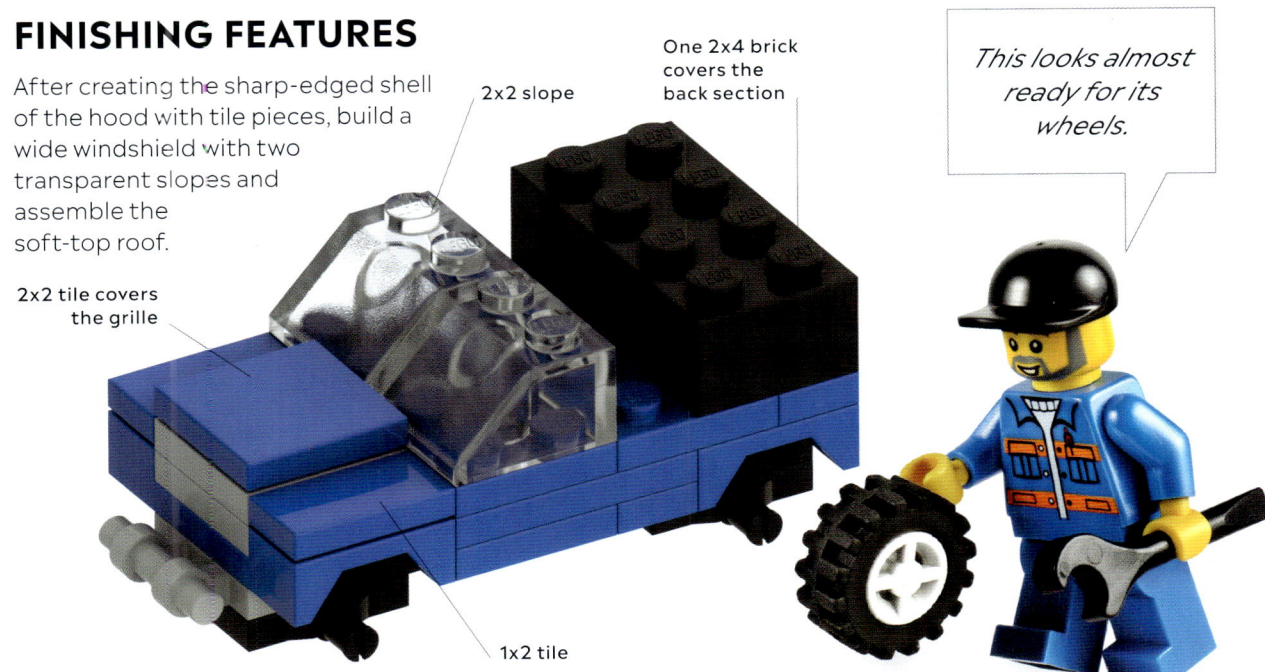

2x2 tile covers the grille

2x2 slope

One 2x4 brick covers the back section

This looks almost ready for its wheels.

1x2 tile

ROCKET CAR

This is a car that will never be caught in traffic. At the slightest sign of a jam, it can engage its rocket engines and propel itself through the skies and beyond. The rocket car's nose cone and small, curved wings (or fins) give it a pointed shape that allows it to blast off in seconds.

I've traveled through 20 galaxies, tracking you down. Now pay your library fine!

But my book isn't due back until Tuesday.

Lever pieces are joysticks for steering

Flaming rocket engine

Two 1x1 slopes make the windshield

Slope with slots forms the front of the fin

Nose cone is a 2x2x2 cone

SKILL LEVEL

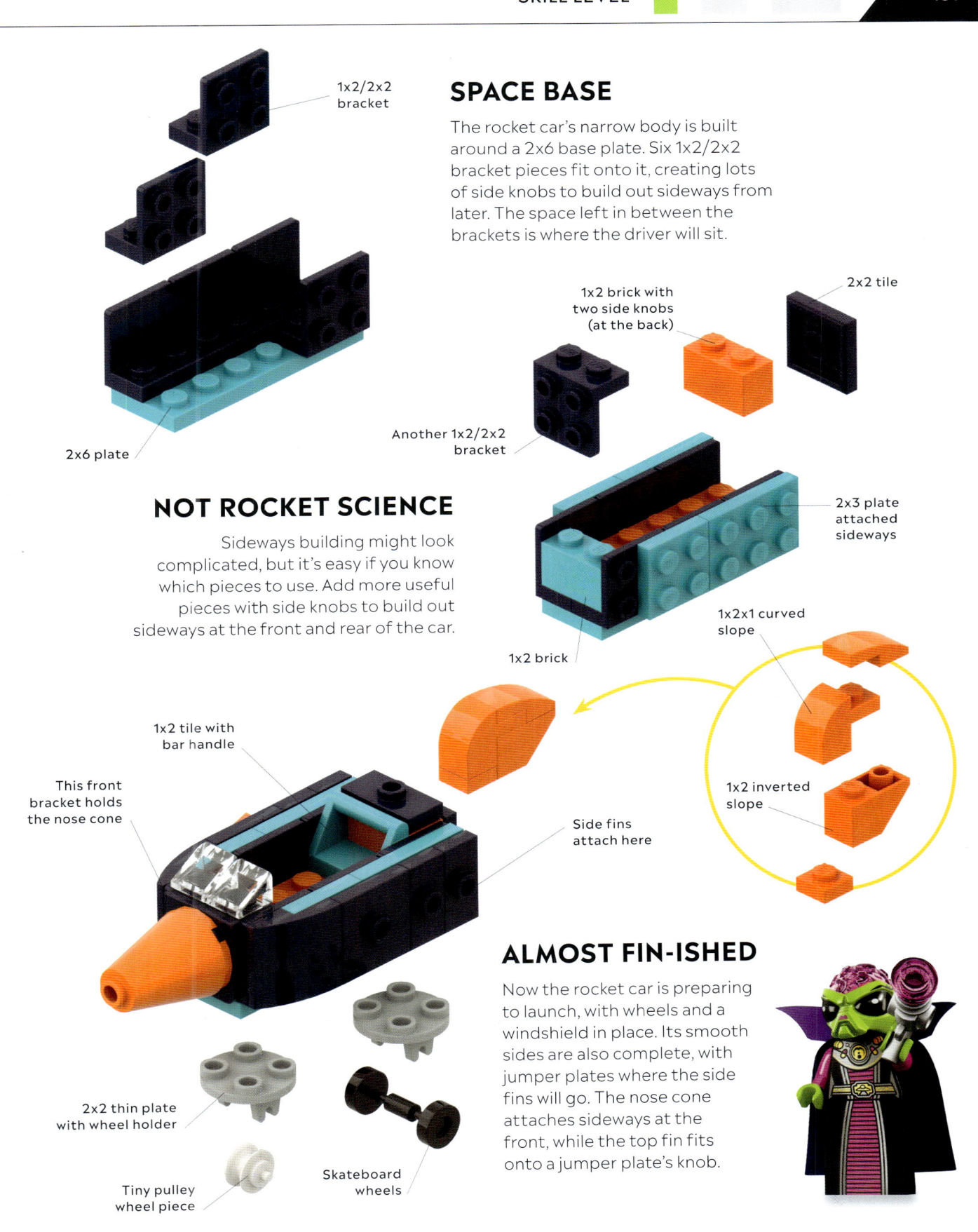

151

SPACE BASE

The rocket car's narrow body is built around a 2x6 base plate. Six 1x2/2x2 bracket pieces fit onto it, creating lots of side knobs to build out sideways from later. The space left in between the brackets is where the driver will sit.

1x2/2x2 bracket

2x6 plate

1x2 brick with two side knobs (at the back)

2x2 tile

Another 1x2/2x2 bracket

2x3 plate attached sideways

1x2 brick

1x2x1 curved slope

1x2 inverted slope

NOT ROCKET SCIENCE

Sideways building might look complicated, but it's easy if you know which pieces to use. Add more useful pieces with side knobs to build out sideways at the front and rear of the car.

1x2 tile with bar handle

This front bracket holds the nose cone

Side fins attach here

2x2 thin plate with wheel holder

Tiny pulley wheel piece

Skateboard wheels

ALMOST FIN-ISHED

Now the rocket car is preparing to launch, with wheels and a windshield in place. Its smooth sides are also complete, with jumper plates where the side fins will go. The nose cone attaches sideways at the front, while the top fin fits onto a jumper plate's knob.

DRAGSTER

Start your engines! This long, narrow car is designed to be first across the finish line on a drag racing circuit. It has a big engine and very light bodywork. A rear wing controls the amount of air (or "drag") moving past it, allowing the dragster to reach incredibly fast speeds.

Create your own racetrack with extra models like this start light

A smooth 2x6 tile forms the top of the rear wing

1x2 tile with handle headrest

Sometimes I drive it to the store to get milk.

Pistol pieces are engine exhausts

Two 1x6 curved slopes create the front of the frame

Small, smooth tires at the front

SKILL LEVEL   153

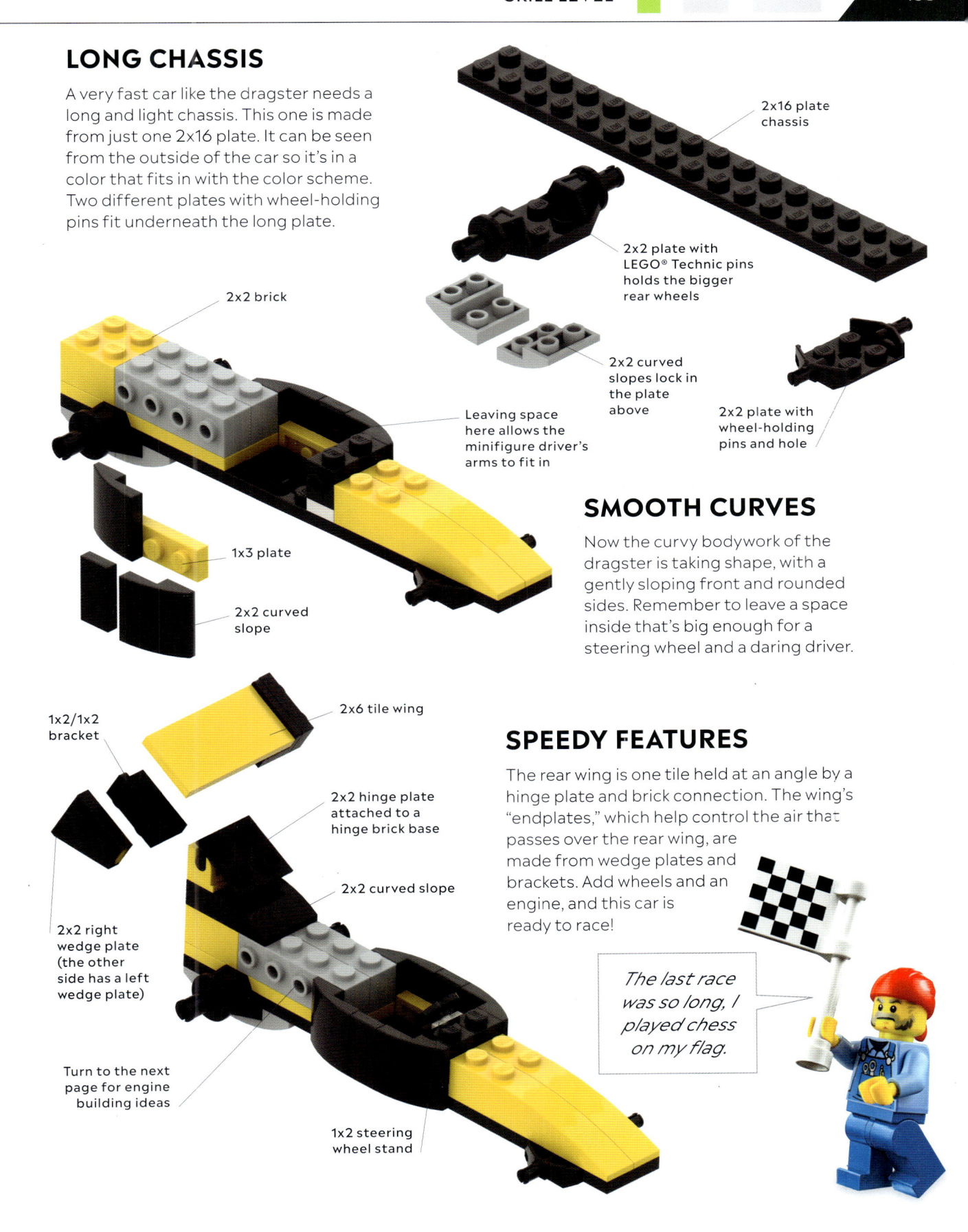

LONG CHASSIS

A very fast car like the dragster needs a long and light chassis. This one is made from just one 2x16 plate. It can be seen from the outside of the car so it's in a color that fits in with the color scheme. Two different plates with wheel-holding pins fit underneath the long plate.

2x16 plate chassis

2x2 plate with LEGO® Technic pins holds the bigger rear wheels

2x2 curved slopes lock in the plate above

2x2 plate with wheel-holding pins and hole

2x2 brick

Leaving space here allows the minifigure driver's arms to fit in

1x3 plate

2x2 curved slope

SMOOTH CURVES

Now the curvy bodywork of the dragster is taking shape, with a gently sloping front and rounded sides. Remember to leave a space inside that's big enough for a steering wheel and a daring driver.

1x2/1x2 bracket

2x6 tile wing

SPEEDY FEATURES

The rear wing is one tile held at an angle by a hinge plate and brick connection. The wing's "endplates," which help control the air that passes over the rear wing, are made from wedge plates and brackets. Add wheels and an engine, and this car is ready to race!

2x2 hinge plate attached to a hinge brick base

2x2 curved slope

2x2 right wedge plate (the other side has a left wedge plate)

Turn to the next page for engine building ideas

1x2 steering wheel stand

The last race was so long, I played chess on my flag.

ENGINES

That engine looks just like my pet dog!

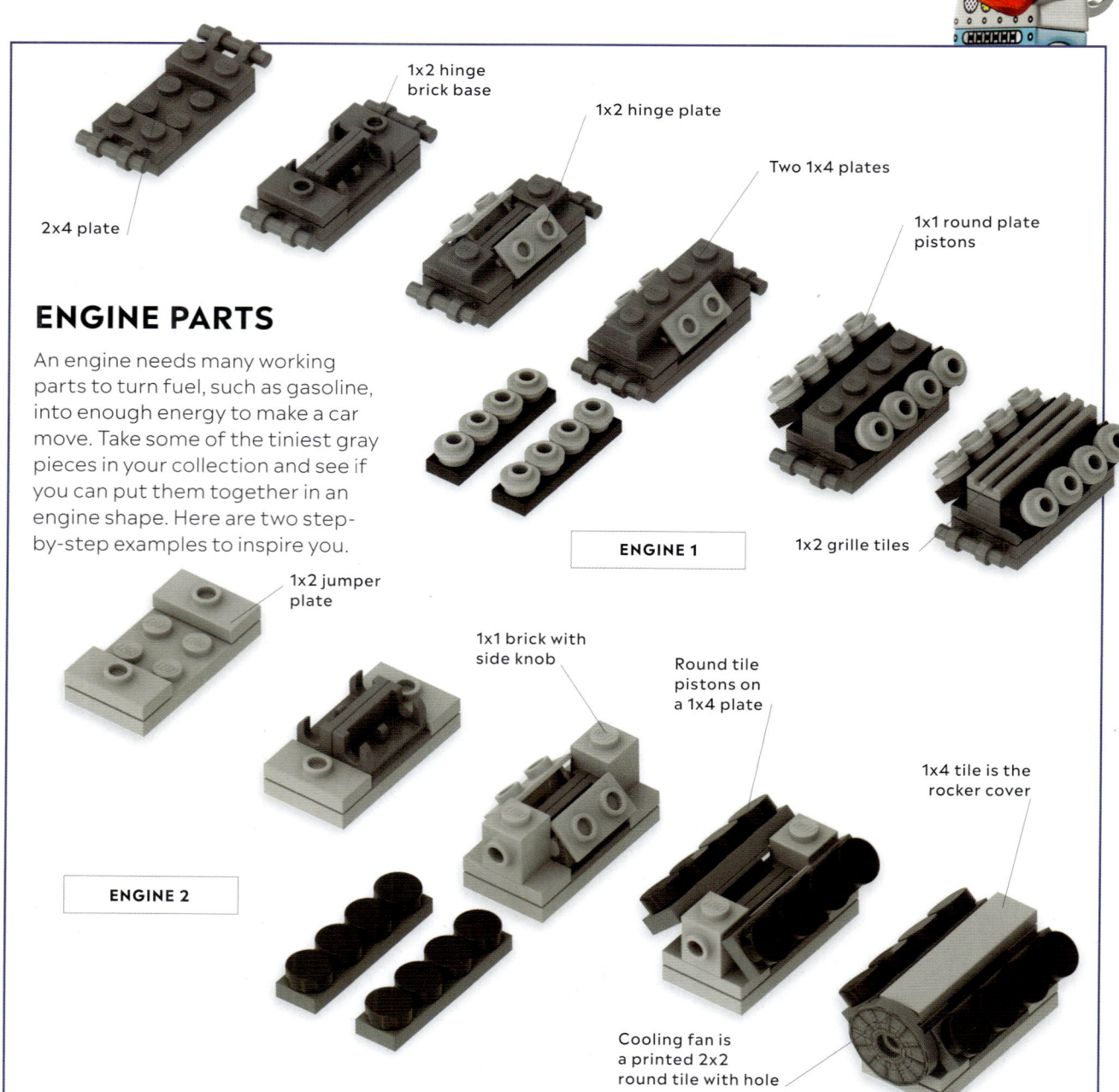

The engine is the roaring heart of a car. Without it, a car wouldn't have the power to move. These amazing machines are mostly hidden under the hoods of cars, but all their metallic pistons, cylinders, and pipes make them interesting to build.

2x4 plate

1x2 hinge brick base

1x2 hinge plate

Two 1x4 plates

1x1 round plate pistons

1x2 grille tiles

ENGINE PARTS

An engine needs many working parts to turn fuel, such as gasoline, into enough energy to make a car move. Take some of the tiniest gray pieces in your collection and see if you can put them together in an engine shape. Here are two step-by-step examples to inspire you.

ENGINE 1

1x2 jumper plate

1x1 brick with side knob

Round tile pistons on a 1x4 plate

1x4 tile is the rocker cover

ENGINE 2

Cooling fan is a printed 2x2 round tile with hole

IDEAS GALLERY   155

ENGINE PIECES

All kinds of small LEGO® pieces can be fitted together in different ways to create mechanical parts and machinery. Do you have any of these useful pieces in your collection?

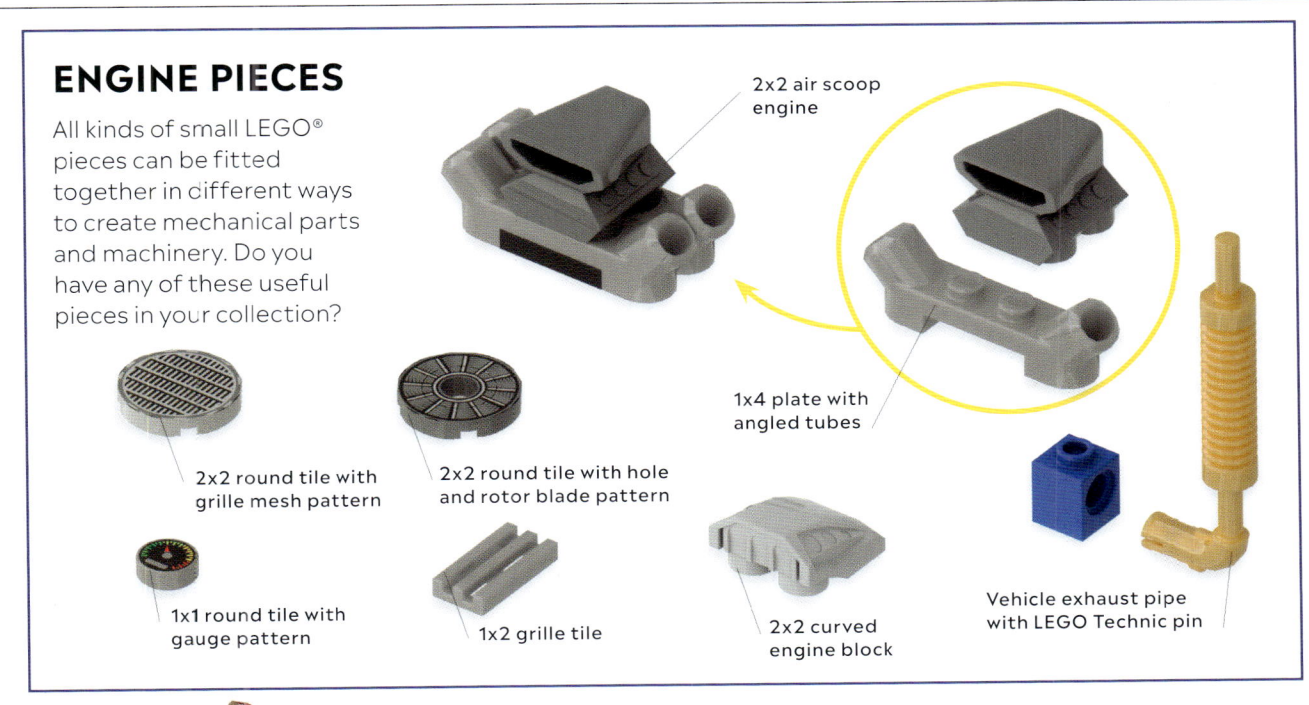

2x2 air scoop engine

1x4 plate with angled tubes

2x2 round tile with grille mesh pattern

2x2 round tile with hole and rotor blade pattern

1x1 round tile with gauge pattern

1x2 grille tile

2x2 curved engine block

Vehicle exhaust pipe with LEGO Technic pin

HOT ROD ENGINE

The hot rod is a stripped-back car with an engine that isn't hidden by bodywork. The flaming "exposed" engine is made from layers of small pieces. The side details are held at an angle using plates with clips attached to bars. Learn more about this car on pages 194-195.

Plates with bar handles

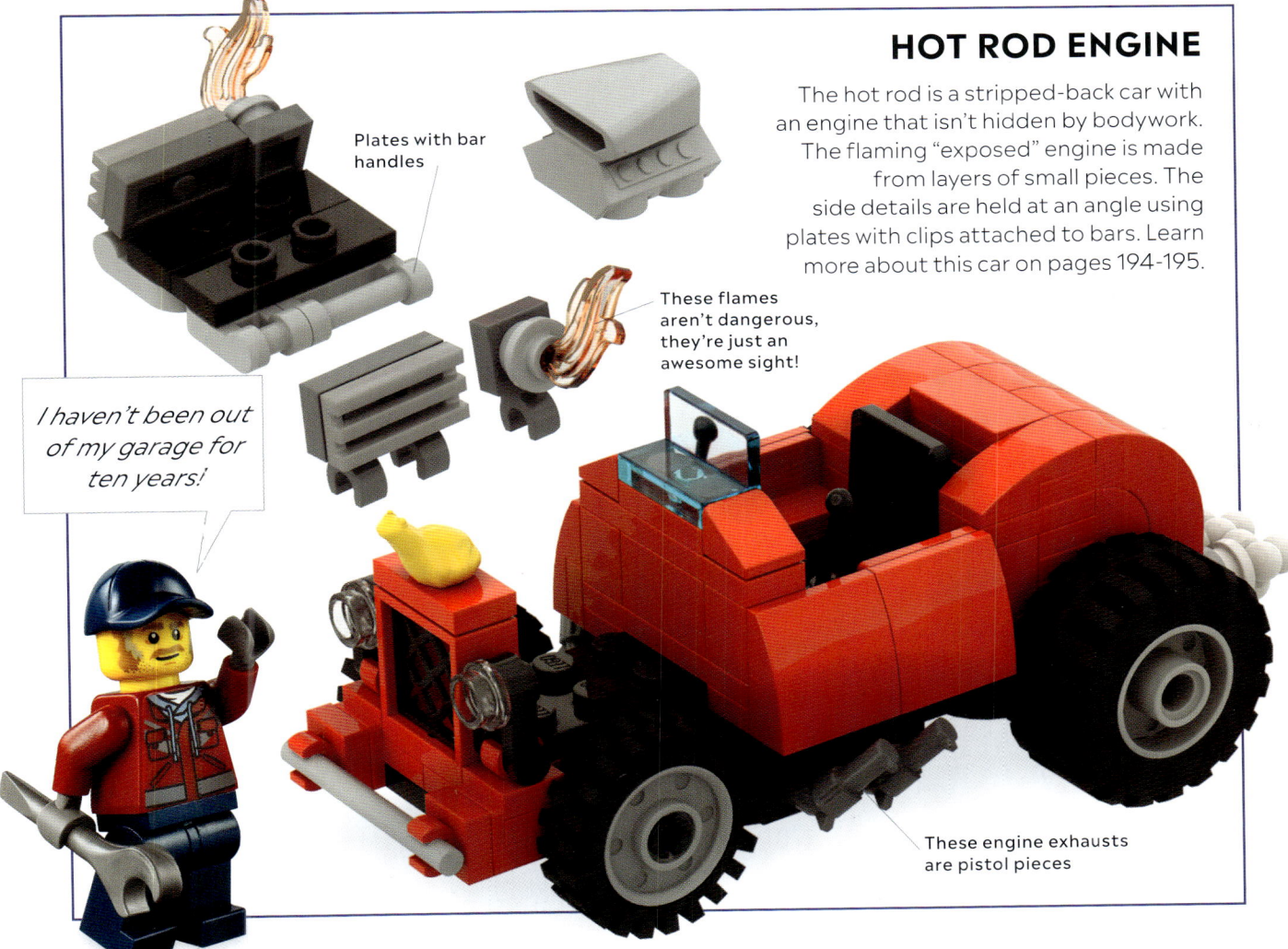

These flames aren't dangerous, they're just an awesome sight!

I haven't been out of my garage for ten years!

These engine exhausts are pistol pieces

ANIMAL CARS

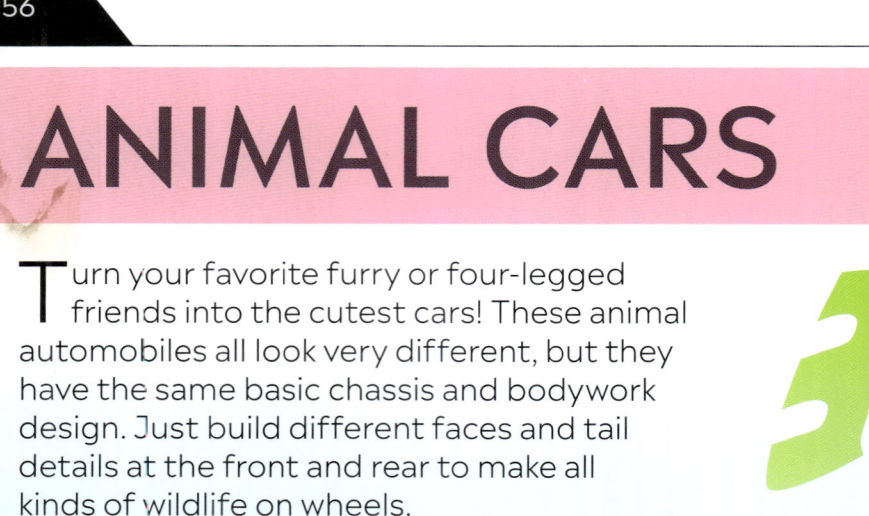

Turn your favorite furry or four-legged friends into the cutest cars! These animal automobiles all look very different, but they have the same basic chassis and bodywork design. Just build different faces and tail details at the front and rear to make all kinds of wildlife on wheels.

Only an elephant car has its trunk at the front!

1x4 double curved slope rear end

Curly monkey tail

3x3 corner plate elephant ear

Jumper plate monkey mouth

1x1 quarter tile rabbit nose

1x1 slopes make a shaggy lion's mane

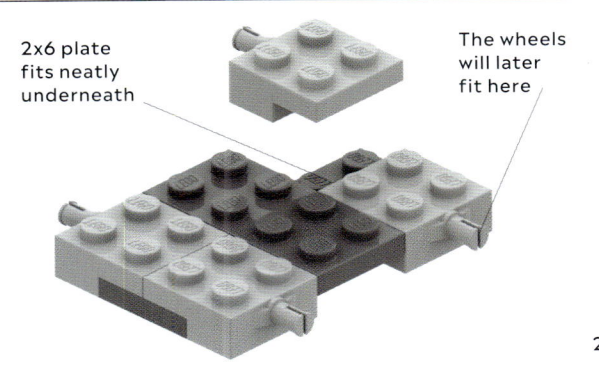

SKILL LEVEL 157

BASIC BODY

The animal cars' bodies are all made from the same pieces. Their small chassis are built around 2x6 plates with six more plates on top: two regular 2x2 plates and four 2x2 plates with wheel-holding pins.

- 2x6 plate fits neatly underneath
- The wheels will later fit here
- 1x4 brick with four side knobs
- 2x2 brick
- 1x2 brick with two side knobs
- There's a 1x2 plate here now too

SIDE STEP

Now there are lots of bricks on the plate chassis. Build in bricks with side knobs on the front, back, and sides of the animal car body so more pieces can be added sideways at the next stage.

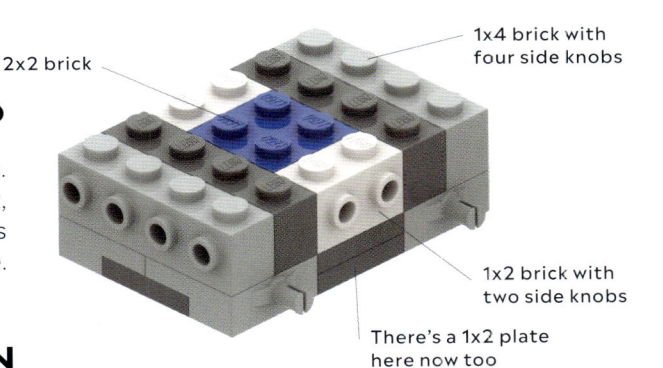

COZY CABIN

It's time to build the driver's cabin. The seat and steering wheel sit low within layers of plates, bricks, and panels.

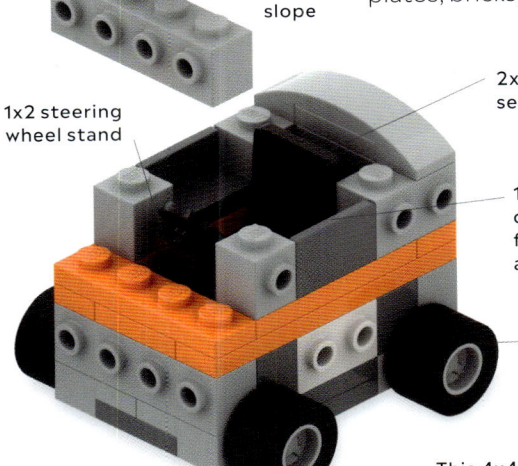

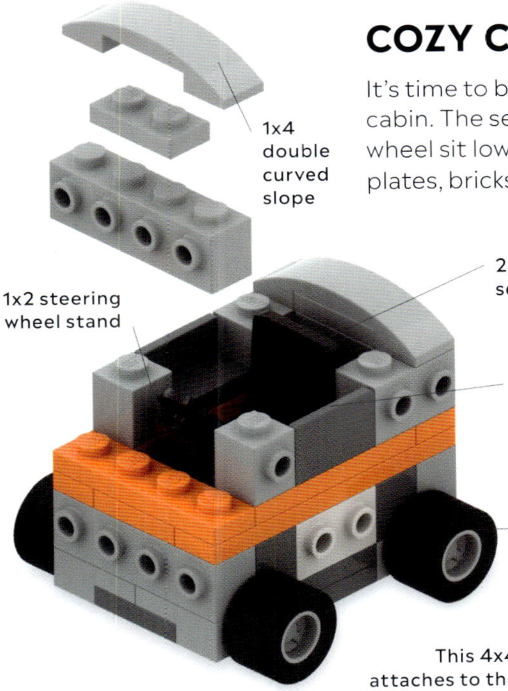

- 1x4 double curved slope
- 1x2 steering wheel stand
- 2x2 driving seat
- 1x2x1 panels create space for minifigure arms
- Small wheels are now attached to the pins

MAKING FACES

Attach all kinds of animal features to the side knobs on the bodywork. This elephant has big ears and a trumpeting trunk at the front, rounded sides, and a swishing tail at the back!

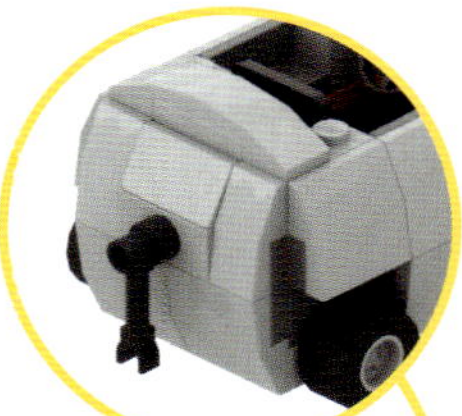

- The back part is built around a 2x4 plate
- This 4x4 plate attaches to the body
- 1x1 round plate
- 1x3x2 arch brick trunk
- Horn piece tusk
- 1x1 tile eye
- 2x2 curved slopes attached to plates

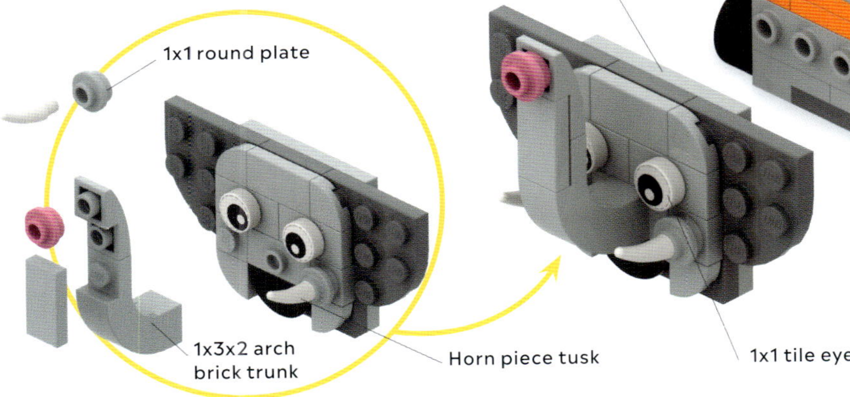

BUMPER CAR

Your minifigures better prepare for a bumpy ride in this fairground favorite. Small, electrically powered bumper cars have rubber bumpers around their bases so riders can race around and crash into each other, creating car-based chaos!

Electricity cable is an antenna piece

I haven't bumped you yet, darling.

Aaaaargh!

1x2 steering wheel stand and steering wheel

Transparent red 1x1 plate tail light

Build more than one bumper car for maximum chaos

Black tiles and curved bricks look like smooth rubber

SKILL LEVEL

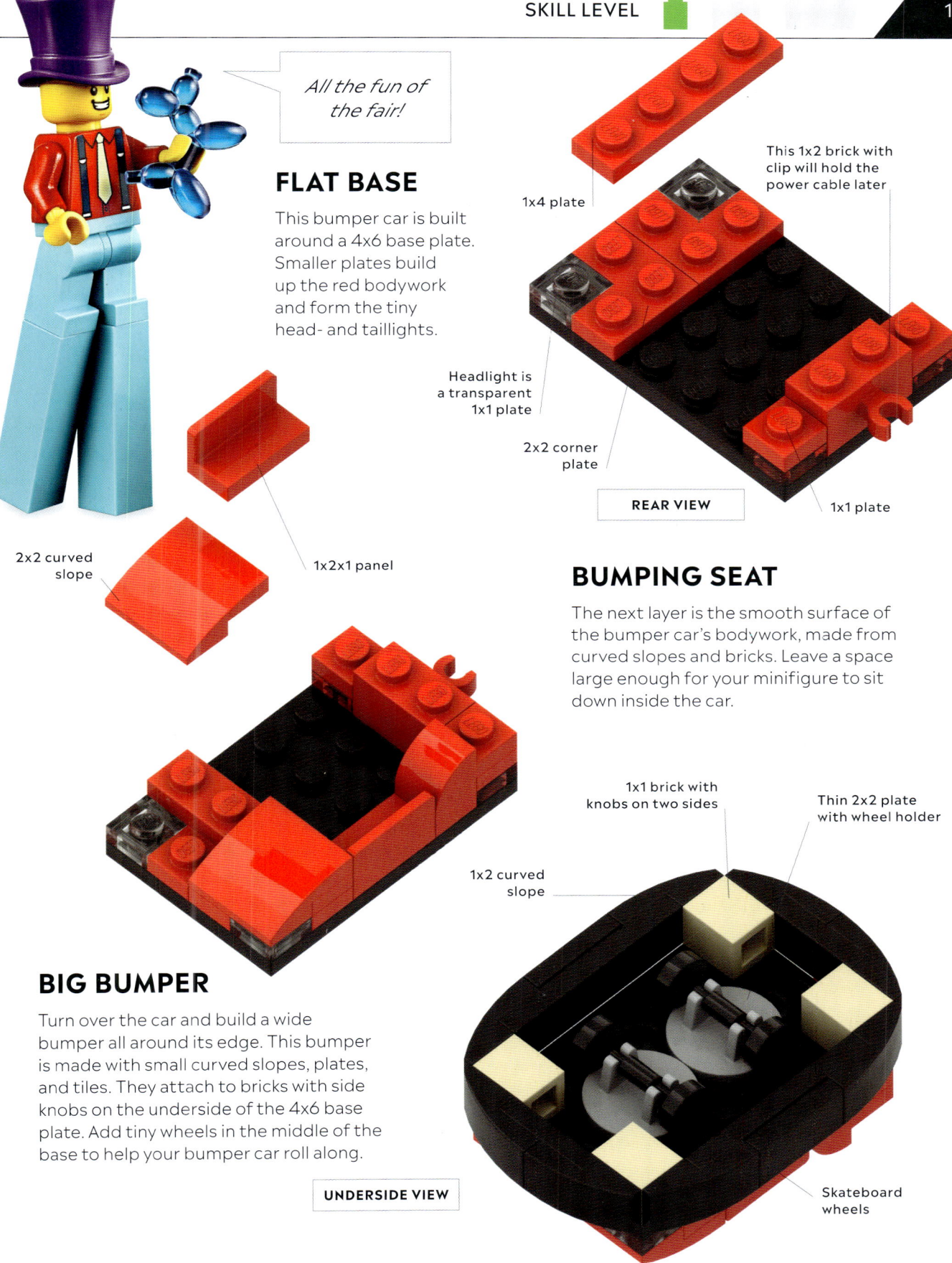

159

All the fun of the fair!

FLAT BASE

This bumper car is built around a 4x6 base plate. Smaller plates build up the red bodywork and form the tiny head- and taillights.

1x4 plate

This 1x2 brick with clip will hold the power cable later

Headlight is a transparent 1x1 plate

2x2 corner plate

REAR VIEW

1x1 plate

2x2 curved slope

1x2x1 panel

BUMPING SEAT

The next layer is the smooth surface of the bumper car's bodywork, made from curved slopes and bricks. Leave a space large enough for your minifigure to sit down inside the car.

1x1 brick with knobs on two sides

Thin 2x2 plate with wheel holder

1x2 curved slope

BIG BUMPER

Turn over the car and build a wide bumper all around its edge. This bumper is made with small curved slopes, plates, and tiles. They attach to bricks with side knobs on the underside of the 4x6 base plate. Add tiny wheels in the middle of the base to help your bumper car roll along.

UNDERSIDE VIEW

Skateboard wheels

FAMILY CAR

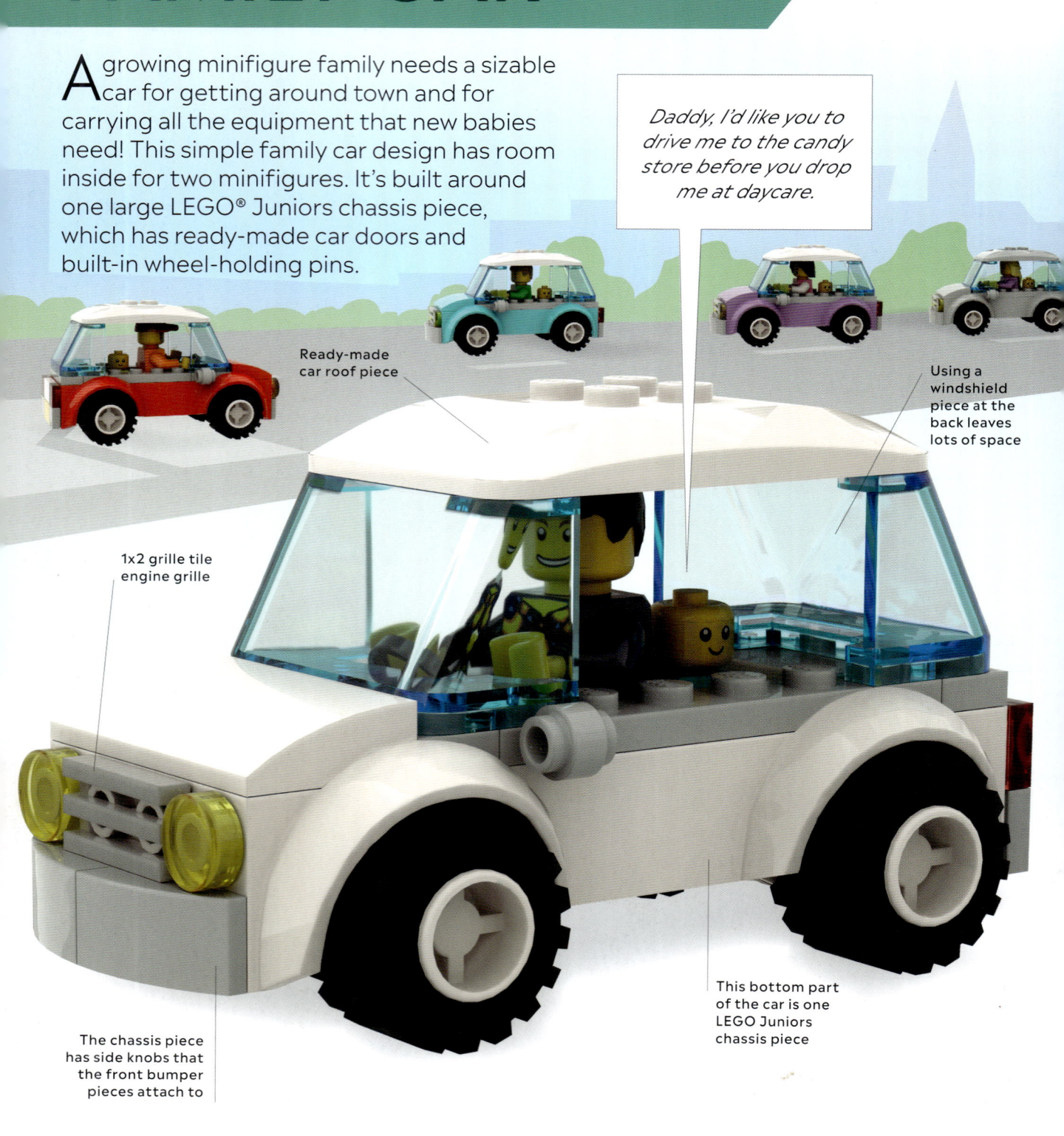

A growing minifigure family needs a sizable car for getting around town and for carrying all the equipment that new babies need! This simple family car design has room inside for two minifigures. It's built around one large LEGO® Juniors chassis piece, which has ready-made car doors and built-in wheel-holding pins.

Daddy, I'd like you to drive me to the candy store before you drop me at daycare.

Ready-made car roof piece

Using a windshield piece at the back leaves lots of space

1x2 grille tile engine grille

The chassis piece has side knobs that the front bumper pieces attach to

This bottom part of the car is one LEGO Juniors chassis piece

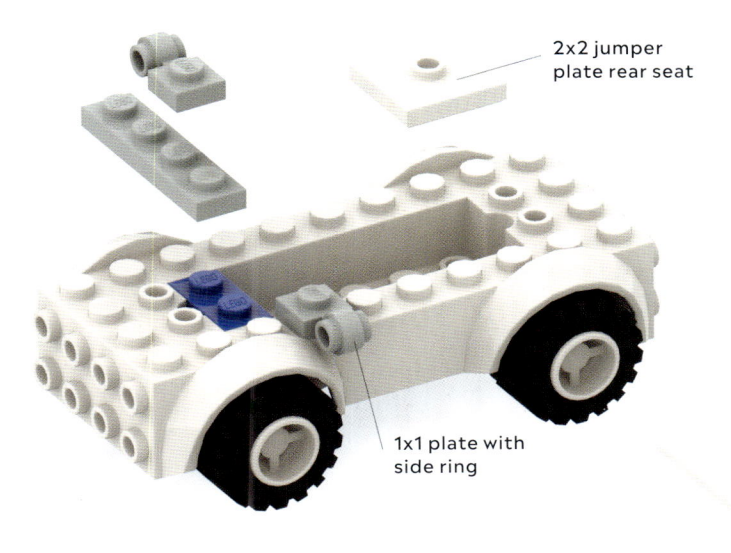

SKILL LEVEL

161

2x2 jumper plate rear seat

ALMOST READY ALREADY

This car's ready-made chassis piece means it can be made LEGO roadworthy in minutes. The wheels, side mirrors, dashboard, and rear seat are the first pieces placed on the chassis.

1x1 plate with side ring

2x4 curved slope hood

HOOD AND BUMPER

After adding more gray plates on top of the chassis, place the one-piece hood at the front. Then build a curved bumper onto the sideways-facing knobs below.

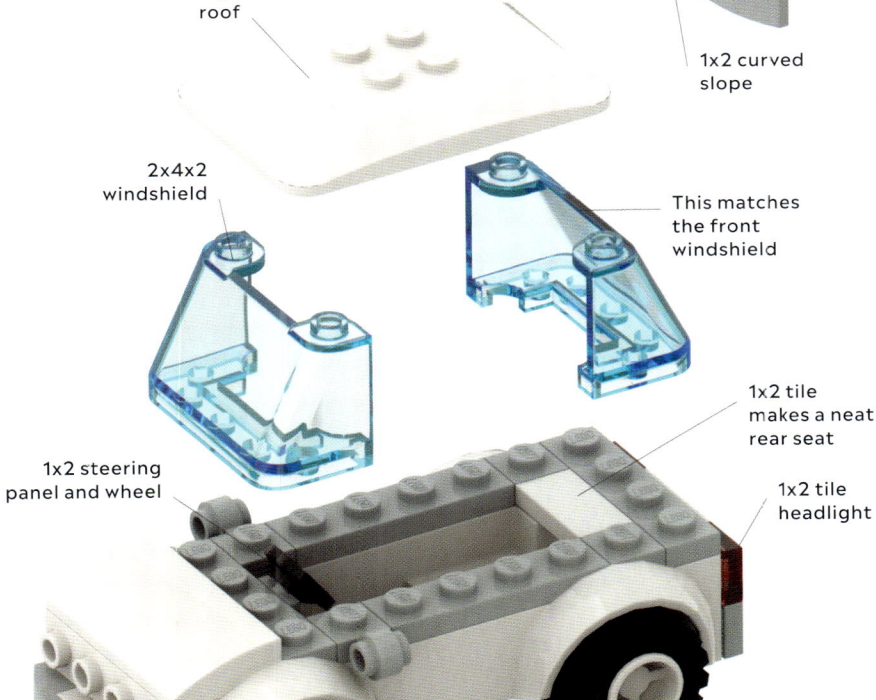

1x4 plate side stripe

4x6x2/3 wedge roof

1x2 curved slope

2x4x2 windshield

This matches the front windshield

Mommy, let's skip the milk and eat that doughnut.

1x2 tile makes a neat rear seat

1x2 tile headlight

1x2 steering panel and wheel

SIMPLE ROOF

Now the steering wheel is in place and this car is almost ready to hit the road. Add a windshield, rear window, and roof, or skip the roof and make it a family-size convertible!

Tires have thick treads

AUTO RICKSHAW

Beep beep! Three-wheeled auto rickshaws can be found in many warm countries around the world, such as India. This one is small and speedy, so it can whiz along narrow city streets. Its open sides allow passengers to enjoy a cooling breeze as they ride along.

Does it have room for my luggage?

1x4x2 windshield

This is a window piece built into the canopy

1x1 round tile headlight

This front wheel is usually found on LEGO airplanes

Handrail is a bar piece between two round plates with open knobs

SKILL LEVEL

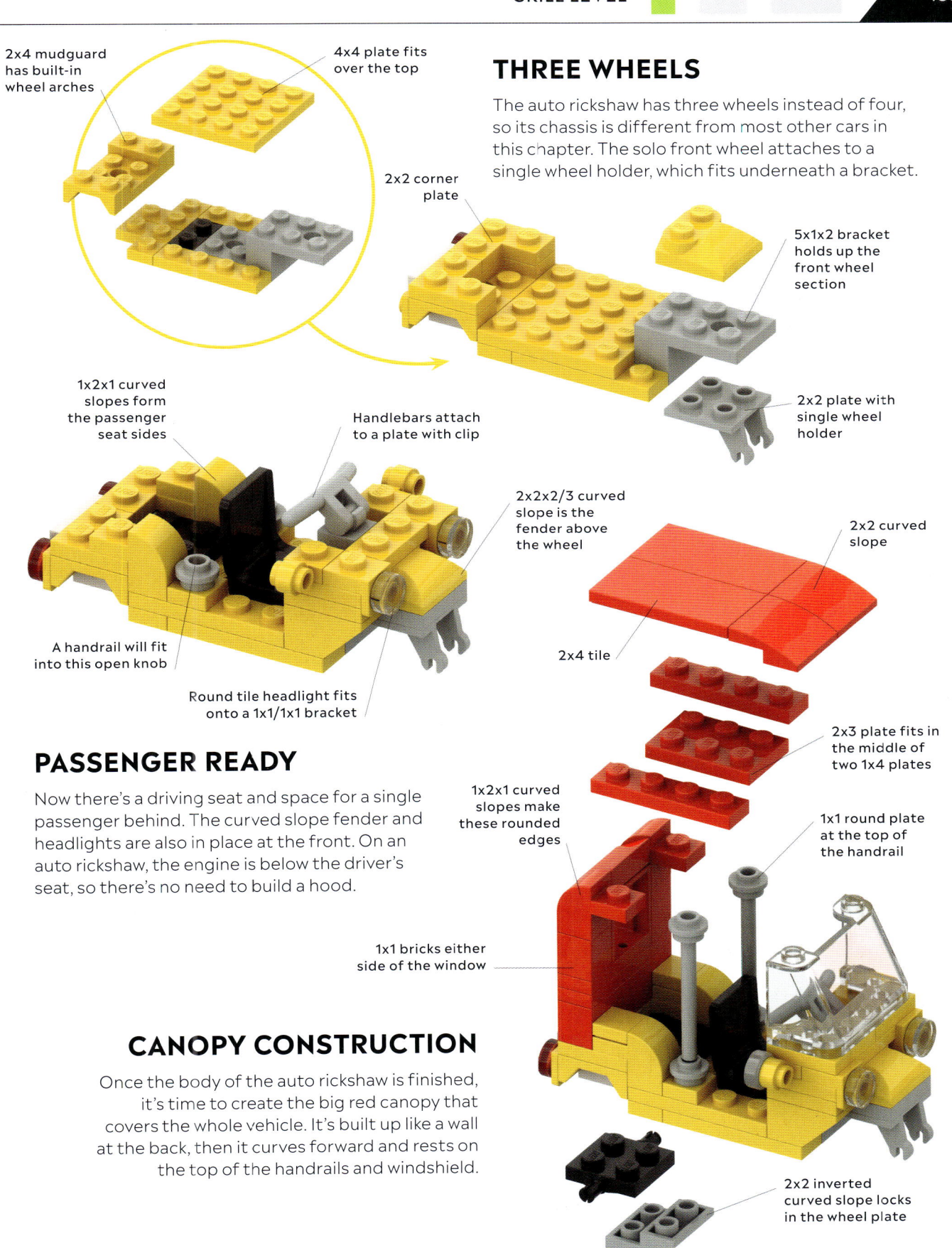

THREE WHEELS

The auto rickshaw has three wheels instead of four, so its chassis is different from most other cars in this chapter. The solo front wheel attaches to a single wheel holder, which fits underneath a bracket.

- 2x4 mudguard has built-in wheel arches
- 4x4 plate fits over the top
- 2x2 corner plate
- 5x1x2 bracket holds up the front wheel section
- 2x2 plate with single wheel holder
- 1x2x1 curved slopes form the passenger seat sides
- Handlebars attach to a plate with clip
- 2x2x2/3 curved slope is the fender above the wheel
- 2x2 curved slope
- 2x4 tile
- A handrail will fit into this open knob
- Round tile headlight fits onto a 1x1/1x1 bracket
- 2x3 plate fits in the middle of two 1x4 plates
- 1x2x1 curved slopes make these rounded edges
- 1x1 round plate at the top of the handrail

PASSENGER READY

Now there's a driving seat and space for a single passenger behind. The curved slope fender and headlights are also in place at the front. On an auto rickshaw, the engine is below the driver's seat, so there's no need to build a hood.

- 1x1 bricks either side of the window

CANOPY CONSTRUCTION

Once the body of the auto rickshaw is finished, it's time to create the big red canopy that covers the whole vehicle. It's built up like a wall at the back, then it curves forward and rests on the top of the handrails and windshield.

- 2x2 inverted curved slope locks in the wheel plate

VINTAGE CAR

This boldly colored, compact classic car is a little piece of history. Designed in the style of practical family cars that were very popular from the 1960s, it has an iconic curved shape. It's perhaps even cooler today than it has ever been!

It's not old, it's a classic.

Four 1x3 curved slopes form the rounded rear end

2x4 tile roof

Two 1x1 transparent plates form a tiny rear window

Large, round mudguards add to the curvy design

Simple round tile headlight

Bumper is one 1x4 double curved slope piece

SKILL LEVEL

START THE CAR

The vintage car has a simple chassis of a 2x10 plate with wheel-holding plates attached underneath. A gray 2x4 plate also fits widthwise under the chassis plate—this will form part of the car's gray running boards or trim.

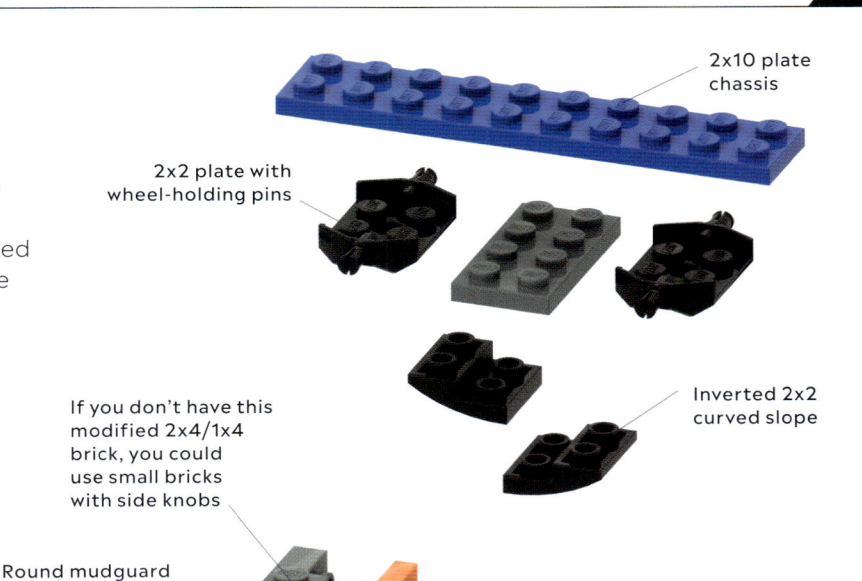

2x10 plate chassis

2x2 plate with wheel-holding pins

Inverted 2x2 curved slope

If you don't have this modified 2x4/1x4 brick, you could use small bricks with side knobs

Round mudguard piece

1x2 slope

1x4 double curved slope

COMPACT CURVES

Now the vintage car has inverted slope side doors, which fit snugly behind the round arches of its mudguards. The lower sections of the front and rear bumpers add more curves to the vintage car's bodywork.

1x2 inverted slope brick

1x3 curved slope

Longer curved slope for the middle of the hood

ROUND ROOF

The distinctive round shape of the vintage car's roof is mostly made from curved slopes in different sizes. Transparent blue plates and bricks then fit neatly into the small spaces underneath to form the rear windows. Add the hood and a roof tile on top, and this classic car has a new lease on life!

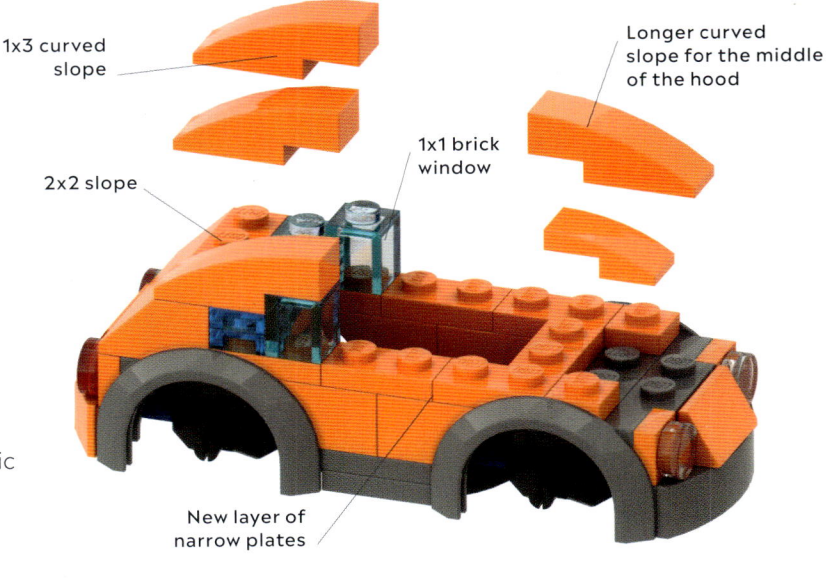

2x2 slope

1x1 brick window

New layer of narrow plates

CITY CAR

Not everyone needs spoilers and fancy features on their cars—sometimes less is more! This nippy, compact car has everything a busy city driver might need: a sturdy chassis, wide windows, and neat little wheels. As an added bonus, it will never struggle for a parking space!

Roof is one 4x4 tile with four knobs on its edge

Windshield pieces form the front and back windows

We are both small and sophisticated.

1x4 double curved slope hood

Simple side stripe is a 1x3 plate

Smooth wheels for city roads

Me too!

SKILL LEVEL

167

SIMPLE BASE

The city car is built up from a ready-made vehicle base that has built-in wheel-holding pins and a 3x4 plate center section. If you don't have the piece, you could use regular plates instead.

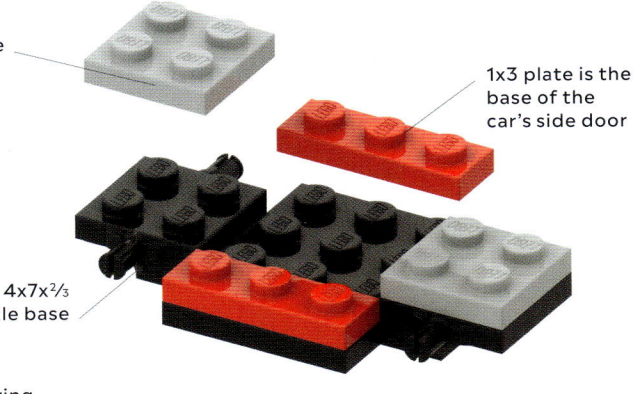

2x2 plate

1x3 plate is the base of the car's side door

4x7x2/3 vehicle base

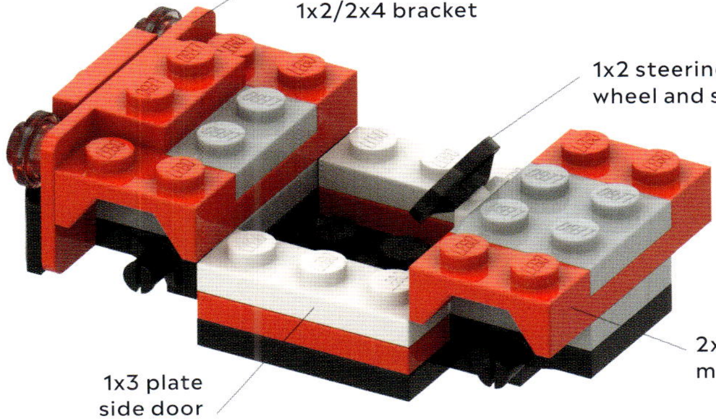

The back bumper is built on a 1x2/2x4 bracket

1x2 steering wheel and stand

SIDE STRIPE

Now the city car's side doors are taking shape. A white 1x3 plate adds a "go-faster" stripe to the door design. There are also tiny mudguards, a rear bumper, and a steering wheel.

1x3 plate side door

2x4 vehicle mudguard

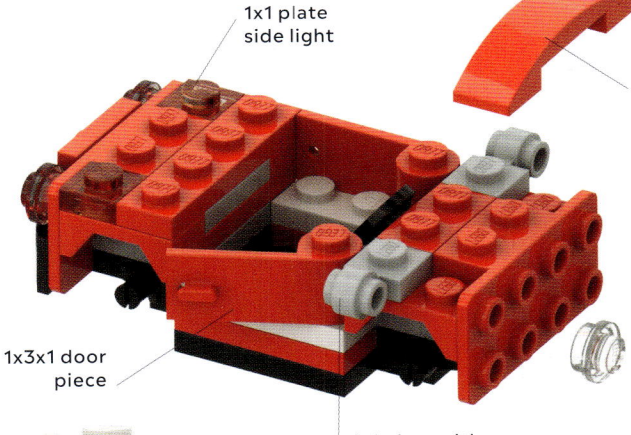

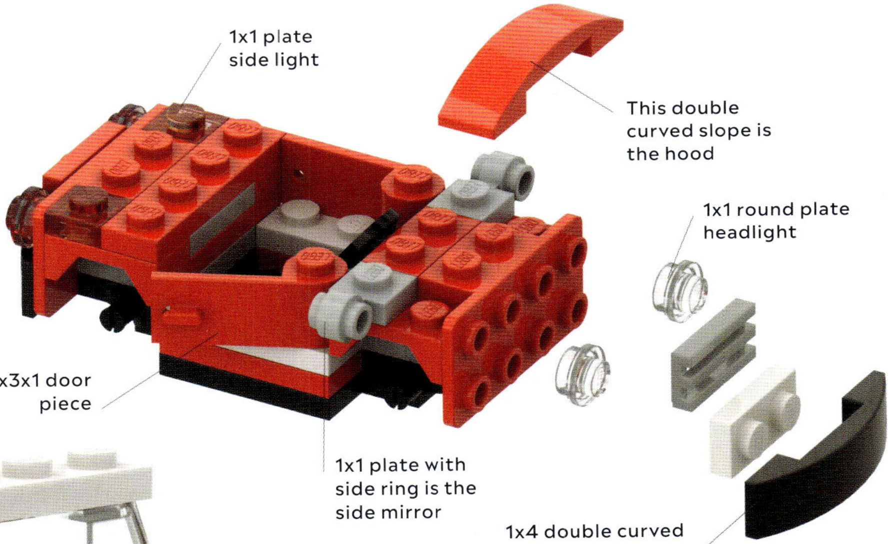

1x1 plate side light

This double curved slope is the hood

BUMPER BRACKET

Like the rear bumper, the front bumper is built onto a bracket piece that has eight sideways-facing knobs on it. The headlights, grille, and curved bumper all fit onto it.

1x1 round plate headlight

This is a modified tile, but an ordinary one would work too

1x3x1 door piece

1x1 plate with side ring is the side mirror

1x4 double curved slope splash shield

2x4x2 windshield

Jumper plate trunk door

WHEELY DONE

Add a windshield at the front and a matching piece at the back for the rear windows. A modified 4x4 tile for the roof tops off the build. Secure the small wheels on their wheel pins, then head off to explore the city!

Simple wheel trim

REAR VIEW

MOON BUGGY

Sam to Base... Can you put my dinner in the oven?

2x2 inverted dish is a satellite dish

Drill piece in the tool storage area

A 1x1 round plate with bar holds this dish

Wide mudguards shield the driver from moon dust

This moon buggy is built to rove across the dusty, rocky surface of the moon so astronauts can learn more about the landscape there. It has four heavy tires to stop it from floating away into space, and two satellite-dish antennas for communicating with other astronauts or people back on Earth.

SKILL LEVEL

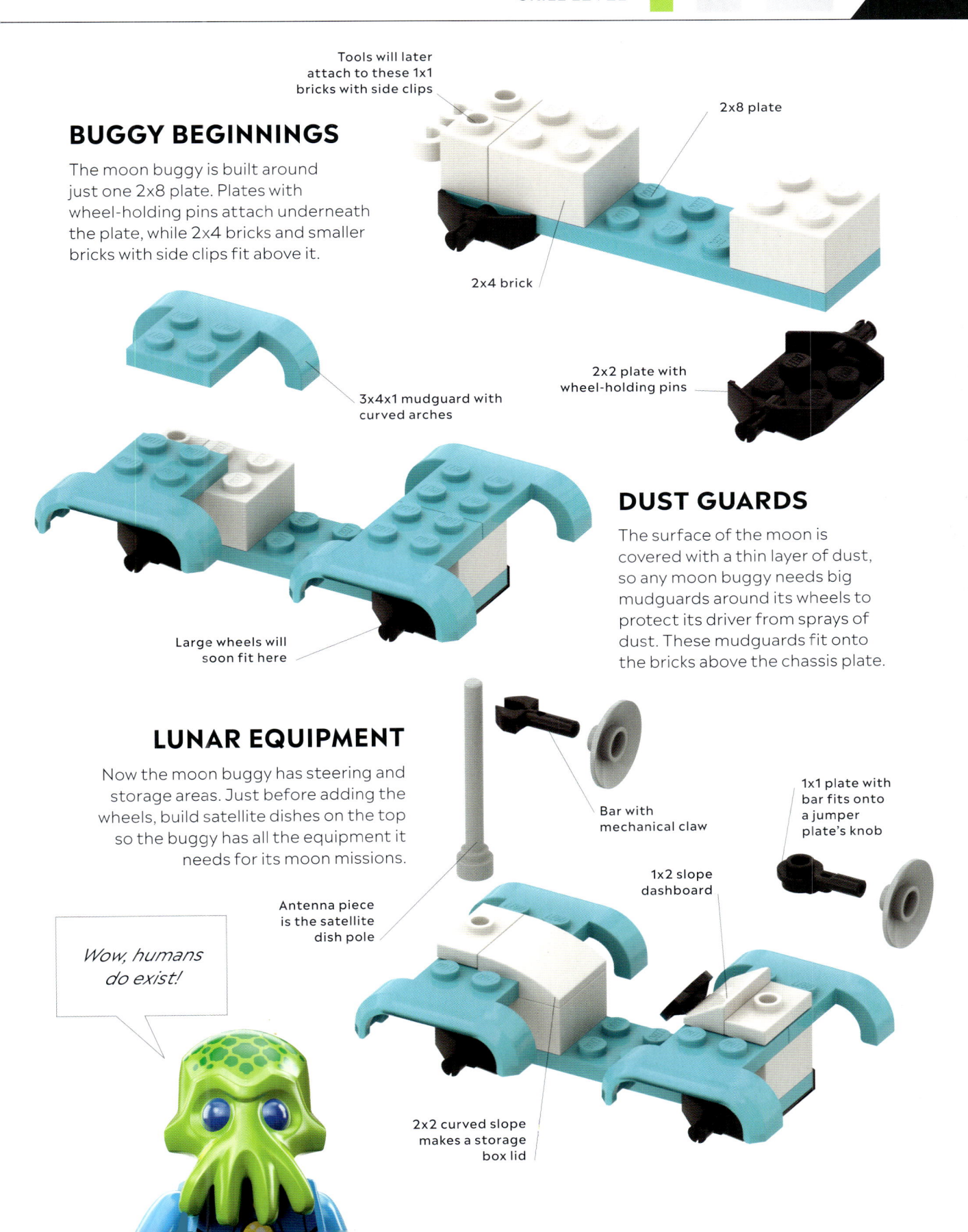

169

BUGGY BEGINNINGS

The moon buggy is built around just one 2x8 plate. Plates with wheel-holding pins attach underneath the plate, while 2x4 bricks and smaller bricks with side clips fit above it.

Tools will later attach to these 1x1 bricks with side clips

2x8 plate

2x4 brick

3x4x1 mudguard with curved arches

2x2 plate with wheel-holding pins

Large wheels will soon fit here

DUST GUARDS

The surface of the moon is covered with a thin layer of dust, so any moon buggy needs big mudguards around its wheels to protect its driver from sprays of dust. These mudguards fit onto the bricks above the chassis plate.

LUNAR EQUIPMENT

Now the moon buggy has steering and storage areas. Just before adding the wheels, build satellite dishes on the top so the buggy has all the equipment it needs for its moon missions.

Bar with mechanical claw

1x1 plate with bar fits onto a jumper plate's knob

1x2 slope dashboard

Antenna piece is the satellite dish pole

2x2 curved slope makes a storage box lid

Wow, humans do exist!

FOUR WHEELER

This heavy-duty all-terrain vehicle can power through places most cars can't. Thick mud, steep hills, and rocky terrain are no trouble for a four wheeler, but it can also join regular cars on roads. This ATV has four extra-thick tires and protective bars on its front bumper.

My costume not only looks cool—it also provides padding in case I fall off!

Handlebar controls the steering

Bumper bar is one grille guard piece

Seat and mudguards are one piece

Four front headlights

Wide off-road tire

SKILL LEVEL

ALL-BLACK BASE

The four wheeler has a simple chassis built around one 2x6 plate. There are 2x2 plates with wheel-holding pins underneath it and plates with rails and clips above it. Unlike many car chassis, all of the pieces can be seen on the final build, so they're all black.

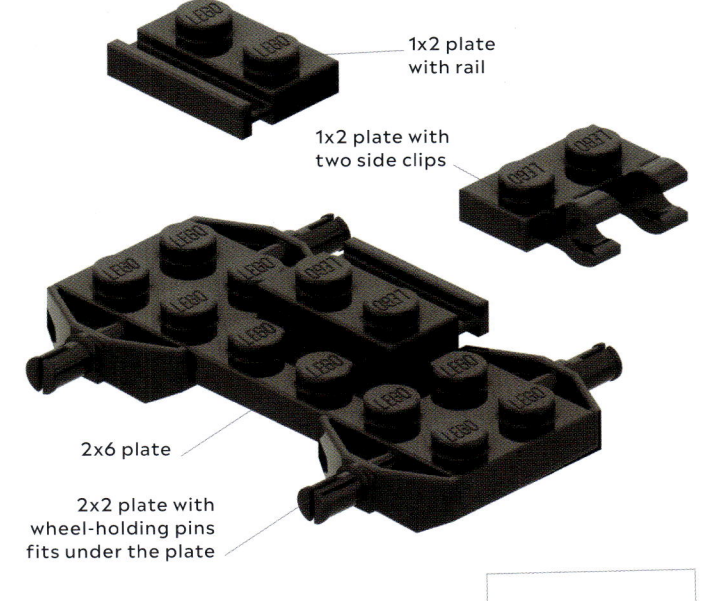

1x2 plate with rail

1x2 plate with two side clips

2x4 mudguard with overhang

2x6 plate

2x2 plate with wheel-holding pins fits under the plate

MUD READY

Top the chassis plates with the lime-green parts of the bodywork. The extra-large, curved mudguards are two matching 2x4 pieces at the front and rear of the ATV.

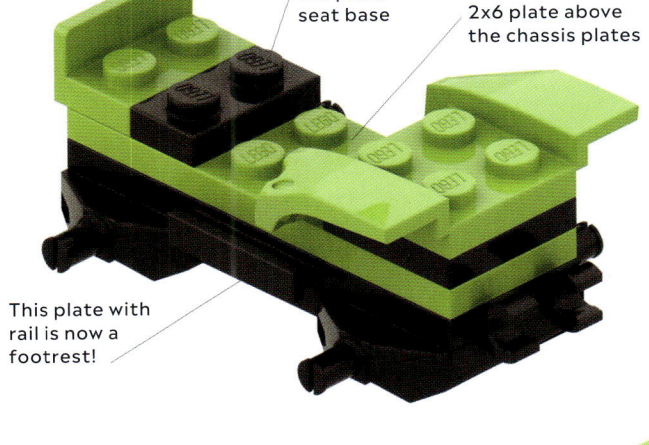

1x2 plate seat base

2x6 plate above the chassis plates

This plate with rail is now a footrest!

Handlebar piece

1x1 tile with clip

A-maize-ing!

1x2/1x4 bracket holds the four headlights

Round tile taillights attached to a 1x2/1x2 inverted bracket

BUMPER BARS

Add the handlebar steering and this four wheeler is almost ready to go off-road. Before it does, add four powerful headlights, and bumper bars to protect the ATV in case it flips or crashes.

1x2 jumper plate

The grille guard piece attaches here

GOLF CART

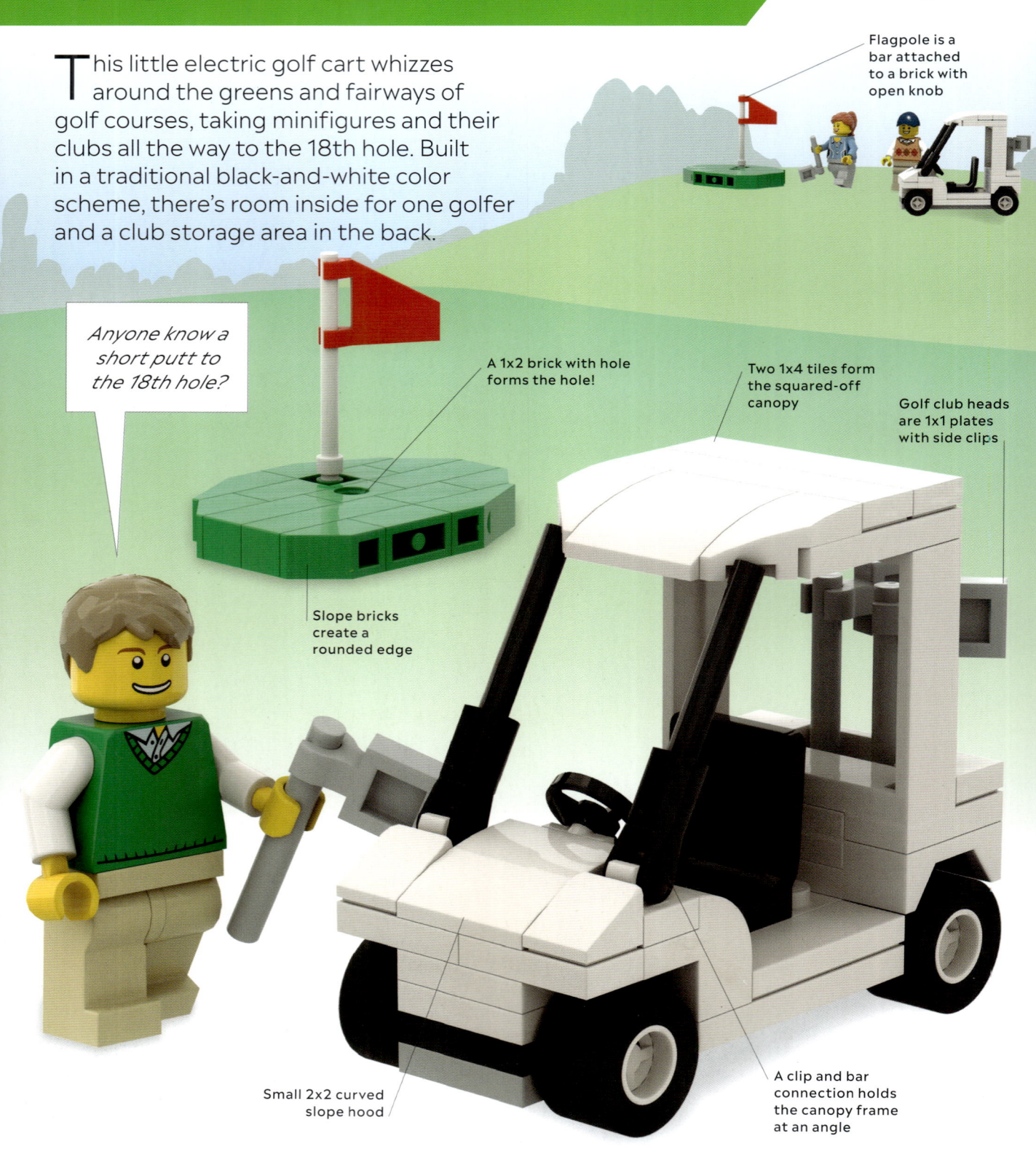

This little electric golf cart whizzes around the greens and fairways of golf courses, taking minifigures and their clubs all the way to the 18th hole. Built in a traditional black-and-white color scheme, there's room inside for one golfer and a club storage area in the back.

Flagpole is a bar attached to a brick with open knob

Anyone know a short putt to the 18th hole?

A 1x2 brick with hole forms the hole!

Two 1x4 tiles form the squared-off canopy

Golf club heads are 1x1 plates with side clips

Slope bricks create a rounded edge

Small 2x2 curved slope hood

A clip and bar connection holds the canopy frame at an angle

SKILL LEVEL   173

TEEING OFF

Golf carts mostly drive around on smooth surfaces, so their chassis can be low to the ground. A gray 2x8 plate locks the plates with wheel-holding pins on the bottom part of the chassis in place and also makes the base of the cart lower.

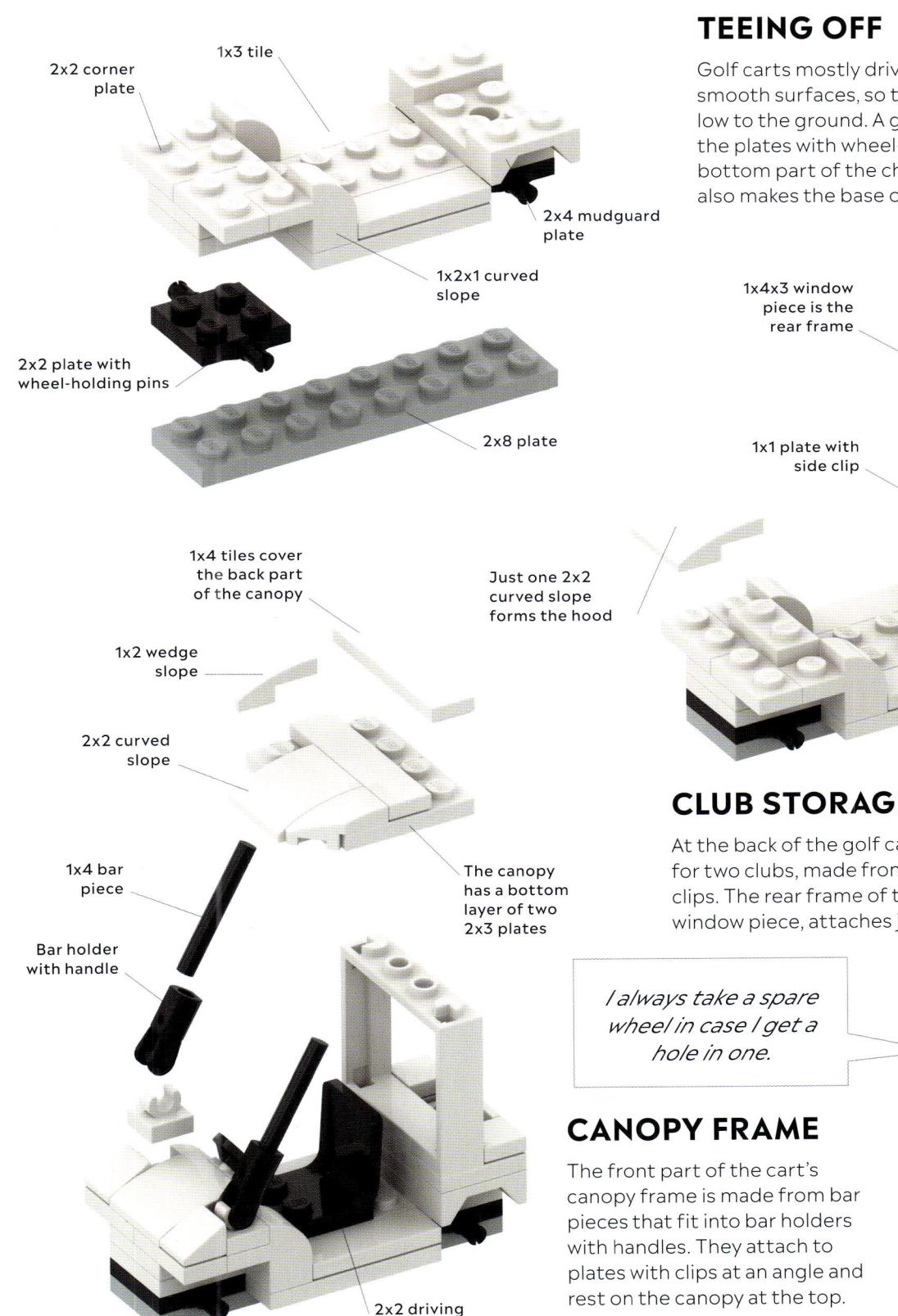

2x2 corner plate

1x3 tile

2x4 mudguard plate

1x2x1 curved slope

2x2 plate with wheel-holding pins

2x8 plate

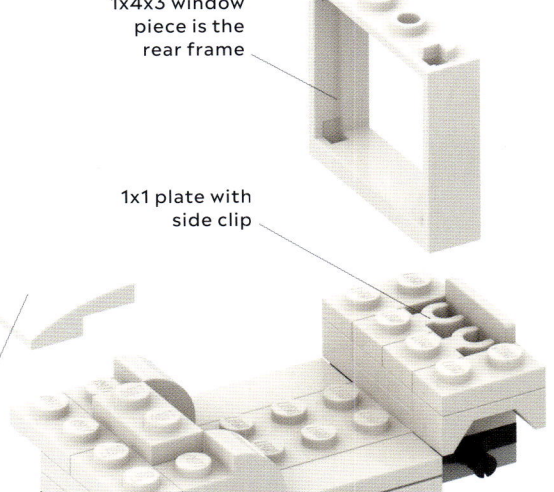

1x4x3 window piece is the rear frame

1x1 plate with side clip

1x4 tiles cover the back part of the canopy

Just one 2x2 curved slope forms the hood

1x2 wedge slope

2x2 curved slope

1x4 bar piece

Bar holder with handle

The canopy has a bottom layer of two 2x3 plates

2x2 driving seat

CLUB STORAGE

At the back of the golf cart is a storage area for two clubs, made from 1x1 plates with side clips. The rear frame of the cart, which is a window piece, attaches just in front of it.

I always take a spare wheel in case I get a hole in one.

CANOPY FRAME

The front part of the cart's canopy frame is made from bar pieces that fit into bar holders with handles. They attach to plates with clips at an angle and rest on the canopy at the top.

RACE CAR

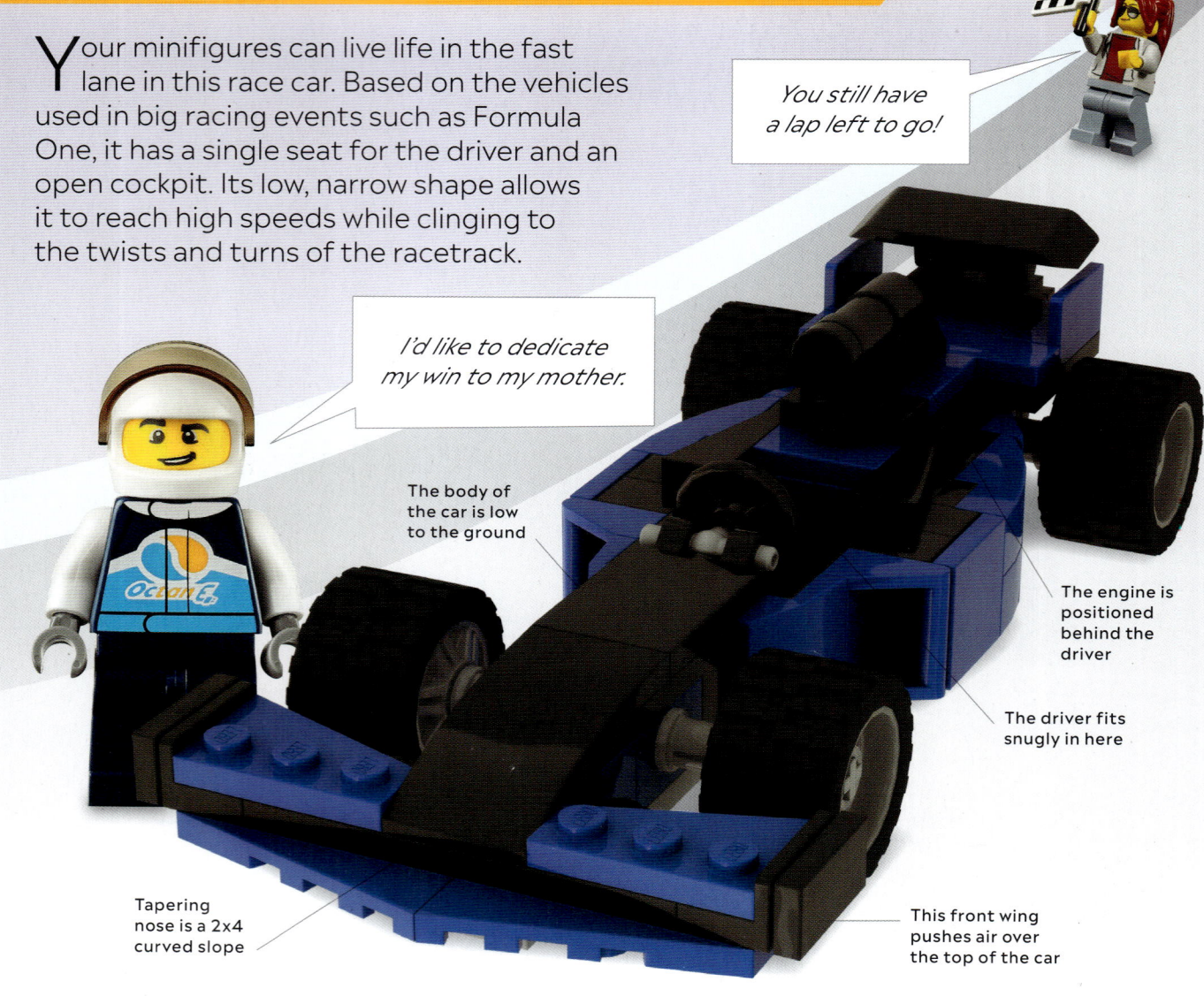

Your minifigures can live life in the fast lane in this race car. Based on the vehicles used in big racing events such as Formula One, it has a single seat for the driver and an open cockpit. Its low, narrow shape allows it to reach high speeds while clinging to the twists and turns of the racetrack.

You still have a lap left to go!

I'd like to dedicate my win to my mother.

The body of the car is low to the ground

The engine is positioned behind the driver

The driver fits snugly in here

Tapering nose is a 2x4 curved slope

This front wing pushes air over the top of the car

2x5x1 bracket

1x2 LEGO Technic brick with hole

LEGO Technic axle

2x16 plate

LEGO Technic bushes stop the axles from moving

LIGHT AND FAST

A super-speedy car like this one needs a light chassis, with no heavy or unnecessary pieces. This minimal chassis is built around a 2x16 plate. Long LEGO Technic axles slide through bricks with holes on the chassis plate to provide places to attach the wheels on the final model.

SKILL LEVEL

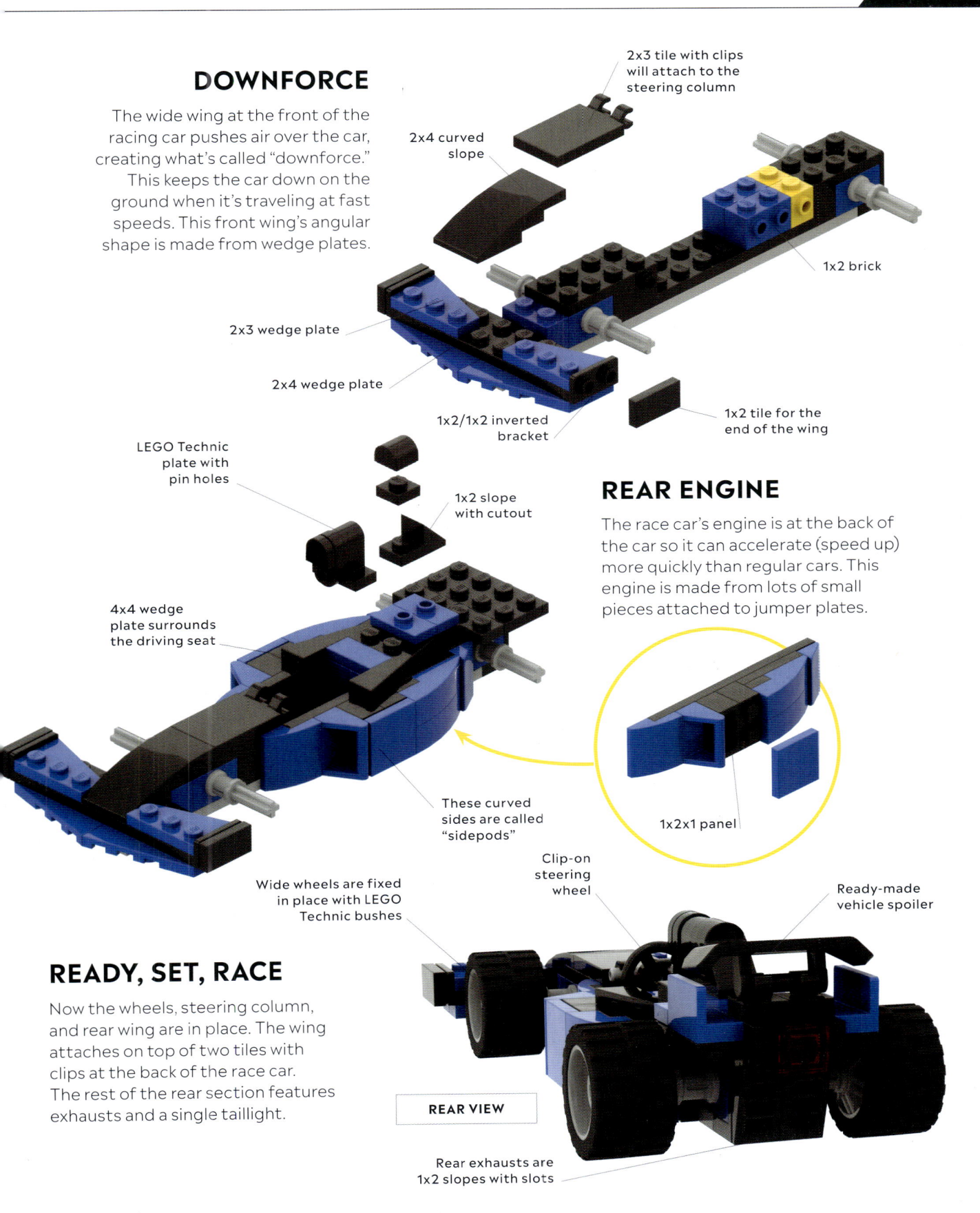

DOWNFORCE

The wide wing at the front of the racing car pushes air over the car, creating what's called "downforce." This keeps the car down on the ground when it's traveling at fast speeds. This front wing's angular shape is made from wedge plates.

2x3 tile with clips will attach to the steering column

2x4 curved slope

1x2 brick

2x3 wedge plate

2x4 wedge plate

1x2/1x2 inverted bracket

1x2 tile for the end of the wing

LEGO Technic plate with pin holes

1x2 slope with cutout

REAR ENGINE

The race car's engine is at the back of the car so it can accelerate (speed up) more quickly than regular cars. This engine is made from lots of small pieces attached to jumper plates.

4x4 wedge plate surrounds the driving seat

These curved sides are called "sidepods"

1x2x1 panel

Wide wheels are fixed in place with LEGO Technic bushes

Clip-on steering wheel

Ready-made vehicle spoiler

READY, SET, RACE

Now the wheels, steering column, and rear wing are in place. The wing attaches on top of two tiles with clips at the back of the race car. The rest of the rear section features exhausts and a single taillight.

REAR VIEW

Rear exhausts are 1x2 slopes with slots

HOT-DOG CAR

Many people love hot dogs, but the owner of this car really, *really* loves hot dogs. With a curved bun base, a mustard-streaked sausage center, and ketchup and mustard exhausts, this car is no old banger!

It's not the wurst car I've owned.

Orange sausage pieces are mustard streaks!

Yellow 1x2 steering column and wheel

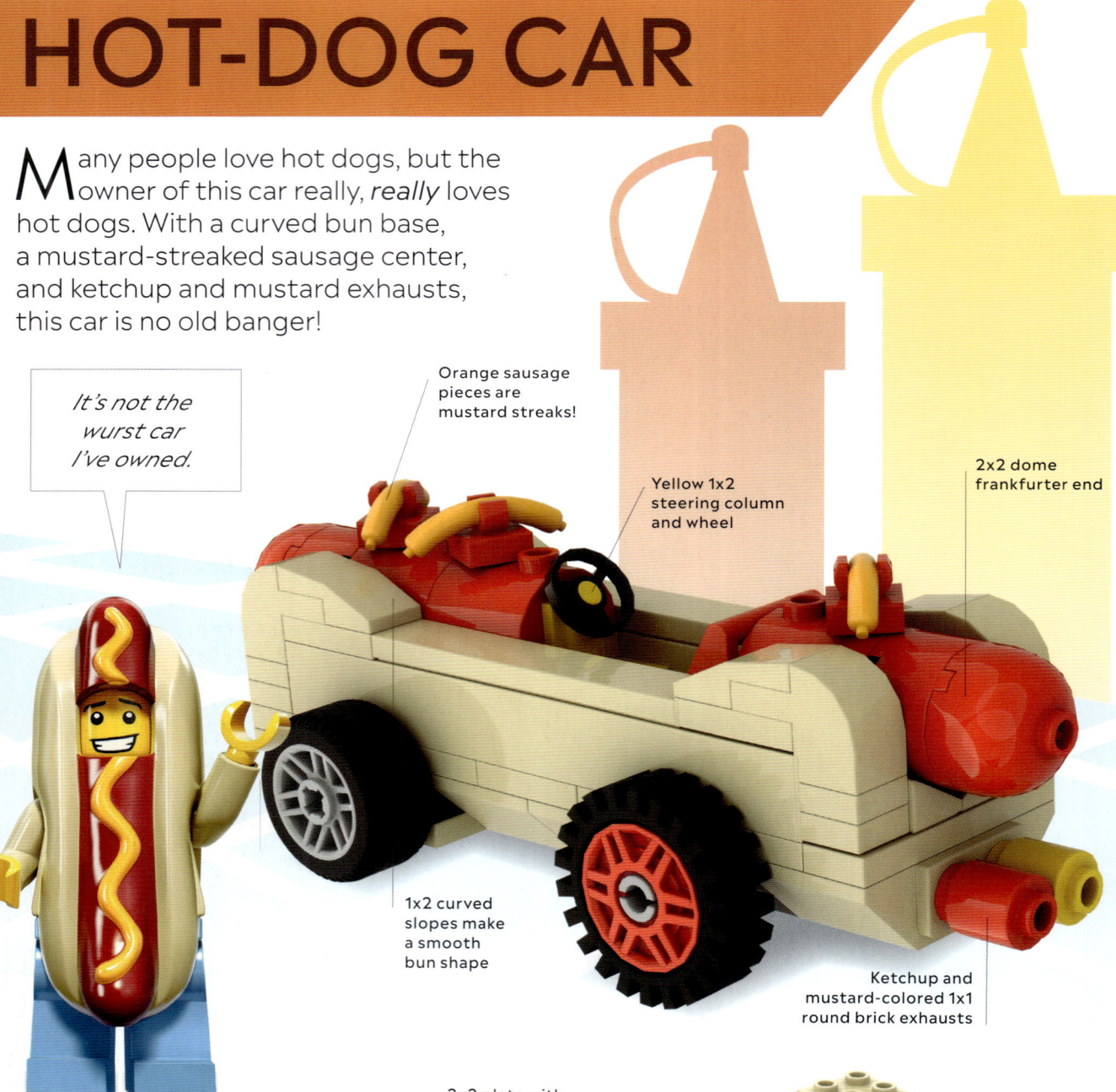

2x2 dome frankfurter end

1x2 curved slopes make a smooth bun shape

Ketchup and mustard-colored 1x1 round brick exhausts

BEGIN THE BUN

A car shaped like a long hot-dog bun needs a lengthy chassis base. The hot-dog car is built around a 2x10 plate. Double inverted slopes attached underneath the chassis plate give the bun base a curved look.

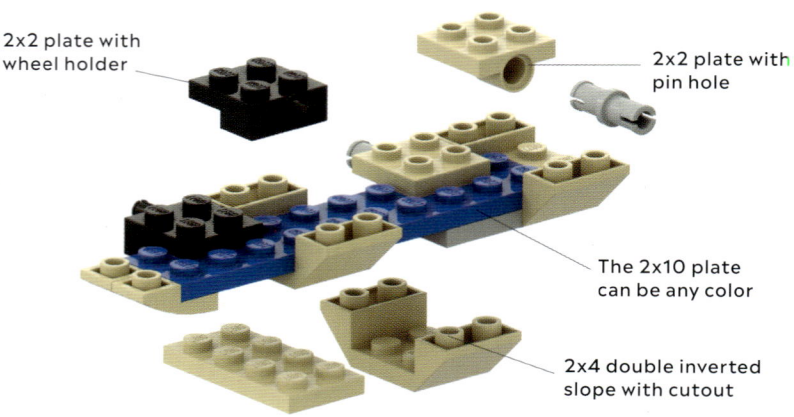

2x2 plate with wheel holder

2x2 plate with pin hole

The 2x10 plate can be any color

2x4 double inverted slope with cutout

SKILL LEVEL

RISING DOUGH

Build more plates on top of the lower part of the car to add height to the bun body. Then attach inverted curved slopes underneath the plates at the front of the car to give the bun its distinctive rounded end.

4x4 plate

The twin sauce exhausts will attach to this 1x2/1x2 bracket

1x2 inverted curved slope

2x4 plate

2x4 double inverted slope with cutout

BUILD A BANGER

Once the bun bodywork is complete, it's time to build the sausage center of the hot-dog car. It's made from double curved slopes in the middle, and round bricks and domes at the ends.

Leave four knobs for the minifigure driver

2x2x1 double curved slope

2x2x²⁄₃ plate with two side knobs

2x2 dome brick

An extra plate layer makes the sausage higher

I'm sorry but I really mustard go.

1x2x1 curved slope

1x1 tile with clip

Steering wheel

FOOD TO GO

Add some more curves to the top edges of the car's bun body using curved slopes, then add a steering column. Serve up the hot-dog car with mustard stripes aptly made from mini hot-dog pieces attached to 1x1 tiles with clips.

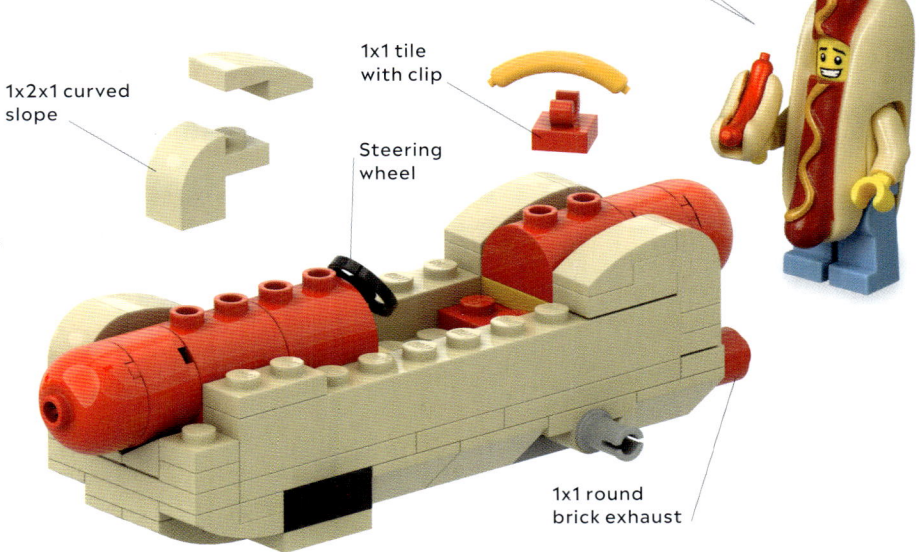

1x1 round brick exhaust

177

178

BLACK CAB

Taxi! This black cab covers miles and miles of city streets each day, picking up passengers and taking them wherever they need to go. Its timeless black-and-chrome design features a light at the top that tells people the cab is available for hire.

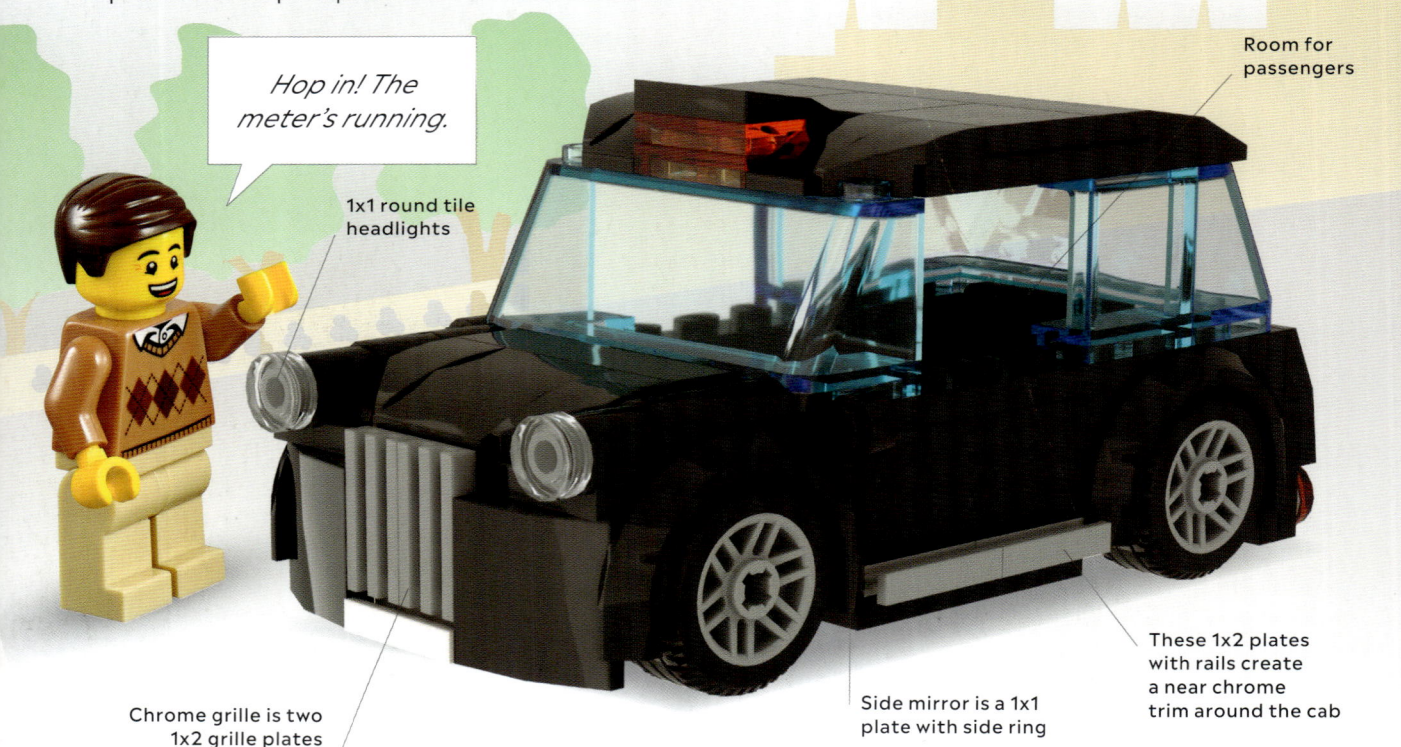

Hop in! The meter's running.

Room for passengers

1x1 round tile headlights

Chrome grille is two 1x2 grille plates attached vertically

Side mirror is a 1x1 plate with side ring

These 1x2 plates with rails create a near chrome trim around the cab

2x2 seat

1x2 steering column and steering wheel

ROOM FOR TWO

Taxi passengers usually sit behind the driver, so it's important to build two rows of seats inside. Place the front and rear seats inside your build at an early stage to ensure there's plenty of minifigure headroom and leg space.

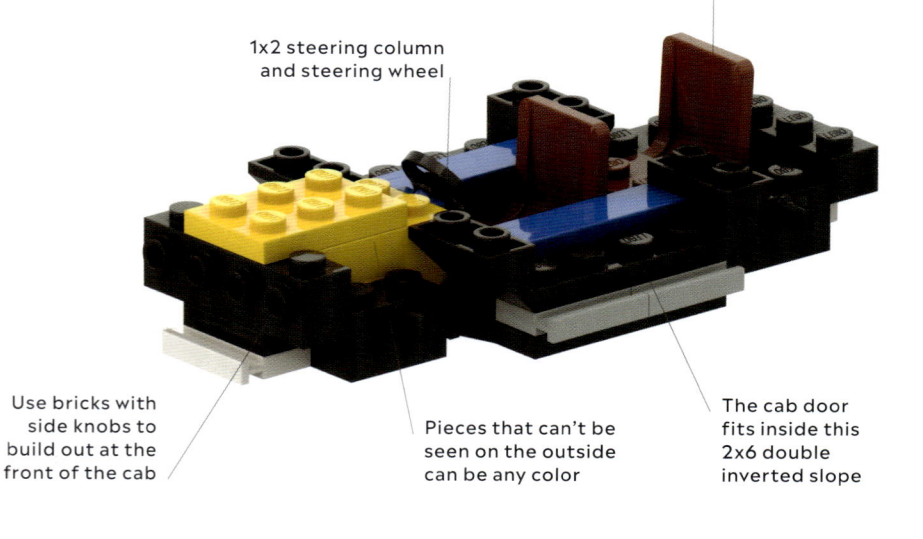

Use bricks with side knobs to build out at the front of the cab

Pieces that can't be seen on the outside can be any color

The cab door fits inside this 2x6 double inverted slope

SKILL LEVEL

BUILDING IN BLACK

The bodywork of the black cab is taking shape, with sleek mudguards and pins for the wheels that are attached underneath the chassis. There are 1x4 brick doors and blue tile armrests for the driver.

Take me to LEGO City, please.

2x4 mudguard

1x4 tile armrest

1x4 brick door

This modified 2x4 brick with wheel pins fits under the chassis

2x2 curved slope

DISTINCTIVE DESIGN

The front section of a classic black cab has a unique shape. The cab's long, curved wings are made from small curved slopes and round bricks with holes, which the headlights slot inside. At the rear, the bumper is tall and flat.

This half pin is the base of the headlight

2x2 curved slope hood

1x1 double curved slope

2x2 curved slope

The roof has a bottom layer of plates

1x2 tile

2x6x2 windshield

READY FOR SERVICE

Taxi passengers love to watch the city streets whiz by on their journeys. Build a wide windshield and lots of windows around the seats before adding the roof. The taxi light is a 1x2 transparent orange tile on top of a plate in the same size and color.

1x2x2 panel

179

PICKUP TRUCK

With an enclosed driver's cabin at the front and an open cargo area at the back, pickup trucks like this are practical vehicles. They are often found on farms or in places of work because they can carry large or heavy loads at the back. But they are also used as regular cars by people who appreciate the extra storage space!

We're about to pick up some gnarly waves!

Large mudguards protect the truck from dirty roads

Curved door panels made from two curved slopes and a tile

Wide wheel-holding plate with LEGO Technic pins

6x6 plate

2x16 plate

Large grille made from four 1x2 grille tiles attached vertically

2x2 inverted curved slope

SECURE START

The pickup truck's chassis build begins with one long 2x16 plate, which supports the whole length of the vehicle. Plates with wheel-holding pins fit under it, along with inverted curved slopes to better secure the truck's wheels from underneath.

SKILL LEVEL

181

OPEN AND CLOSED

Now it's possible to see where the two areas of the pickup truck will be built up. A 4x6 plate at the back will become the open cargo area. The smaller plates at the front form the base of the enclosed driver's cabin.

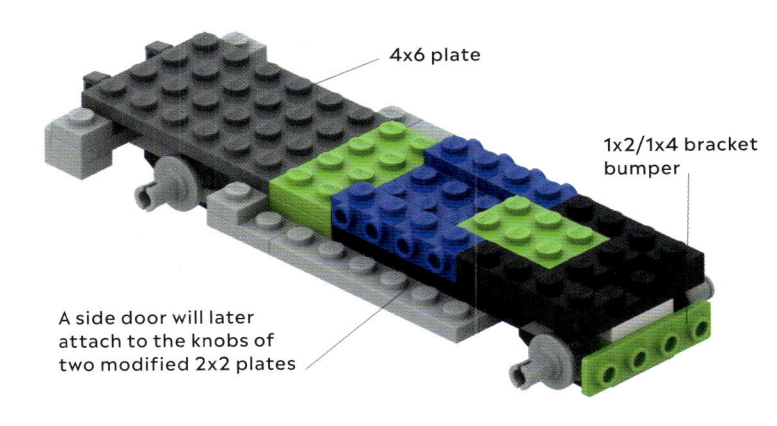

4x6 plate

1x2/1x4 bracket bumper

A side door will later attach to the knobs of two modified 2x2 plates

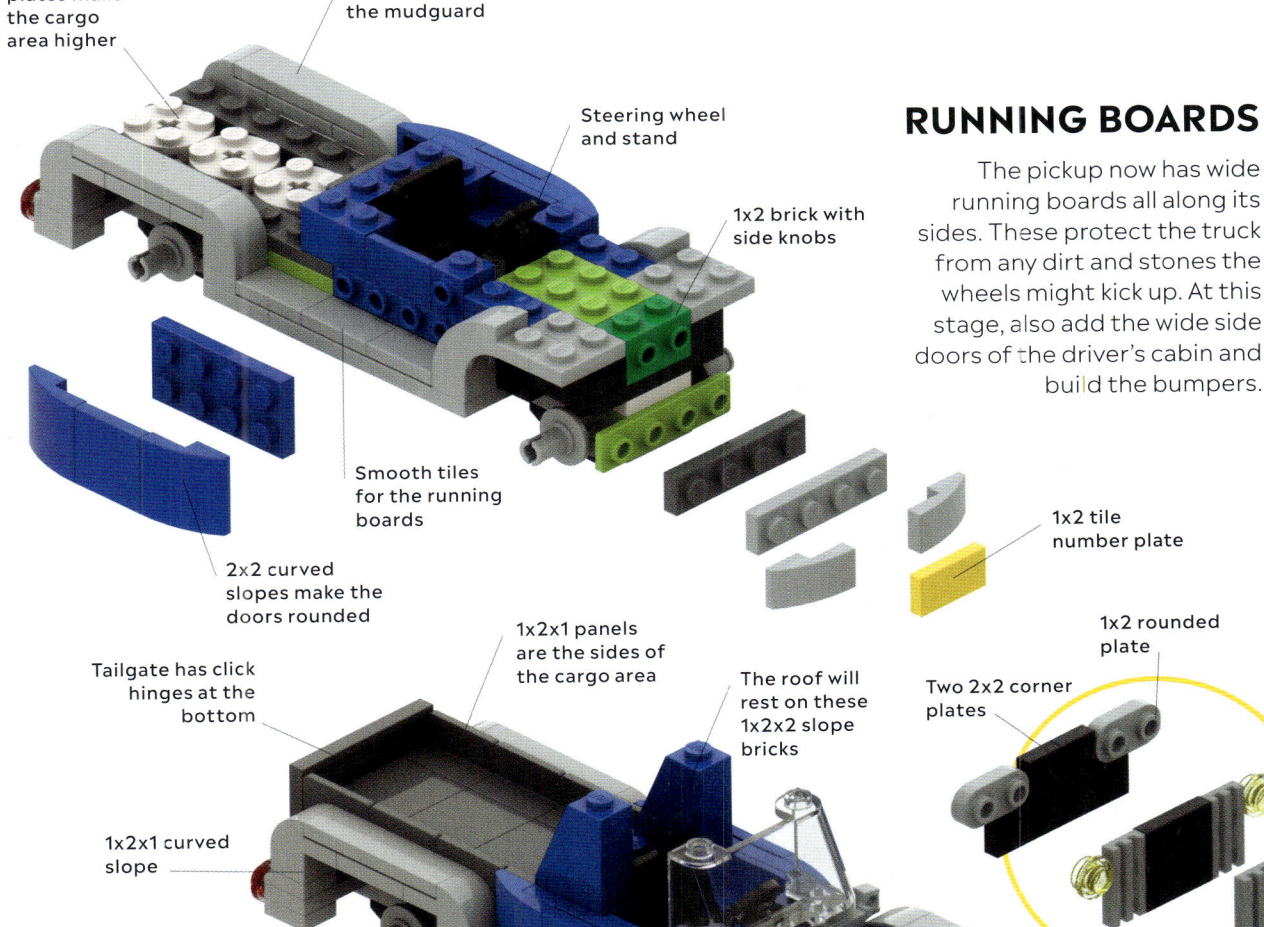

2x2 round plates make the cargo area higher

1x4 tile tops the mudguard

Steering wheel and stand

1x2 brick with side knobs

RUNNING BOARDS

The pickup now has wide running boards all along its sides. These protect the truck from any dirt and stones the wheels might kick up. At this stage, also add the wide side doors of the driver's cabin and build the bumpers.

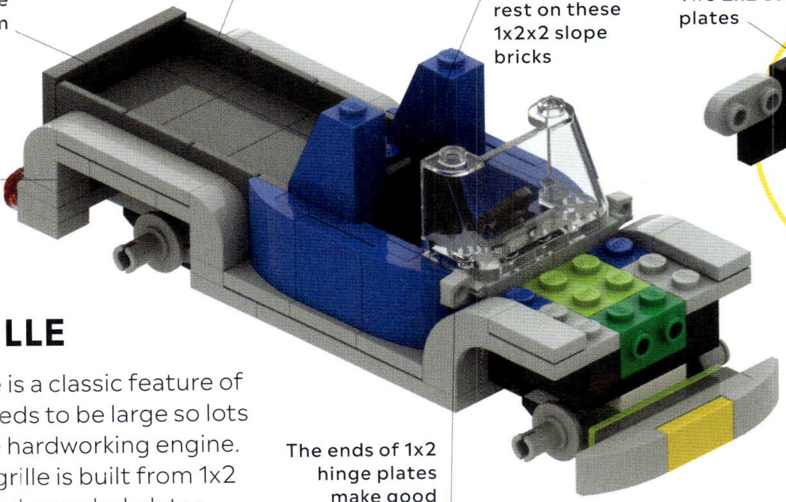

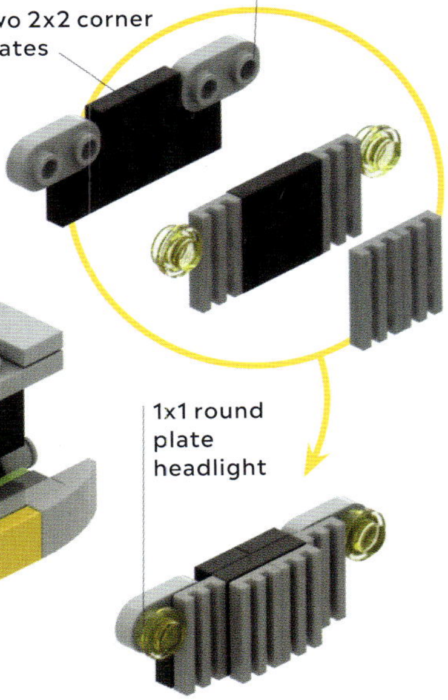

Smooth tiles for the running boards

2x2 curved slopes make the doors rounded

1x2 tile number plate

1x2 rounded plate

Tailgate has click hinges at the bottom

1x2x1 panels are the sides of the cargo area

The roof will rest on these 1x2x2 slope bricks

Two 2x2 corner plates

1x2x1 curved slope

1x1 round plate headlight

ENGINE GRILLE

A giant engine grille is a classic feature of a pickup truck. It needs to be large so lots of air can get to the hardworking engine. This pickup truck's grille is built from 1x2 grille tiles, plates, and rounded plates.

The ends of 1x2 hinge plates make good side mirrors!

SWITCHED-UP PICKUPS

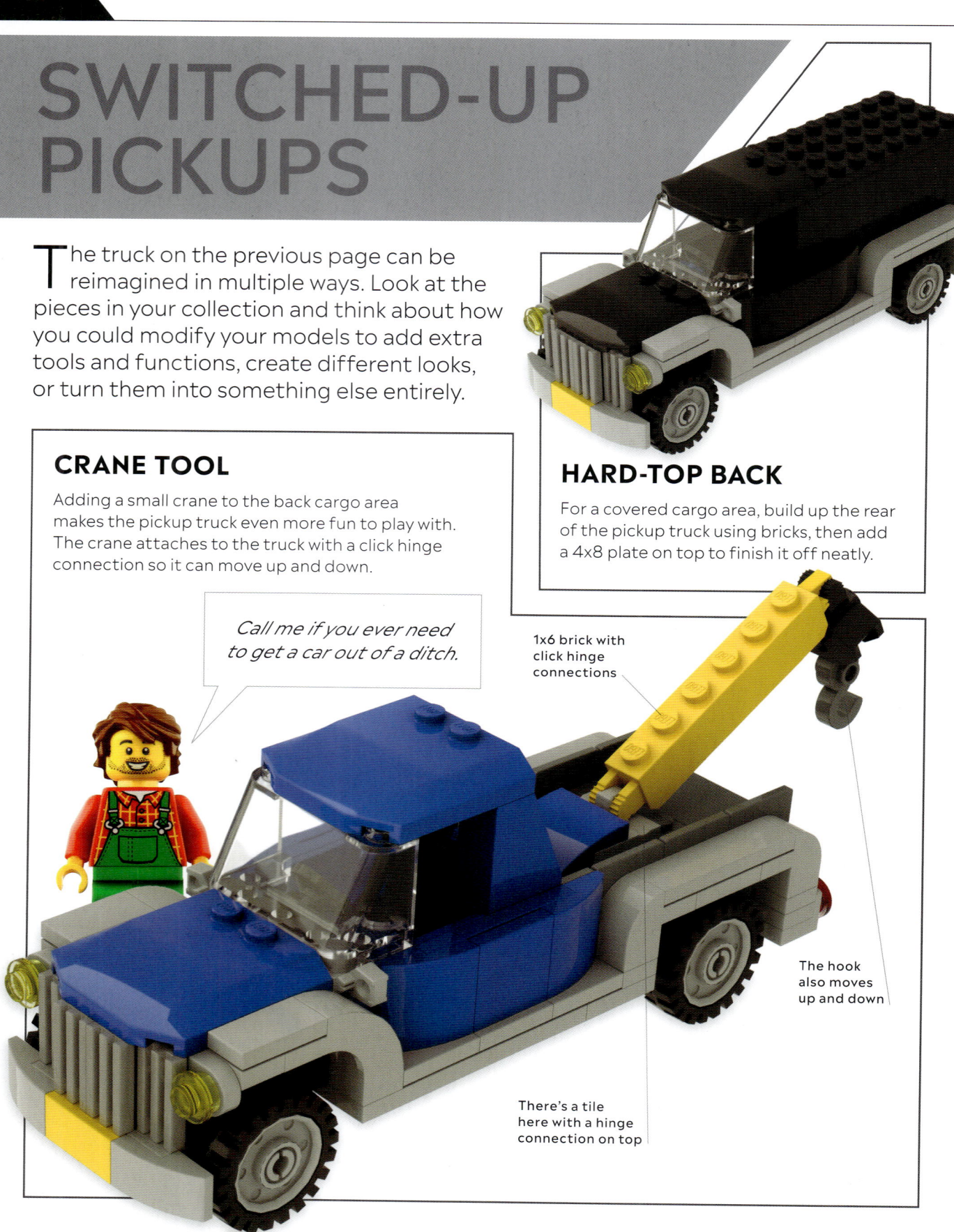

The truck on the previous page can be reimagined in multiple ways. Look at the pieces in your collection and think about how you could modify your models to add extra tools and functions, create different looks, or turn them into something else entirely.

CRANE TOOL

Adding a small crane to the back cargo area makes the pickup truck even more fun to play with. The crane attaches to the truck with a click hinge connection so it can move up and down.

HARD-TOP BACK

For a covered cargo area, build up the rear of the pickup truck using bricks, then add a 4x8 plate on top to finish it off neatly.

Call me if you ever need to get a car out of a ditch.

1x6 brick with click hinge connections

The hook also moves up and down

There's a tile here with a hinge connection on top

IDEAS GALLERY 183

SOUND SYSTEM

This pickup truck has been turned into a stage on wheels, for concerts on the go! Removing the sides creates room for a minifigure performer as well as two large speakers and a microphone.

Speaker base made from three plates

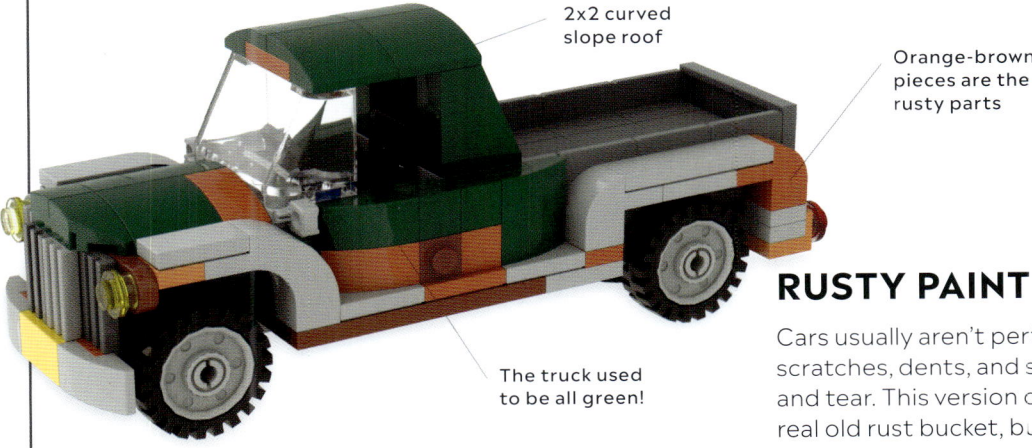

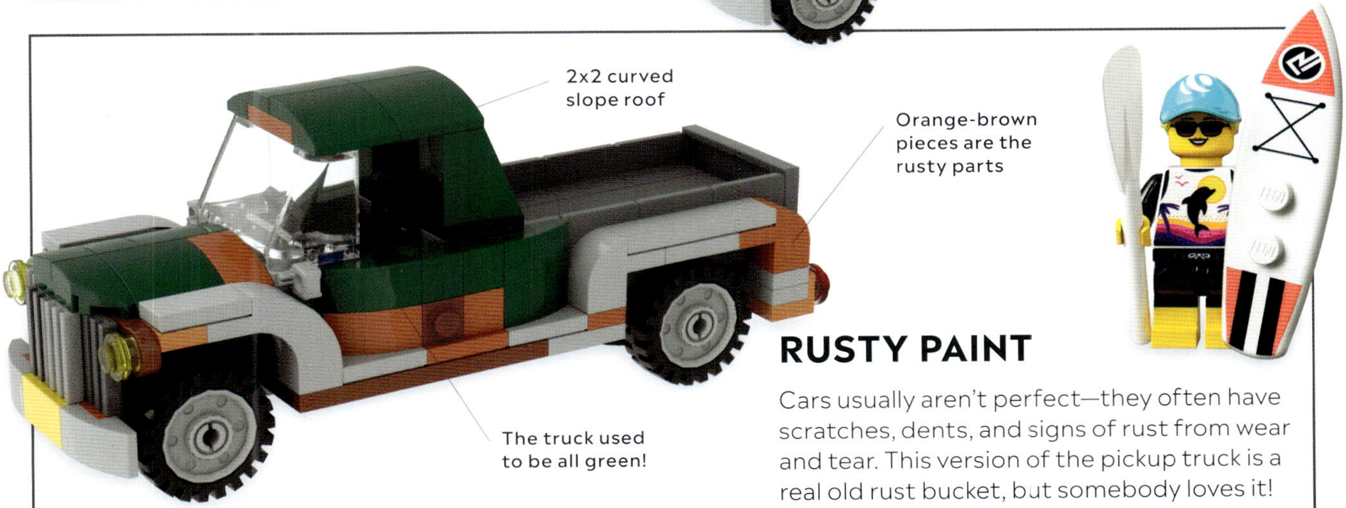

RUSTY PAINT

Cars usually aren't perfect—they often have scratches, dents, and signs of rust from wear and tear. This version of the pickup truck is a real old rust bucket, but somebody loves it!

2x2 curved slope roof

Orange-brown pieces are the rusty parts

The truck used to be all green!

HOT ROD

Like the hot rod car on pages 194–195, the pickup truck has now been stripped back and rebuilt with a roaring exposed engine and flaming and lightning sides.

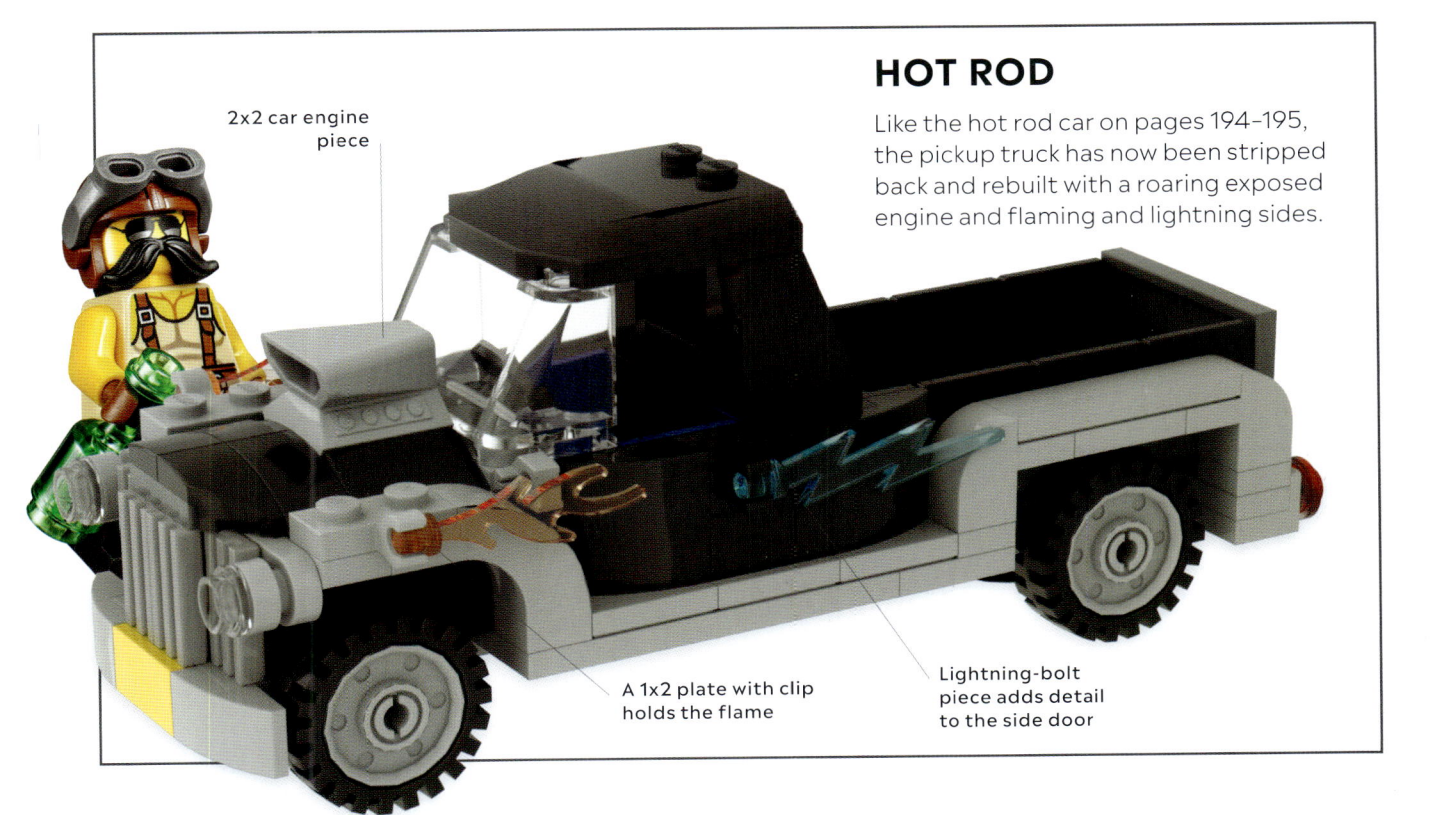

2x2 car engine piece

A 1x2 plate with clip holds the flame

Lightning-bolt piece adds detail to the side door

BRICK CAR

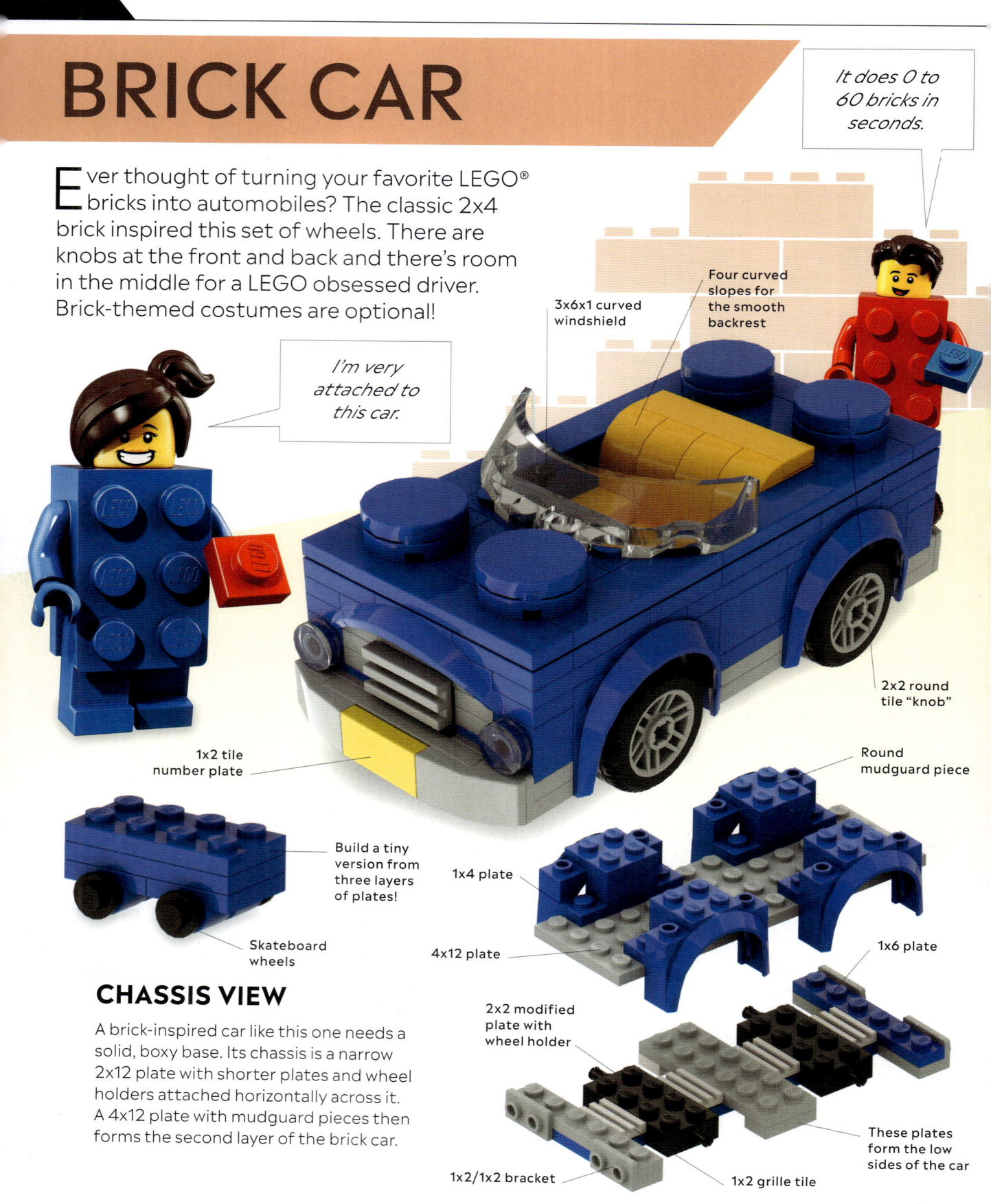

It does 0 to 60 bricks in seconds.

Ever thought of turning your favorite LEGO® bricks into automobiles? The classic 2x4 brick inspired this set of wheels. There are knobs at the front and back and there's room in the middle for a LEGO obsessed driver. Brick-themed costumes are optional!

I'm very attached to this car.

- 3x6x1 curved windshield
- Four curved slopes for the smooth backrest
- 1x2 tile number plate
- 2x2 round tile "knob"
- Round mudguard piece
- 1x4 plate
- 4x12 plate
- 1x6 plate
- 2x2 modified plate with wheel holder
- 1x2/1x2 bracket
- 1x2 grille tile
- These plates form the low sides of the car
- Build a tiny version from three layers of plates!
- Skateboard wheels

CHASSIS VIEW

A brick-inspired car like this one needs a solid, boxy base. Its chassis is a narrow 2x12 plate with shorter plates and wheel holders attached horizontally across it. A 4x12 plate with mudguard pieces then forms the second layer of the brick car.

SKILL LEVEL

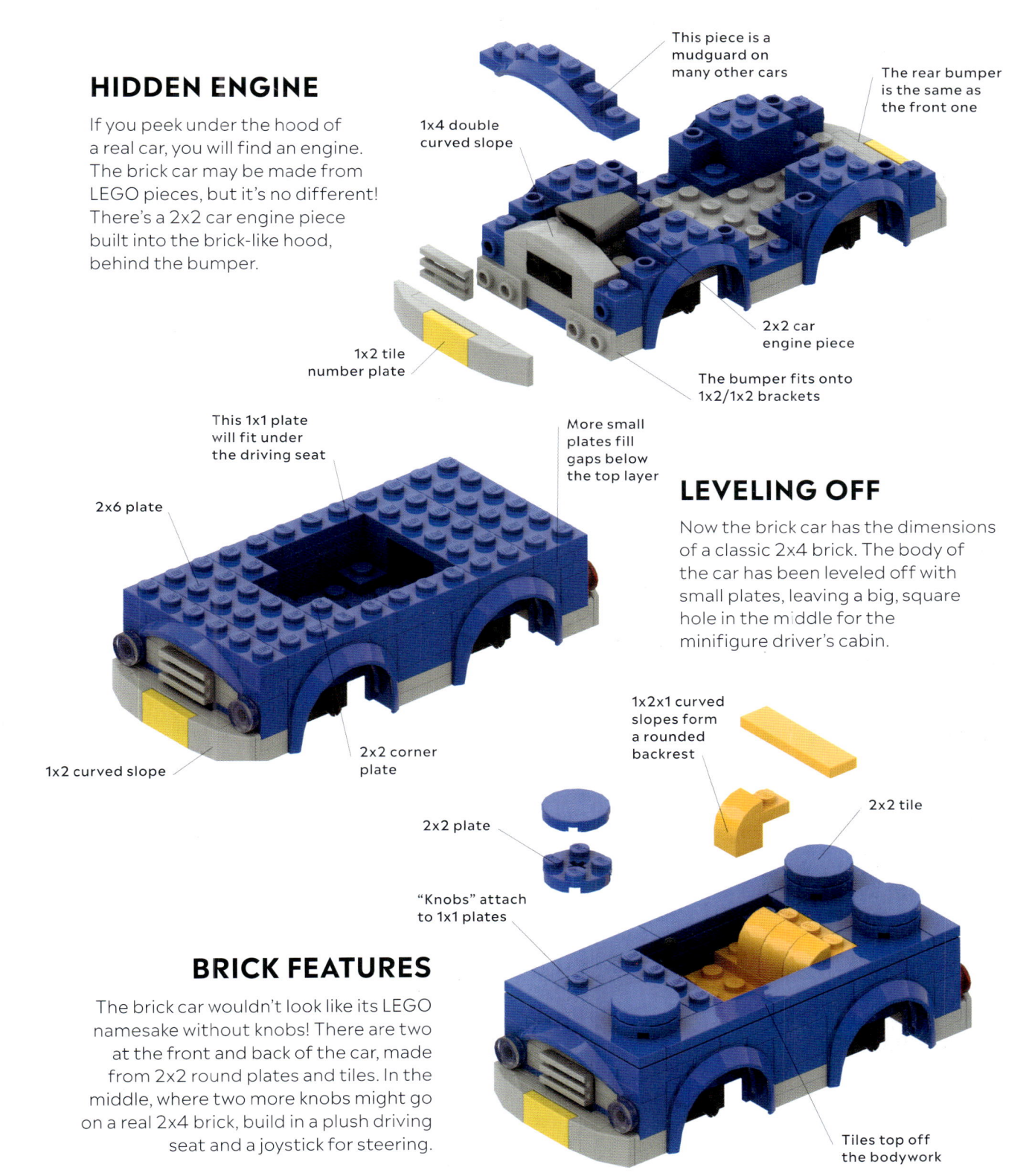

185

HIDDEN ENGINE

If you peek under the hood of a real car, you will find an engine. The brick car may be made from LEGO pieces, but it's no different! There's a 2x2 car engine piece built into the brick-like hood, behind the bumper.

This piece is a mudguard on many other cars

1x4 double curved slope

The rear bumper is the same as the front one

1x2 tile number plate

2x2 car engine piece

The bumper fits onto 1x2/1x2 brackets

This 1x1 plate will fit under the driving seat

More small plates fill gaps below the top layer

2x6 plate

LEVELING OFF

Now the brick car has the dimensions of a classic 2x4 brick. The body of the car has been leveled off with small plates, leaving a big, square hole in the middle for the minifigure driver's cabin.

1x2 curved slope

2x2 corner plate

1x2x1 curved slopes form a rounded backrest

2x2 plate

2x2 tile

"Knobs" attach to 1x1 plates

BRICK FEATURES

The brick car wouldn't look like its LEGO namesake without knobs! There are two at the front and back of the car, made from 2x2 round plates and tiles. In the middle, where two more knobs might go on a real 2x4 brick, build in a plush driving seat and a joystick for steering.

Tiles top off the bodywork

SPORTS CAR

Your minifigures can feel the wind in their LEGO hair in this convertible sports car! Its soft-top roof, made from black tiles and slopes, can be removed for top-down driving on sunny days. Built for speed, this sleek vehicle is small and low to the ground, with white racing stripes on the hood.

- 2x4 tiles complete the soft-top roof
- Two 2x3 slopes form the base of the roof
- 1x2 curved slope stripe
- Smoked glass 3x4x1 windshield piece
- 1x2 curved slope bumper
- 1x4 brick door
- Transparent 1x1 slope is a headlight lens

That's one sturdy-looking chassis.

BODY BASE

The sports car body is built around one chassis piece. Plates and bricks are then added on top, underneath, and on the sides. If you don't have a chassis piece like this, you could make your own using small plates.

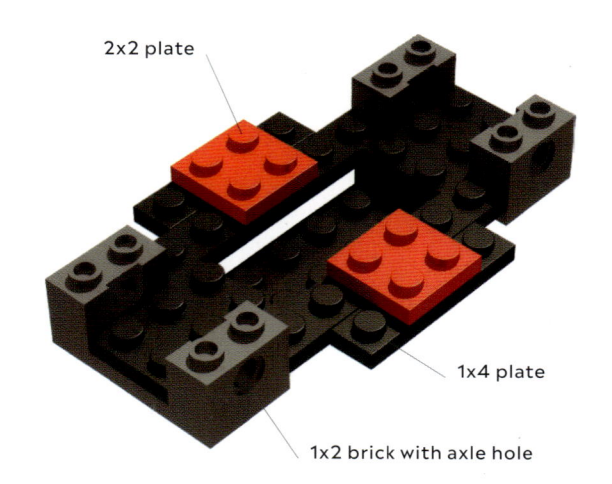

- 2x2 plate
- 1x4 plate
- 1x2 brick with axle hole

SKILL LEVEL

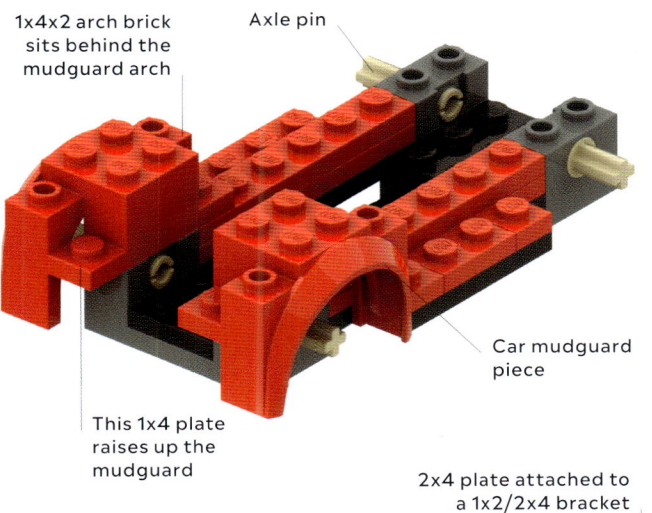

187

1x4x2 arch brick sits behind the mudguard arch

Axle pin

HIGH WHEELS

A speedy sports car needs a body that's low to the ground so it can move fast and make sharp turns with ease. To keep the chassis low, attach the wheel axles and mudguards slightly above it.

Car mudguard piece

This 1x4 plate raises up the mudguard

2x4 plate attached to a 1x2/2x4 bracket

BUMPER DETAILS

The headlights and license plates at the front and rear of the sports car are attached using sideways-building techniques. Bracket and headlight bricks are especially useful for sideways building.

1x1 headlight brick

Yellow 1x2 tile license plate

1x1 transparent orange plate is a glowing headlight

1x1 tile extends the racing stripe

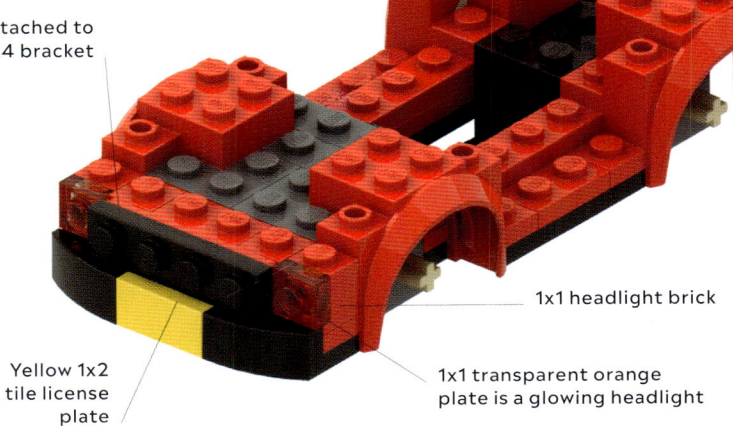

The windshield attaches here

SMOOTH RIDE

Curved slopes and small tiles form the sports car's smooth, racing-striped hood. Several curved slope pieces above the back bumper continue the sleek curves at the rear.

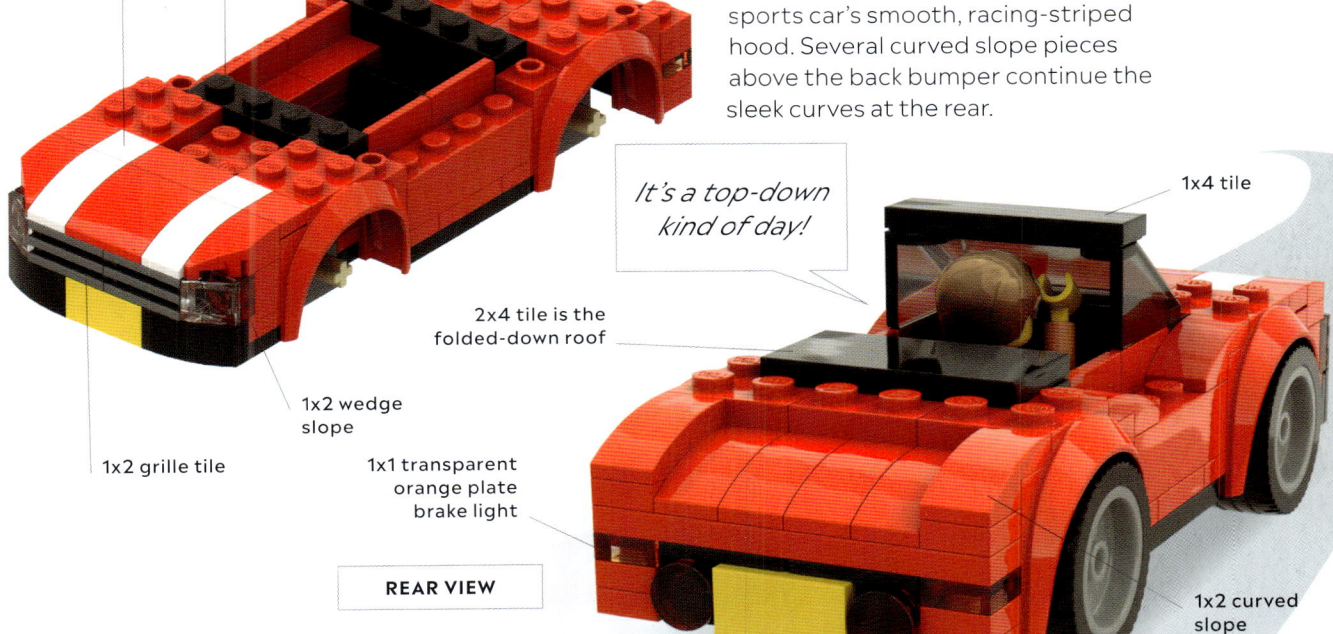

It's a top-down kind of day!

1x4 tile

2x4 tile is the folded-down roof

1x2 wedge slope

1x2 grille tile

1x1 transparent orange plate brake light

REAR VIEW

1x2 curved slope

UNDERWATER CAR

There are roads all over the land, but much of the ocean is unexplored—until now! This propeller-powered aquatic automobile is built to roam the ocean floor. The underwater car's two watertight seats have big bubble windows so its driver and his small passenger can get a 360-degree view of the passing marine life.

I bought this car for shellfish reasons, but the boy seems to like it too.

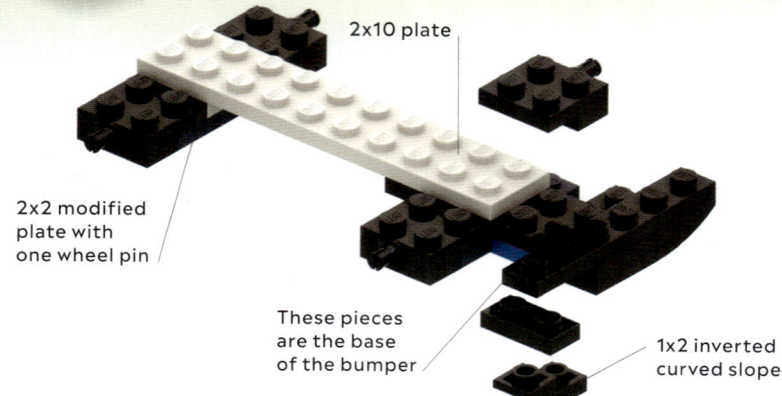

- Smaller rear seat and window
- Thick tread tires for moving through sand
- Bubble window is a half sphere windshield
- Lifebuoy and flipper door handle
- Snorkel bumper with bubbles

- 2x2 modified plate with one wheel pin
- 2x10 plate
- These pieces are the base of the bumper
- 1x2 inverted curved slope

SANDY CHASSIS

The underwater car has a high chassis because it spends most of its time navigating deep-water sand dunes. It's built around a 2x10 plate. Plates with wheel-holding pins fit to smaller plates underneath it, so the chassis plate is higher than the wheel pins.

SKILL LEVEL

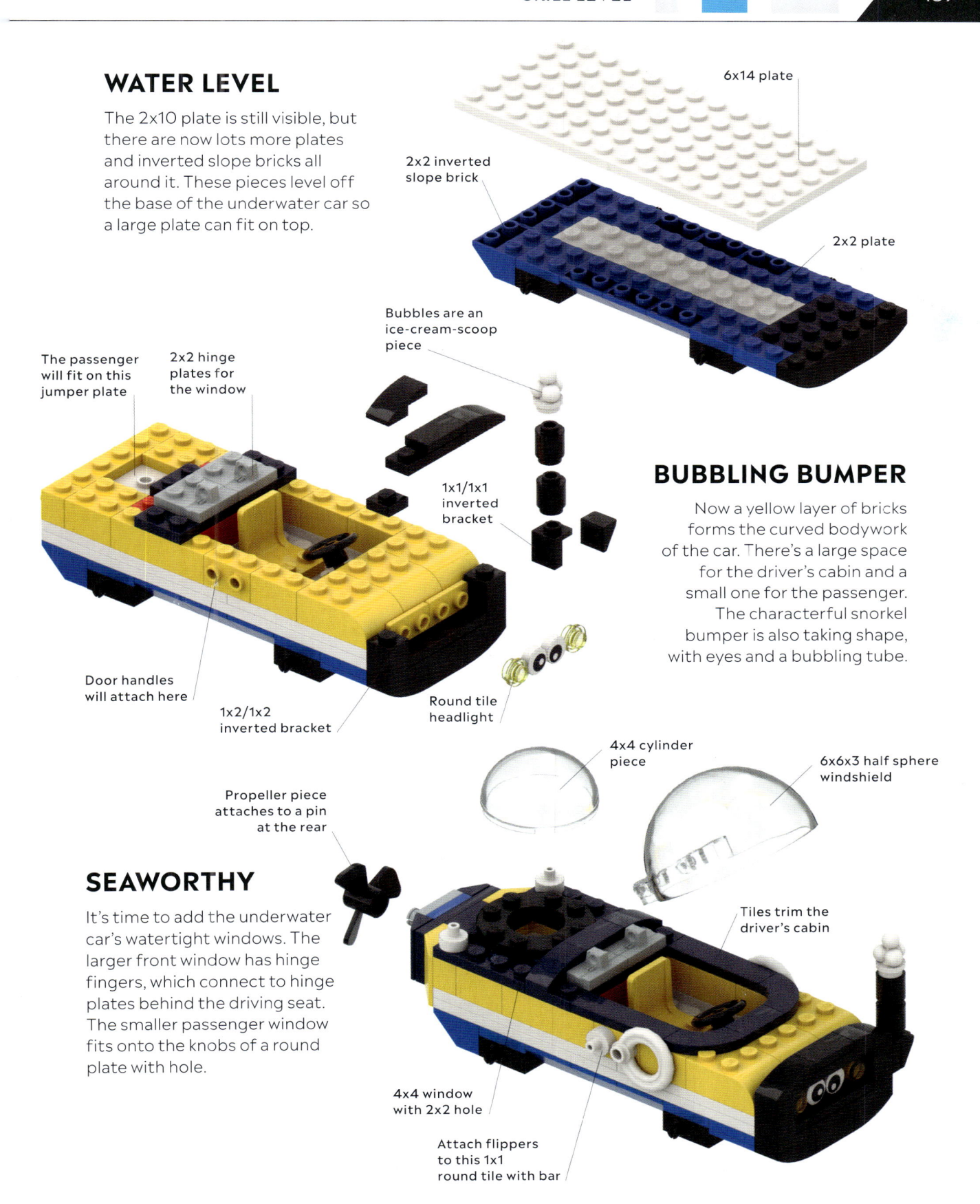

189

WATER LEVEL

The 2x10 plate is still visible, but there are now lots more plates and inverted slope bricks all around it. These pieces level off the base of the underwater car so a large plate can fit on top.

6x14 plate

2x2 inverted slope brick

2x2 plate

The passenger will fit on this jumper plate

2x2 hinge plates for the window

Bubbles are an ice-cream-scoop piece

1x1/1x1 inverted bracket

BUBBLING BUMPER

Now a yellow layer of bricks forms the curved bodywork of the car. There's a large space for the driver's cabin and a small one for the passenger. The characterful snorkel bumper is also taking shape, with eyes and a bubbling tube.

Door handles will attach here

1x2/1x2 inverted bracket

Round tile headlight

Propeller piece attaches to a pin at the rear

4x4 cylinder piece

6x6x3 half sphere windshield

SEAWORTHY

It's time to add the underwater car's watertight windows. The larger front window has hinge fingers, which connect to hinge plates behind the driving seat. The smaller passenger window fits onto the knobs of a round plate with hole.

Tiles trim the driver's cabin

4x4 window with 2x2 hole

Attach flippers to this 1x1 round tile with bar

DRIVERLESS CAR

This is one clever car! Driverless or self-driving cars like this don't need a driver because they can plan a driving route, read road signs, and sense the environment around them. This build has an access ramp and plenty of room inside for someone in a wheelchair.

The large back door opens at the top

Look, no hands!

Six curved slopes form the short hood

1x1 cheese slope side mirror

Simple 1x1 round tile headlights

Small, smooth tires

BUILT-IN RAMP

The low base of the driverless car has a first layer of a 4x8 plate with lots of smaller plates on top to make it six knobs wide. A row of curved slopes at one end forms the ramp for getting in and out of the vehicle.

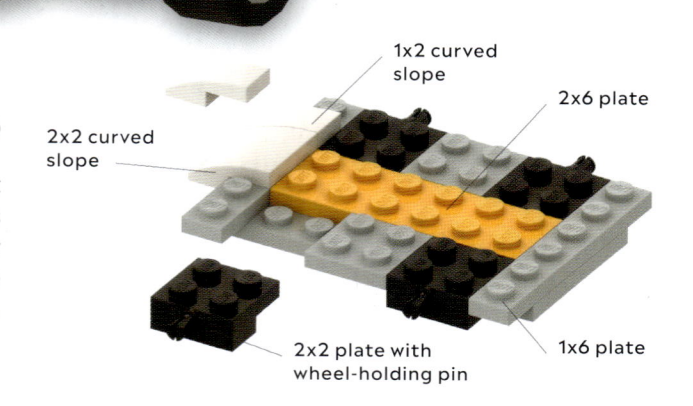

1x2 curved slope

2x6 plate

2x2 curved slope

2x2 plate with wheel-holding pin

1x6 plate

SKILL LEVEL

191

CAR FLOOR

Add tiles on top of the chassis layers to create a smooth car floor. The cheese slopes in the floor will stop a wheelchair from rolling too far forward when the car is moving.

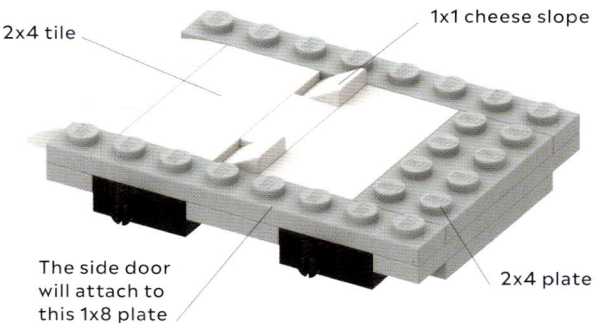

2x4 tile

1x1 cheese slope

The side door will attach to this 1x8 plate

2x4 plate

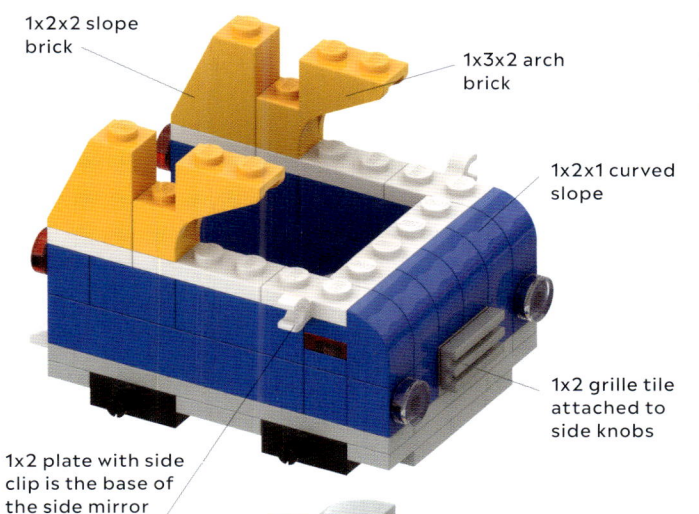

1x2x2 slope brick

1x3x2 arch brick

1x2x1 curved slope

1x2 grille tile attached to side knobs

1x2 plate with side clip is the base of the side mirror

SOLID DESIGN

The steeply sloping hood and bumper details are now in place at the front of the driverless car, and bricks of different sizes form the solid sides of the bodywork. To build the top part of the car's frame, start with slope bricks and arches.

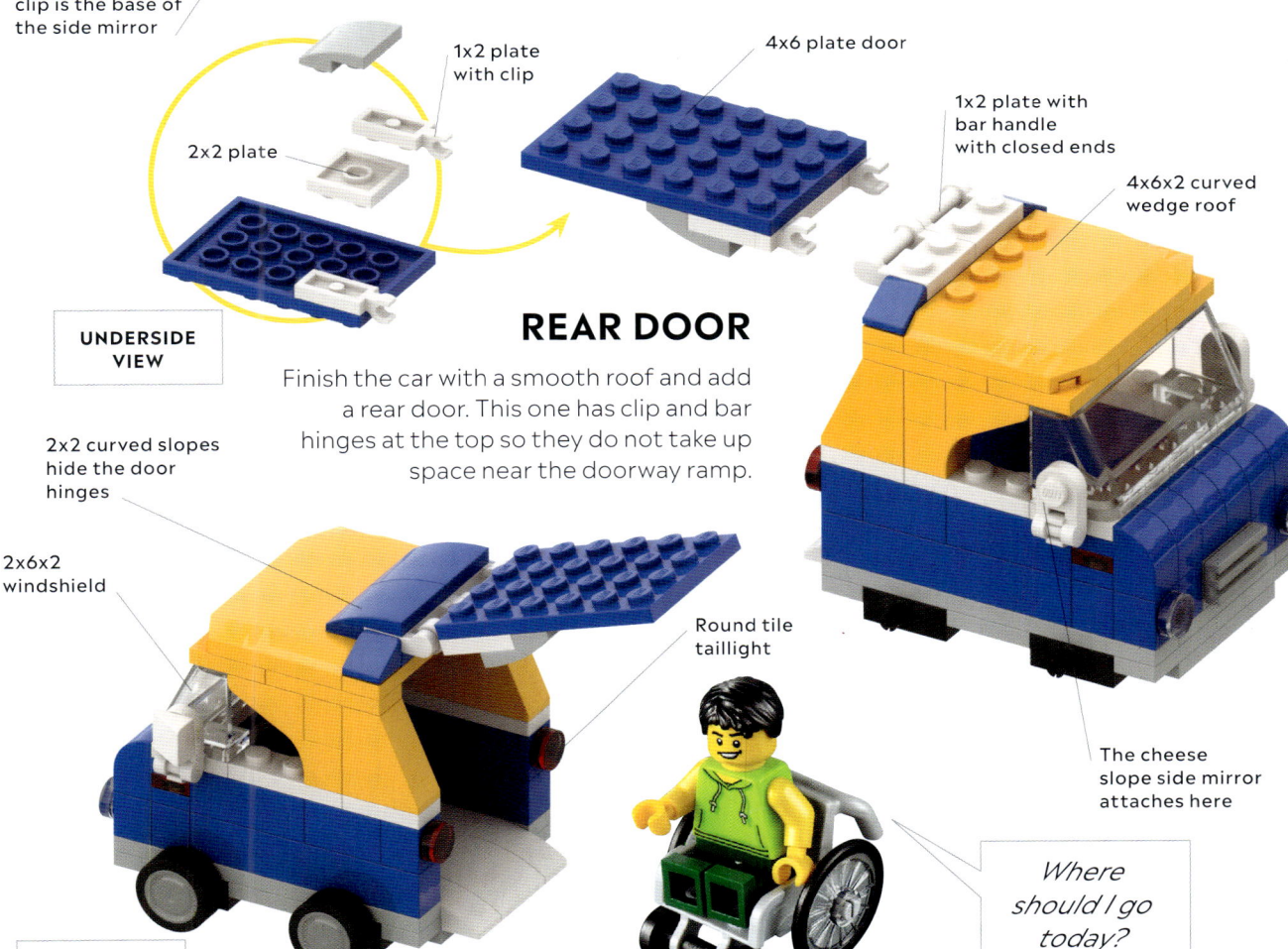

2x2 plate

1x2 plate with clip

4x6 plate door

1x2 plate with bar handle with closed ends

4x6x2 curved wedge roof

UNDERSIDE VIEW

REAR DOOR

Finish the car with a smooth roof and add a rear door. This one has clip and bar hinges at the top so they do not take up space near the doorway ramp.

2x2 curved slopes hide the door hinges

2x6x2 windshield

Round tile taillight

The cheese slope side mirror attaches here

Where should I go today?

REAR VIEW

CLOUD CAR

Not all cars need to be built for the road. This imaginary flying car floats high among the clouds, spreading color and cheer instead of gas fumes from its rainbow exhaust! It looks just like a fluffy cloud thanks to its rounded shape and white-and-aqua color scheme.

Perhaps I can let go of this balloon now...

Transparent 1x2x1 panel is a tiny windshield

Tiles and plates in rainbow colors for the exhaust

2x2 round tiles attached on the side enhance the round, fluffy shape

1x1 double curved slopes make a curved bumper

4x8 plate

1x2 curved slope

2x2 curved slope

Thin 2x2 plate with tiny wheels

FIRST CURVES

The base of the cloud car isn't as fluffy-looking as the exterior—it's one rectangular 4x8 plate. But the cloud curves are already starting to appear, with the addition of curved slopes underneath the front and back of the chassis.

GATHERING CLOUDS

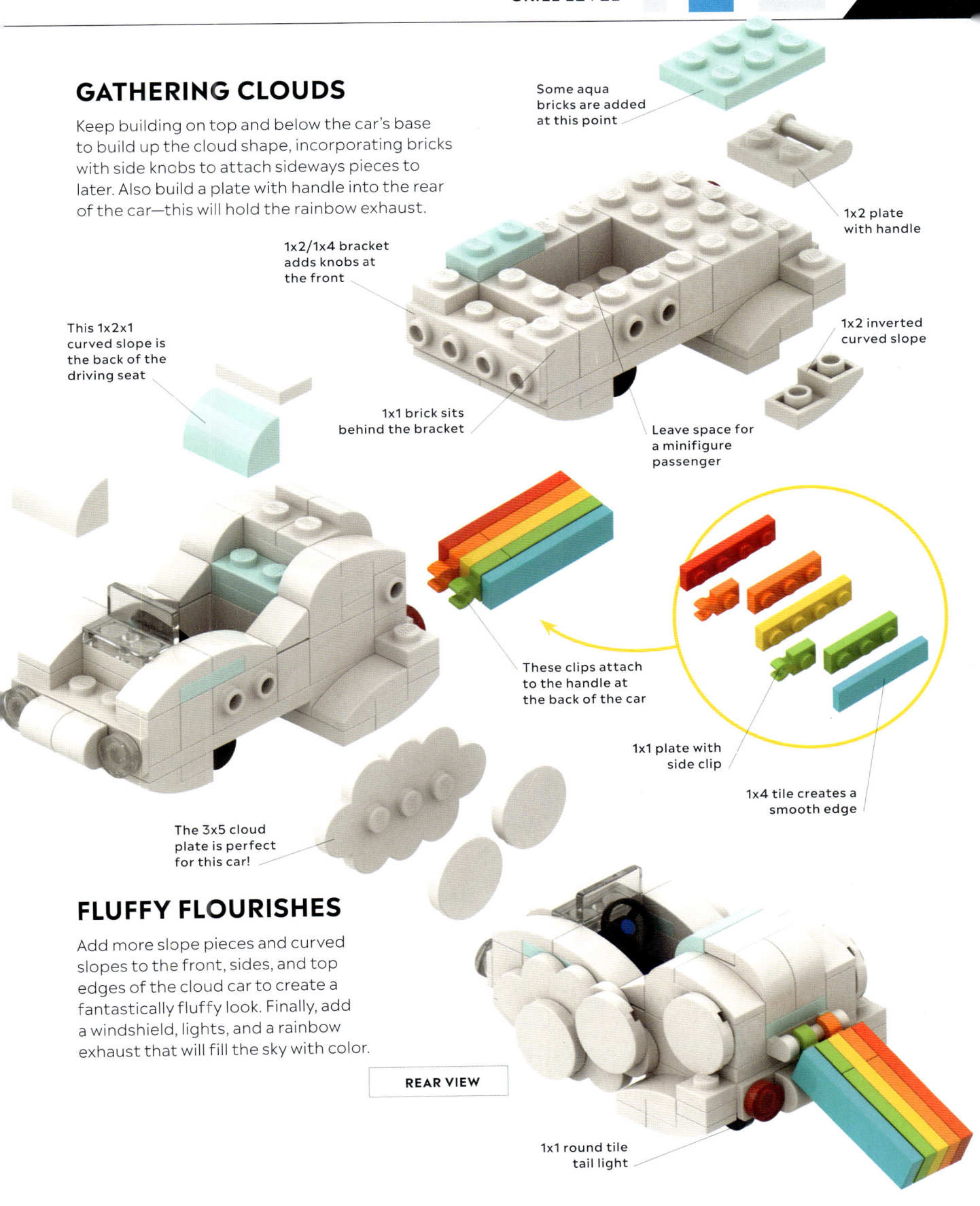

Keep building on top and below the car's base to build up the cloud shape, incorporating bricks with side knobs to attach sideways pieces to later. Also build a plate with handle into the rear of the car—this will hold the rainbow exhaust.

Some aqua bricks are added at this point

1x2 plate with handle

1x2/1x4 bracket adds knobs at the front

1x2 inverted curved slope

This 1x2x1 curved slope is the back of the driving seat

1x1 brick sits behind the bracket

Leave space for a minifigure passenger

These clips attach to the handle at the back of the car

1x1 plate with side clip

1x4 tile creates a smooth edge

The 3x5 cloud plate is perfect for this car!

FLUFFY FLOURISHES

Add more slope pieces and curved slopes to the front, sides, and top edges of the cloud car to create a fantastically fluffy look. Finally, add a windshield, lights, and a rainbow exhaust that will fill the sky with color.

REAR VIEW

1x1 round tile tail light

HOT ROD

A hot rod is usually an older car that's been stripped back and rebuilt to be even faster and cooler than it was before. This one looks revved up and ready to race with its huge engine, flames, and fearsome froggy figurehead!

I use any extra engine oil on my hair.

- Small panel windshield
- Frog piece attached to a jumper plate
- "Nerf bar" bumper made from a bar piece attached to clips
- Twin side exhausts are pistol pieces
- Smaller wheels at the front
- LEGO Technic pin
- LEGO Technic 2x4 brick with holes
- Inverted curved slopes lock the wheel holders in place
- 2x2 plate with wide wheel holders

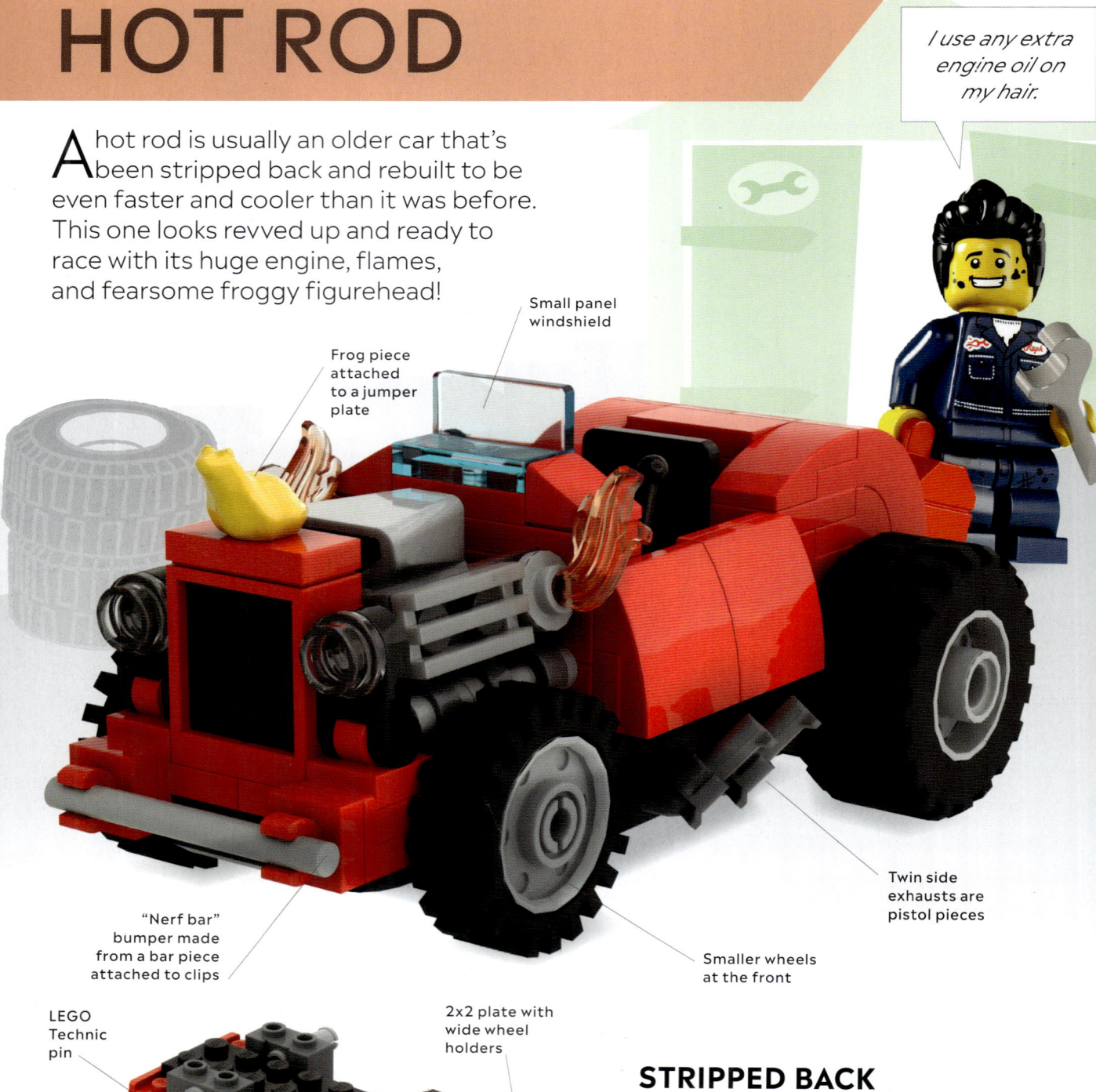

STRIPPED BACK

Hot-rod-owning minifigures love to strip cars right back to their bases like this. The hot rod's chassis is built around a 2x12 plate. There are different wheel-holding parts at the front and the back to accommodate the different-size wheels.

SKILL LEVEL

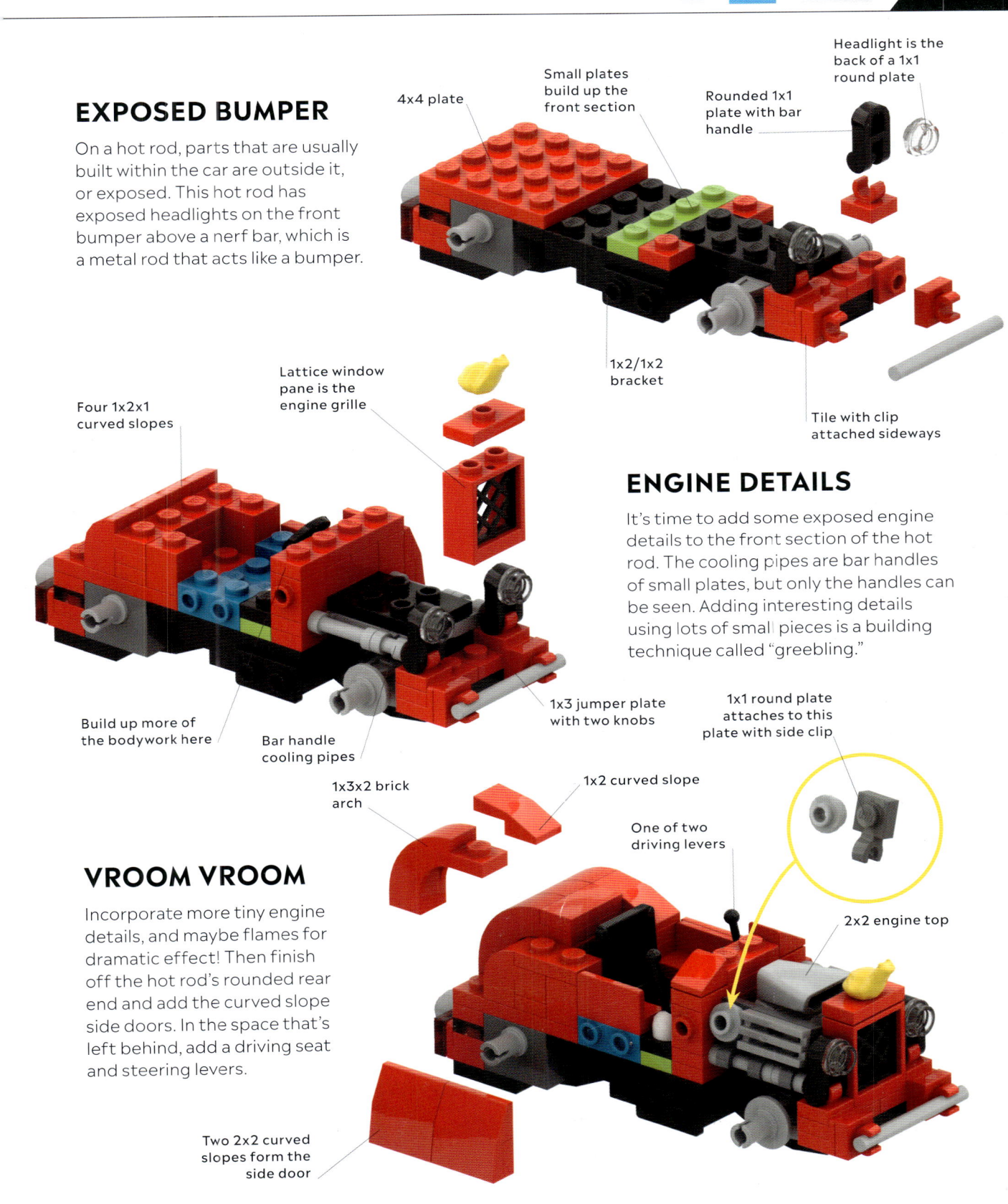

195

EXPOSED BUMPER

On a hot rod, parts that are usually built within the car are outside it, or exposed. This hot rod has exposed headlights on the front bumper above a nerf bar, which is a metal rod that acts like a bumper.

- 4x4 plate
- Small plates build up the front section
- Rounded 1x1 plate with bar handle
- Headlight is the back of a 1x1 round plate
- 1x2/1x2 bracket
- Tile with clip attached sideways

- Four 1x2x1 curved slopes
- Lattice window pane is the engine grille
- Build up more of the bodywork here
- Bar handle cooling pipes
- 1x3x2 brick arch

ENGINE DETAILS

It's time to add some exposed engine details to the front section of the hot rod. The cooling pipes are bar handles of small plates, but only the handles can be seen. Adding interesting details using lots of small pieces is a building technique called "greebling."

- 1x3 jumper plate with two knobs
- 1x2 curved slope
- One of two driving levers
- 1x1 round plate attaches to this plate with side clip
- 2x2 engine top

VROOM VROOM

Incorporate more tiny engine details, and maybe flames for dramatic effect! Then finish off the hot rod's rounded rear end and add the curved slope side doors. In the space that's left behind, add a driving seat and steering levers.

- Two 2x2 curved slopes form the side door

PIRATE CAR

Shiver me timbers! Is it a car? Is it a boat? It's both! A pirate's life may be mostly spent at sea, but when a salty sea dog needs to be a landlubber like the rest of us, this pirate car has everything they need to feel at home. There's a wide deck, ship's wheel steering, and a scary-looking sail.

1x1 round brick forms the top of the mast

The poop deck is the car's trunk

Skull on the mainsail

Do you have any treasure to pay for parking?

The ship's wheel can spin

The boat is one premade piece

Thick-tread tires for rolling on sand

SKILL LEVEL

HULL CHASSIS

The base of a boat is a hull and the base of a car is a chassis, so this build is a bit of both! It begins with a layer of inverted slope bricks attached to a plate. More plates with pins fit underneath it. A boat piece then goes on top of the base.

- 5x14x2 boat piece
- 2x2 inverted slope bricks
- 2x4 plate with pins
- This 2x8 plate locks in the plates above it
- 1x2 jumper plate
- 1x1 tile with clip

WALK THE ... STEPS

The next stage involves building inside the boat. Stepped jumper plates and 1x4 plates form the steps leading up to the poop deck. At the front, add a tile with clip to attach the figurehead to later.

- 1x2/1x2 brackets hold up the sail
- 1x1x6 pillar mast
- 1x1 round tile skeleton eye
- 1x1 half circle tile jawbone
- 2x2 macaroni tile is part of a gold trim
- A bracket piece here holds the rear details

RAISE THE SAIL

After polishing the poop deck, it's time to build the tall mast and then the skeleton mainsail. Its skull face is made from small tiles and jumper plates attached to a 4x6 plate.

- The car's figurehead is a 4x4 trapezoid flag
- The ship's wheel spins on a pin attached here
- 2x2 slope raises up the wheel
- 1x1/1x1 inverted bracket

197

YELLOW TAXI

You can't miss this bright yellow taxi cab. Hail one of these and it'll take you anywhere you need to go. Yellow taxis like this one are associated with New York City, but they can be found in many cities around the world.

Will you be my getaway vehicle?

Two double slopes make a rooftop advertising sign

Flat trunk formed from plates on their sides

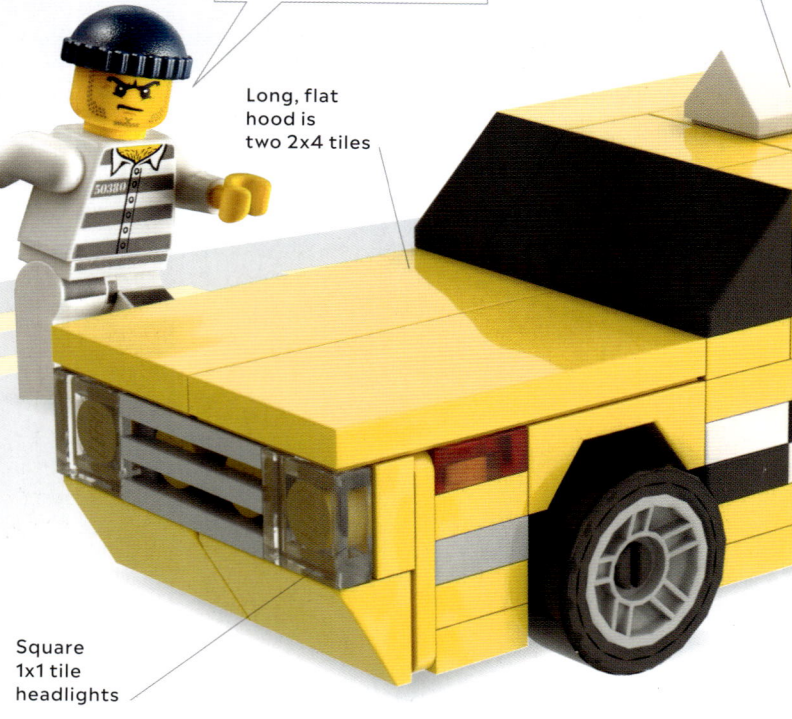

Long, flat hood is two 2x4 tiles

Square 1x1 tile headlights

Checkered door made from narrow plates in alternating colors

2x4 mudguard piece

2x2 corner plate

2x2 plate with wheel-holding pins

1x1 plate check pattern

4x4 plate

LOW CHASSIS

The yellow taxi's body is wide and low to the ground, so there are two extra layers of plates underneath the 2x8 plate that the wheels attach to. More plate layers fit above that plate, including two 2x4 mudguard pieces.

SKILL LEVEL

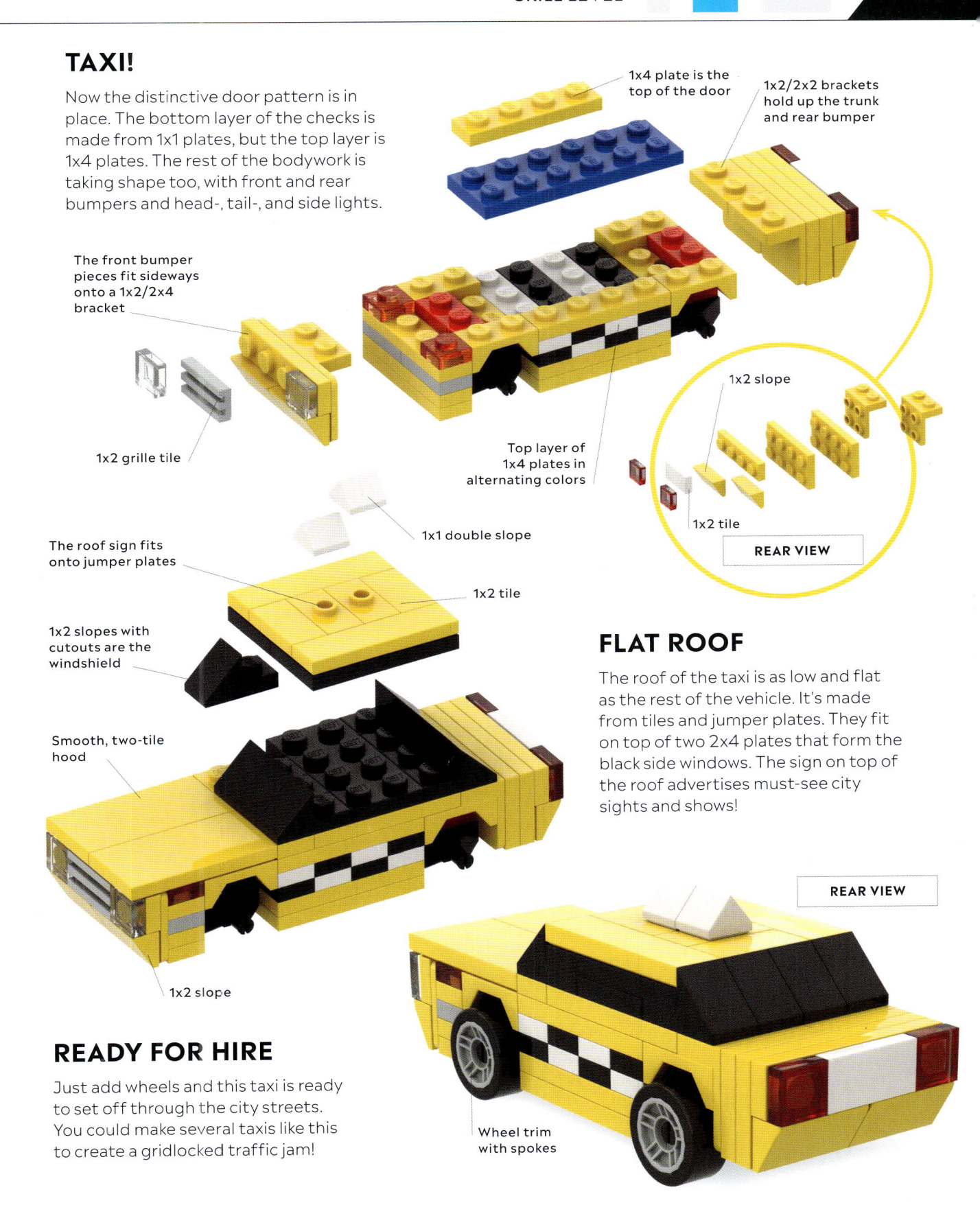

TAXI!

Now the distinctive door pattern is in place. The bottom layer of the checks is made from 1x1 plates, but the top layer is 1x4 plates. The rest of the bodywork is taking shape too, with front and rear bumpers and head-, tail-, and side lights.

1x4 plate is the top of the door

1x2/2x2 brackets hold up the trunk and rear bumper

The front bumper pieces fit sideways onto a 1x2/2x4 bracket

1x2 grille tile

Top layer of 1x4 plates in alternating colors

1x2 slope

1x2 tile

REAR VIEW

The roof sign fits onto jumper plates

1x2 slopes with cutouts are the windshield

1x1 double slope

1x2 tile

FLAT ROOF

The roof of the taxi is as low and flat as the rest of the vehicle. It's made from tiles and jumper plates. They fit on top of two 2x4 plates that form the black side windows. The sign on top of the roof advertises must-see city sights and shows!

Smooth, two-tile hood

1x2 slope

REAR VIEW

READY FOR HIRE

Just add wheels and this taxi is ready to set off through the city streets. You could make several taxis like this to create a gridlocked traffic jam!

Wheel trim with spokes

STREET FURNITURE

There are lots of objects on or near roads that are there to keep us safe, whether we're in a car or walking or working near them. These objects, known as street furniture, are often overlooked in real life, but they can make a LEGO road more interesting.

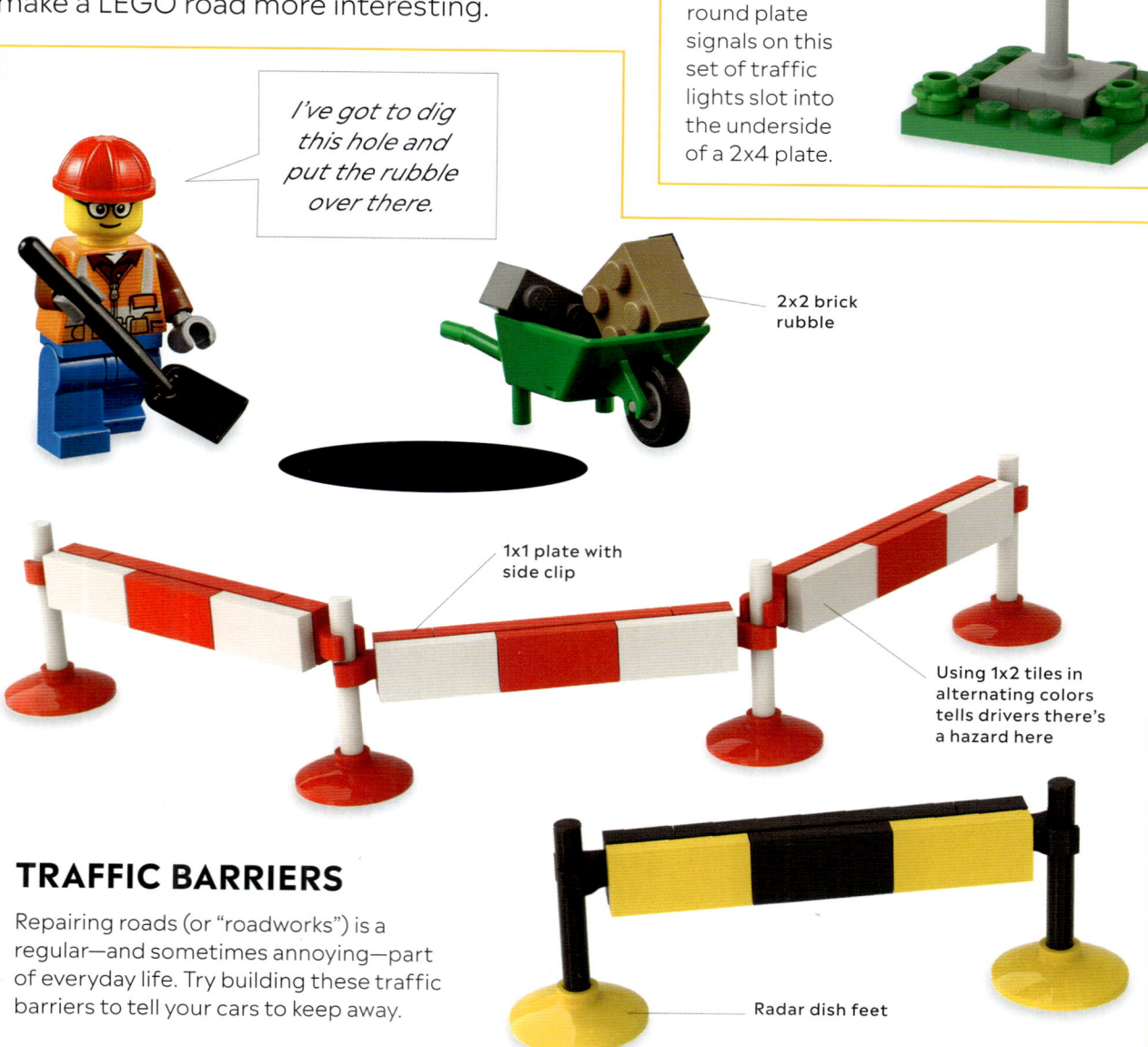

TRAFFIC LIGHTS

This is the tube part of a 1x1 round plate

The red, amber, and green 1x1 round plate signals on this set of traffic lights slot into the underside of a 2x4 plate.

I've got to dig this hole and put the rubble over there.

2x2 brick rubble

1x1 plate with side clip

Using 1x2 tiles in alternating colors tells drivers there's a hazard here

TRAFFIC BARRIERS

Repairing roads (or "roadworks") is a regular—and sometimes annoying—part of everyday life. Try building these traffic barriers to tell your cars to keep away.

Radar dish feet

IDEAS GALLERY 201

1x1 brick with side knob

Pole is a bar piece

STOP SIGN

This traffic-stopping road sign is made from a 2x2 round tile with an open knob. A T-piece slots into the knob to make the white "stop" line.

TRAFFIC CONES

Make lots of tiny traffic cones to show your cars where they can and can't go on a road! These build ideas use two different sizes of the LEGO cone piece.

2x2 plate base

1x1 cone

2x2 cone

SPEED BUMP

These bumps in the road force cars to slow down on streets where there are lots of minifigures around. They are easy to make with curved bricks.

1x4 double curved slope

There is one 2x8 plate underneath

I' have to take this rubble and fill up that hole over there.

GINGERBREAD CAR

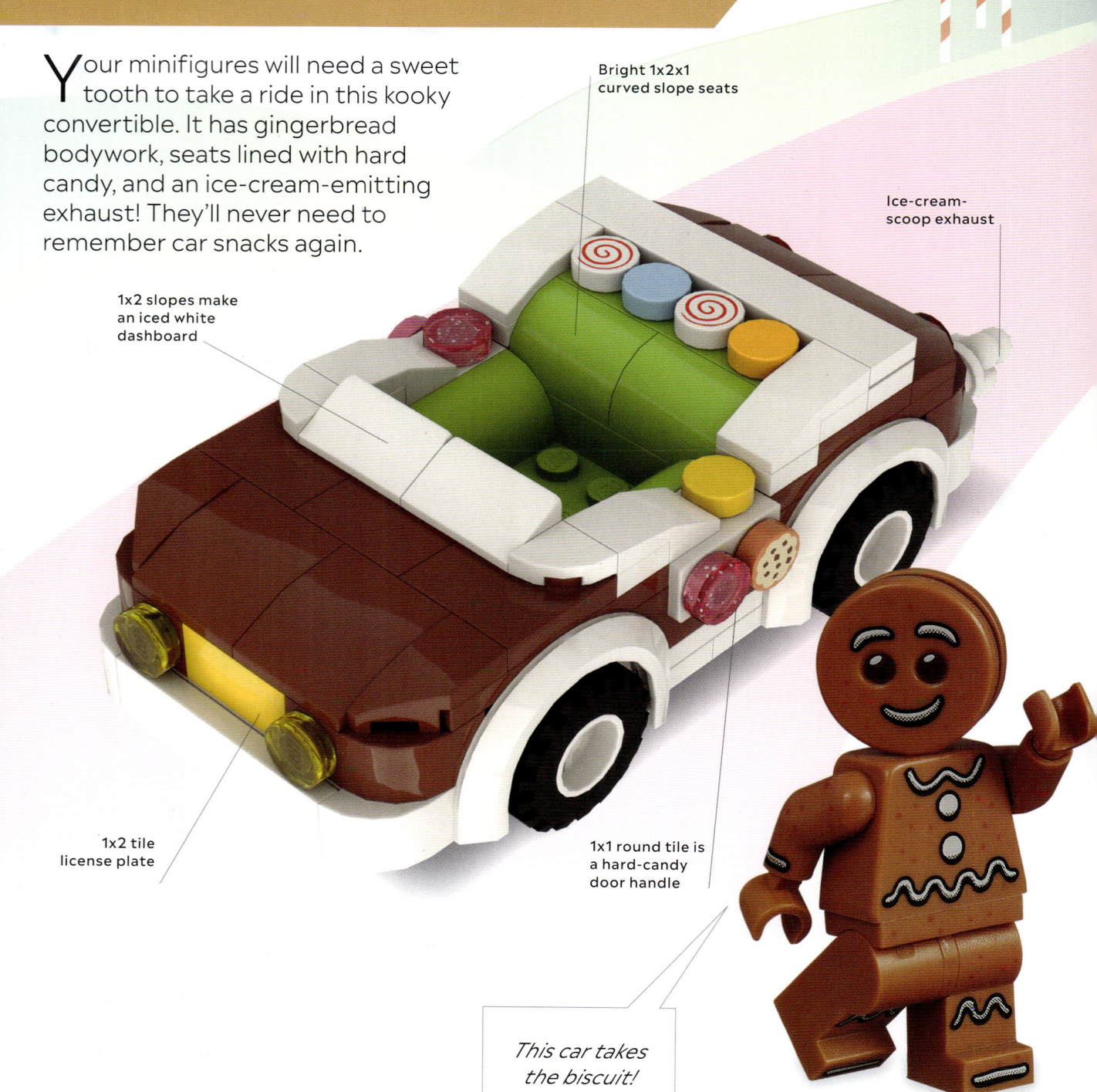

Your minifigures will need a sweet tooth to take a ride in this kooky convertible. It has gingerbread bodywork, seats lined with hard candy, and an ice-cream-emitting exhaust! They'll never need to remember car snacks again.

Bright 1x2x1 curved slope seats

Ice-cream-scoop exhaust

1x2 slopes make an iced white dashboard

1x2 tile license plate

1x1 round tile is a hard-candy door handle

This car takes the biscuit!

SKILL LEVEL

INGREDIENTS

The base of the gingerbread car is made from bricks and plates in any color, because it can't be seen on the final build. There's a blue 2x10 base plate, with more plates fitted horizontally across it to add width. Bracket pieces at each end hold the front and rear bumper pieces.

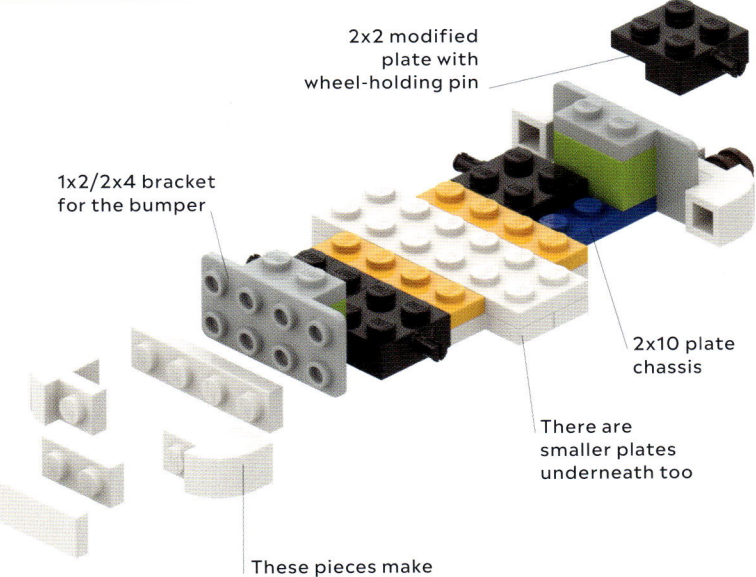

2x2 modified plate with wheel-holding pin

1x2/2x4 bracket for the bumper

2x10 plate chassis

There are smaller plates underneath too

These pieces make the bumper rounded

1x2 inverted slope brick

2x2 bricks fill gaps in the middle

SHAPING THE DOUGH

Now the car has round, icing-colored mudguards for the wheels to whir inside and gingerbread side doors made from inverted slope bricks. The front and back bumpers are in place too, with headlight and license plate details.

Round mudguard piece

Round tile headlight attached to a 1x1 plate

Leave this space for the seats

1x2x1 curved slope armrest

2x2 curved slope

Add a tile on top of this 1x6 plate for a smooth finish

FRESHLY BAKED

Build up the gingerbread hood with brown plates and tiles, then add curved slopes and wedges for a rounded finish. Next, add the wheels and hard candy details, and this gingerbread car will be ready to serve to your minifigures!

The door handle will fit onto this 1x2/1x2 bracket

White pieces create an iced trim

1x2 wedge on the hood edge

CAMPER

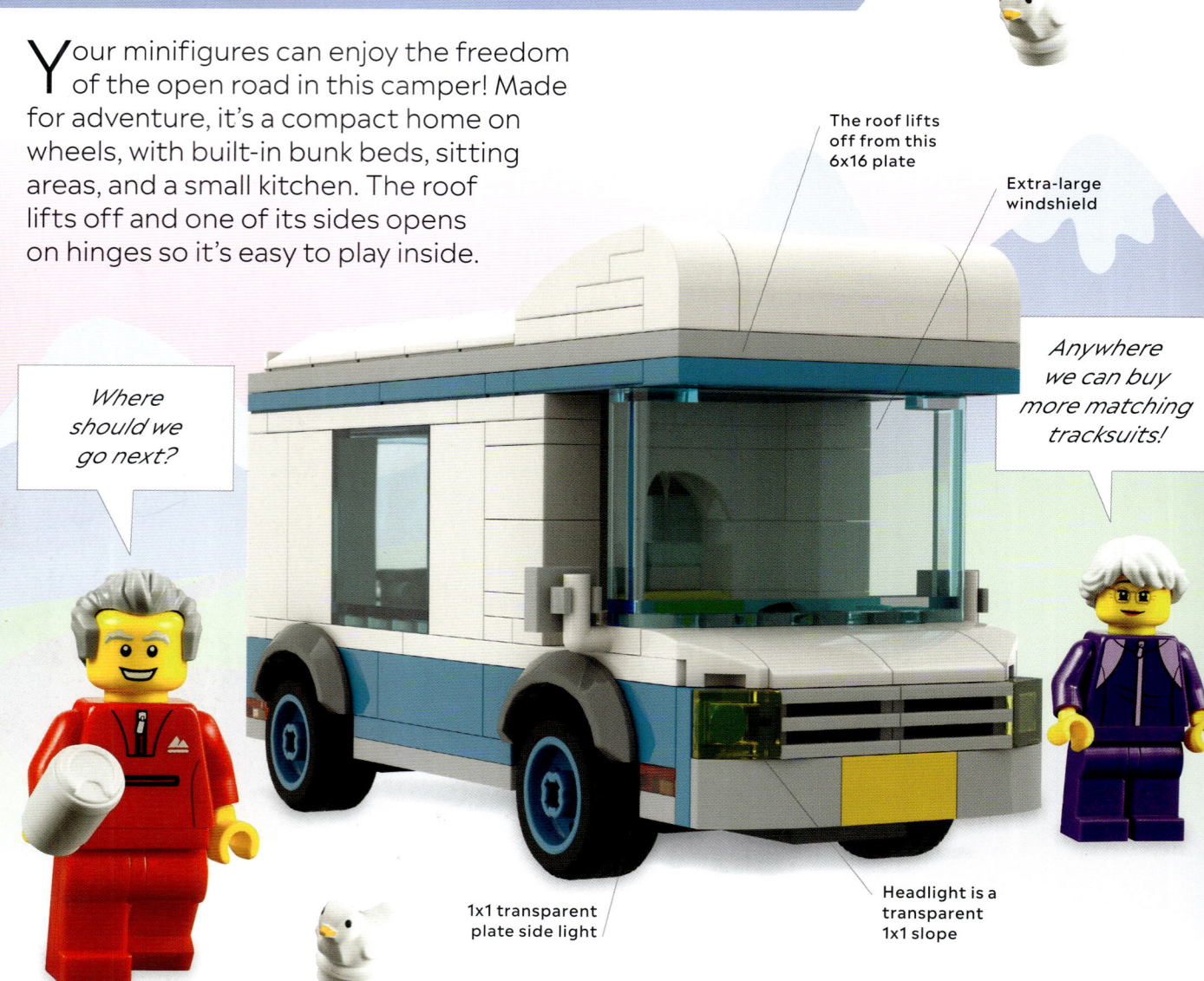

Your minifigures can enjoy the freedom of the open road in this camper! Made for adventure, it's a compact home on wheels, with built-in bunk beds, sitting areas, and a small kitchen. The roof lifts off and one of its sides opens on hinges so it's easy to play inside.

The roof lifts off from this 6x16 plate

Extra-large windshield

Where should we go next?

Anywhere we can buy more matching tracksuits!

1x1 transparent plate side light

Headlight is a transparent 1x1 slope

HOME SWEET HOME

To build the foundations of this mobile home, start with two 4x8 plates. Attach mudguards above them and wheel-holding plates below, locking them in with another plate underneath. Then build up the sides and bumpers of the camper with more plates and bricks.

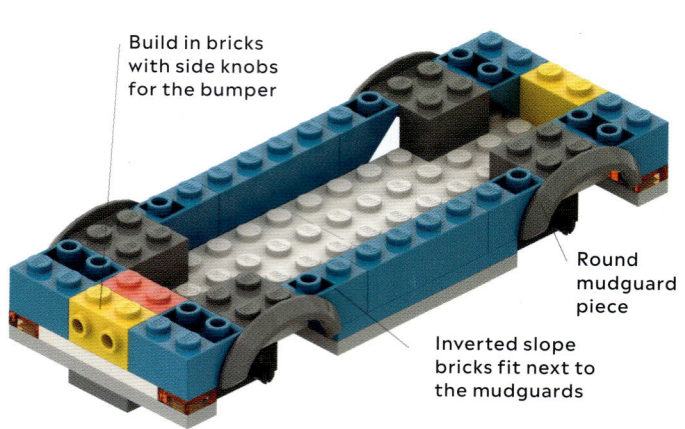

Build in bricks with side knobs for the bumper

Round mudguard piece

Inverted slope bricks fit next to the mudguards

SKILL LEVEL

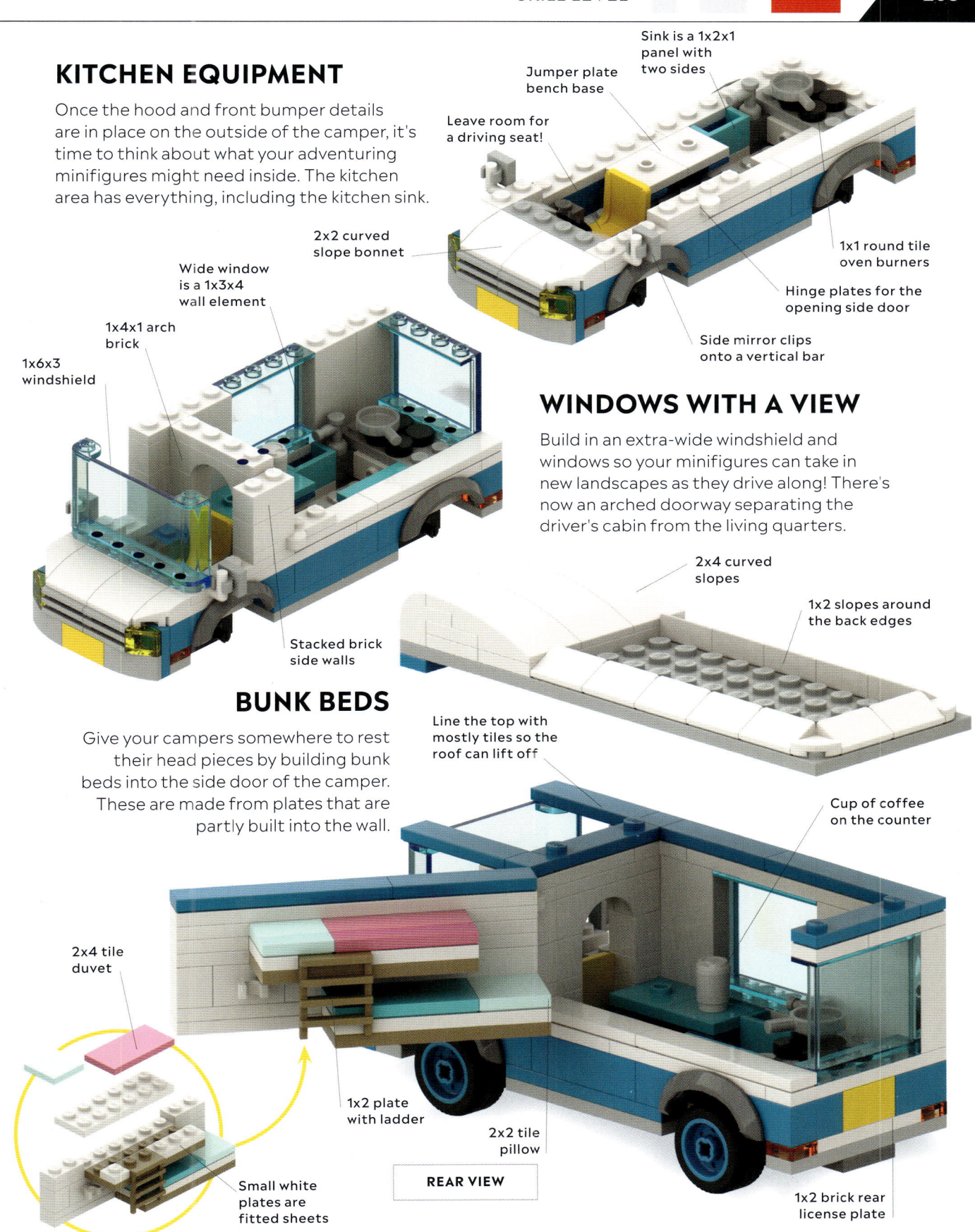

205

KITCHEN EQUIPMENT

Once the hood and front bumper details are in place on the outside of the camper, it's time to think about what your adventuring minifigures might need inside. The kitchen area has everything, including the kitchen sink.

- Sink is a 1x2x1 panel with two sides
- Jumper plate bench base
- Leave room for a driving seat!
- 2x2 curved slope bonnet
- 1x1 round tile oven burners
- Hinge plates for the opening side door
- Side mirror clips onto a vertical bar

- Wide window is a 1x3x4 wall element
- 1x4x1 arch brick
- 1x6x3 windshield
- Stacked brick side walls

WINDOWS WITH A VIEW

Build in an extra-wide windshield and windows so your minifigures can take in new landscapes as they drive along! There's now an arched doorway separating the driver's cabin from the living quarters.

- 2x4 curved slopes
- 1x2 slopes around the back edges
- Line the top with mostly tiles so the roof can lift off

BUNK BEDS

Give your campers somewhere to rest their head pieces by building bunk beds into the side door of the camper. These are made from plates that are partly built into the wall.

- 2x4 tile duvet
- 1x2 plate with ladder
- 2x2 tile pillow
- Cup of coffee on the counter
- 1x2 brick rear license plate
- Small white plates are fitted sheets

REAR VIEW

ICE-CREAM TRUCK

What's that sound? Could it be …? The ice-cream truck's here! Hearing the cheerful melodies of this vehicle brings joy to children and adults alike on a warm summer day. A quirky and colorful shop on wheels, the ice-cream truck drives around residential streets selling delicious cold treats from its serving hatch.

2x2 dome with hole is a scoop of ice cream

A gold 2x2x2 cone makes a tasty-looking wafer cone

Striped roof made from curved slopes and tiles

1x1 half round tiles form the curved ends of this shop sign

I've had a sundae every Sunday since I was a kid.

Bright colors to attract attention from all angles

Vanilla ice cream on the serving counter

SKILL LEVEL

207

WIDE LOAD

The ice-cream truck needs lots of room inside for freezers and ice-cream-making equipment, so its chassis must be very wide. The bottom layer is actually a narrow 4x12 plate, but more plates on top increase its width. At the front, a brick with side knobs allows the bumper parts to attach sideways.

4x6 plate fits into the middle section

2x5x1 bracket

Modified 1x2 brick has four side knobs

Modified 2x6 plate

Two 1x2 grille tiles form one engine grille

2x1 curved slope bumper

2x2 plate

These pieces can be any color

Faucet piece is the vanilla nozzle

Handle made from a round plate with bar

1x2 plate with pin hole holds the headlight

2x2 curved slope hood

1x2 grille tile drainer

ICE-CREAM MACHINE

After building up the lower sides of the truck, create some of the inside details, such as a chocolate and vanilla soft-serve ice-cream machine and ice pops on a counter. Leave space for a driving seat too.

Modified 1x4 tile with two knobs

2x6x2 windshield

REMOVABLE ROOF

The ice-cream truck needs an extra-wide windshield for spotting hungry customers and a large serving counter for handing over its wares. The last part of the build is the striped roof, which can easily be lifted off the six knobs at the top of the truck.

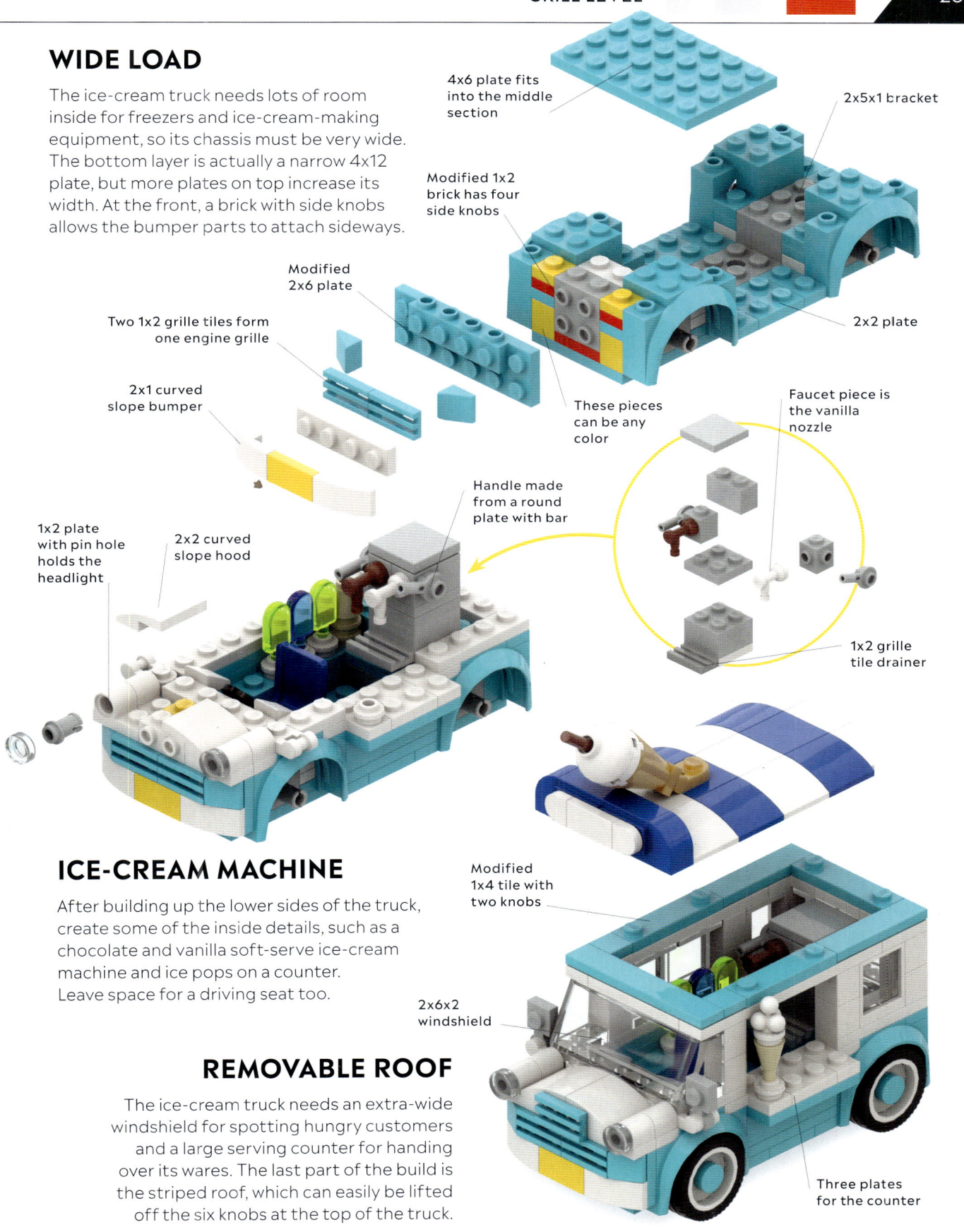

Three plates for the counter

ROYAL CAR

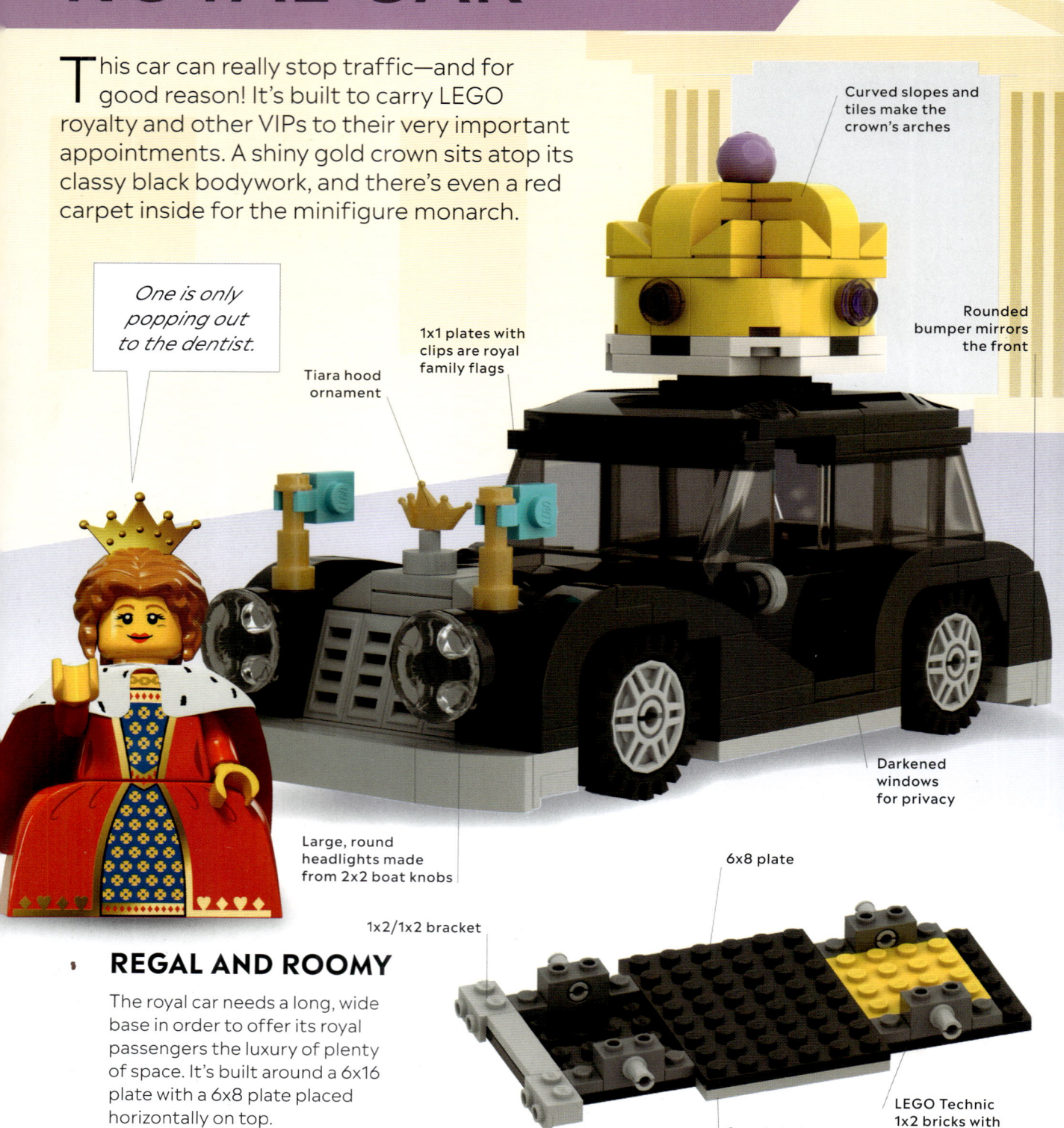

This car can really stop traffic—and for good reason! It's built to carry LEGO royalty and other VIPs to their very important appointments. A shiny gold crown sits atop its classy black bodywork, and there's even a red carpet inside for the minifigure monarch.

One is only popping out to the dentist.

Curved slopes and tiles make the crown's arches

1x1 plates with clips are royal family flags

Tiara hood ornament

Rounded bumper mirrors the front

Darkened windows for privacy

Large, round headlights made from 2x2 boat knobs

1x2/1x2 bracket

6x8 plate

Gray 1x6 plate underneath

LEGO Technic 1x2 bricks with holes hold the wheel pins

REGAL AND ROOMY

The royal car needs a long, wide base in order to offer its royal passengers the luxury of plenty of space. It's built around a 6x16 plate with a 6x8 plate placed horizontally on top.

SKILL LEVEL

CROWN CURVES

The royal car now has its seats and red carpet on the inside. Its shiny curved sides and silver trim on the outside make it stand out as a top-quality car.

1x4 curved slope

1x3x2 arch brick

Comfortable rear seat

T-piece holds the wheel

Four slotted slopes form the front grille

1x2 tile armrests

Build up the hood with bricks and plates

Tile red carpet wraps around the driving seat

STEER CLEAR

Instead of a steering wheel panel, the royal car has steering built into a dashboard made from a bracket and two plates with clips.

This plate with ring is the base of the side mirror

The windows are topped with mostly tiles so the roof can lift off

1x2x2 panel window

The headlights will attach here

ROYAL PROTECTION

Now the royal car has a sloping, smooth hood made from curved slopes to protect its engine. It also has a smoked glass windshield and windows to protect its passengers' privacy.

The hood ornament fits here

1x1 round tile with bar

2x2 plates and boat knobs make curved headlights

MONSTER TRUCK

Smaller automobiles cower at the sight of this beast of a car—it can crush them in seconds! A monster truck is a regular pickup truck or car that has been modified to have enormous wheels and intimidating features, such as an exposed engine, bright lights, bumper bars, and even chomping teeth.

Floodlights are the undersides of 1x1 round plates

2x2x3 slope bricks support the roof

Bumper bars made from prison window bars

A monster truck this terrifying is a rare sight.

These tires are often seen on LEGO tractors

Tooth plates hang from the bumper

3x4x1 curved mudguard

2x5x1 bracket

1x14 LEGO Technic brick with holes

2x2 brick with pins and axle hole

Enormous wheels hang from these two pins

MONSTROUS BASE

This is a chassis built around two LEGO Technic bricks with holes, which attach to three 4x4 plates below. Wide, arched mudguard pieces fit onto the bricks. Bricks with pin holes and large bracket pieces sit inside the space between the longer bricks with holes.

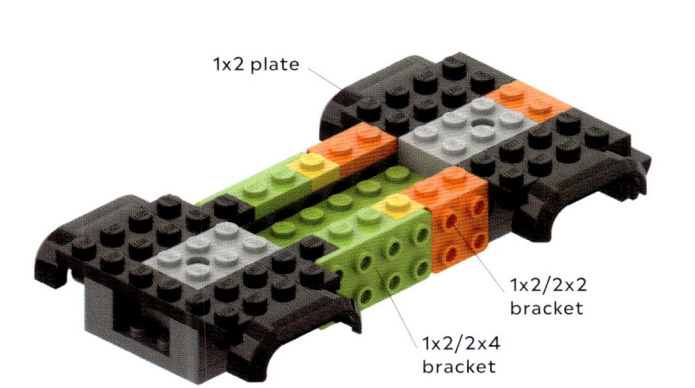

SKILL LEVEL   211

1x2 plate

BEFORE THE BODYWORK

Add plates in different sizes on top of the chassis base, along with pieces with side knobs on the outer edges. The smoother bodywork pieces will then attach to those side knobs in the next stages of the build.

1x2/2x2 bracket

1x2/2x4 bracket

The red taillights are now in place

2x2 driving seat

1x2/2x2 bracket

TERRIFYING TEETH

Monsters have sharp teeth, so monster trucks should have them too! These teeth are attached before the truck's front bumper details are. They fit onto bracket pieces with sideways knobs.

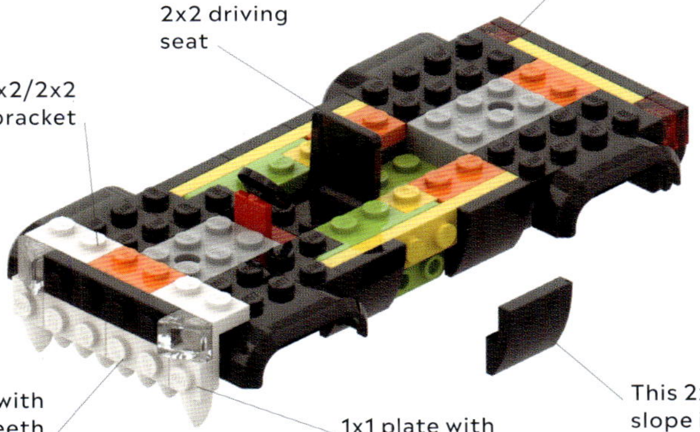

1x2 plate with three teeth

1x1 plate with one horizontal tooth

This 2x2 curved slope is part of the side door

These plates are the cargo area

1x2x1 panel pieces are armrests

1x4 plate with angled tubes

HUGE ENGINE

If the monster truck's teeth aren't enough to scare passing motorists, its exposed engine might be! Its made from plates with angled tubes and a large "air-scoop" piece, which supplies air to the engine.

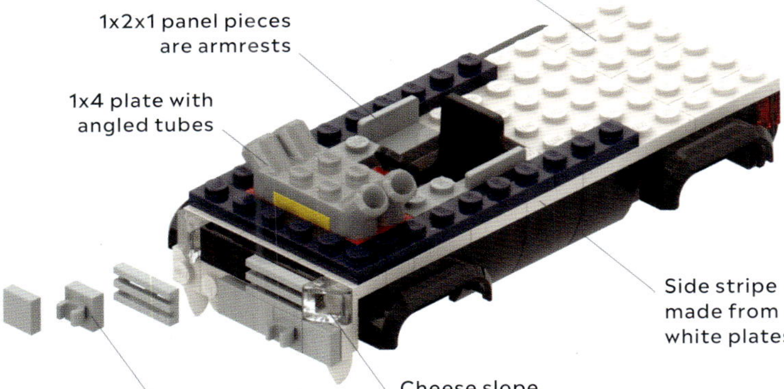

Side stripe made from white plates

4x6 plate roof

The bumper bars will attach here

Cheese slope headlight

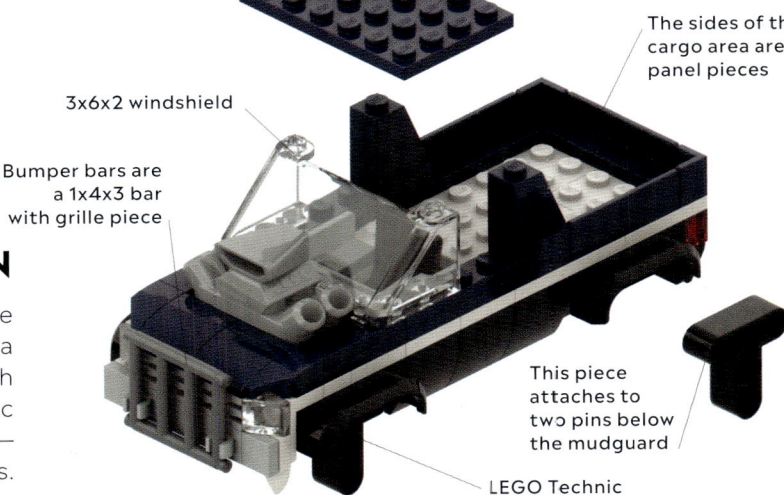

The sides of the cargo area are panel pieces

3x6x2 windshield

Bumper bars are a 1x4x3 bar with grille piece

ENCLOSED CABIN

The daring minifigure driver of the monster truck needs the protection of a strong roof, so cover the driver's cabin with a 4x6 plate. Finally, attach LEGO Technic liftarm pieces below the mudguards—these will hold the thundering wheels.

This piece attaches to two pins below the mudguard

LEGO Technic liftarm

FOX
PAGES 126–127

I hope that bot knows what it's doing.

ZZZ...PFFFT... programming error!

SWITCHED-UP PICKUPS
PAGES 182–183

Hey, I only asked for a tire change!

MONSTER MECHANIC
PAGES 300–301

HOT ROD
PAGES 194–195

SUBURBAN STREET
PAGES 34–35

MODEL MASH-UP

So you say your engine is making funny noises...

MONSTER TRUCK
PAGES 210–211

LIMOUSINE

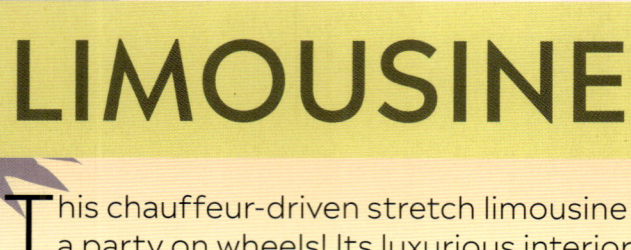

This chauffeur-driven stretch limousine is a party on wheels! Its luxurious interior has a swanky sofa, its own karaoke machine, and best of all a hot tub at the back. Just don't make any sharp turns!

This is great if you ignore the honking!

- 2x6x3 windshield
- Three 2x4 tiles create a sharp-edged hood
- Small curved slopes form the rounded bumper
- 4x10 chassis piece with recessed center
- 2x4 plate
- 2x2 modified plate with wheel-holding pin
- The rear bumper will attach to these modified plates with side knobs
- Opening side door has two hinge-plate connections
- Transparent blue 1x2 plates and tiles are whirling hot-tub water
- 2x2 corner plate
- These small plates won't be seen in the final model so they can be any color
- 2x4 mudguard with arch

STRETCHED CHASSIS

An extra-long car needs a double-length chassis. The limousine's base is made from two premade chassis pieces. There are two plates with wheel pins at either end of the chassis and white plates underneath and around them.

CHAUFFEUR'S SEAT

After building up the chassis with plates, add crisp white mudguards above the wheel-holding pins and put in a seat and steering wheel for the chauffeur. At the back, add bricks and more plates to create a raised section for the hot tub.

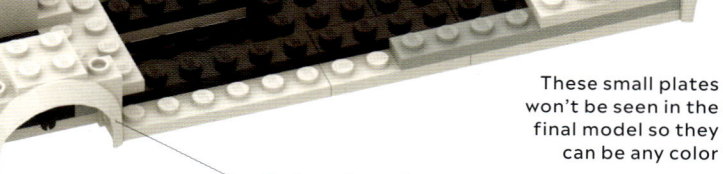

SKILL LEVEL 215

BUMPER BUILDS

Both the limousine's bumpers are built separately from the main build. The front bumper has a pointed silver grille and dual headlights for a glitzy look. There are matching dual taillights on the rear bumper that are made using the same technique.

- Dual headlights made from a 1x2 rounded plate and two 1x1 round tiles
- 1x1 plate with ring side mirror
- 1x4 brick with side knobs
- The door fits onto this swiveling hinge-plate connection
- Four 1x2 slopes with slots make a pointed grille
- Transparent black 1x2x2 panel is a smoked-glass driver's window

CAR KARAOKE

Add the smoked-glass windshield and blacked-out windows of the limousine, then get ready to party! Build pieces with side knobs into the door to attach the karaoke sound system and other fun details to.

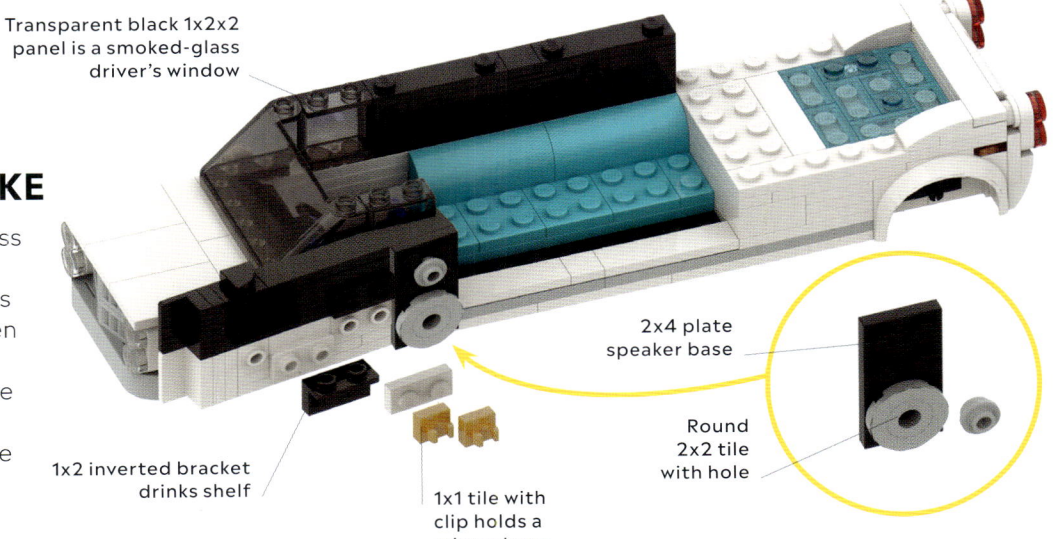

- 1x2 inverted bracket drinks shelf
- 1x1 tile with clip holds a microphone
- 2x4 plate speaker base
- Round 2x2 tile with hole

RAISING THE ROOF

The long, white roof has a base layer of plates topped with tiles and curved slopes. It attaches to four knobs on one side of the limousine so it can be removed easily.

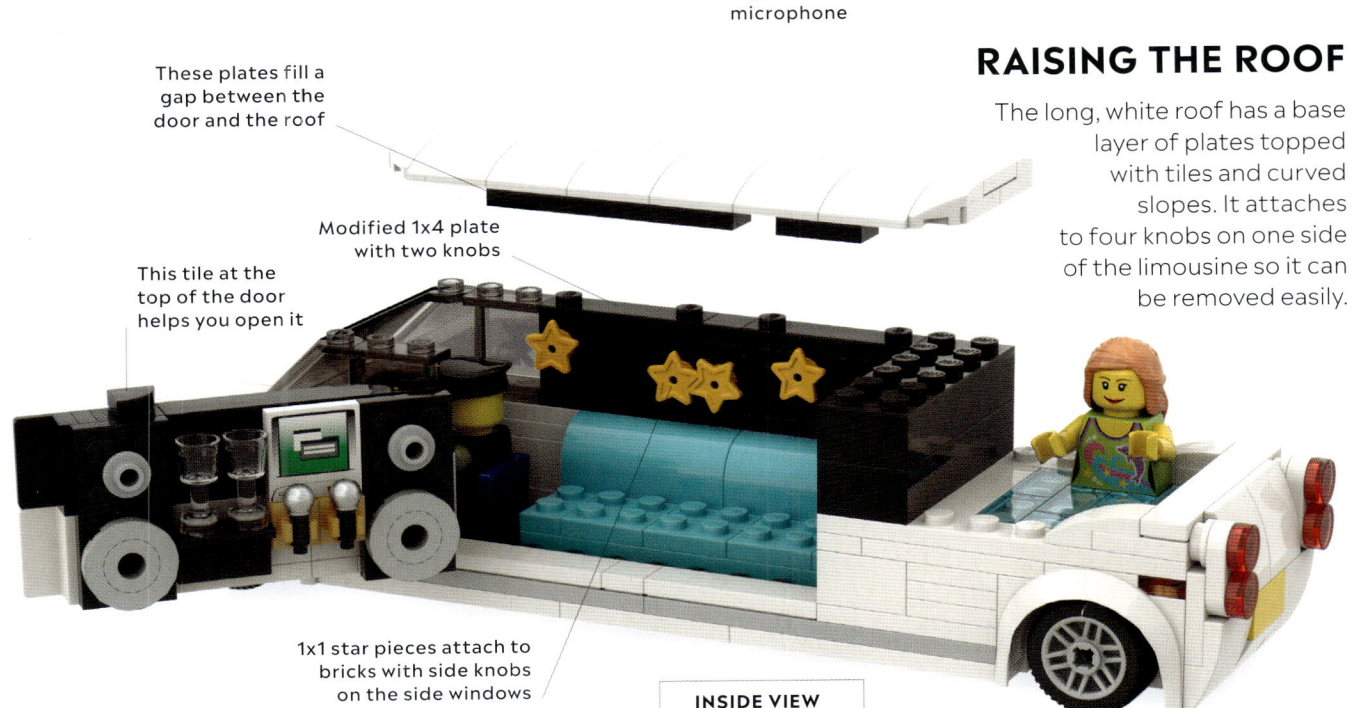

- These plates fill a gap between the door and the roof
- Modified 1x4 plate with two knobs
- This tile at the top of the door helps you open it
- 1x1 star pieces attach to bricks with side knobs on the side windows

INSIDE VIEW

Dinosaurs

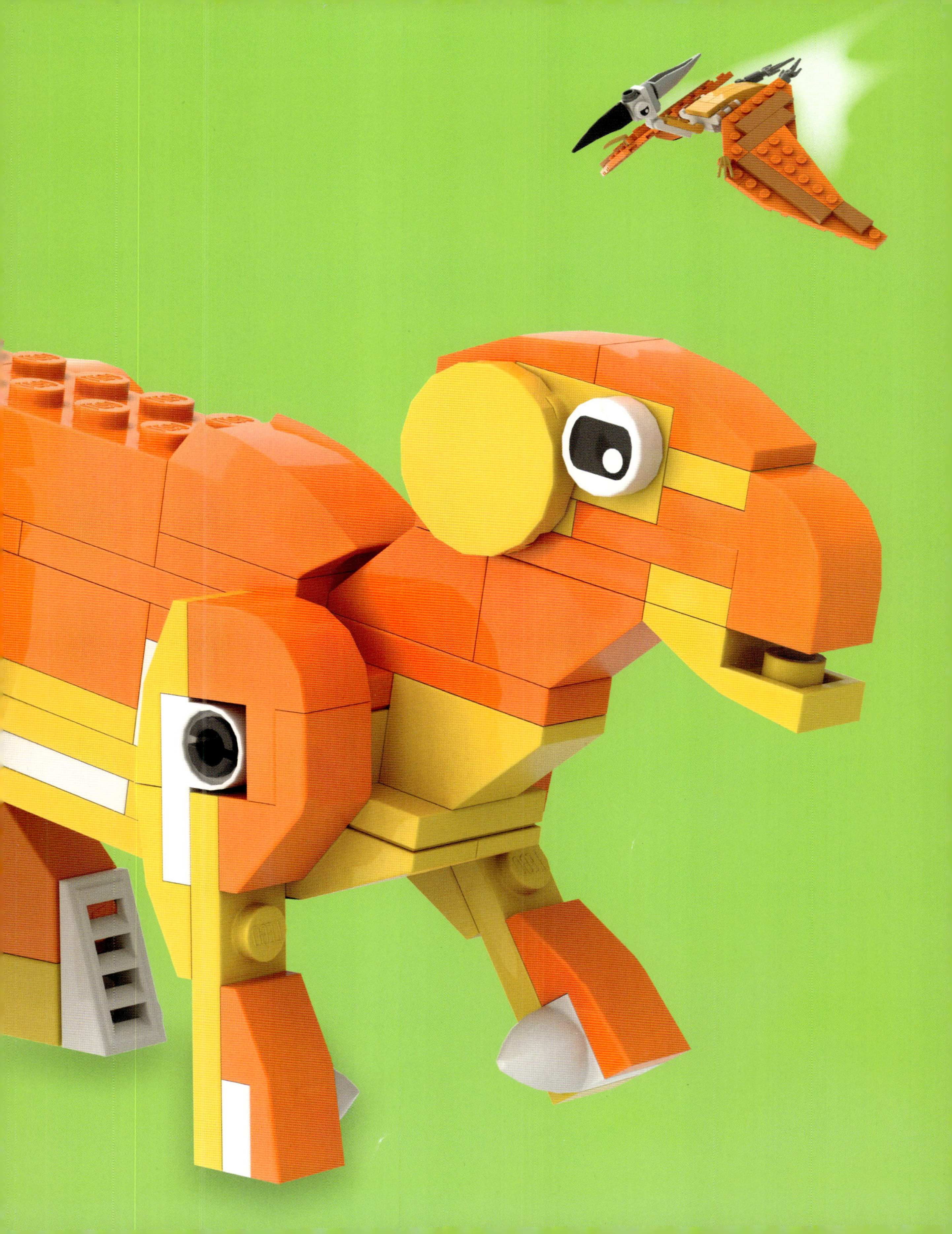

BRACHIOSAURUS

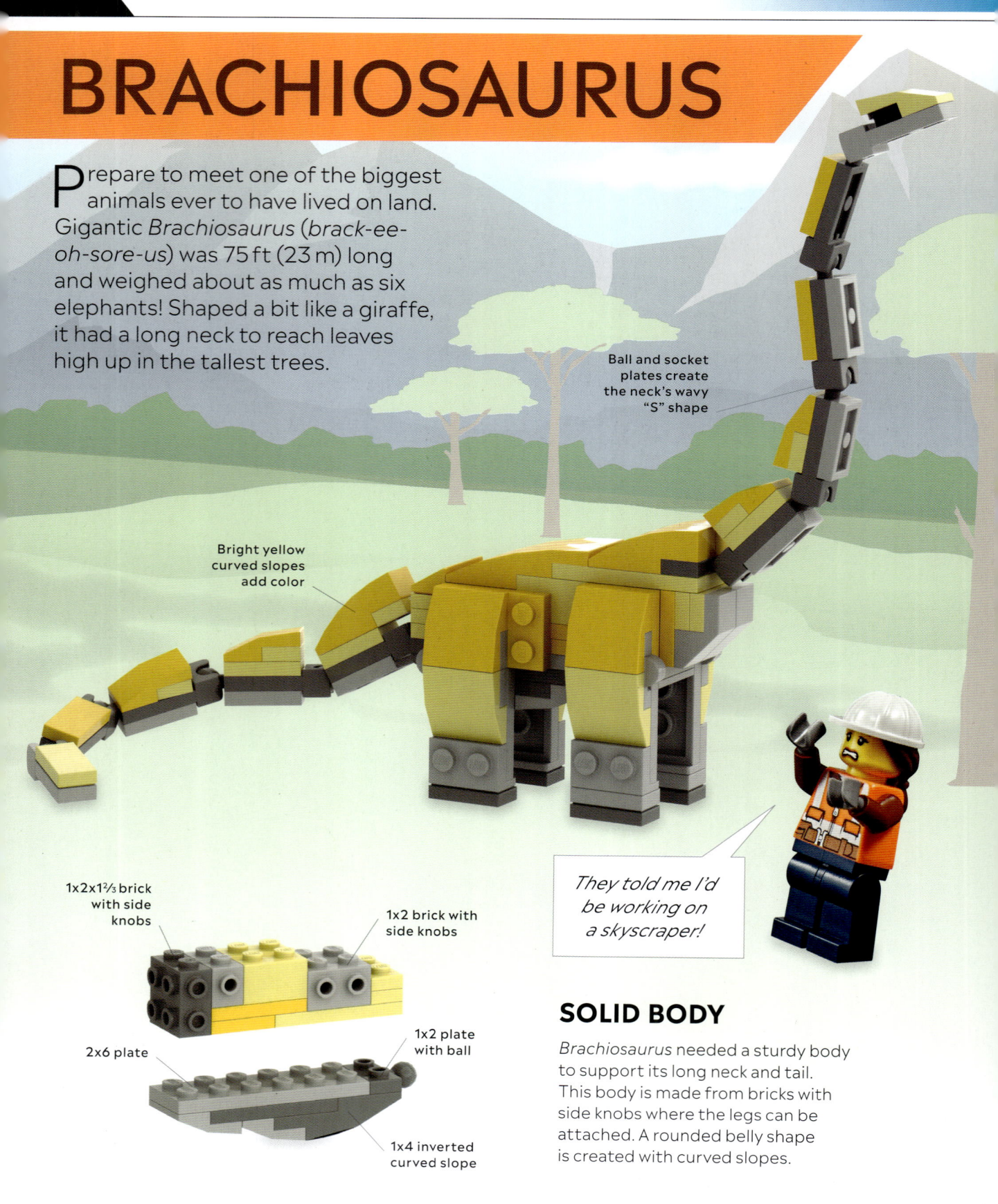

JURASSIC PERIOD

Prepare to meet one of the biggest animals ever to have lived on land. Gigantic *Brachiosaurus* (brack-ee-oh-sore-us) was 75 ft (23 m) long and weighed about as much as six elephants! Shaped a bit like a giraffe, it had a long neck to reach leaves high up in the tallest trees.

Ball and socket plates create the neck's wavy "S" shape

Bright yellow curved slopes add color

1x2x1⅔ brick with side knobs

1x2 brick with side knobs

2x6 plate

1x2 plate with ball

1x4 inverted curved slope

They told me I'd be working on a skyscraper!

SOLID BODY

Brachiosaurus needed a sturdy body to support its long neck and tail. This body is made from bricks with side knobs where the legs can be attached. A rounded belly shape is created with curved slopes.

SKILL LEVEL

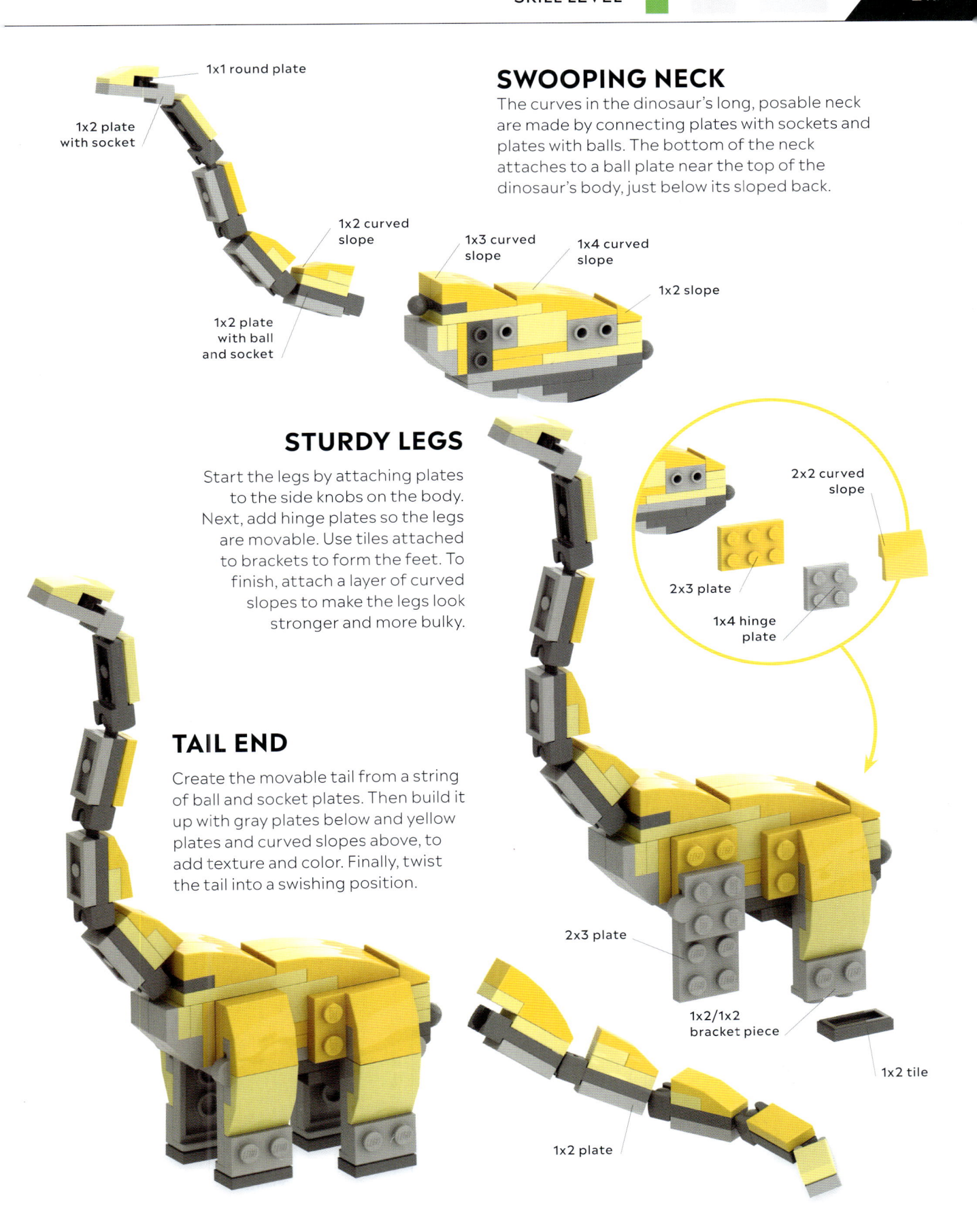

219

SWOOPING NECK

The curves in the dinosaur's long, posable neck are made by connecting plates with sockets and plates with balls. The bottom of the neck attaches to a ball plate near the top of the dinosaur's body, just below its sloped back.

- 1x1 round plate
- 1x2 plate with socket
- 1x2 curved slope
- 1x2 plate with ball and socket
- 1x3 curved slope
- 1x4 curved slope
- 1x2 slope

STURDY LEGS

Start the legs by attaching plates to the side knobs on the body. Next, add hinge plates so the legs are movable. Use tiles attached to brackets to form the feet. To finish, attach a layer of curved slopes to make the legs look stronger and more bulky.

- 2x2 curved slope
- 2x3 plate
- 1x4 hinge plate

TAIL END

Create the movable tail from a string of ball and socket plates. Then build it up with gray plates below and yellow plates and curved slopes above, to add texture and color. Finally, twist the tail into a swishing position.

- 2x3 plate
- 1x2/1x2 bracket piece
- 1x2 tile
- 1x2 plate

LONGISQUAMA

TRIASSIC PERIOD

This little lizard-like reptile gets its name from the long, narrow structures on its back. *Longisquama* (long-ge-skwa-mah) means "long scales" in Greek. Scientists still aren't sure quite what these structures were used for, but they do know that *Longisquama* was very small and lived in trees.

- Horn pieces form the shorter plumes
- Longer plumes made from minifigure hockey sticks
- Printed 1x1 round tile eye
- A clip and bar connection here allows the leg to move
- 1x1 rounded plates with bars make little feet

- 1x1 brick with hole
- 1x2 brick with two side knobs
- LEGO® Technic pin
- 1x6 plate
- 1x4 curved slope is the rounded chest

TINY BODY

The body of *Longisquama* was roughly the size of a mouse, so it doesn't take many LEGO® pieces to make it! Its lizard-like legs will later attach to the blue pins built into the body.

SKILL LEVEL

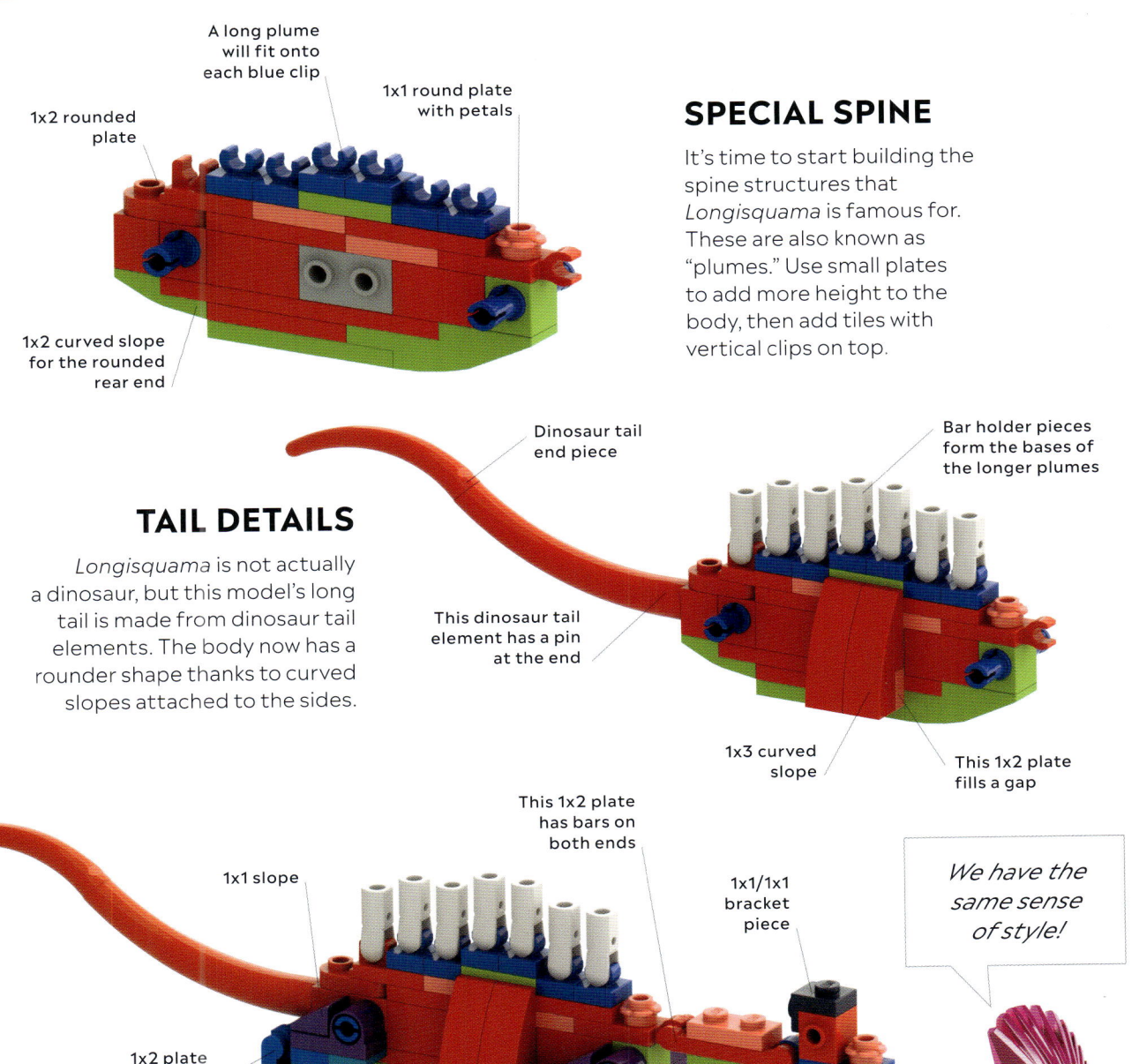

SPECIAL SPINE

It's time to start building the spine structures that *Longisquama* is famous for. These are also known as "plumes." Use small plates to add more height to the body, then add tiles with vertical clips on top.

- 1x2 rounded plate
- A long plume will fit onto each blue clip
- 1x1 round plate with petals
- 1x2 curved slope for the rounded rear end

TAIL DETAILS

Longisquama is not actually a dinosaur, but this model's long tail is made from dinosaur tail elements. The body now has a rounder shape thanks to curved slopes attached to the sides.

- Dinosaur tail end piece
- This dinosaur tail element has a pin at the end
- Bar holder pieces form the bases of the longer plumes
- 1x3 curved slope
- This 1x2 plate fills a gap

LEGS AND HEAD

Now this creature has four matching, posable legs made from curved slopes, small slopes, and plates with clips and holes. Its neck and head are also taking shape—it just needs eyes, a mouth, and long plumes, then it's ready to roam!

- 1x1 slope
- This 1x2 plate has bars on both ends
- 1x1/1x1 bracket piece
- 1x2 plate with bar
- 1x2 curved slope
- 1x2 plate with hole

We have the same sense of style!

FUN FACT
Lizard-like creatures like *Longisquama* have long tails for balance and, sometimes, for grabbing onto things.

221

PTERANODON

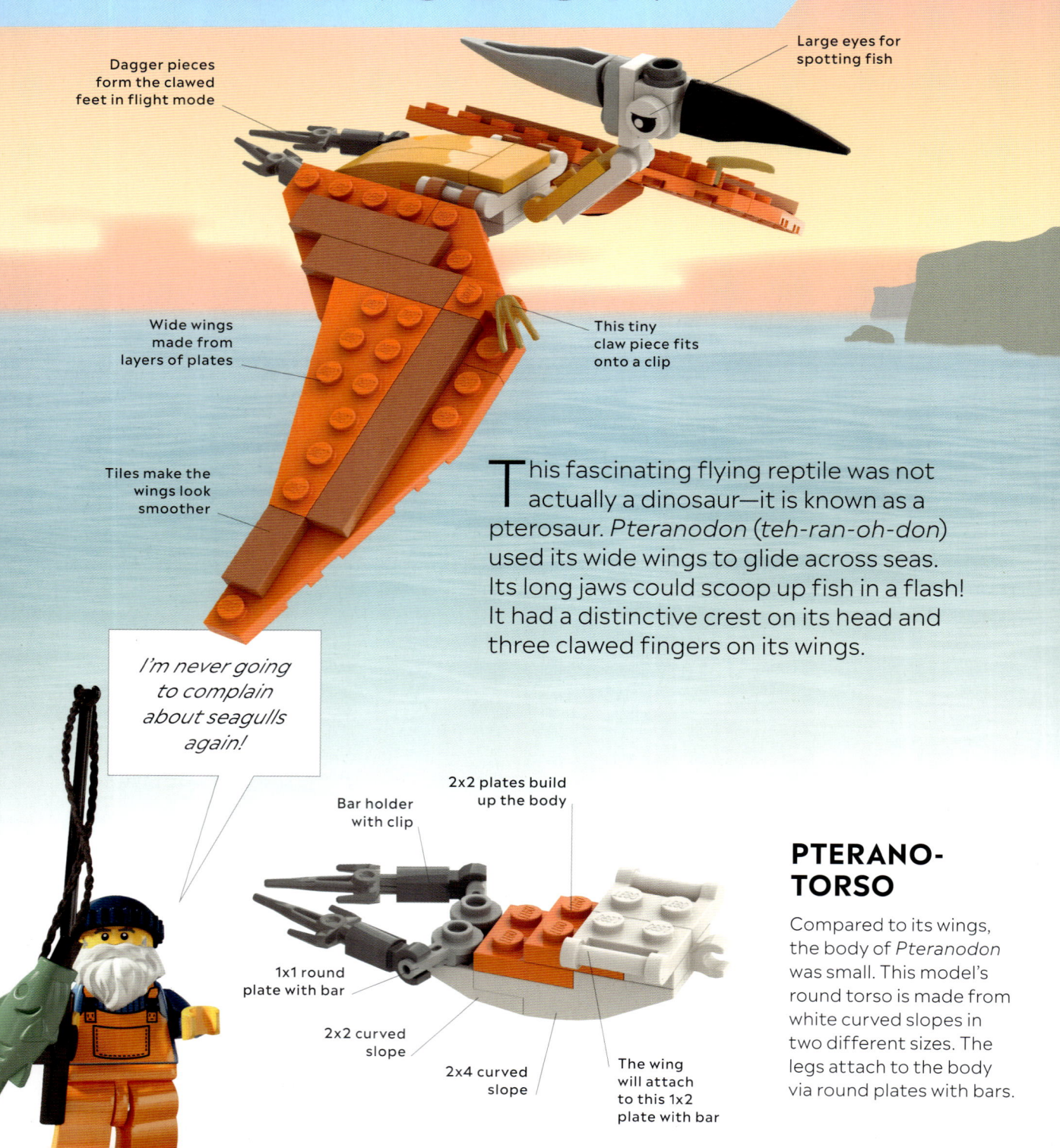

CRETACEOUS PERIOD

- Large eyes for spotting fish
- Dagger pieces form the clawed feet in flight mode
- Wide wings made from layers of plates
- This tiny claw piece fits onto a clip
- Tiles make the wings look smoother

I'm never going to complain about seagulls again!

This fascinating flying reptile was not actually a dinosaur—it is known as a pterosaur. *Pteranodon* (teh-ran-oh-don) used its wide wings to glide across seas. Its long jaws could scoop up fish in a flash! It had a distinctive crest on its head and three clawed fingers on its wings.

- 2x2 plates build up the body
- Bar holder with clip
- 1x1 round plate with bar
- 2x2 curved slope
- 2x4 curved slope
- The wing will attach to this 1x2 plate with bar

PTERANO-TORSO

Compared to its wings, the body of *Pteranodon* was small. This model's round torso is made from white curved slopes in two different sizes. The legs attach to the body via round plates with bars.

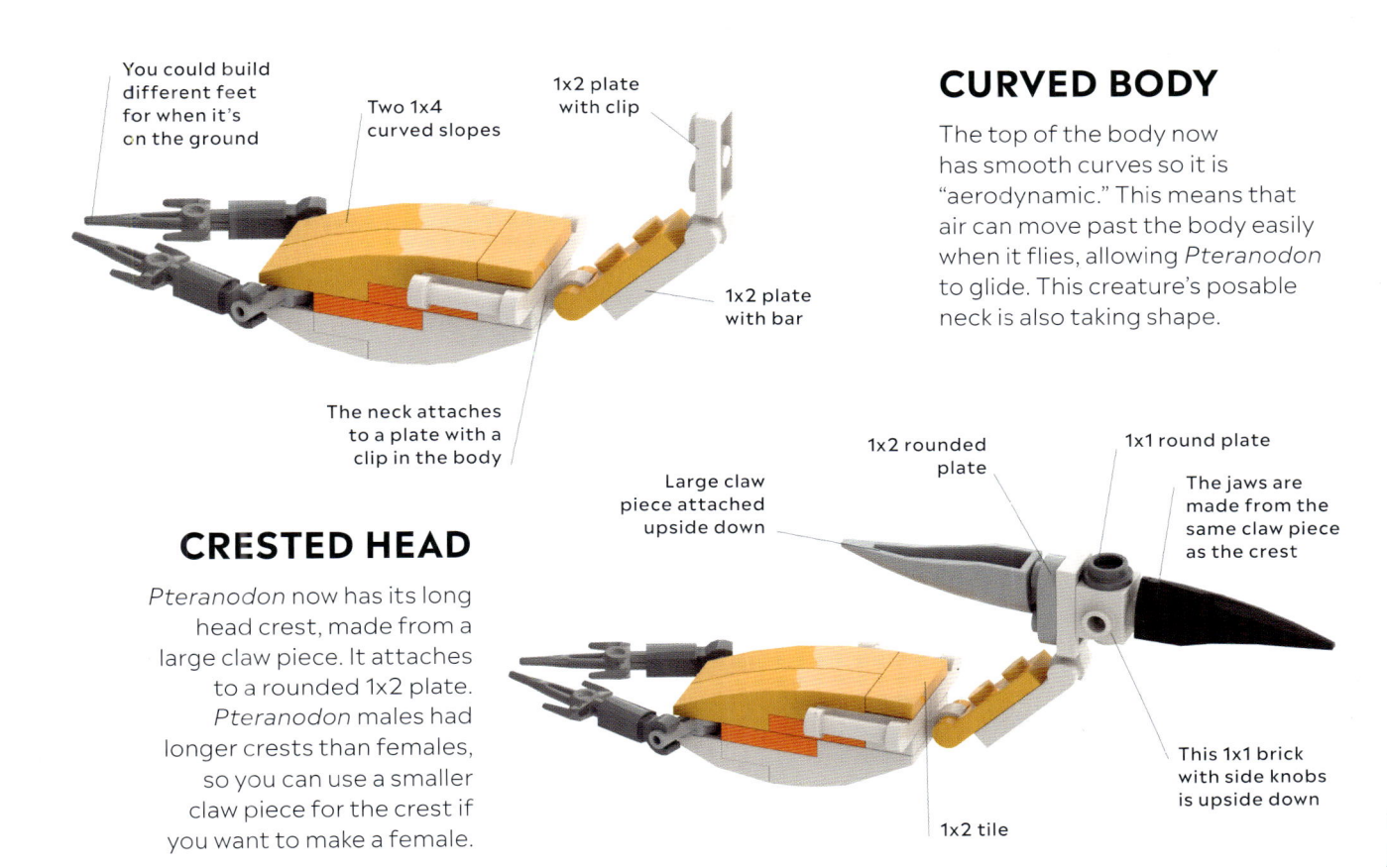

SKILL LEVEL

223

CURVED BODY

The top of the body now has smooth curves so it is "aerodynamic." This means that air can move past the body easily when it flies, allowing *Pteranodon* to glide. This creature's posable neck is also taking shape.

- You could build different feet for when it's on the ground
- Two 1x4 curved slopes
- 1x2 plate with clip
- 1x2 plate with bar
- The neck attaches to a plate with a clip in the body

CRESTED HEAD

Pteranodon now has its long head crest, made from a large claw piece. It attaches to a rounded 1x2 plate. *Pteranodon* males had longer crests than females, so you can use a smaller claw piece for the crest if you want to make a female.

- Large claw piece attached upside down
- 1x2 rounded plate
- 1x1 round plate
- The jaws are made from the same claw piece as the crest
- This 1x1 brick with side knobs is upside down
- 1x2 tile

TAKING FLIGHT

The gliding wings of *Pteranodon* are built from wedge plates in various sizes with small tiles on top. There are plates with horizontal clips underneath the wings that allow them to attach to plates with bars on the body.

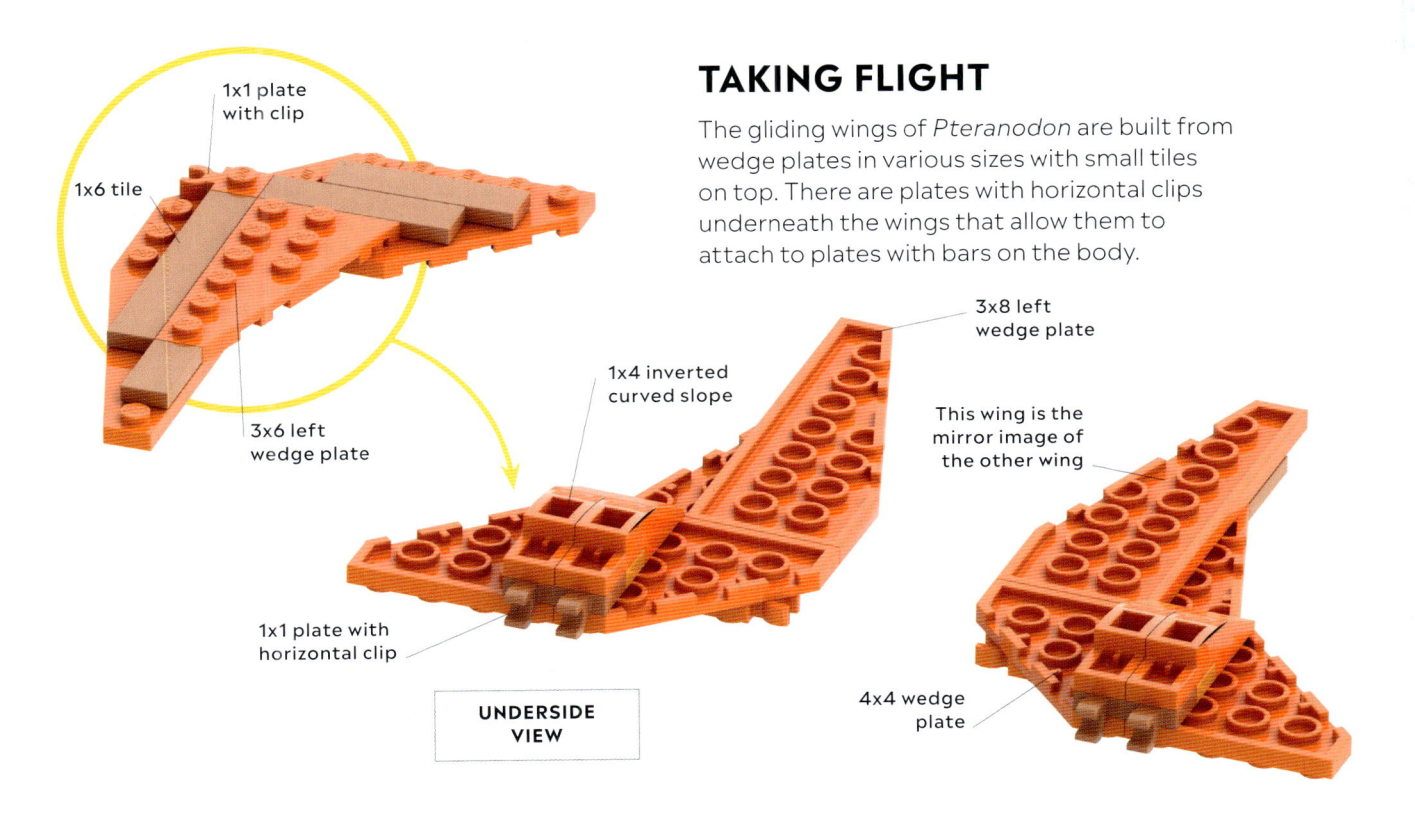

- 1x1 plate with clip
- 1x6 tile
- 3x6 left wedge plate
- 1x4 inverted curved slope
- 3x8 left wedge plate
- This wing is the mirror image of the other wing
- 1x1 plate with horizontal clip
- 4x4 wedge plate

UNDERSIDE VIEW

COMPSOGNATHUS

JURASSIC PERIOD

Don't underestimate this petite predator! One of the smallest dinosaurs ever to be discovered, *Compsognathus* (comp-sog-naith-us) was only about the size of a chicken, but it was a fierce theropod and skillful hunter. It had sharp teeth and claws, excellent eyesight, and it could run very fast.

- Narrow head and jaws
- A long tail helped with balance
- "S"-shaped neck
- 1x1 round bricks lengthen the tail
- Short arms with hooked claws
- Curved, clawed toes

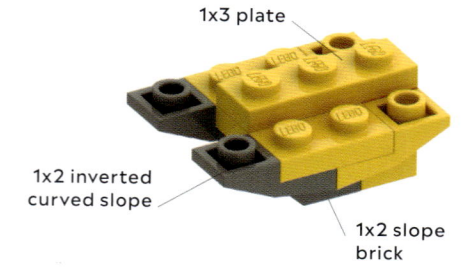

Hey, want to play fetch?

DINO FACT
Compsognathus would have hunted for small prey, such as lizards and insects. It probably also scavenged meat from the bones of creatures killed by much bigger predators.

- 1x3 plate
- 1x2 inverted curved slope
- 1x2 slope brick

BOTTOM FIRST

Most of this *Compsognathus* model is built sideways, so instead of starting with the dinosaur's belly build, this time you can begin with its bottom. Bricks form the base layer, with plates on top. Curved slopes are the rear sides of the hind legs.

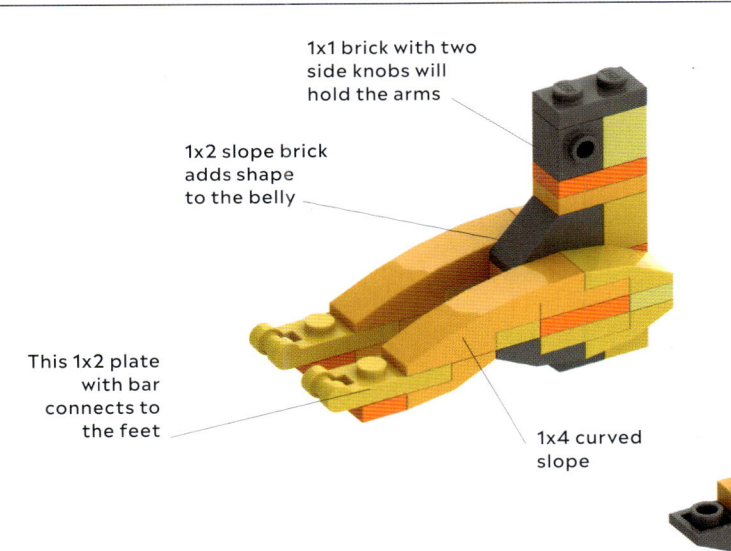

SKILL LEVEL

225

- **1x1 brick with two side knobs will hold the arms**
- **1x2 slope brick adds shape to the belly**
- **This 1x2 plate with bar connects to the feet**
- **1x4 curved slope**

SLIM LIMBS

Compsognathus had long, slim legs that allowed it to stride toward its fleeing prey and pounce on it. Still working upward from the bottom, widen your model's legs using narrow plates, adding more curved slopes for the strong upper thighs and calves.

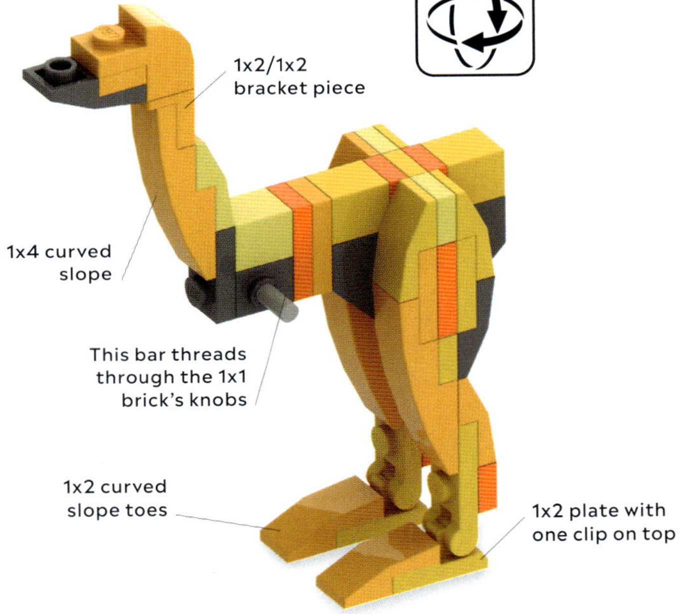

NECK CURVES

Theropods like *Compsognathus* usually had an "S"-shaped neck. You can get the right shape by combining a regular curved slope with a smaller inverted one at the bottom of the neck. Once the neck is in place, build the head on top using more curved slopes.

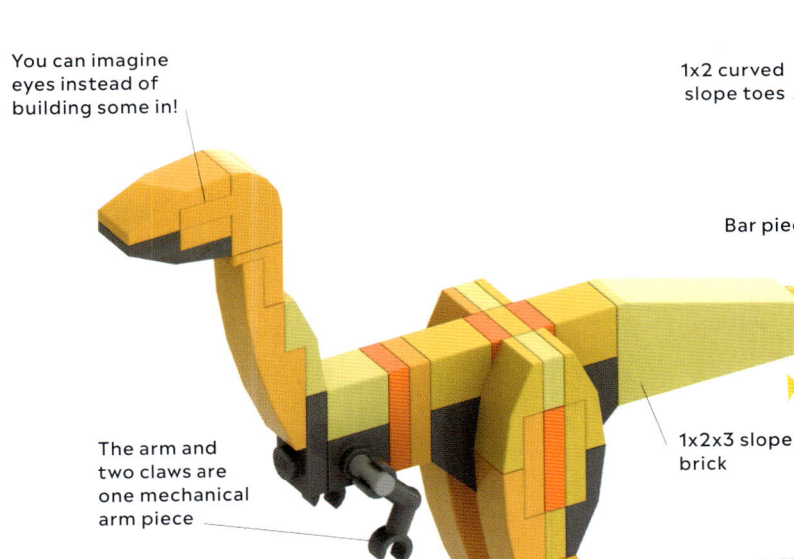

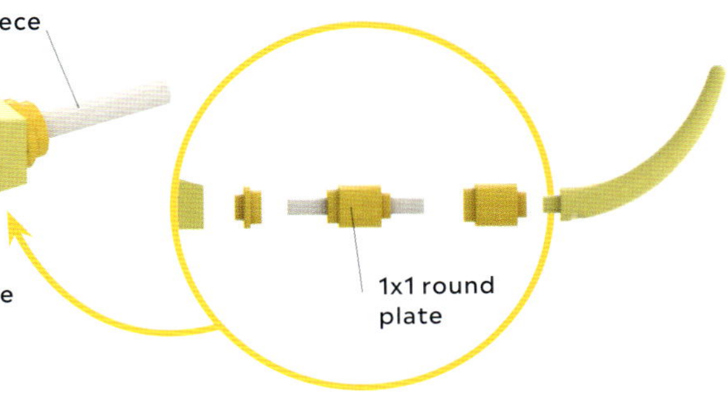

- **1x2/1x2 bracket piece**
- **1x4 curved slope**
- **This bar threads through the 1x1 brick's knobs**
- **1x2 curved slope toes**
- **1x2 plate with one clip on top**

- **You can imagine eyes instead of building some in!**
- **The arm and two claws are one mechanical arm piece**
- **Bar piece**
- **1x2x3 slope brick**
- **1x1 round plate**

LITTLE AND LONG

Now *Compsognathus* has small yet powerful arms, and sharp claws for grasping its prey. Next, bring some balance to this theropod's LEGO life by building a long tail to help it stay upright on two legs, especially while running at fast speeds. Ready, set ... run!

DIMORPHODON

JURASSIC PERIOD

This toothy pterosaur was one of the first flying reptiles in the Jurassic period. *Dimorphodon* (die-mor-foe-don) had a large head and beak; leathery wings like a bat; and a long, stiff tail that it used to steer itself through the air. It could be found flying over land and sea looking for small animals or fish, or climbing on cliffs and tree branches.

Smooth wings are built sideways

Stiff tail acts like a rudder in the air

This part of the wing can flap up and down

Wing claw is a 1x1 tile with clip

Toothy jaws

It eats small animals? I'm off!

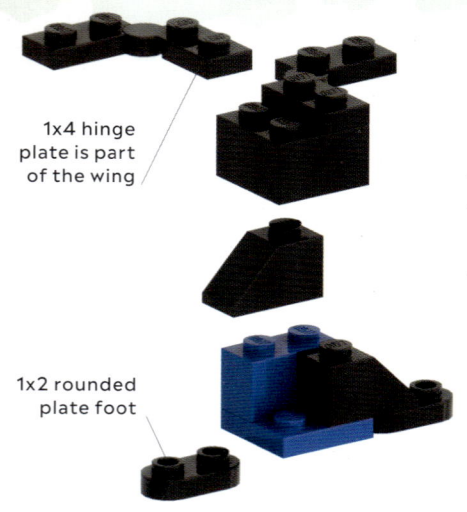

1x4 hinge plate is part of the wing

1x2 rounded plate foot

LEGS FIRST

Dimorphodon had two hind legs that it used for walking on land and scurrying up trees. This model's hind legs are made from slope bricks, with rounded plate feet underneath. The hinge plates that connect the wings are also in place.

UNDERSIDE VIEW

Another 1x4 hinge plate

WINGING IT

Build underneath the hinge plates at the top of the body to start filling out the wings. There are two more hinge plates underneath the first ones—these connections allow the end of each wing to flap independently.

1x2 slope brick

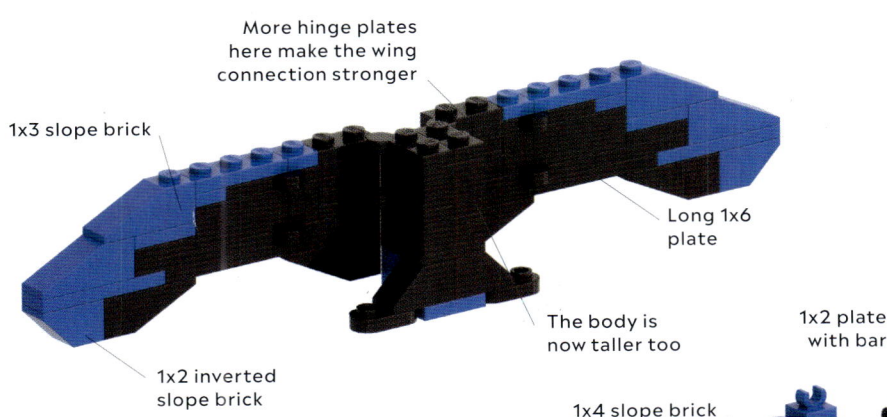

SKILL LEVEL

WIDEN THE WINGS

Give your *Dimorphodon* model a wider wingspan by building long, narrow plates and bricks into your model. Slope bricks in various sizes give the ends of the wings a streamlined shape.

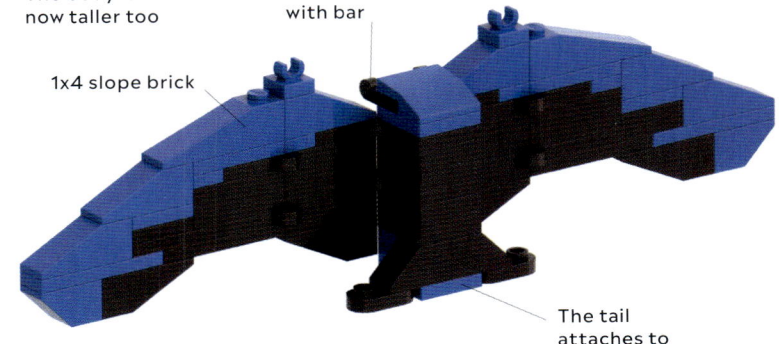

- More hinge plates here make the wing connection stronger
- 1x3 slope brick
- Long 1x6 plate
- The body is now taller too
- 1x2 inverted slope brick
- 1x2 plate with bar
- 1x4 slope brick

CLAWS AND CONNECTORS

Finish off the wings with more slope bricks, then add tiny, sharp wing claws. *Dimorphodon* used these to walk and to capture its prey. Also add a plate with bar at the top of the body for the head to attach to.

- The tail attaches to this plate

BONY TAIL

The long tail of *Dimorphodon* is one of its defining features. This model's tail is made from a balloon bottom piece and a bar. The round brick at the end of the tail is a tiny flap of skin that helped to steer the flying reptile.

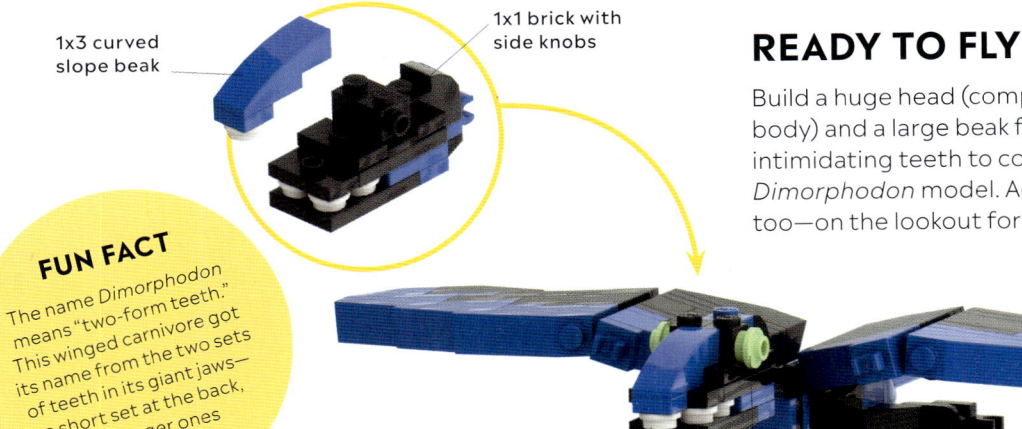

- Wings in flight mode
- REAR VIEW
- 2x2 balloon bottom
- Hind claws are 1x1 plates with horizontal clips

READY TO FLY

Build a huge head (compared to the body) and a large beak filled with intimidating teeth to complete your *Dimorphodon* model. Add big eyes too—on the lookout for a hearty meal!

- 1x3 curved slope beak
- 1x1 brick with side knobs

FUN FACT

The name *Dimorphodon* means "two-form teeth." This winged carnivore got its name from the two sets of teeth in its giant jaws—a short set at the back, and longer ones around the front.

- Long, protruding front teeth
- 1x1 round plate eye
- Legs in flying position

228 | TRIASSIC PERIOD

MELANOROSAURUS

The heavy feet of *Melanorosaurus* (meh-lan-or-oh-sore-us), which means "black mountain lizard," stomped the Earth at what is known as the very beginning of the Age of Dinosaurs—the Triassic period. One of the first super sauropods, it walked on four legs to support its bulky body and used its long neck to reach tall trees.

DINO FACT
Melanorosaurus didn't have the body armor that some other herbivores had, or sharp teeth like carnivores, so it probably lived in a large herd for protection from predators.

Tiny head compared to its body

Long, humped back

Long, curved neck

Long tail balances out the body weight

This pin connection allows the legs to move

What a lovely long neck!

Thick legs and feet, like an elephant's

This dinosaur probably wasn't red, so choose any color you like!

SKILL LEVEL

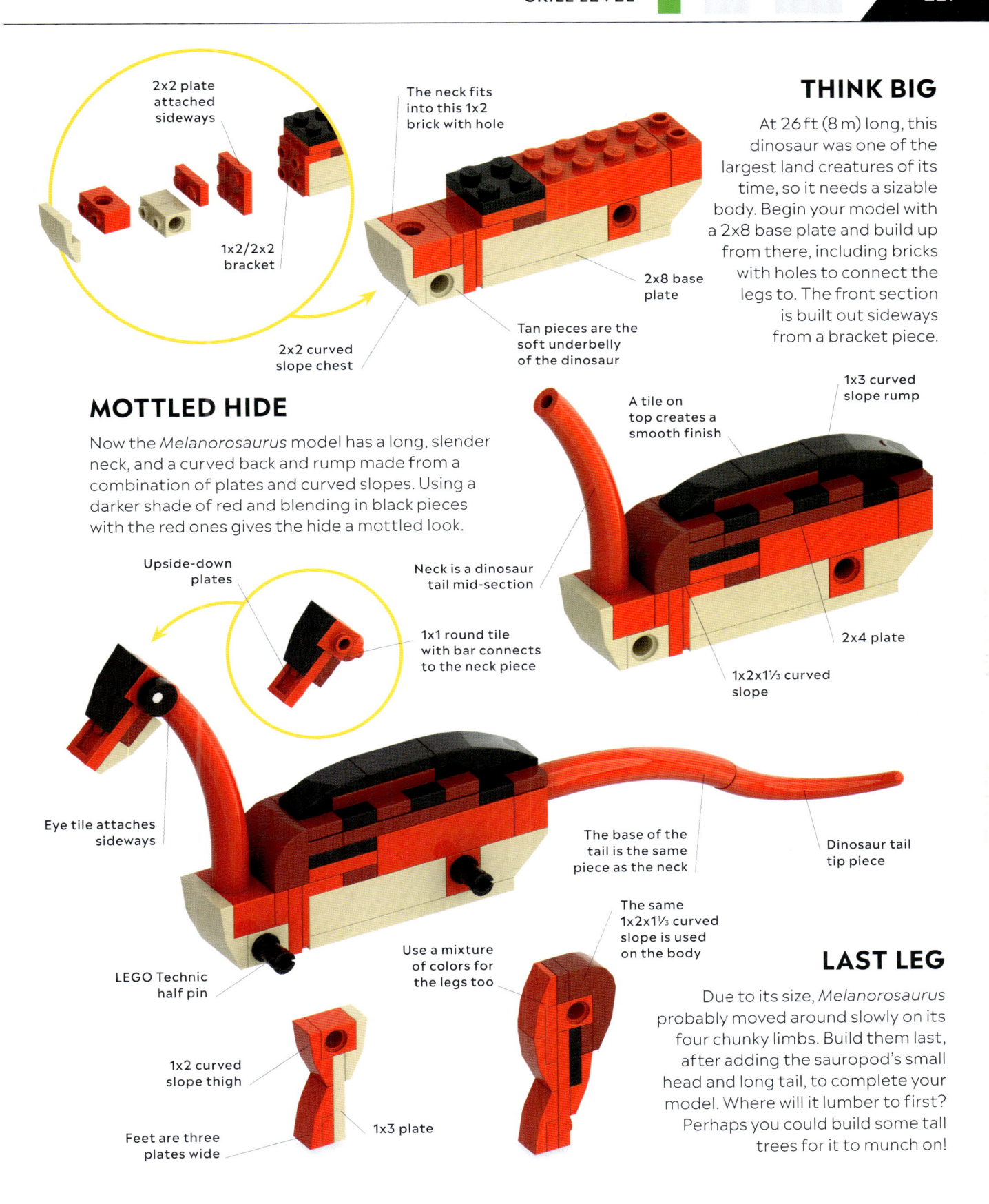

229

THINK BIG

At 26 ft (8 m) long, this dinosaur was one of the largest land creatures of its time, so it needs a sizable body. Begin your model with a 2x8 base plate and build up from there, including bricks with holes to connect the legs to. The front section is built out sideways from a bracket piece.

- 2x2 plate attached sideways
- 1x2/2x2 bracket
- The neck fits into this 1x2 brick with hole
- 2x2 curved slope chest
- 2x8 base plate
- Tan pieces are the soft underbelly of the dinosaur
- A tile on top creates a smooth finish
- 1x3 curved slope rump
- 2x4 plate
- 1x2x1⅓ curved slope

MOTTLED HIDE

Now the *Melanorosaurus* model has a long, slender neck, and a curved back and rump made from a combination of plates and curved slopes. Using a darker shade of red and blending in black pieces with the red ones gives the hide a mottled look.

- Upside-down plates
- Neck is a dinosaur tail mid-section
- 1x1 round tile with bar connects to the neck piece
- Eye tile attaches sideways
- LEGO Technic half pin
- 1x2 curved slope thigh
- Feet are three plates wide
- 1x3 plate
- Use a mixture of colors for the legs too
- The base of the tail is the same piece as the neck
- Dinosaur tail tip piece
- The same 1x2x1⅓ curved slope is used on the body

LAST LEG

Due to its size, *Melanorosaurus* probably moved around slowly on its four chunky limbs. Build them last, after adding the sauropod's small head and long tail, to complete your model. Where will it lumber to first? Perhaps you could build some tall trees for it to munch on!

CLAWS AND TEETH

It's time to take a closer look at the sharpest parts of dinosaurs ... No, not their brains—their teeth and claws! Fossils of these body parts tell us a lot about what they ate, how they ate it, and how they lived. If you want to build realistic-looking dinosaurs, it's a good thing to get the size, shape, and position of their teeth and claws right.

Look more closely at this raptor on pages 232-233

KILLER CLAWS

This *Velociraptor* might look like it needs a pedicure, but that upright toe claw is an important part of its anatomy. Some raptor dinosaurs had giant hooked claws on the second toe of each foot for stabbing and gripping prey.

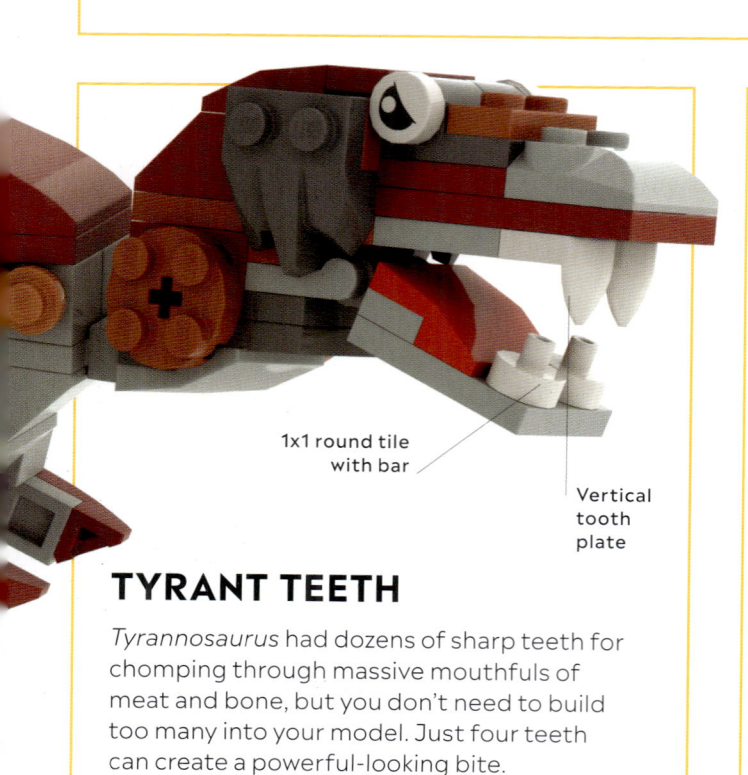

This small horn piece fits into an open knob

1x1 round tile with bar

Vertical tooth plate

TYRANT TEETH

Tyrannosaurus had dozens of sharp teeth for chomping through massive mouthfuls of meat and bone, but you don't need to build too many into your model. Just four teeth can create a powerful-looking bite.

THUMB SPIKE

Some dinosaurs had a long, sharp thumb claw on each hand. This unusual feature, which can be seen on *Iguanodon* and *Baryonyx*, has baffled scientists for a very long time.

Sharp-looking horn piece

Horizontal tooth plate

IDEAS GALLERY   231

SABER TEETH

That's one memorable smile! Or is it a snarl? *Smilodon* was famous for its two elongated canine teeth. This model's are made from two horn pieces. They clip onto plates with clips built into the roof of the mouth. The lower teeth, made from smaller horns, fit neatly behind them when the beast bites down.

Horn pieces have thin ends that fit into clips

This creature actually had four toe claws, but three works fine!

UNDERSIDE VIEW

Slide plates hold the claws in place

Find upright angles of this model on pages 258-259

UNDERSIDE VIEW

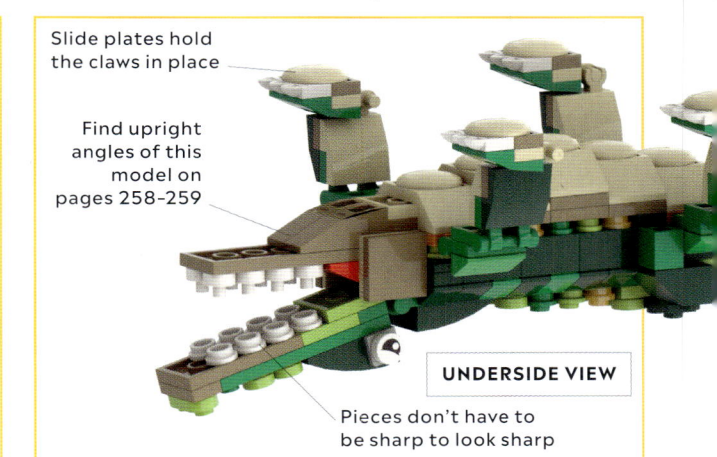

Pieces don't have to be sharp to look sharp

OPEN WIDE

Don't be tempted to give this *Deinosuchus* a tummy tickle—it won't end well! This upside-down angle of the toothy terror shows the rows of round plate teeth in its upper jaws.

TUSKS

Were a mammoth's curved tusks claws or teeth? The answer is ... teeth. They were like huge incisors (front teeth) that grew out from the mammoth's skull, and they could reach up to 13 ft (4 m) in length! On this model, they're made from just two LEGO pieces: a candle and a dinosaur tail tip.

Time to brush your tusks!

This candle piece fits into a clip

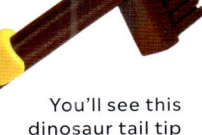

You'll see this dinosaur tail tip a lot in this chapter

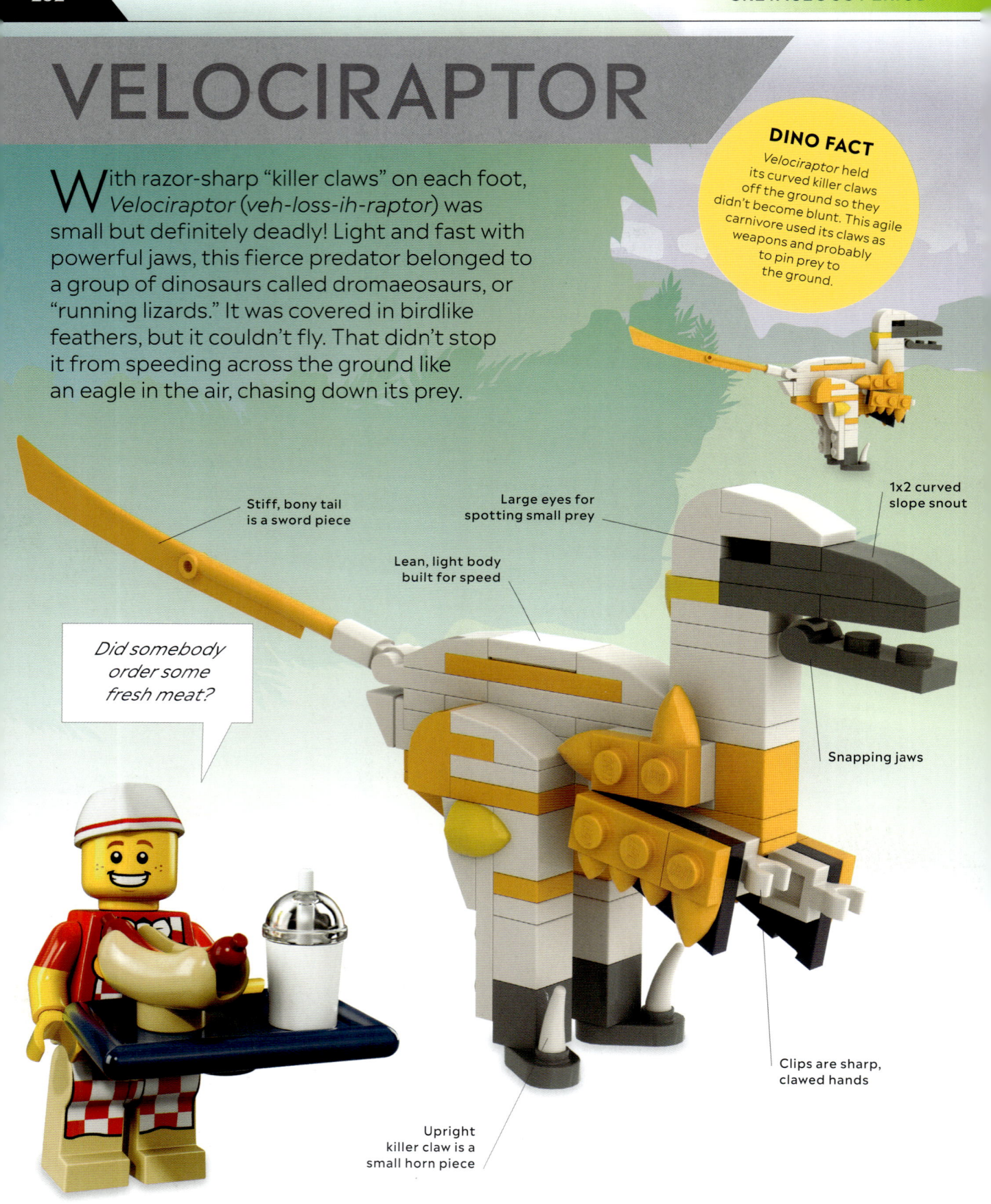

VELOCIRAPTOR

CRETACEOUS PERIOD

With razor-sharp "killer claws" on each foot, *Velociraptor* (veh-loss-ih-raptor) was small but definitely deadly! Light and fast with powerful jaws, this fierce predator belonged to a group of dinosaurs called dromaeosaurs, or "running lizards." It was covered in birdlike feathers, but it couldn't fly. That didn't stop it from speeding across the ground like an eagle in the air, chasing down its prey.

DINO FACT
Velociraptor held its curved killer claws off the ground so they didn't become blunt. This agile carnivore used its claws as weapons and probably to pin prey to the ground.

Stiff, bony tail is a sword piece

Large eyes for spotting small prey

1x2 curved slope snout

Lean, light body built for speed

Snapping jaws

Did somebody order some fresh meat?

Clips are sharp, clawed hands

Upright killer claw is a small horn piece

LIGHTWEIGHT BODY

Velociraptor had a lean body because it needed to be an agile hunter that could run quickly. Begin your model with a narrow body built from lots of small plates and bricks. Include curved slopes at the bottom to give it a rounded shape, and add in tail and arm connections.

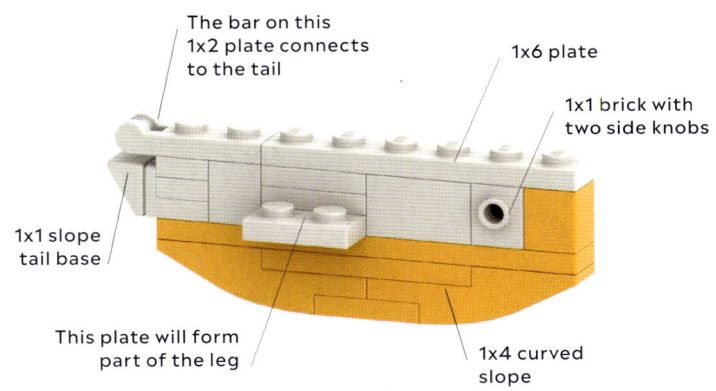

The bar on this 1x2 plate connects to the tail

1x6 plate

1x1 brick with two side knobs

1x1 slope tail base

This plate will form part of the leg

1x4 curved slope

This bar holder with clip is part of the tail

Include tooth plates for a feathered look

Dark gray plates and slopes form the snout

1x2 curved slope

These knobs look like teeth

HUNTING HEAD

Finish off the curved top of the body, then construct the head, snout, and fearsome jaws. They're all built up from the body except for the lower jaw piece, which attaches to a plate with clip in the neck. This connection allows the lower jaw to open and close.

More leg pieces are now in place too

2x3 wedge plate widens the wing

1x1 horizontal tooth plate

Just add the long tail here to finish

1x4 plate connected sideways

WINGED LIMBS

Now this model has legs and its all-important killer claws. Scientists know that *Velociraptor* couldn't fly because its winged arms were too short, but they were filled with long, thick feathers for warmth and for show. Build them using lots of plates with teeth, for a feathery effect.

1x1 vertical tooth plate is more feathers

1x2 rounded plate foot

SKILL LEVEL

233

PARASAUROLOPHUS

CRETACEOUS PERIOD

You might have been able to hear this handsome herbivore coming—scientists think it used its large, tubelike head crest to make loud, booming sounds like a trumpet! *Parasaurolophus* (pa-ra-sore-oh-loaf-us) belonged to a group of dinosaurs called hadrosaurs. Unlike most herbivores, it could walk on two or four legs. *Parasaurolophus* traveled in herds, probably making a racket!

Head crest is a 1x2 curved slope

Small beak for cropping plants

Thinner, shorter forearms

Long, stiff tail helped with balance

Muscular hind legs

Give me some trumpets!

DINO FACT
The head crest of *Parasaurolophus* grew longer as the dinosaur matured. The small crest of a young *Parasaurolophus* would probably have made a different sound than an adult's.

SQUARE START

Begin this *Parasaurolophus* model with a neatly squared-off middle section. It has a 2x6 plate base, with three bricks on top. The bracket piece at the front of the body will provide side knobs for the chest pieces.

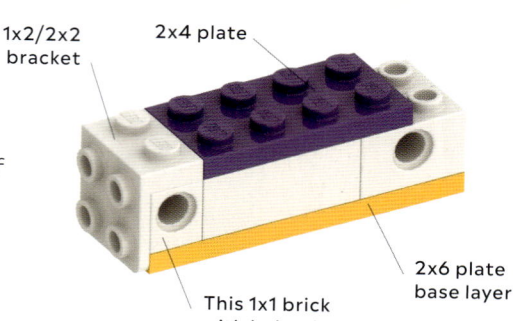

1x2/2x2 bracket

2x4 plate

This 1x1 brick with hole will hold a forearm

2x6 plate base layer

SKILL LEVEL

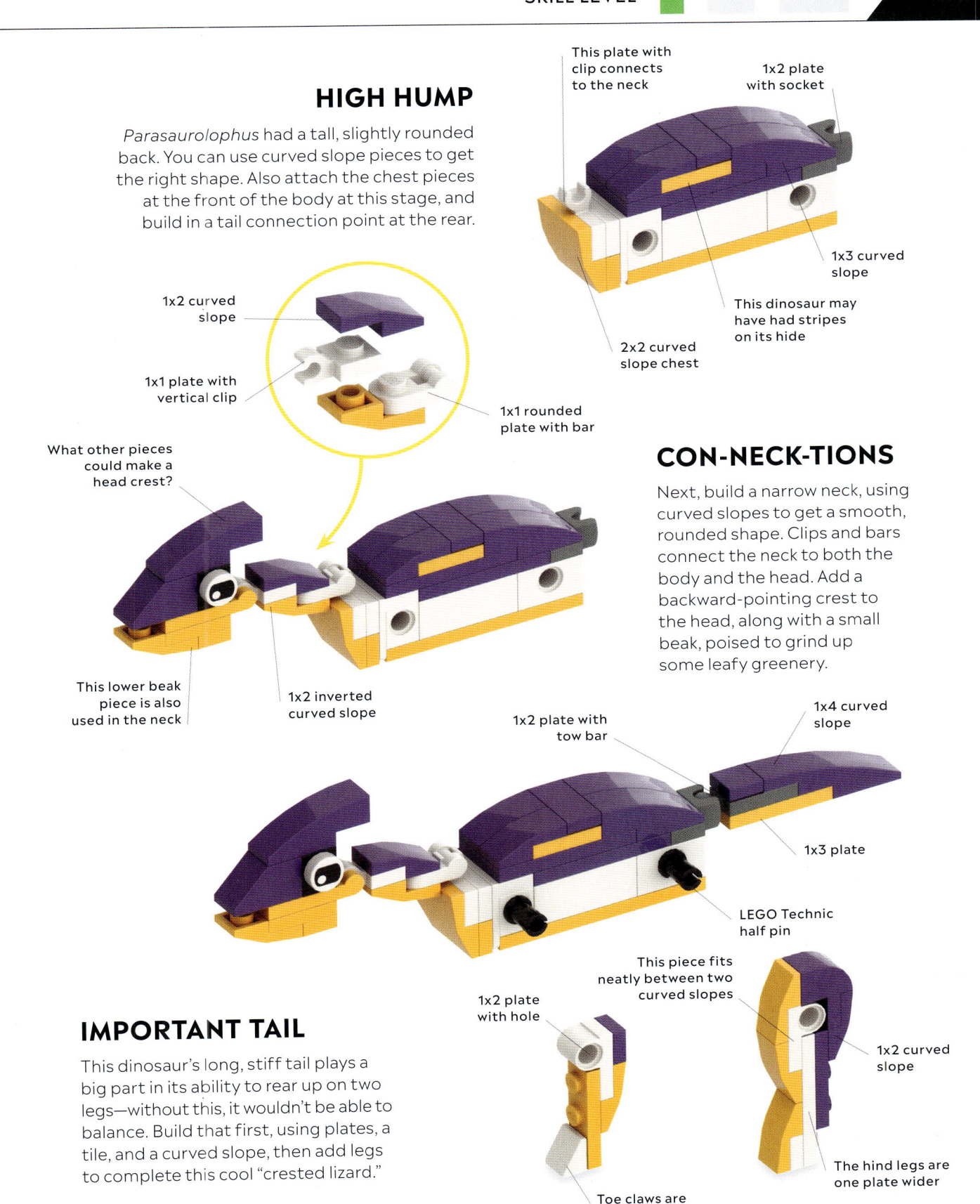

235

HIGH HUMP

Parasaurolophus had a tall, slightly rounded back. You can use curved slope pieces to get the right shape. Also attach the chest pieces at the front of the body at this stage, and build in a tail connection point at the rear.

This plate with clip connects to the neck

1x2 plate with socket

1x3 curved slope

This dinosaur may have had stripes on its hide

2x2 curved slope chest

1x2 curved slope

1x1 plate with vertical clip

1x1 rounded plate with bar

What other pieces could make a head crest?

CON-NECK-TIONS

Next, build a narrow neck, using curved slopes to get a smooth, rounded shape. Clips and bars connect the neck to both the body and the head. Add a backward-pointing crest to the head, along with a small beak, poised to grind up some leafy greenery.

This lower beak piece is also used in the neck

1x2 inverted curved slope

1x2 plate with tow bar

1x4 curved slope

1x3 plate

LEGO Technic half pin

This piece fits neatly between two curved slopes

1x2 plate with hole

1x2 curved slope

IMPORTANT TAIL

This dinosaur's long, stiff tail plays a big part in its ability to rear up on two legs—without this, it wouldn't be able to balance. Build that first, using plates, a tile, and a curved slope, then add legs to complete this cool "crested lizard."

Toe claws are a 1x1 slope

The hind legs are one plate wider

236 | TRIASSIC PERIOD

CYNOGNATHUS

This creature might look a bit like a pet dog, but don't be fooled! *Cynognathus* (*sigh-nog-nay-thus*) was a ferocious predator with wide, powerful jaws built for chomping meat. It was the same size as a wolf, with a heavy body, strong legs, and a short tail. Just like a wolf or dog, scientists think that its body was covered in hair.

- Half circle tile nose
- Stacked plates form the stocky body
- Short tail topped with a 1x3 tile
- Ball connection allows the tail to move around
- Plates widen the thighs of the hind legs
- Wide 2x3 plate lower jaw
- Small leg is a 1x2x1⅓ curved slope
- Paws are 1x1 rounded plates with bars

My goodness, Daphne, is that yours?

FUN FACT
Cynognathus belongs to a group of reptiles called cynodonts. Scientists think they may have been related to modern-day mammals, including humans!

SKILL LEVEL   237

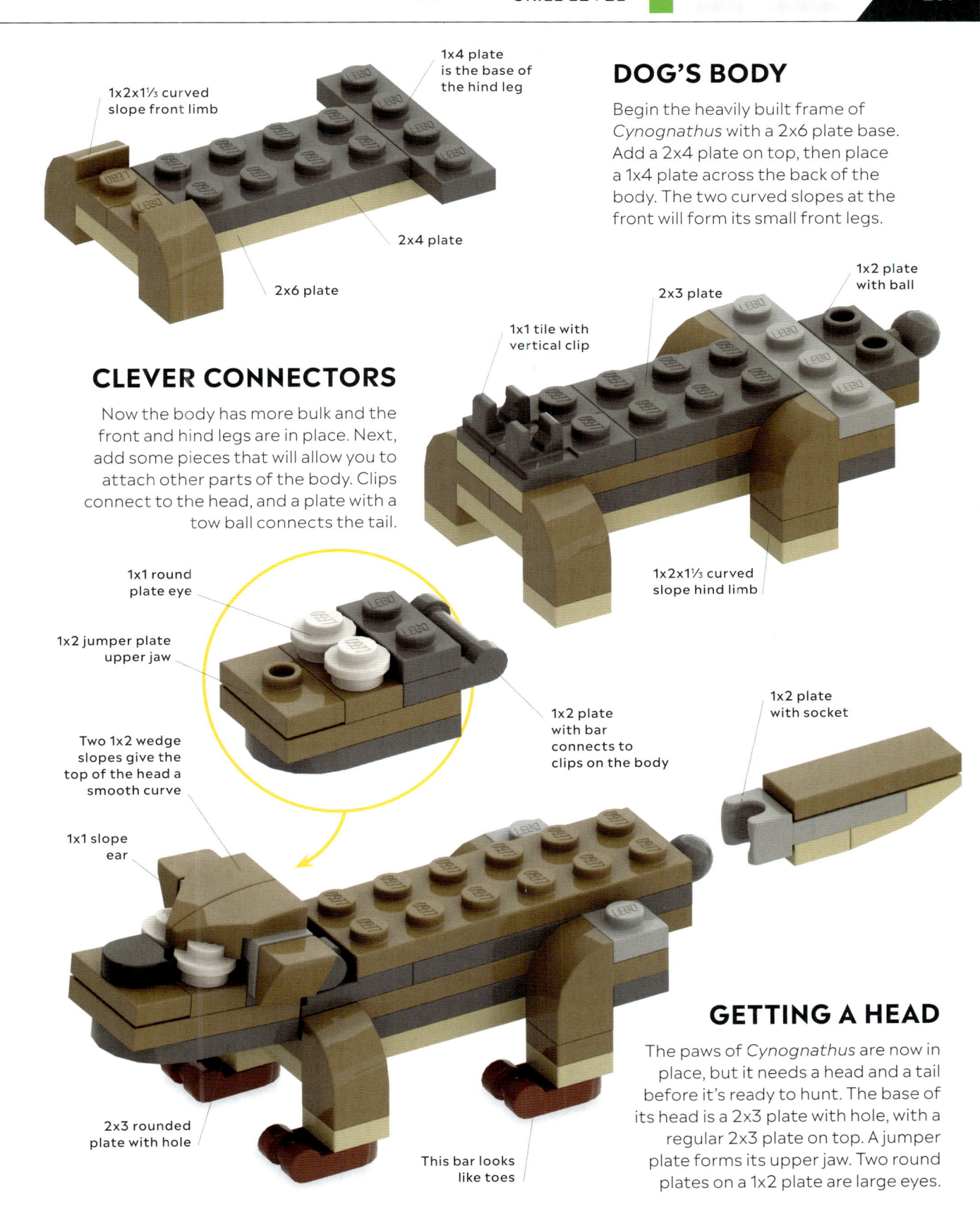

DOG'S BODY

Begin the heavily built frame of *Cynognathus* with a 2x6 plate base. Add a 2x4 plate on top, then place a 1x4 plate across the back of the body. The two curved slopes at the front will form its small front legs.

- 1x2x1⅓ curved slope front limb
- 1x4 plate is the base of the hind leg
- 2x4 plate
- 2x6 plate

CLEVER CONNECTORS

Now the body has more bulk and the front and hind legs are in place. Next, add some pieces that will allow you to attach other parts of the body. Clips connect to the head, and a plate with a tow ball connects the tail.

- 1x1 tile with vertical clip
- 2x3 plate
- 1x2 plate with ball
- 1x2x1⅓ curved slope hind limb
- 1x1 round plate eye
- 1x2 jumper plate upper jaw
- Two 1x2 wedge slopes give the top of the head a smooth curve
- 1x2 plate with bar connects to clips on the body
- 1x2 plate with socket
- 1x1 slope ear
- 2x3 rounded plate with hole
- This bar looks like toes

GETTING A HEAD

The paws of *Cynognathus* are now in place, but it needs a head and a tail before it's ready to hunt. The base of its head is a 2x3 plate with hole, with a regular 2x3 plate on top. A jumper plate forms its upper jaw. Two round plates on a 1x2 plate are large eyes.

CRETACEOUS PERIOD

PROTOCERATOPS

Dinky *Protoceratops* (*pro-toe-seh-rah-tops*) was a bit like a prehistoric sheep. It was about the same size, and it roamed around in large herds for protection from any predators looking for a snack. Like its larger ceratopsian relative *Triceratops* (pages 280-281), it had a neck frill and a parrot-like beak.

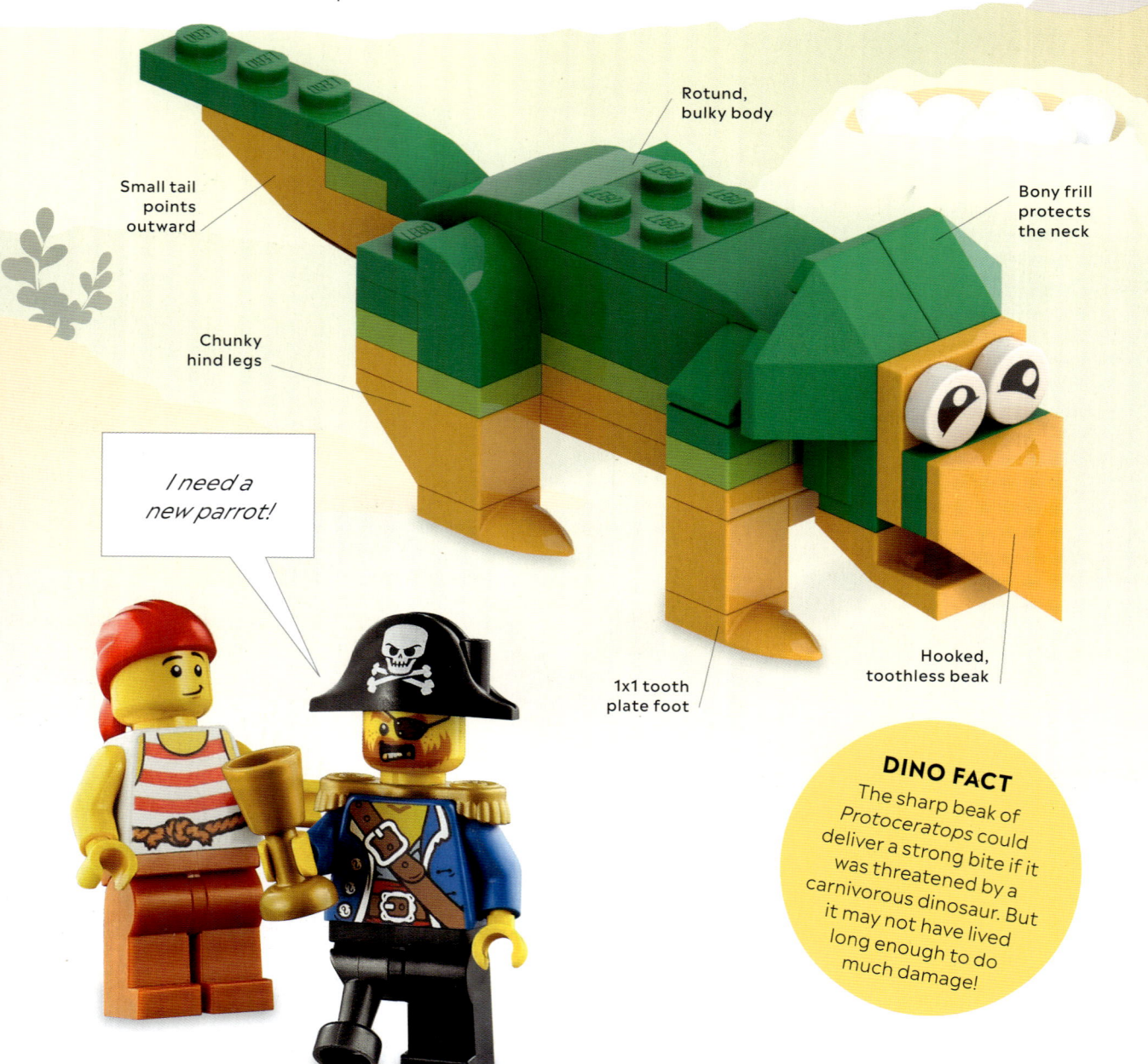

- Small tail points outward
- Rotund, bulky body
- Bony frill protects the neck
- Chunky hind legs
- 1x1 tooth plate foot
- Hooked, toothless beak

I need a new parrot!

DINO FACT
The sharp beak of *Protoceratops* could deliver a strong bite if it was threatened by a carnivorous dinosaur. But it may not have lived long enough to do much damage!

SKILL LEVEL 239

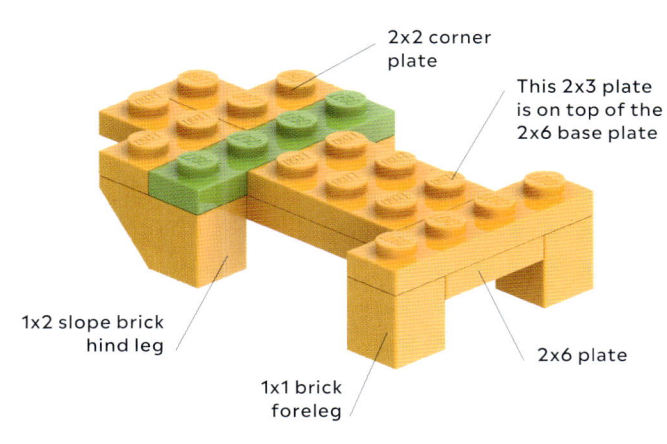

SHORT STUFF

You don't need many pieces to create a pint-size *Protoceratops*. This model's barrel-shaped body is built up from one 2x6 plate, with smaller plates on top. Add small bricks underneath the second plate layer to form the legs.

- 2x2 corner plate
- This 2x3 plate is on top of the 2x6 base plate
- 1x2 slope brick hind leg
- 1x1 brick foreleg
- 2x6 plate

TAIL TO TOE

Now the base of the small tail is attached to the back of the body. It's made from an inverted curved slope, which makes the tail look like it's curving upward. This tiny ceratopsian also has feet, and its legs have more shape.

- 1x4 inverted curved slope
- 1x2x1⅓ curved slope thigh
- 1x1 slope shoulder

FINISHING FRILLS

Smooth off the back and tail of *Protoceratops* with curved slopes, leaving a couple of plates with knobs for texture. Then build its beaked head and neck frill. They attach to the side knobs of a bracket piece on the body.

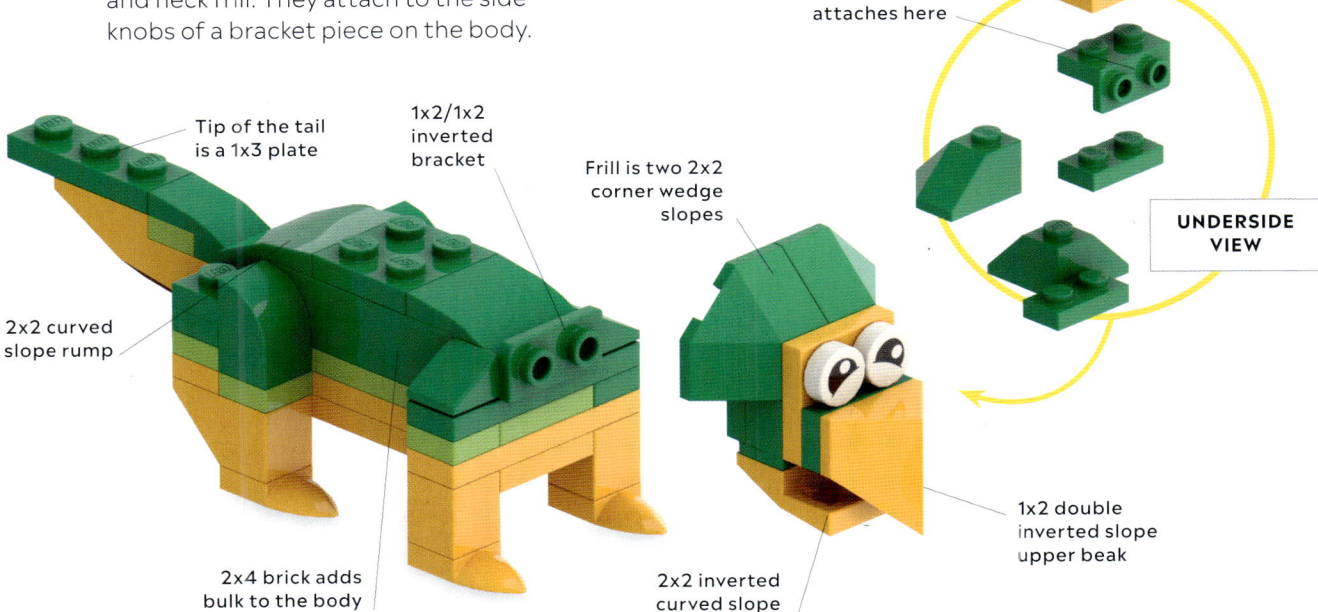

- 2x2 plate
- The lower part of the beak attaches here
- Tip of the tail is a 1x3 plate
- 1x2/1x2 inverted bracket
- Frill is two 2x2 corner wedge slopes
- 2x2 curved slope rump
- 2x4 brick adds bulk to the body
- 2x2 inverted curved slope lower beak
- 1x2 double inverted slope upper beak

UNDERSIDE VIEW

MICRO DINOSAURS

Many dinosaurs are known for their size, but yours don't have to be enormous. Try building LEGO dinosaurs at microscale, which is smaller than minifigure size. Think of the most noticeable physical features of your favorite ones and figure out how to create them using the smallest pieces in your collection. Build their other body parts at the same scale so they look in proportion.

- 1x3 curved slope back
- 1x1 brick with side knobs
- 1x2 plate with clip tail

LITTLE FRILL

This dinosaur's tough neck frill and three horns make it easy to identify. The chunky legs of this microscale *Triceratops* are made from three small pieces, and its tiny tail is just one piece.

- This open knob is the snout horn
- 1x1 slope beak
- 3x3 round corner plate neck frill
- 1x2 curved slope hind leg

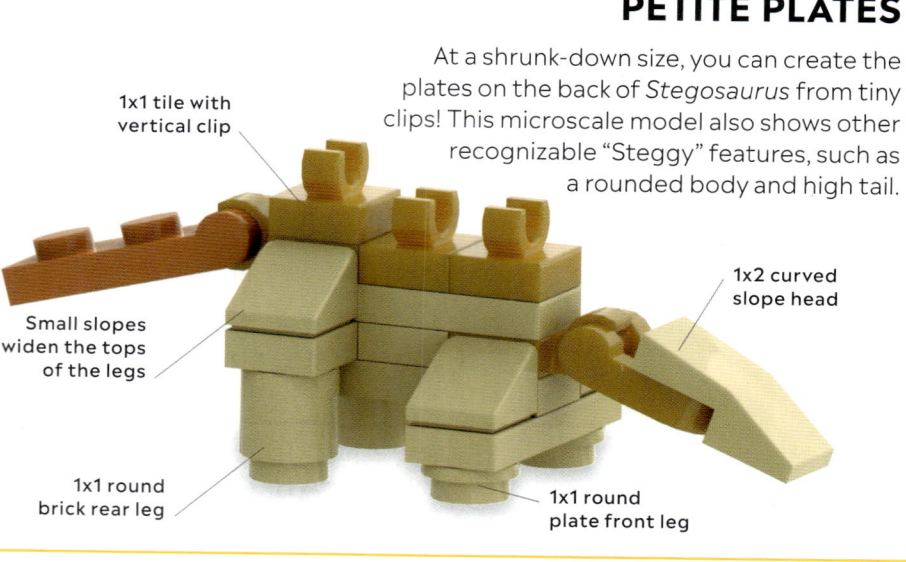

These look small enough to catch!

PETITE PLATES

At a shrunk-down size, you can create the plates on the back of *Stegosaurus* from tiny clips! This microscale model also shows other recognizable "Steggy" features, such as a rounded body and high tail.

- 1x1 tile with vertical clip
- Small slopes widen the tops of the legs
- 1x1 round brick rear leg
- 1x2 curved slope head
- 1x1 round plate front leg

IDEAS GALLERY

241

TINY T. REX

The king of the dinosaurs looks a little less scary at microscale. It's so small, we can't even see its long, sharp teeth! Its distinctive short arms and claws are now even tinier than usual. Despite its size, this T. rex looks poised to run fast after its prey.

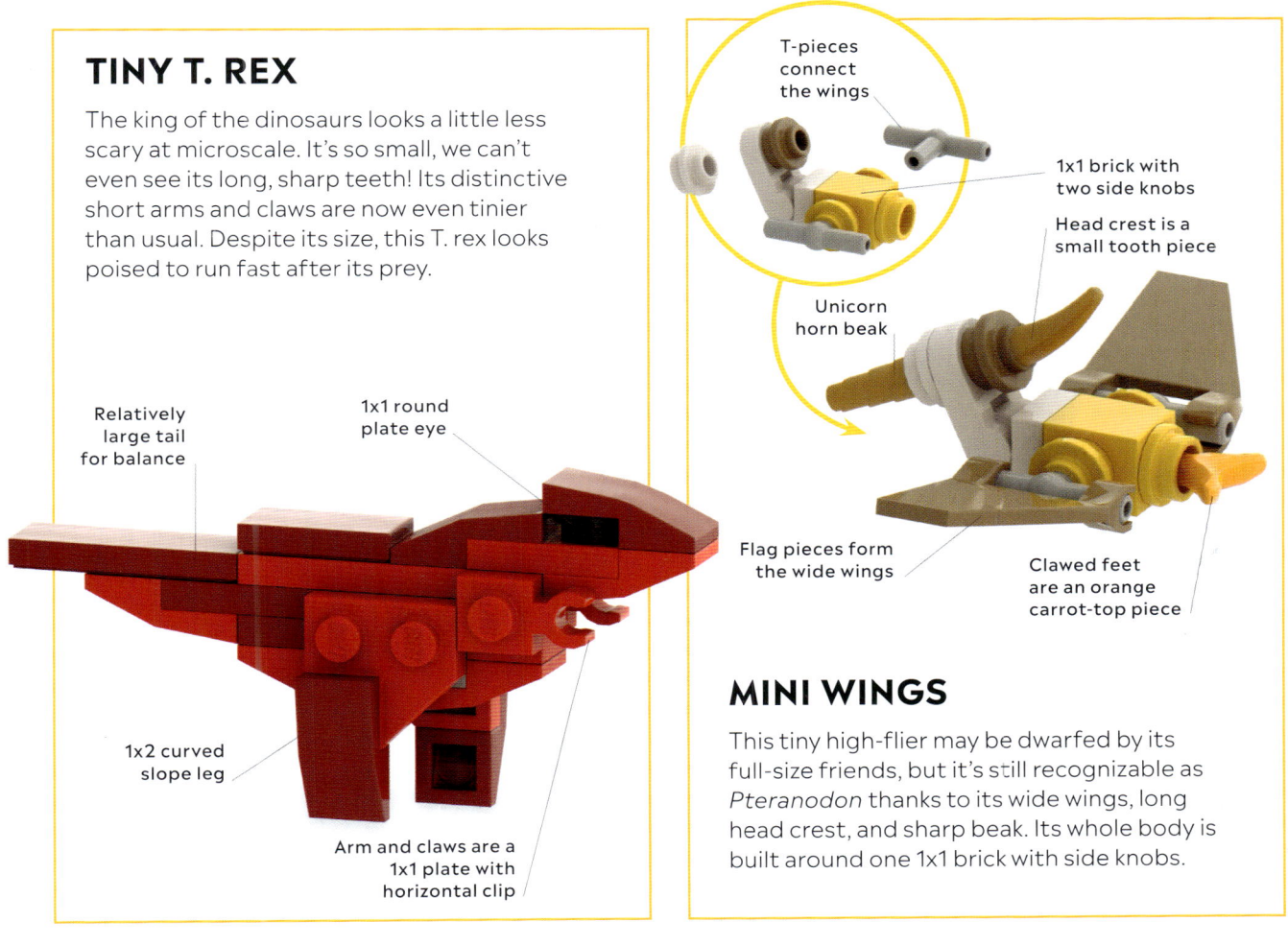

Relatively large tail for balance

1x1 round plate eye

1x2 curved slope leg

Arm and claws are a 1x1 plate with horizontal clip

T-pieces connect the wings

1x1 brick with two side knobs

Head crest is a small tooth piece

Unicorn horn beak

Flag pieces form the wide wings

Clawed feet are an orange carrot-top piece

MINI WINGS

This tiny high-flier may be dwarfed by its full-size friends, but it's still recognizable as *Pteranodon* thanks to its wide wings, long head crest, and sharp beak. Its whole body is built around one 1x1 brick with side knobs.

DINKY DIPPY

One of the longest land animals that ever lived, this dinosaur is known for its very long neck and tail. Both of these physical features are made from the same LEGO piece—a dinosaur tail tip—on this microscale *Diplodocus*.

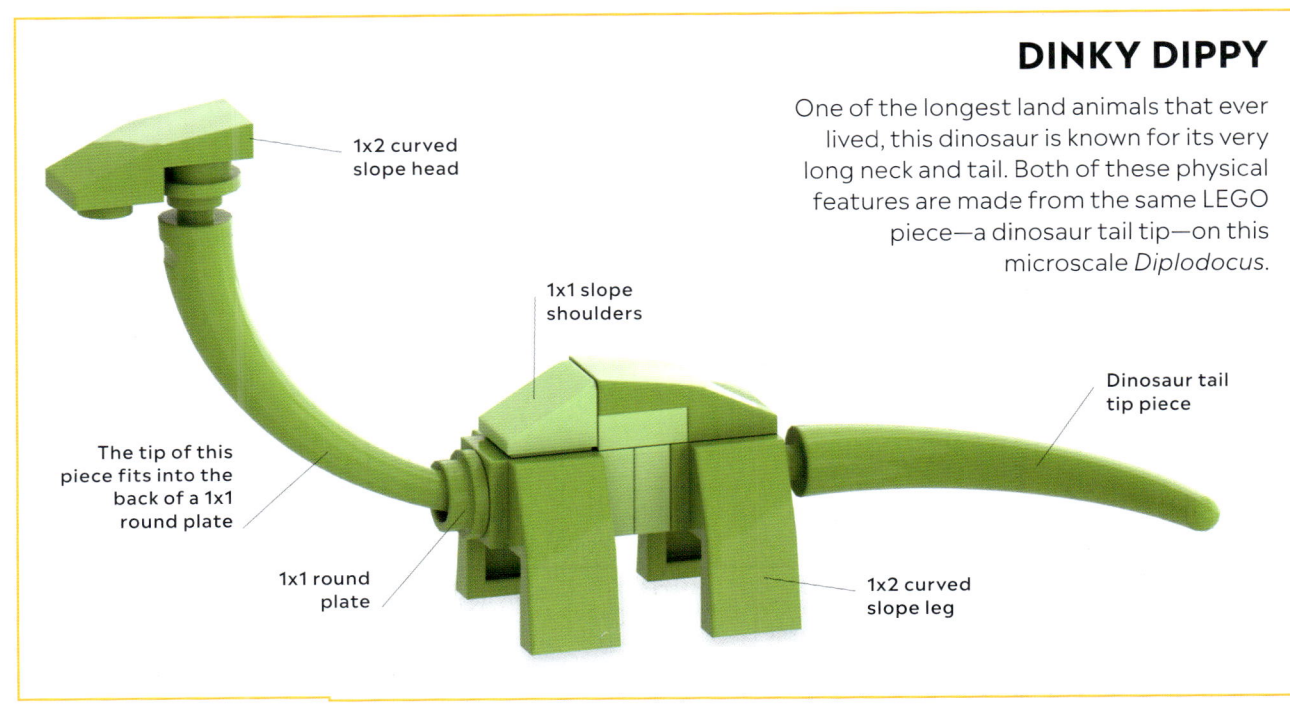

1x2 curved slope head

1x1 slope shoulders

Dinosaur tail tip piece

The tip of this piece fits into the back of a 1x1 round plate

1x1 round plate

1x2 curved slope leg

DOYOUTHINKHESAURUS

If you spot Doyouthinkhesaurus before it spots you ... RUN! With huge, dinner-plate eyes; a dangerous-looking underbite; and a squat, muscular body that looks poised to pounce, this imaginary dinosaur is not to be messed with. Most of the dinosaurs in this chapter are based on those we know existed, but prehistoric creatures designed by you can look however you like!

FUN FACT
Doyouthinkhesaurus looks a bit like a prehistoric frog. The biggest frog that ever lived, the fantastically named Cretaceous Beelzebufo, could be found in swamps at the time of the dinosaurs.

Large eyes on the lookout for passers-by

Smooth, hunched back made from curved slopes

Mottled pink and purple skin

Protruding lower jaw is built sideways

Wide, clawed feet, like a crocodile's

Do you think he saw us?

START SOMEWHERE
When designing an imaginary dinosaur, you can visualize your creature before you begin building, or just make a start and see where your imagination takes you! This model starts off with two layers of plates.

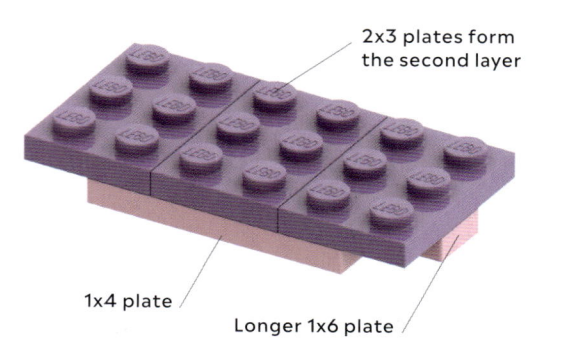

2x3 plates form the second layer

1x4 plate

Longer 1x6 plate

SKILL LEVEL

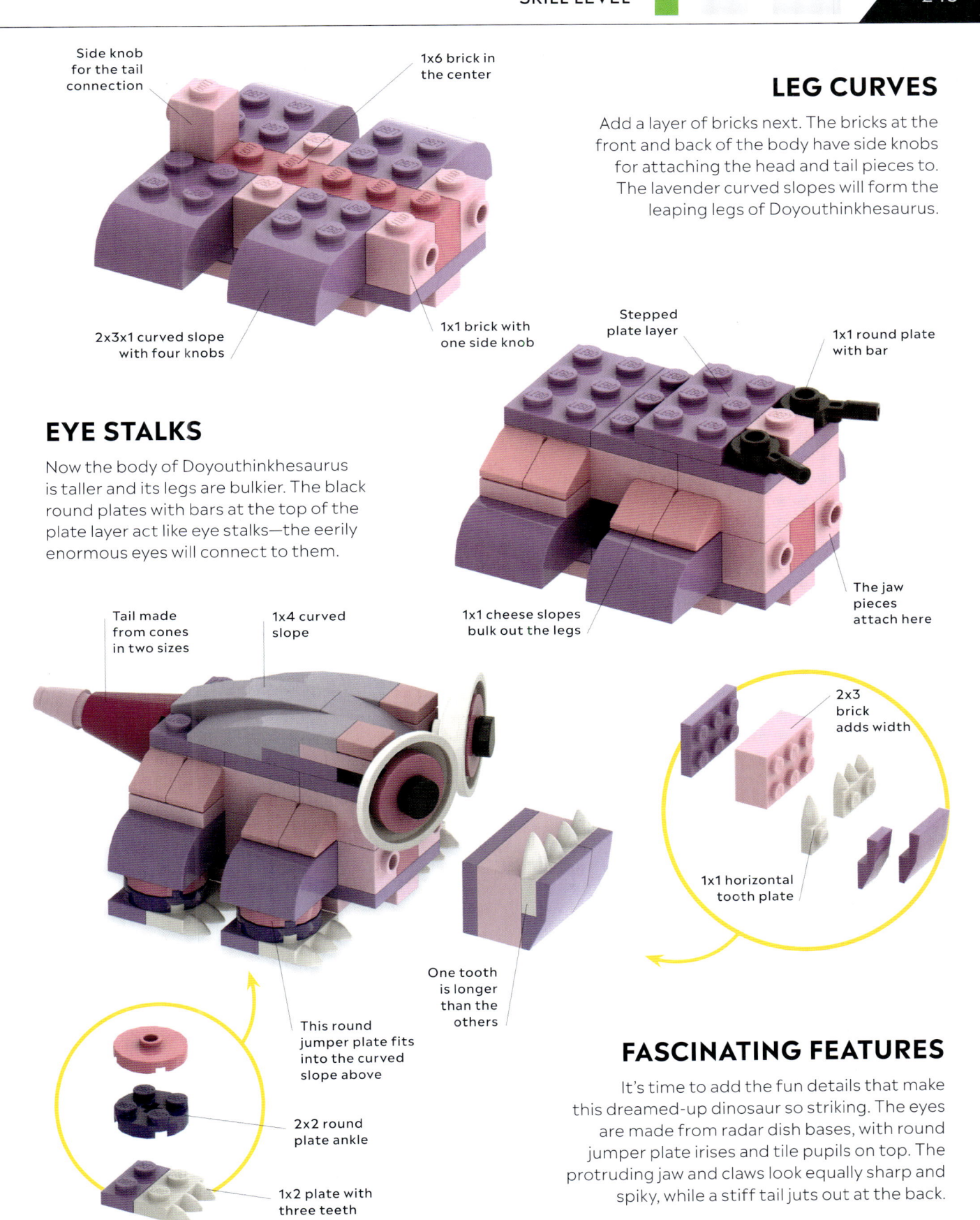

243

- Side knob for the tail connection
- 1x6 brick in the center
- 2x3x1 curved slope with four knobs
- 1x1 brick with one side knob

LEG CURVES

Add a layer of bricks next. The bricks at the front and back of the body have side knobs for attaching the head and tail pieces to. The lavender curved slopes will form the leaping legs of Doyouthinkhesaurus.

- Stepped plate layer
- 1x1 round plate with bar
- 1x1 cheese slopes bulk out the legs
- The jaw pieces attach here

EYE STALKS

Now the body of Doyouthinkhesaurus is taller and its legs are bulkier. The black round plates with bars at the top of the plate layer act like eye stalks—the eerily enormous eyes will connect to them.

- Tail made from cones in two sizes
- 1x4 curved slope
- 2x3 brick adds width
- 1x1 horizontal tooth plate
- One tooth is longer than the others
- This round jumper plate fits into the curved slope above
- 2x2 round plate ankle
- 1x2 plate with three teeth

FASCINATING FEATURES

It's time to add the fun details that make this dreamed-up dinosaur so striking. The eyes are made from radar dish bases, with round jumper plate irises and tile pupils on top. The protruding jaw and claws look equally sharp and spiky, while a stiff tail juts out at the back.

BARYONYX

CRETACEOUS PERIOD

Imagine a creature with a crocodile-shaped skull; huge, hooklike claws; more teeth than a T. rex; and the power to run on two legs ... Meet *Baryonyx* (*bah-ree-on-icks*), a fish-eating theropod from the Cretaceous period. Scientists think that it used its thumb claws to catch and slash slippery prey.

DINO FACT
Although *Baryonyx* was built for catching and eating fish, it did dine on land-based creatures as well. One *Baryonyx* fossil was found with the bones of a baby *Iguanodon* (pages 266–267) inside its stomach!

- Bony head crest for display
- Long jaws for grabbing fish
- Extra-sharp thumb claw
- This clip is two smaller claws
- Thick, powerful arms made from plates and a curved slope
- Muscular tail for balance
- Pointy toe claws

It's strange—this is the first fish I've caught in days!

FORWARD PLANNING
Like most theropod dinosaurs, *Baryonyx* balanced its whole body weight on two strong legs. When building this first part of your model, think about where you want to attach the two legs. Build in sturdy connecting bricks there, such as bricks with holes.

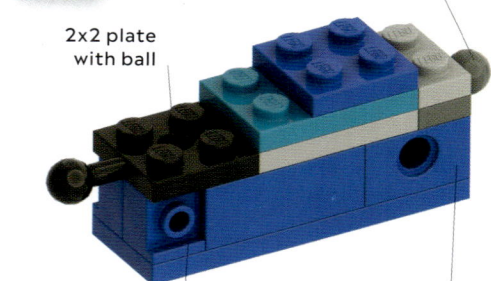

- 2x2 plate with ball
- Plate with ball tail connector
- The arm will attach to this headlight brick
- 1x2 brick with hole leg connector

SKILL LEVEL

245

BALANCING OUT

Without its long, powerful tail, *Baryonyx* would have toppled over on its two legs! This model's tail is connected by plates with balls and sockets, which allow it to swish from side to side as *Baryonyx* strides around.

- Smooth, rounded back
- The tail colors match those on the body
- 1x2 curved slope is a tapering tail tip
- This clever 1x2 plate has a ball at one end and a socket at the other
- LEGO Technic pin for the leg
- Light gray pieces create stripes on the body

POWERFUL LIMBS

Now that the tail is in place and the smooth, heavy body is complete, it's time to give your model some strong limbs. The arms are made from eight small pieces. They attach to the body via a mechanical claw piece. The legs, which are mostly built sideways, have wide, clawed feet at the bottom.

- This part of the mechanical claw slots into the body
- Rounded shin made from curved slopes
- 1x2 wedge slope (right)
- 1x2 wedge slope (left)
- Thumb claw is a small horn piece
- 1x2 slope with cutout
- 2x2 round plate base

LEFT ARM

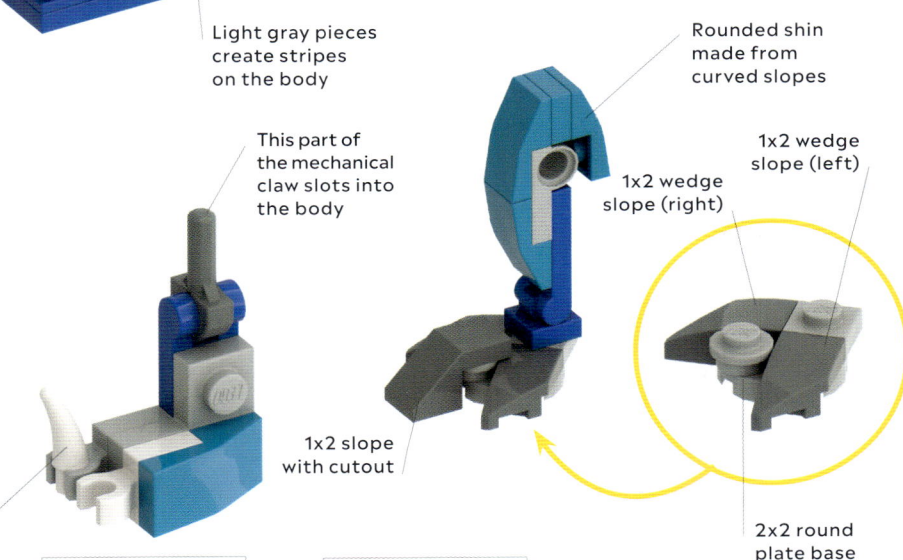

LEFT LEG

- The eye and head crest sit on top of this bracket
- This mechanical claw acts like a clip
- This piece connects to the neck
- This brick has knobs on all of its sides

SNAP, SNAP

All this model needs now is a toothy grin! *Baryonyx* had 96 teeth in its long, narrow jaws—that's twice as many as most other theropods, including *Tyrannosaurus* (pages 282–283). The lower jaw can snap shut via a clip and bar connection when the dinosaur finds some prey.

CRETACEOUS PERIOD

PACHYCEPHALOSAURUS

Look out! This big-headed dinosaur may have used its extremely thick skull to ram into competitors and predators—and it looks poised to attack! *Pachycephalosaurus* (pack-ee-sef-ah-low-sore-us), whose name translates as "thick-headed lizard," was a relative of *Triceratops* (pages 280–281).

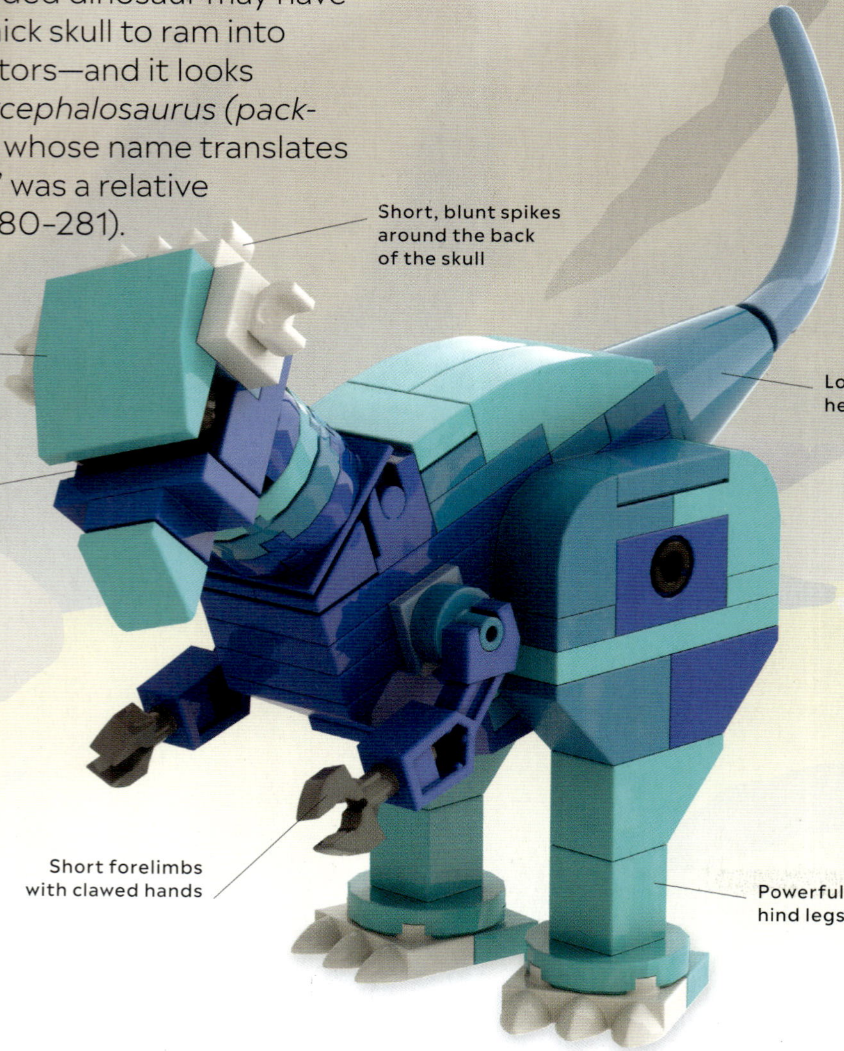

- Short, blunt spikes around the back of the skull
- Thick skull roof is a 2x2 curved slope
- Large 1x1 round plate eyes
- Long, heavy tail
- Short forelimbs with clawed hands
- Powerful hind legs

DINO FACT
Scientists once thought *Pachycephalosaurus* was herbivorous, but recent research shows that it may have eaten both plants and small animals because it had sharper teeth than most herbivores.

Why does that dinosaur look familiar?

BIPEDAL BODY

Begin your *Pachycephalosaurus* model with a round tummy made from inverted curved slopes and a plate. This dinosaur was bipedal, which means it walked on two strong legs, so add bricks with holes that will connect to the two legs on top.

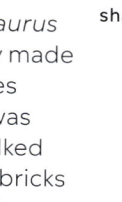

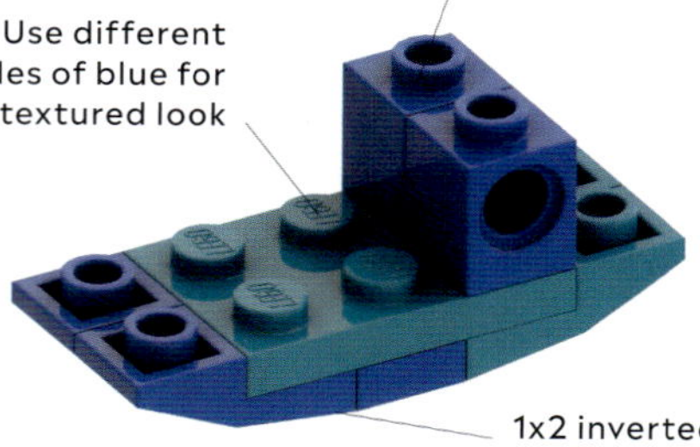

- Use different shades of blue for a textured look
- 1x1 brick with hole will hold a connecting pin
- 1x2 inverted curved slope

SKILL LEVEL

VERTICAL NECK

Now this dinosaur has a bulky, rotund body, mostly made from small bricks and plates. At the front of the body, there's a hinge brick and plate connection that holds the short, thick neck at a vertical angle. The back of the hinge plate rests on slope bricks to make it stable.

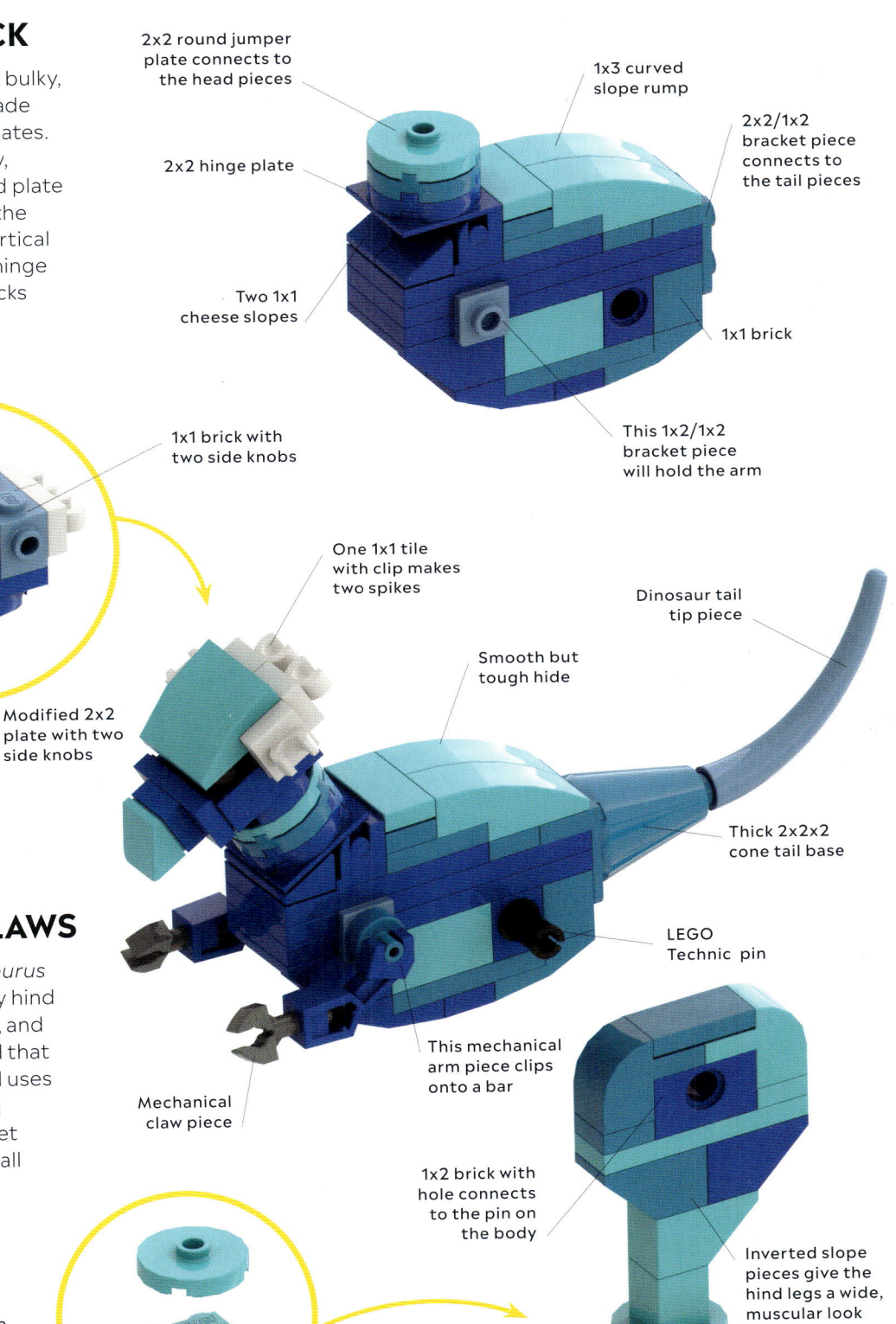

2x2 round jumper plate connects to the head pieces

2x2 hinge plate

Two 1x1 cheese slopes

1x3 curved slope rump

2x2/1x2 bracket piece connects to the tail pieces

1x1 brick

This 1x2/1x2 bracket piece will hold the arm

1x1 brick with two side knobs

The snout and jaws connect sideways

Modified 2x2 plate with two side knobs

One 1x1 tile with clip makes two spikes

Smooth but tough hide

Dinosaur tail tip piece

Thick 2x2x2 cone tail base

LEGO Technic pin

SPIKES AND CLAWS

Get this *Pachycephalosaurus* moving by building hefty hind legs and small forelimbs, and add the bony, spiky head that it's famous for. The head uses lots of sideways building techniques in order to get lots of detail out of a small number of pieces.

Mechanical claw piece

This mechanical arm piece clips onto a bar

1x2 brick with hole connects to the pin on the body

Inverted slope pieces give the hind legs a wide, muscular look

Toe claws are a 1x2 plate with three teeth

1x2 plate is the heel of the foot

247

STEGOSAURUS

JURASSIC PERIOD

The famous *Stegosaurus* (steg-oh-sore-us) was a dinosaur not to be messed with. This placid plant-eater was no fierce predator, but it had some impressive-looking body armor, including a double row of hard plates on its back and a spiked tail. Just the sight of it might have been enough to scare away would-be attackers!

DINO FACT
Even though the large plates on the back of *Stegosaurus* gave the impression they could do serious damage, scientists think they were mainly used for display.

- Brightly colored back plate
- Bricks with side knobs hold up the plates
- Sharp beak is a vertical tooth plate
- Dangerous tail spikes
- Taller back legs
- Round tiles attached sideways look like spotted skin
- 1x1 slope toes

This Steggy selfie is going to get A LOT of likes.

- 1x2 plate with clip
- 2x3 plate
- 1x2 curved slope

ELEPHANT SIZE

A fully grown *Stegosaurus* was as large as an elephant, so begin your build with a broad belly base. This model's is made from three layers of plates and curved slopes.

BIG PLANS

The body of the *Stegosaurus* model is now taller and there are connection points for the neck, legs, and tail. Building in connecting pieces at this early stage helps you plan out your model's proportions.

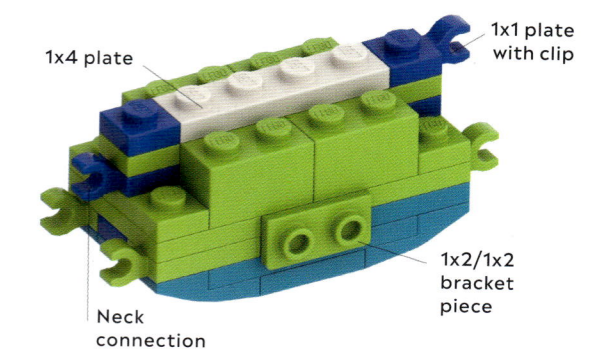

- 1x4 plate
- 1x1 plate with clip
- 1x2/1x2 bracket piece
- Neck connection

SKILL LEVEL 249

BODY COMPLETE

Round off the chunky body of the *Stegosaurus* using curved slopes and tiles. The two jumper plates and small plate at the top of the body will act as a base for the two rows of spiky back plates.

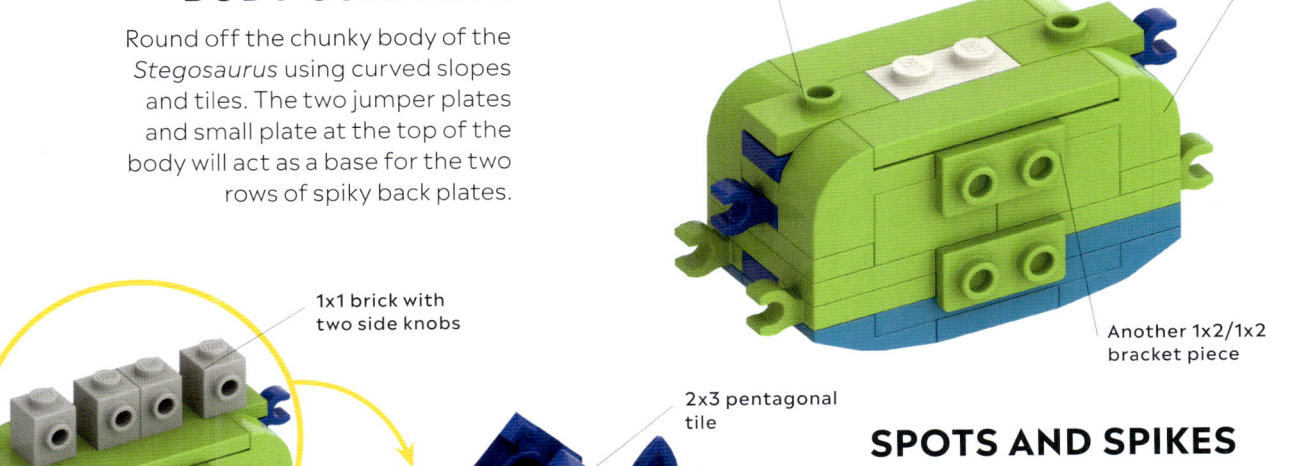

1x2 jumper plate

1x2x1⅓ curved slope

Another 1x2/1x2 bracket piece

1x1 brick with two side knobs

2x3 pentagonal tile

SPOTS AND SPIKES

Attach round tiles to the brackets on the sides of the body to make the body look wider and mottled. Then it's time to build the well-known plates of *Stegosaurus*. This model's plates are attached using a sideways building technique.

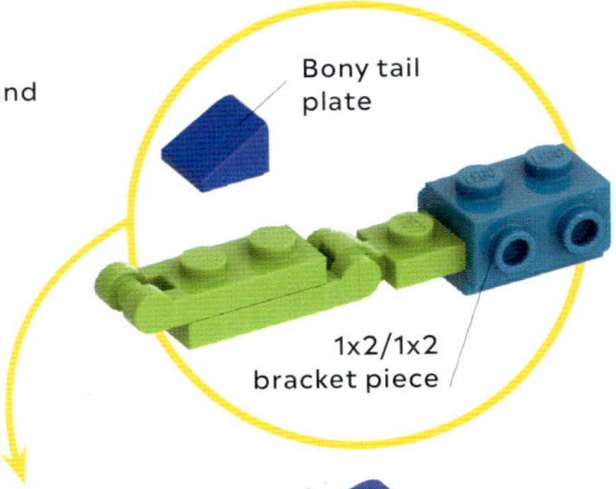

1x1 tooth plate

Alternate plate colors in places to make more spots on the skin

2x2 round tile

1x1 round tile

Bony tail plate

1x2/1x2 bracket piece

DEFENSE MECHANISM

Finish off your model with a small, beaked head and four legs. The hind legs are longer than the front ones so this dinosaur can graze on plants that are low to the ground. Also give your "Steggy" a long, lashing tail with four deadly spikes at the end for defense.

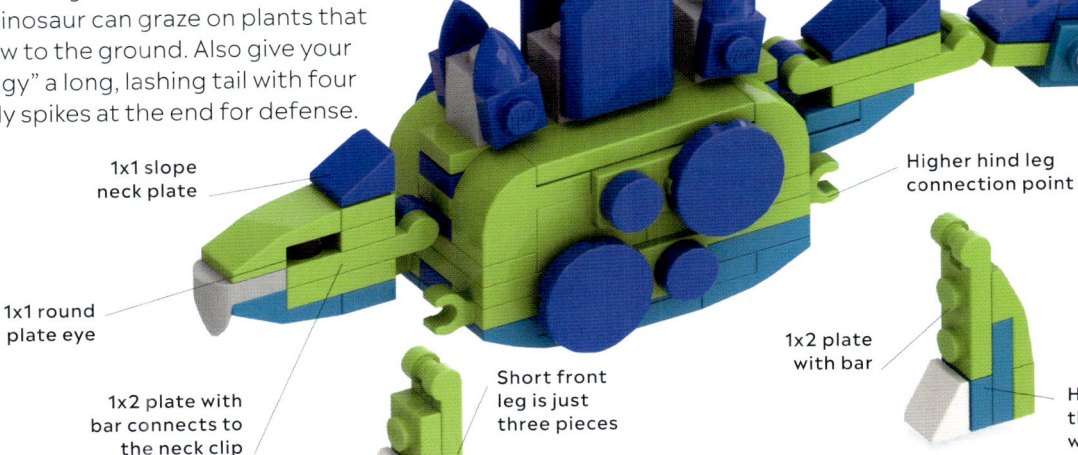

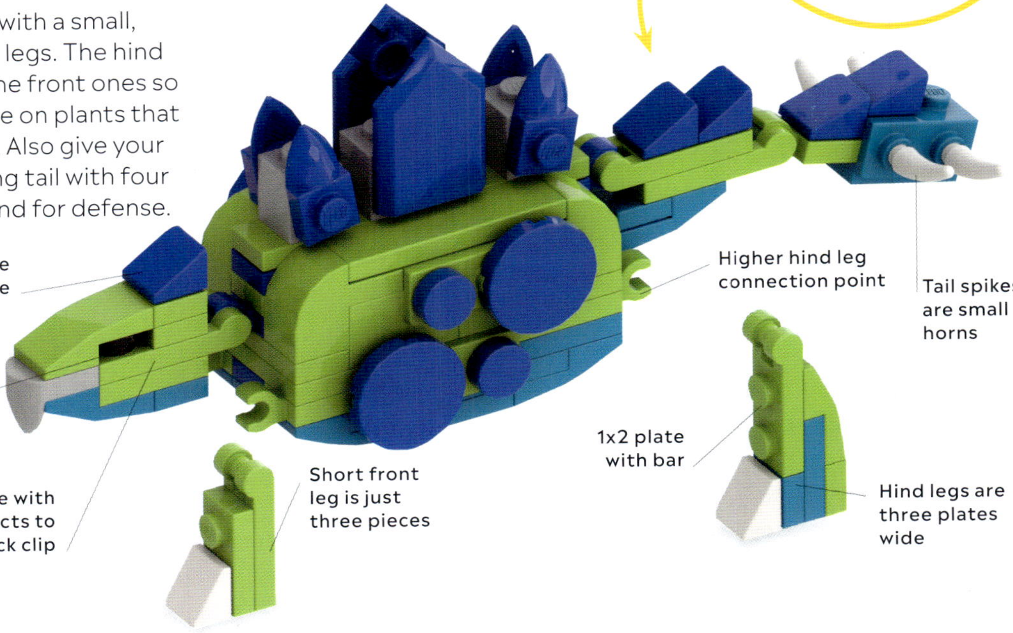

1x1 slope neck plate

1x1 round plate eye

1x2 plate with bar connects to the neck clip

Short front leg is just three pieces

Higher hind leg connection point

Tail spikes are small horns

1x2 plate with bar

Hind legs are three plates wide

HAPPYOSAURUS

This imaginary colorful creature is sure to bring joy and good cheer to any prehistoric landscape! Gather up a bunch of the brightest pieces in your LEGO collection and see if you can put them together to make a dinosaur. Be inspired by the real dinosaurs in this chapter or build something completely new. It's up to you!

DINO FACT
Scientists think that a tiny dinosaur called Caihong had rainbow feathers that shimmered in the light. Its name means "rainbow" in Mandarin.

- Smiling eyes
- Flexible neck made from movable connections
- Rounded snout
- Sparkly rump
- Slide plates make a soft-looking tummy
- Stripy ankles—or are they socks?
- Chunky blue feet
- Dinosaurs that walk on two legs usually have a long tail for balance

Ah, a fellow fan of clashing colors!

BACKFLIP

This Happyosaurus model starts off with a gray plate that sits right in the middle of the body. Lay this plate flat to begin with and build layers of plates and bricks on top. These will form the back of the dinosaur.

- 1x3 curved slope shoulders
- The neck will attach here
- 4x6 base plate
- This clip will hold an arm
- Layers of plates and bricks widen the body
- The legs will attach to these wider plates

SKILL LEVEL

CUDDLY BELLY

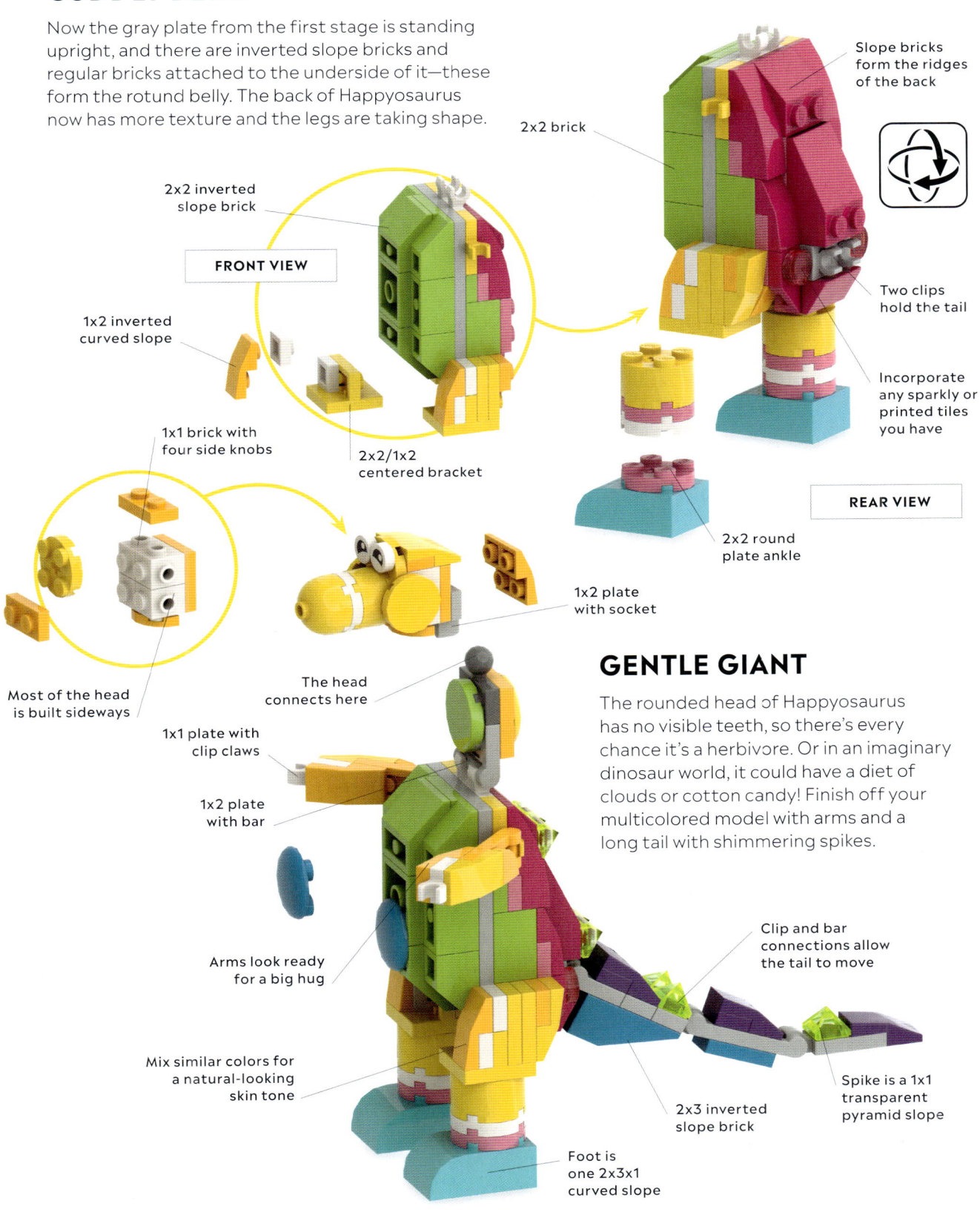

Now the gray plate from the first stage is standing upright, and there are inverted slope bricks and regular bricks attached to the underside of it—these form the rotund belly. The back of Happyosaurus now has more texture and the legs are taking shape.

- Slope bricks form the ridges of the back
- 2x2 brick
- 2x2 inverted slope brick
- FRONT VIEW
- 1x2 inverted curved slope
- Two clips hold the tail
- Incorporate any sparkly or printed tiles you have
- 1x1 brick with four side knobs
- 2x2/1x2 centered bracket
- REAR VIEW
- 2x2 round plate ankle
- 1x2 plate with socket
- Most of the head is built sideways
- The head connects here
- 1x1 plate with clip claws
- 1x2 plate with bar

GENTLE GIANT

The rounded head of Happyosaurus has no visible teeth, so there's every chance it's a herbivore. Or in an imaginary dinosaur world, it could have a diet of clouds or cotton candy! Finish off your multicolored model with arms and a long tail with shimmering spikes.

- Arms look ready for a big hug
- Clip and bar connections allow the tail to move
- Mix similar colors for a natural-looking skin tone
- 2x3 inverted slope brick
- Spike is a 1x1 transparent pyramid slope
- Foot is one 2x3x1 curved slope

JURASSIC PERIOD

ALLOSAURUS

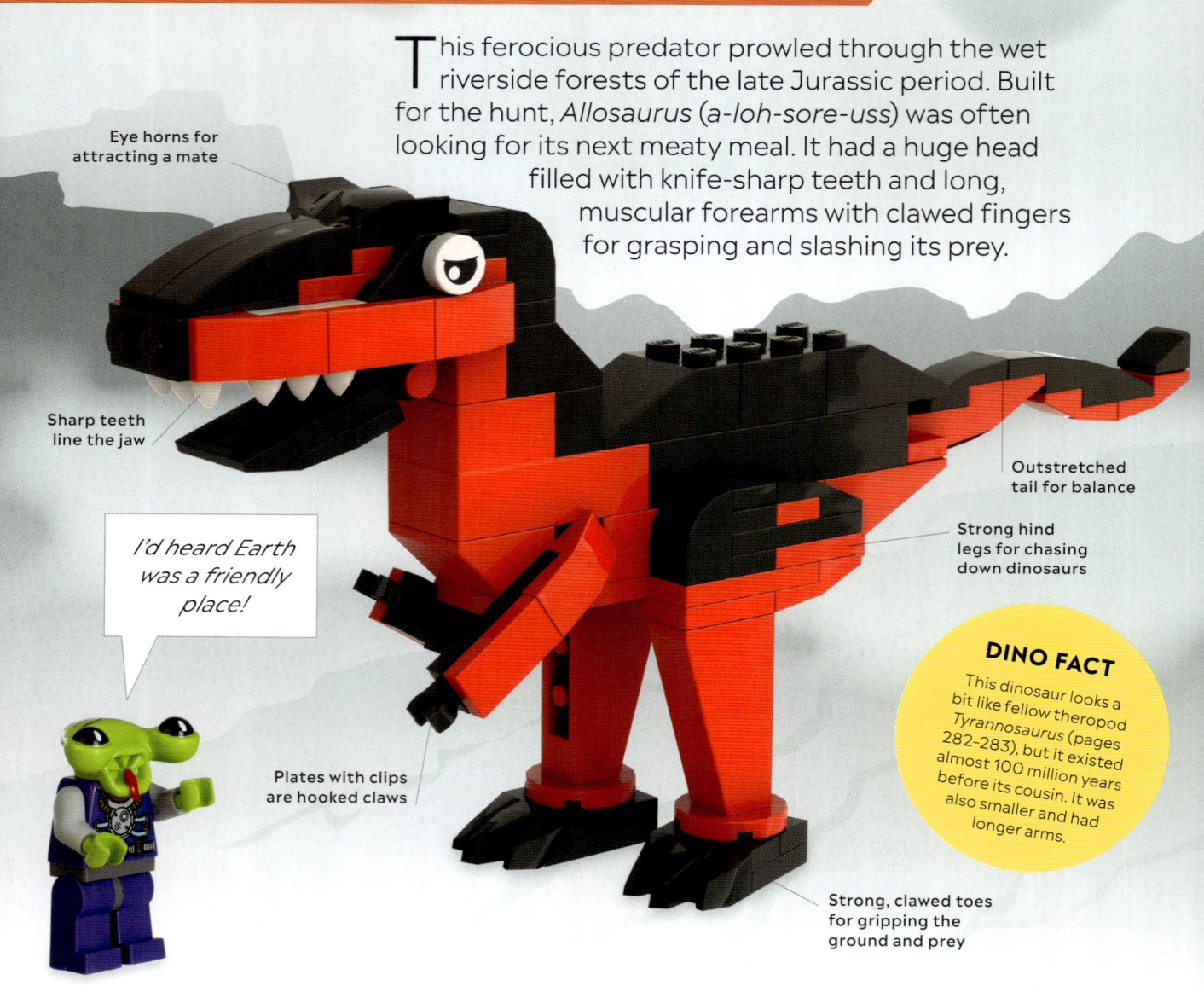

This ferocious predator prowled through the wet riverside forests of the late Jurassic period. Built for the hunt, *Allosaurus* (a-loh-sore-uss) was often looking for its next meaty meal. It had a huge head filled with knife-sharp teeth and long, muscular forearms with clawed fingers for grasping and slashing its prey.

Eye horns for attracting a mate

Sharp teeth line the jaw

Outstretched tail for balance

Strong hind legs for chasing down dinosaurs

I'd heard Earth was a friendly place!

Plates with clips are hooked claws

DINO FACT
This dinosaur looks a bit like fellow theropod *Tyrannosaurus* (pages 282–283), but it existed almost 100 million years before its cousin. It was also smaller and had longer arms.

Strong, clawed toes for gripping the ground and prey

NOT-SO-SCARY START

Even bloodcurdling carnivores like *Allosaurus* have a softer side! Build this model's underbelly and begin the smooth thighs before moving on to the more intimidating body parts. Start with a 2x8 plate and add smaller plates on top and below.

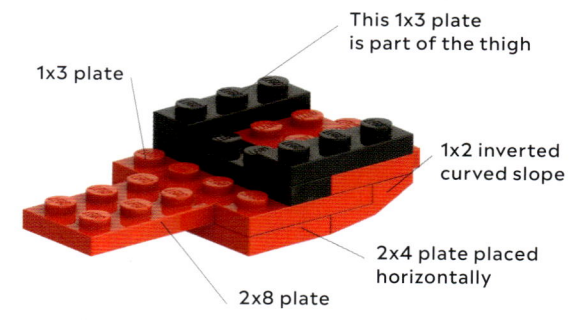

This 1x3 plate is part of the thigh

1x3 plate

1x2 inverted curved slope

2x4 plate placed horizontally

2x8 plate

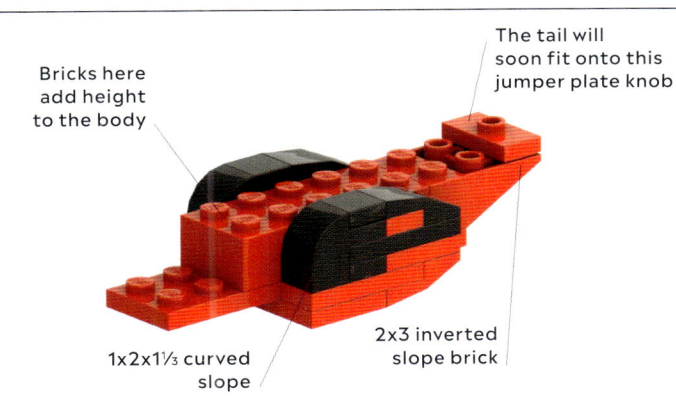

SKILL LEVEL

TAPERED SHAPE

The rear of this dinosaur's body tapers down toward its tail. This model uses an inverted slope brick to create the right shape. The tops of the thighs are now in place too, and the middle of the body is taller.

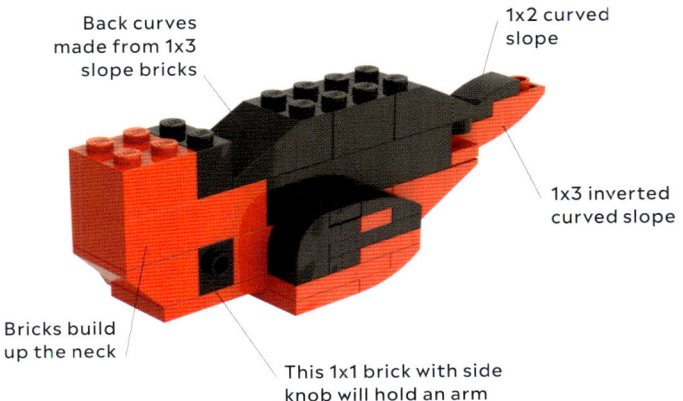

- Bricks here add height to the body
- The tail will soon fit onto this jumper plate knob
- 1x2x1⅓ curved slope
- 2x3 inverted slope brick
- Back curves made from 1x3 slope bricks
- 1x2 curved slope
- 1x3 inverted curved slope
- Bricks build up the neck
- This 1x1 brick with side knob will hold an arm

BIGGER THREAT

Lengthen the *Allosaurus* body by beginning the neck and tail. Many dinosaur models use connecting pieces, such as clips and bars or tow balls, to attach these parts, but this time they're built right into the body of the beast.

- Eye horns are 1x2 curved slopes with bowed ends
- 1x2 plate with handled bar
- 1x2 plate nostrils
- A tooth plate attaches to this 1x1 brick with side knob
- The lower jaw is built upside down

BUILD A BITE

It's time to build the mighty head of this hunter. Include pieces with side knobs in the upper jaws to attach teeth and face details to. There's also a plate with bar built into the back of the head—the lower jaw clips onto this.

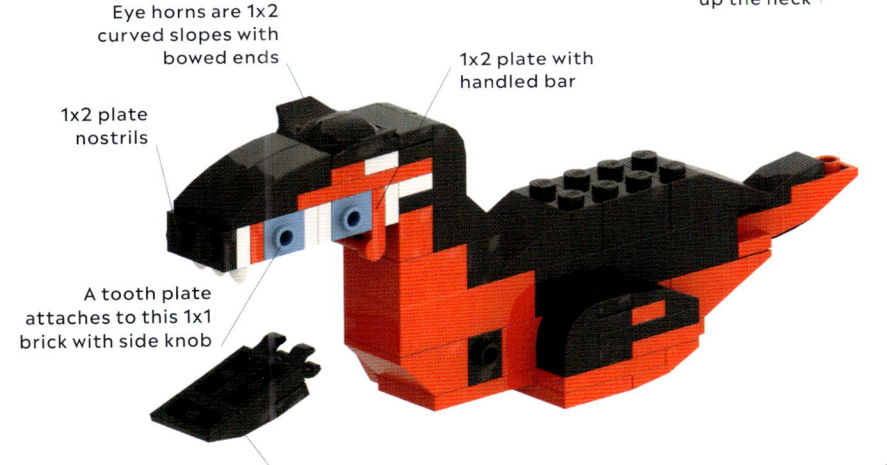

FEARSOME FEATURES

Allosaurus used its clawed feet as well as its hooked hand claws to pin down prey. Add both, along with more face details and the tip of the tail to top off this toothy terror!

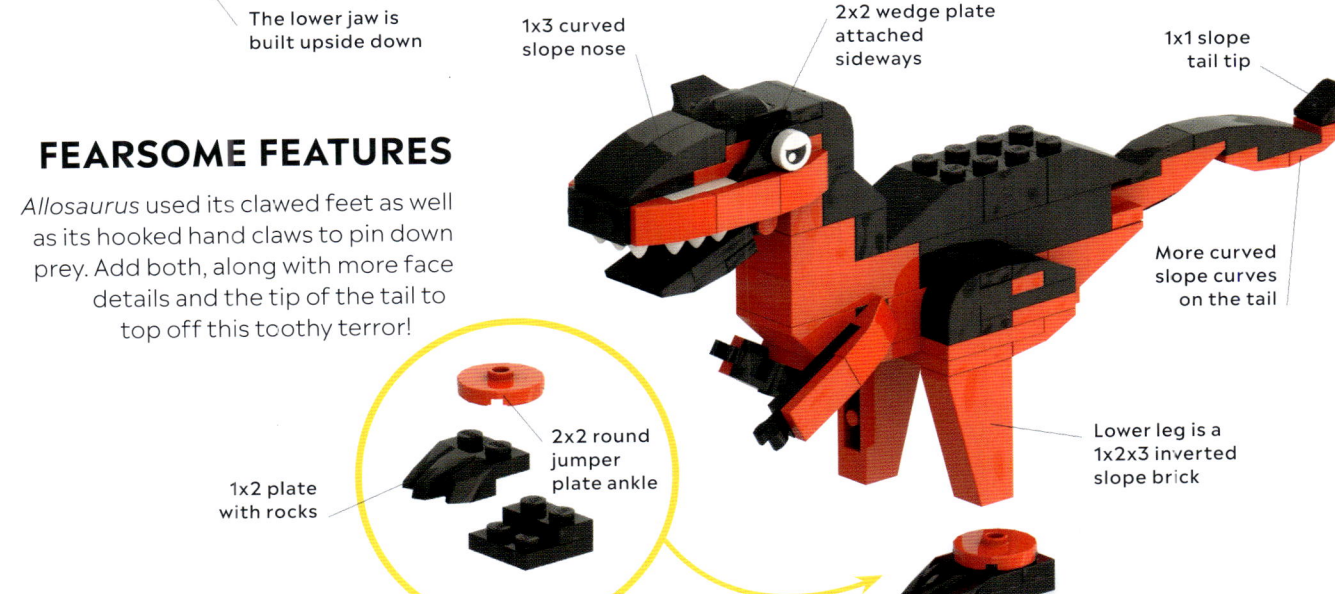

- 1x3 curved slope nose
- 2x2 wedge plate attached sideways
- 1x1 slope tail tip
- 1x2 plate with rocks
- 2x2 round jumper plate ankle
- More curved slope curves on the tail
- Lower leg is a 1x2x3 inverted slope brick

HABITATS

The Earth looked very different when the dinosaurs dominated. There were no cities or famous buildings to visit, but there were all sorts of environments that the dinosaurs shared, called habitats. They lived in forests and deserts, and by rivers and oceans all over the world. Try building some habitats for your LEGO creations.

I'm hoping to fit right in at the watering hole.

MICROSCALE FOREST

Match your habitats to the size of your dinosaur models. This riparian woodland—which is a forest beside a river—is the ideal stomping ground for your microscale creations.

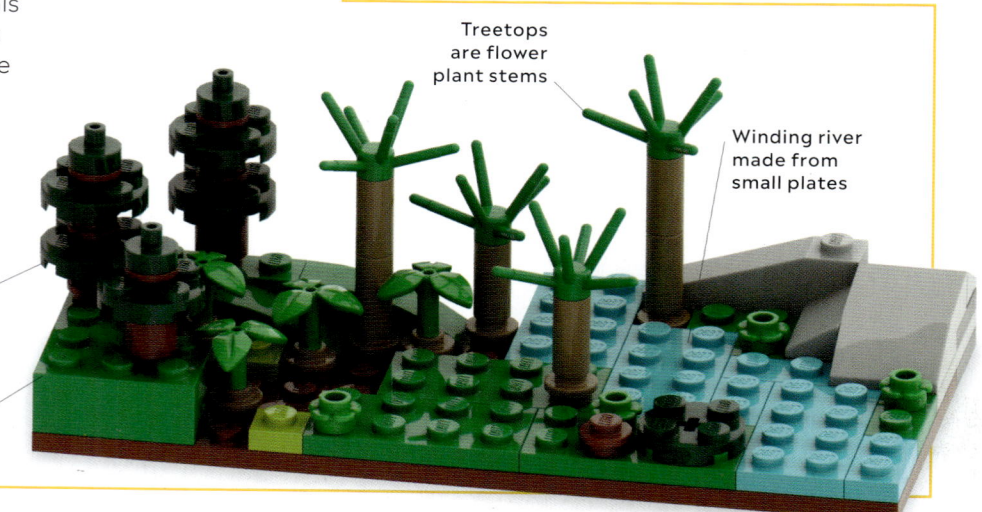

- Build trees in different sizes for a natural look
- Bricks look like hilly spots
- Treetops are flower plant stems
- Winding river made from small plates

OCEAN SHORE

Build just a hint of an ocean and its shoreline as a backdrop for your dinosaurs that love the salty sea air! Many dinosaurs lived close to coastlines and rivers because there were lots of plants and fish to eat there.

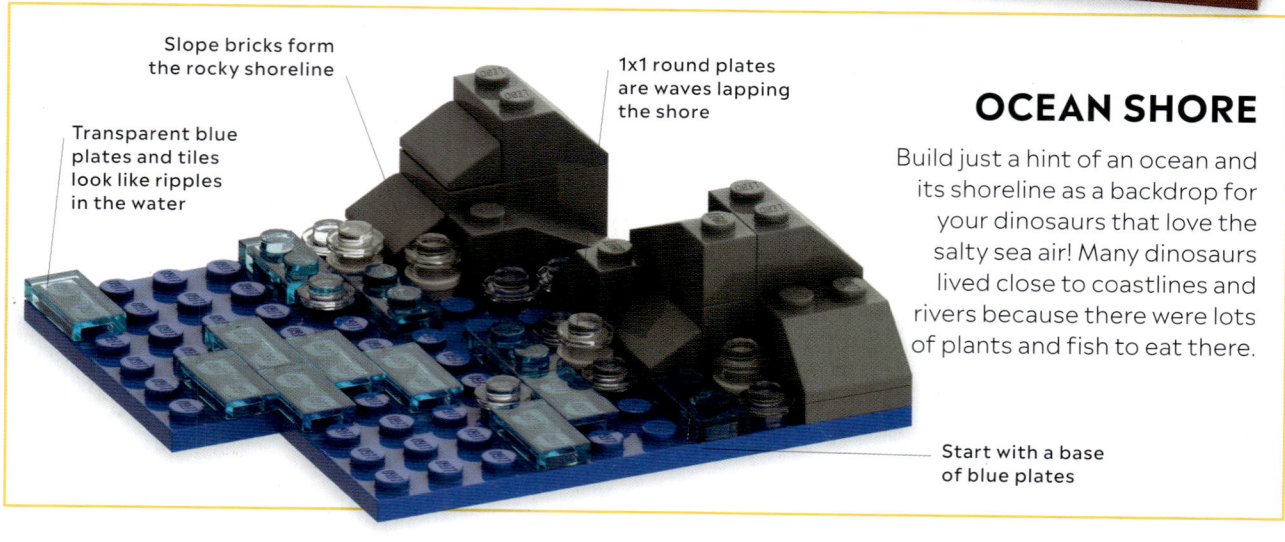

- Slope bricks form the rocky shoreline
- Transparent blue plates and tiles look like ripples in the water
- 1x1 round plates are waves lapping the shore
- Start with a base of blue plates

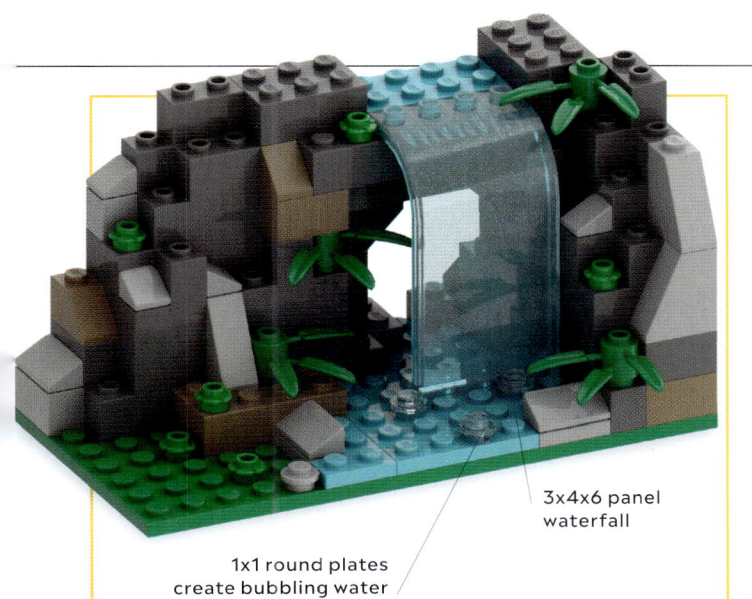

IDEAS GALLERY   255

WATERING HOLE

This is a habitat where dinosaurs big and small and all kinds of prehistoric life forms came to eat, drink, and bathe in dry, desert landscapes.

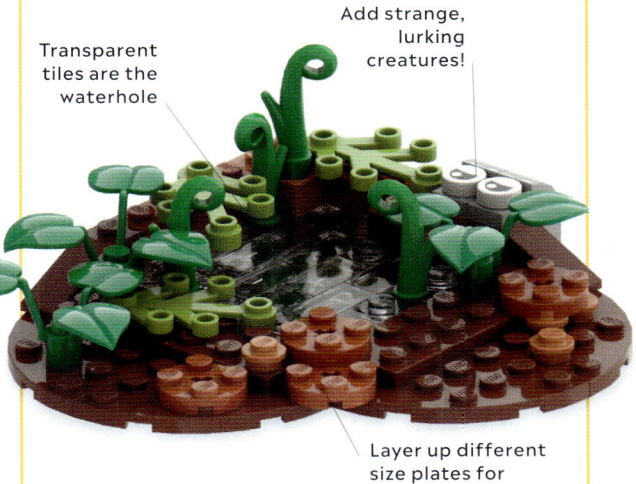

Transparent tiles are the waterhole

Add strange, lurking creatures!

3x4x6 panel waterfall

1x1 round plates create bubbling water

Layer up different size plates for uneven ground

RAVINE

A ravine is a narrow valley with steep sides that usually has water running through it. A lush landscape like this would have been a place where dinosaur herbivores thrived—and carnivores came to snack on them!

SWAMP FOREST

Hordes of dinosaurs made their homes in and around hot and steamy swamps like this, especially during the Cretaceous period. Many dinosaur fossils are found in them today.

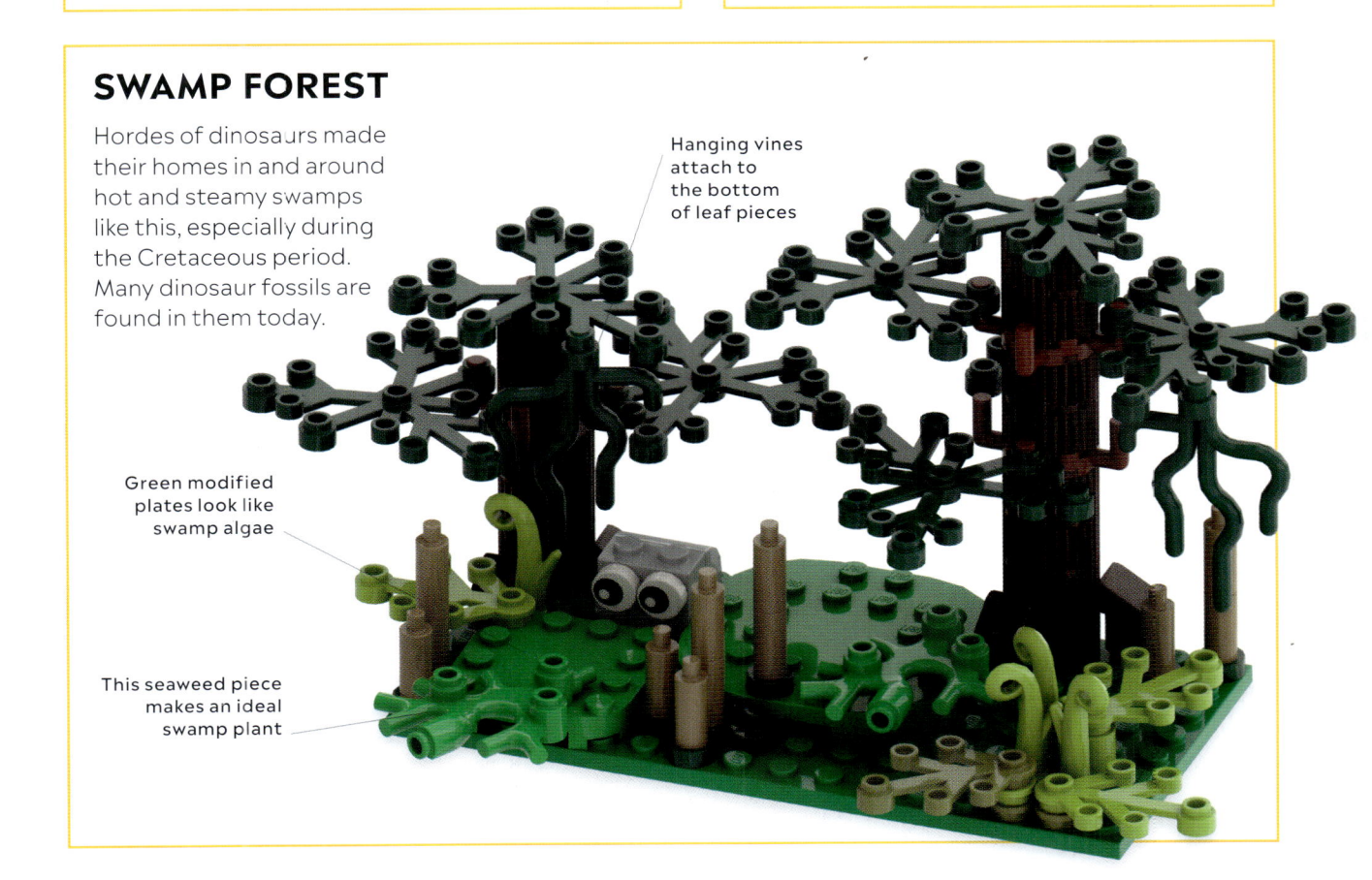

Hanging vines attach to the bottom of leaf pieces

Green modified plates look like swamp algae

This seaweed piece makes an ideal swamp plant

CAUDIPTERYX

CRETACEOUS PERIOD

Hey, good looking! With its dazzling fanned tail and arm feathers, *Caudipteryx* (caw-dip-ter-iks) looked a bit like a prehistoric peacock. This small theropod dinosaur couldn't fly but it could run very fast on its long legs. It had feathers to keep it warm and also to attract a mate.

- Toothless beak for eating plants and insects
- Flexible neck with click hinge connections
- Feathered breast made from vertical tooth plates
- Arm feathers are 3x4 plant pieces
- Clawed feet made from wedge slopes
- Fan-shaped tail feathers
- Long, bony tail

DINO FACT
Paleontologists know that *Caudipteryx* couldn't fly because its wings were too small. In fact, they think that only a few feathered dinosaurs could fly.

Why did the Caudipteryx cross the road?

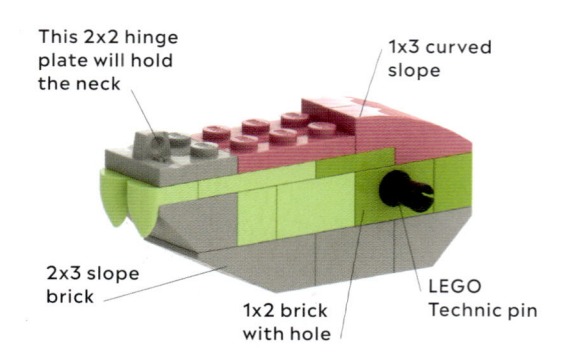

COMPACT BODY

The body of *Caudipteryx* was about the same size and shape as a small turkey. Most of this model's body is made from bricks, including some sloped bricks to give it a rounded shape. Nobody knows what color *Caudipteryx* was, so you can use any colors in your collection.

- This 2x2 hinge plate will hold the neck
- 1x3 curved slope
- 2x3 slope brick
- 1x2 brick with hole
- LEGO Technic pin

SKILL LEVEL 257

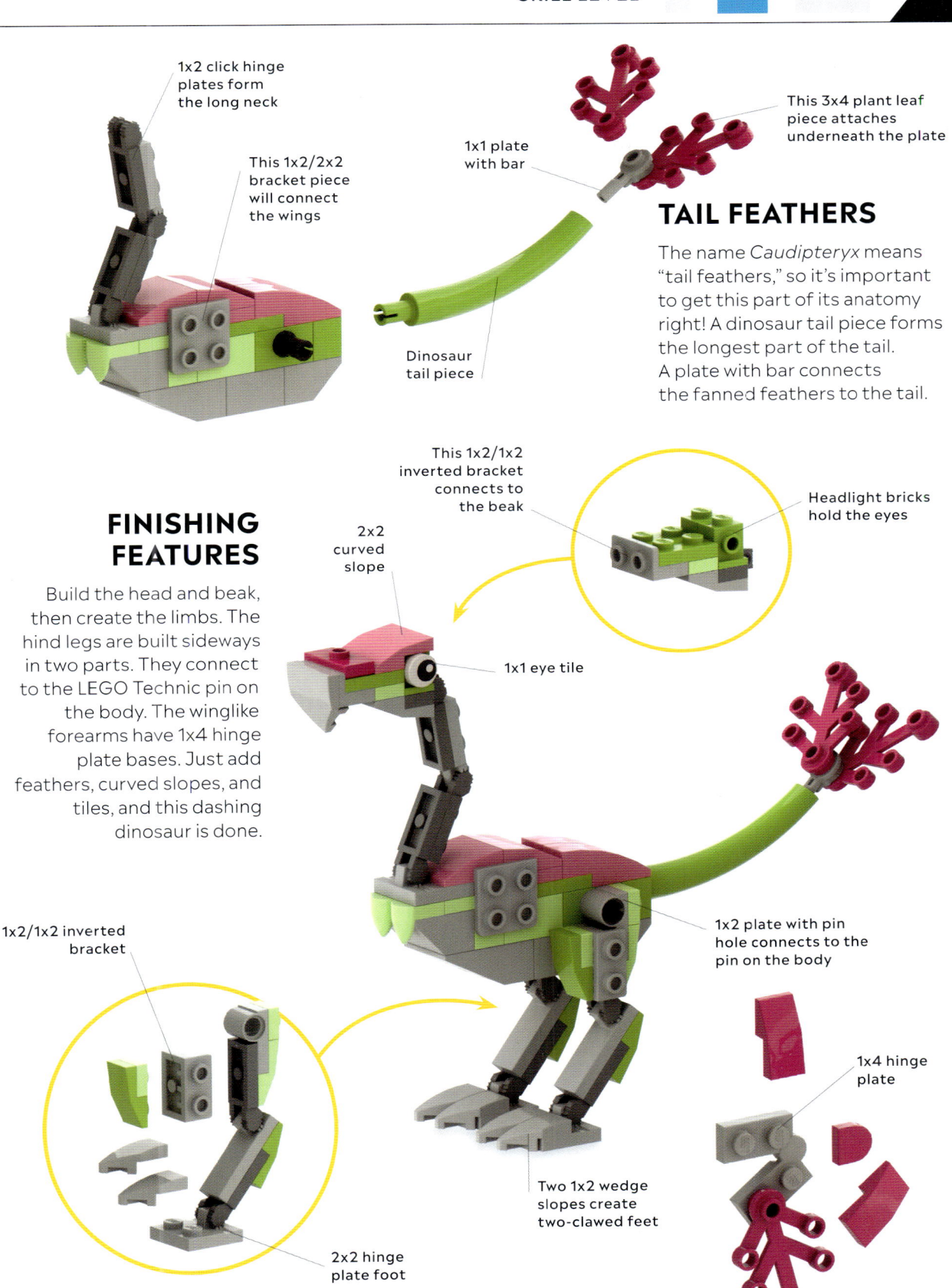

1x2 click hinge plates form the long neck

This 1x2/2x2 bracket piece will connect the wings

1x1 plate with bar

This 3x4 plant leaf piece attaches underneath the plate

TAIL FEATHERS

The name *Caudipteryx* means "tail feathers," so it's important to get this part of its anatomy right! A dinosaur tail piece forms the longest part of the tail. A plate with bar connects the fanned feathers to the tail.

Dinosaur tail piece

FINISHING FEATURES

Build the head and beak, then create the limbs. The hind legs are built sideways in two parts. They connect to the LEGO Technic pin on the body. The winglike forearms have 1x4 hinge plate bases. Just add feathers, curved slopes, and tiles, and this dashing dinosaur is done.

This 1x2/1x2 inverted bracket connects to the beak

2x2 curved slope

Headlight bricks hold the eyes

1x1 eye tile

1x2/1x2 inverted bracket

1x2 plate with pin hole connects to the pin on the body

1x4 hinge plate

2x2 hinge plate foot

Two 1x2 wedge slopes create two-clawed feet

CRETACEOUS PERIOD

DEINOSUCHUS

This prehistoric predator's name translates from Greek as "terrible crocodile," so you might want to give it a wide berth! *Deinosuchus* (die-no-soo-kuss) was an enormous relative of modern-day alligators and crocodiles. It hunted in the waters of rivers and swamps, where it would feast on fish, turtles, and—if it got lucky—dinosaurs.

FUN FACT
An adult *Deinosuchus* could grow up to 39 ft (12 m) in length. From head to tail, that's as long as a T. Rex! *Deinosuchus* was twice as long as a saltwater crocodile, which is the biggest croc in the world today.

- 1x1 round plate "scutes"
- Enormous, scaly body
- Terrifying teeth made from 1x1 round tiles with bars
- Belly is low to the ground
- Small feet with clawed toes
- Long, powerful tail

I've fought off alligators, but this is in another league!

- This 1x2/2x2 inverted bracket will be part of the jaw
- 1x1 plate with clip
- 2x4 double inverted curved slope with cutout
- First of four plate layers

LONG BASE

Start your *Deinosuchus* model right in the middle of its enormous body. There are four layers of plates in this section, with two double inverted slopes at either end of the plates. Build plates with clips and bars on the edges to hold the legs and tail later.

SKILL LEVEL

259

CHOMPING JAWS

Now *Deinosuchus* has tough, bony plates of armor all over its body. Next, it needs powerful jaws built for gripping and crushing its prey. The lower part of the jaw has a 2x8 plate base. The upper part of it is made from layers of smaller plates.

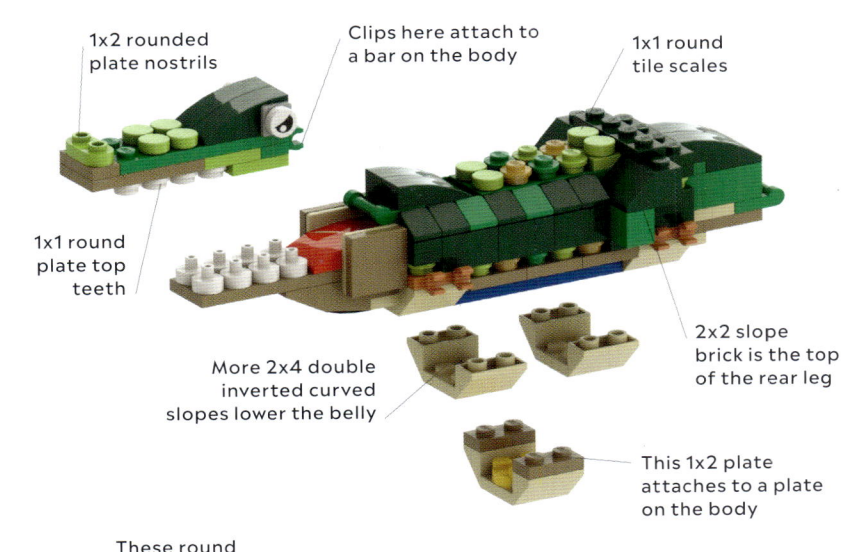

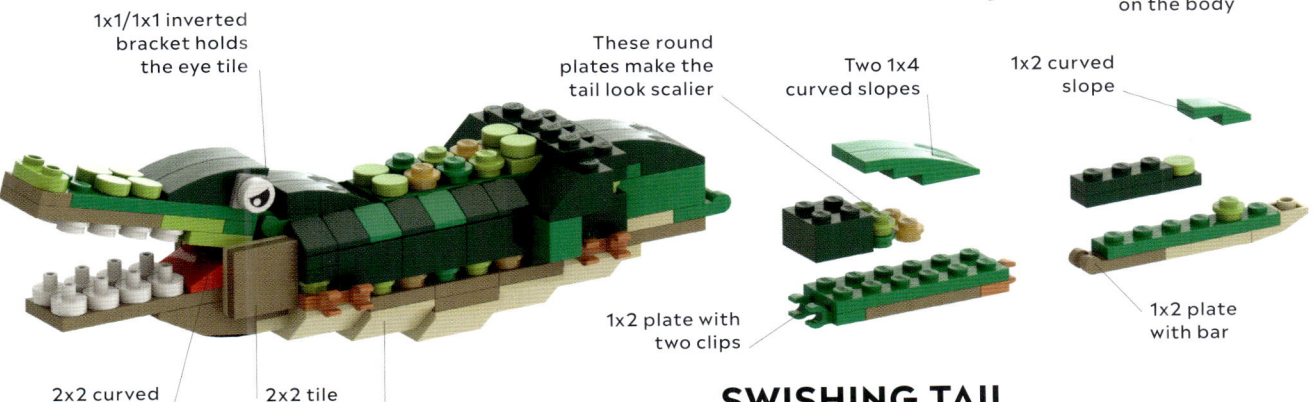

- 1x2 rounded plate nostrils
- Clips here attach to a bar on the body
- 1x1 round tile scales
- 1x1 round plate top teeth
- More 2x4 double inverted curved slopes lower the belly
- 2x2 slope brick is the top of the rear leg
- This 1x2 plate attaches to a plate on the body
- 1x1/1x1 inverted bracket holds the eye tile
- These round plates make the tail look scalier
- Two 1x4 curved slopes
- 1x2 curved slope
- 1x2 plate with two clips
- 1x2 plate with bar
- 2x2 curved slope tongue
- 2x2 tile
- These slopes are now attached at different heights

SWISHING TAIL

Deinosuchus used its massive, muscular tail to propel itself through water. This tail is made in two parts. Thanks to clip and bar connections, it can move up and down in two places.

- 2x2 brick widens the tail
- You could also use round plates and small tooth pieces here
- Add smooth 2x2 slide plates under the tail and belly
- 1x2 plate with bar
- This 1x2 plate with bar attaches to the body
- Slopes add curves to the top of each leg
- 1x2 plate with bar (with open ends)

LITTLE LEGS

It might have been a colossal carnivore, but *Deinosuchus* had relatively small legs for its size. That's because it spent most of its time in the water. Finish off your fearsome crocodilian with four short but sturdy legs.

DIMETRODON

Permian Period

This spine-chilling creature roamed the Earth millions of years before the very first dinosaurs. *Dimetrodon* (*dai-met-roh-don*) belonged to a group of animals called synapsids. With razor-sharp, pointed teeth, it was one of the most dangerous predators of its time. Scientists think it used the tall sail on its back to scare enemies and attract mates.

- Large sail is a hexagonal flag piece
- Curved body supported by four legs
- This 1x2 slope continues the curve of the sail
- Small, lizard-like head
- Long, narrow tail
- Lower jaw is a 1x1 plate with clip
- The limbs move on LEGO Technic half pins
- Legs have a wide, reptilian stance
- Body is low to the ground
- 1x1 slope feet

Hoist the main sail... Oh, you have!

FUN FACT
The sail of a *Dimetrodon* may have been brightly colored to attract attention, so your model's can be any shade you like.

SKILL LEVEL

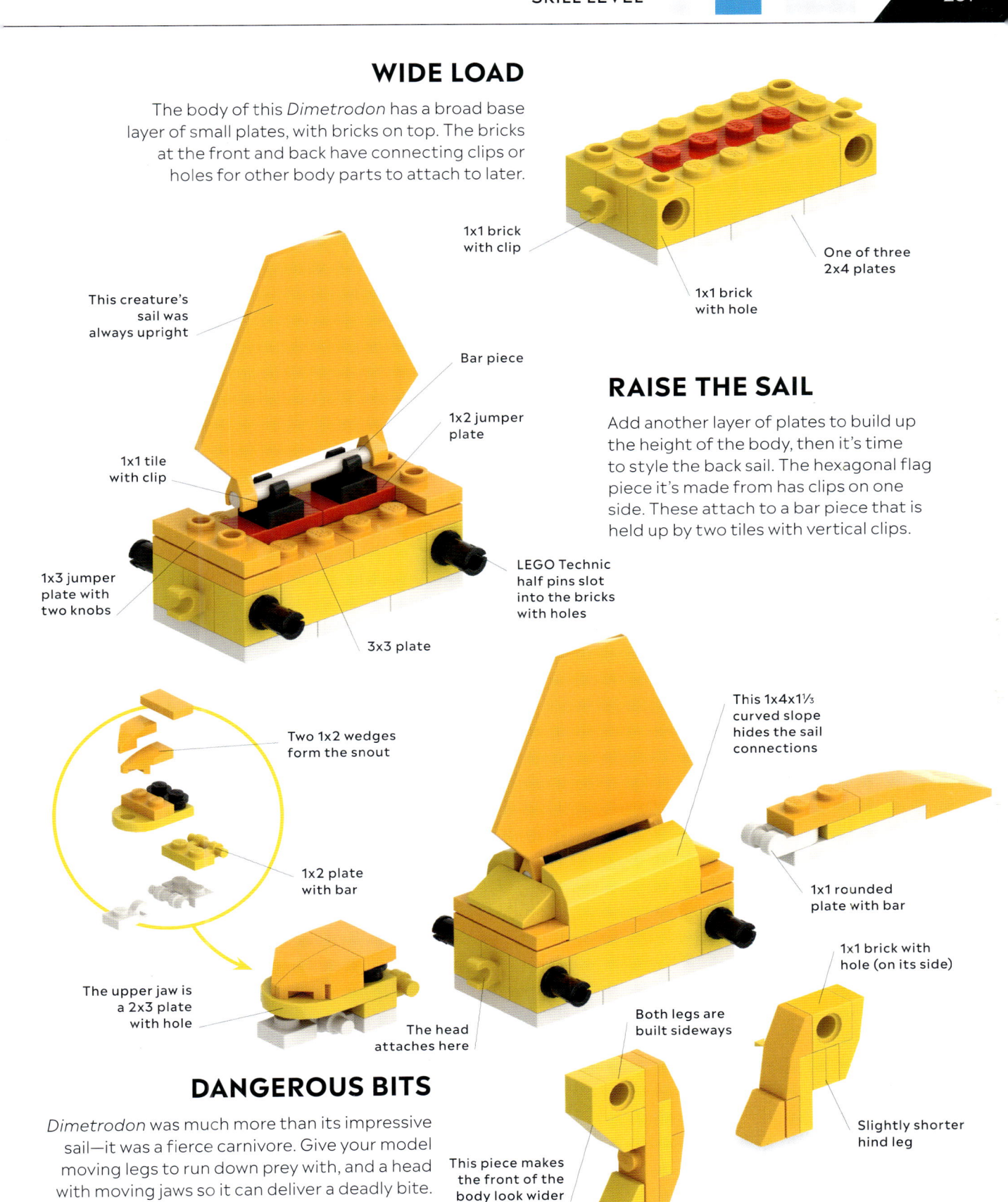

261

WIDE LOAD

The body of this *Dimetrodon* has a broad base layer of small plates, with bricks on top. The bricks at the front and back have connecting clips or holes for other body parts to attach to later.

- 1x1 brick with clip
- One of three 2x4 plates
- 1x1 brick with hole

RAISE THE SAIL

Add another layer of plates to build up the height of the body, then it's time to style the back sail. The hexagonal flag piece it's made from has clips on one side. These attach to a bar piece that is held up by two tiles with vertical clips.

- This creature's sail was always upright
- Bar piece
- 1x2 jumper plate
- 1x1 tile with clip
- 1x3 jumper plate with two knobs
- 3x3 plate
- LEGO Technic half pins slot into the bricks with holes

- Two 1x2 wedges form the snout
- 1x2 plate with bar
- The upper jaw is a 2x3 plate with hole
- The head attaches here
- This 1x4x1⅓ curved slope hides the sail connections
- 1x1 rounded plate with bar
- 1x1 brick with hole (on its side)
- Both legs are built sideways
- This piece makes the front of the body look wider
- Slightly shorter hind leg

DANGEROUS BITS

Dimetrodon was much more than its impressive sail—it was a fierce carnivore. Give your model moving legs to run down prey with, and a head with moving jaws so it can deliver a deadly bite.

TRIASSIC PERIOD

ICHTHYOSAURUS

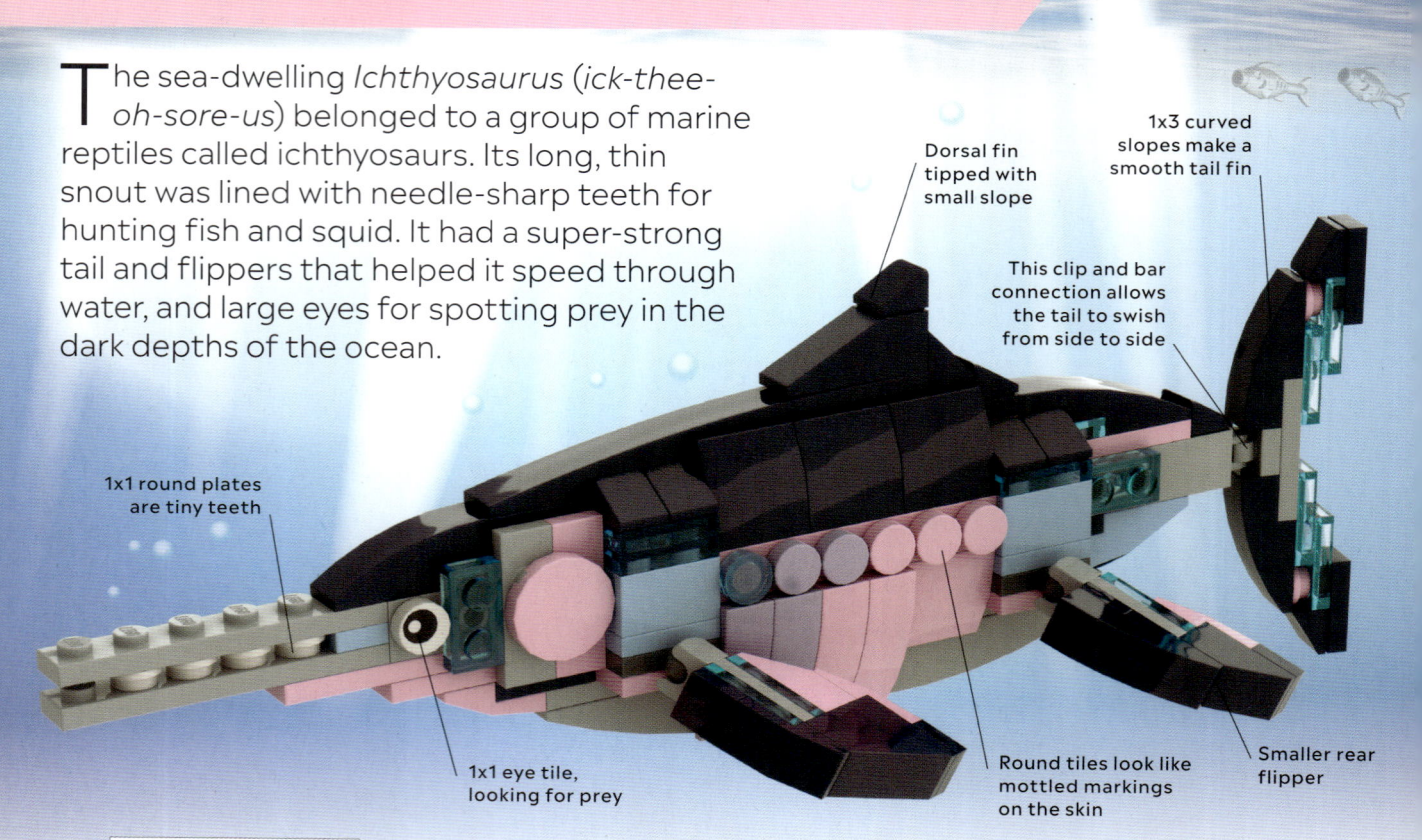

The sea-dwelling *Ichthyosaurus* (ick-thee-oh-sore-us) belonged to a group of marine reptiles called ichthyosaurs. Its long, thin snout was lined with needle-sharp teeth for hunting fish and squid. It had a super-strong tail and flippers that helped it speed through water, and large eyes for spotting prey in the dark depths of the ocean.

- Dorsal fin tipped with small slope
- 1x3 curved slopes make a smooth tail fin
- This clip and bar connection allows the tail to swish from side to side
- 1x1 round plates are tiny teeth
- 1x1 eye tile, looking for prey
- Round tiles look like mottled markings on the skin
- Smaller rear flipper

Who ordered the tuna steak sandwich?

- 1x6 plate
- 2x3 plate
- 1x6 brick
- 1x3 curved slope continues the smooth line of the body
- 1x6 inverted curved slope
- The rear flipper will attach to this 1x2 plate with ball

LONG AND STRONG

The base of this creature's powerful body is made from two inverted curved slopes. There are narrow plates and bricks on top. At this stage, you can also attach connections for the four flippers.

FUN FACT

Ichthyosaurus looked a lot like a dolphin or a shark, and is definitely not a "swimming dinosaur." These animals lived in the seas at the same time as dinosaurs walked on land.

SKILL LEVEL

263

READY FOR SQUID

Now *Ichthyosaurus* has long, thin jaws made from two narrow plates. Its row of four 1x1 plate teeth is sandwiched between the plates. The rest of the body is starting to take shape too.

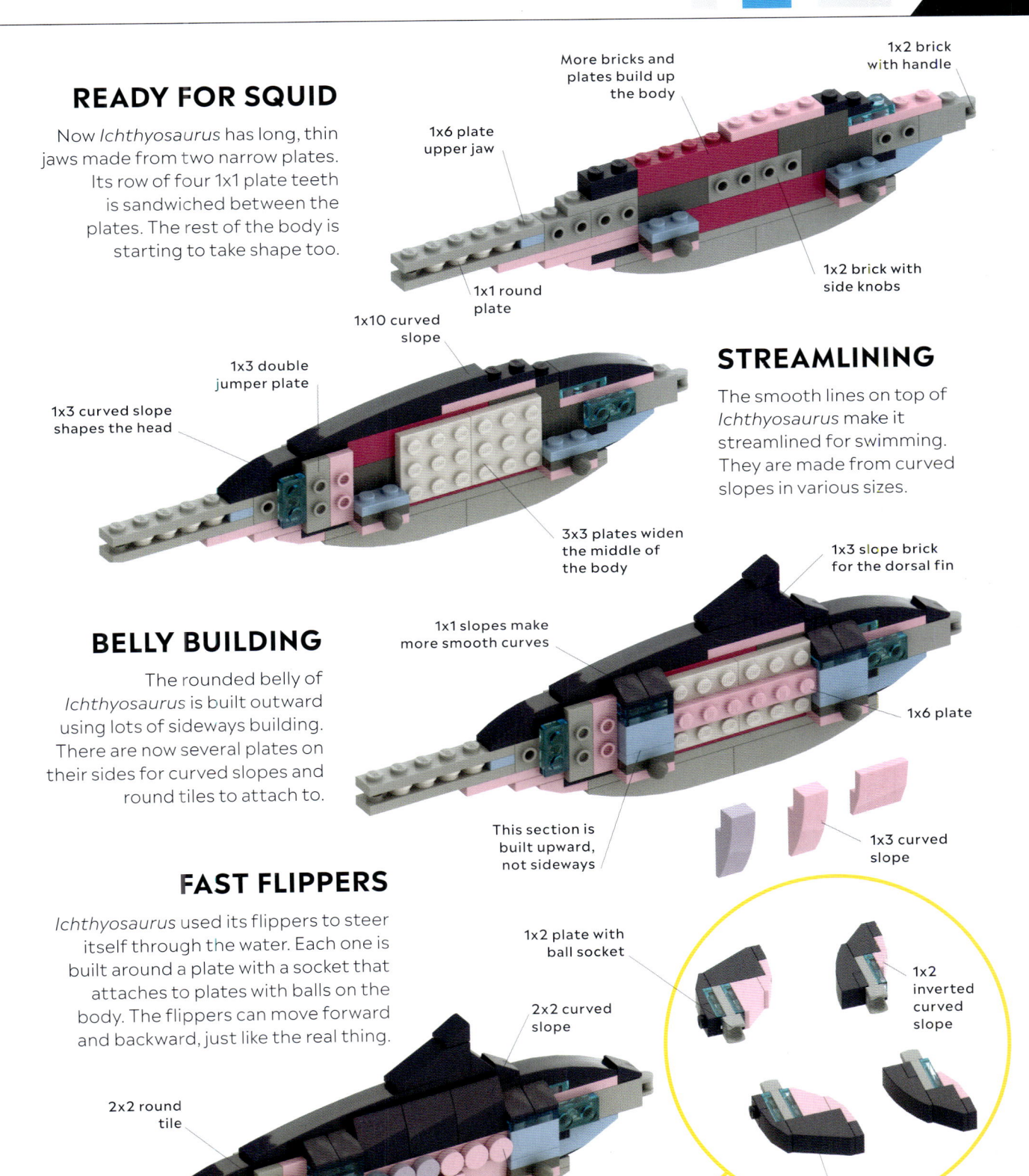

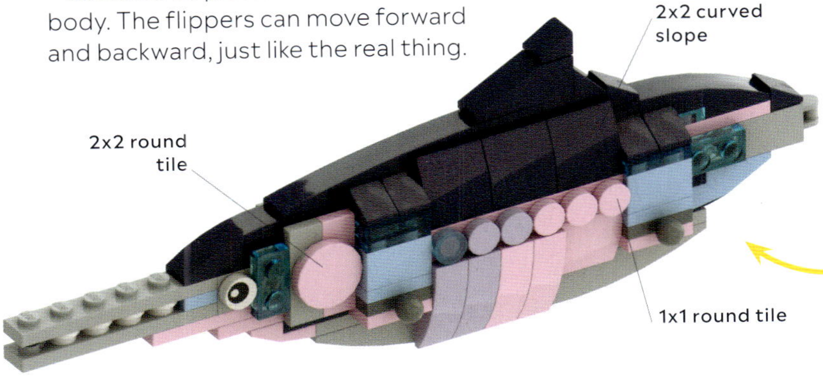

More bricks and plates build up the body

1x6 plate upper jaw

1x2 brick with handle

1x1 round plate

1x2 brick with side knobs

1x10 curved slope

1x3 double jumper plate

1x3 curved slope shapes the head

STREAMLINING

The smooth lines on top of *Ichthyosaurus* make it streamlined for swimming. They are made from curved slopes in various sizes.

3x3 plates widen the middle of the body

1x3 slope brick for the dorsal fin

BELLY BUILDING

The rounded belly of *Ichthyosaurus* is built outward using lots of sideways building. There are now several plates on their sides for curved slopes and round tiles to attach to.

1x1 slopes make more smooth curves

1x6 plate

This section is built upward, not sideways

1x3 curved slope

FAST FLIPPERS

Ichthyosaurus used its flippers to steer itself through the water. Each one is built around a plate with a socket that attaches to plates with balls on the body. The flippers can move forward and backward, just like the real thing.

1x2 plate with ball socket

2x2 curved slope

1x2 inverted curved slope

2x2 round tile

1x1 round tile

1x2x2 slope bricks help make the front fins wider

PLESIOSAURUS

JURASSIC PERIOD

This carnivorous sea reptile swam the prehistoric seas in search of fish, squid, and smaller marine reptiles. *Plesiosaurus* (*plee-see-oh-sore-us*) moved through the water a bit like a turtle, using its paddlelike flippers to propel itself forward. Its long neck could move quickly to catch prey.

- Small head shaped with curved slopes
- Long neck built in sections
- Broad, flat body like a turtle
- Tip of the tail is a 2x2x2 cone
- Tiles on top make the flippers look smooth
- Large flippers made from wedge plates
- 1x2 tile with two clips holds the neck
- 1x1 brick with side knob
- 1x2x1⅔ brick with side knobs
- LEGO Technic 1x2 brick with axle hole
- This modified 2x2 tile has two knobs at one end

Is that the Loch Ness Monster?

SIDEWAYS START

Gather lots of bricks with side knobs to begin your *Plesiosaurus* model. Use them to build up the height of the body, then attach inverted curved slopes below them to form the smooth, streamlined underside.

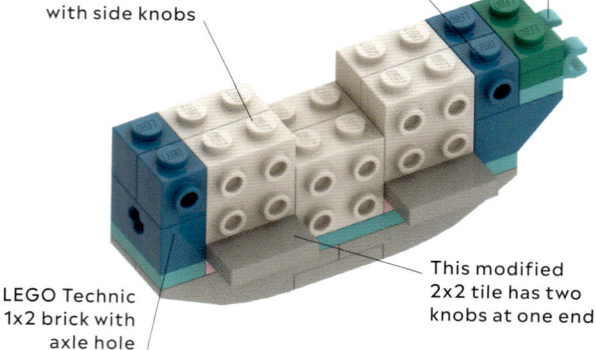

SKILL LEVEL

265

SMOOTH SURFACE

Now there are lots of curved slope pieces attached to the top and sides of the body, making it smoother and wider. Next, add pieces that will allow you to connect the tail and four flippers. There's a LEGO Technic axle at the back for the tail, and plates with balls for the flippers.

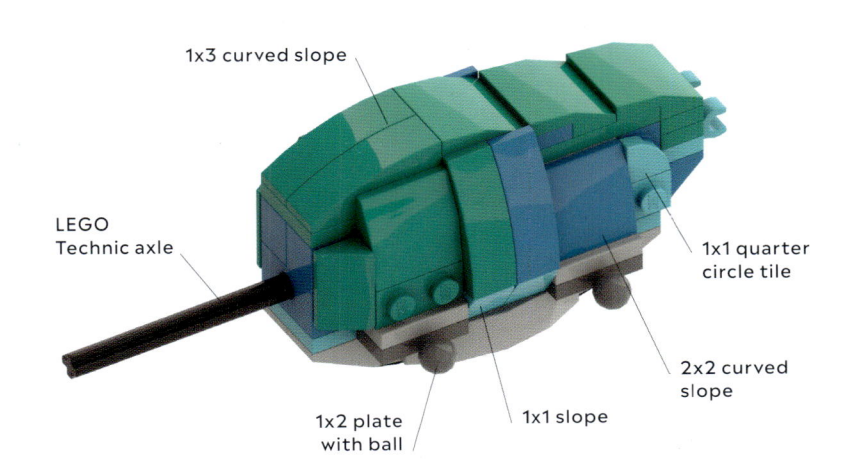

- 1x3 curved slope
- LEGO Technic axle
- 1x1 quarter circle tile
- 2x2 curved slope
- 1x2 plate with ball
- 1x1 slope

- 3x3x2 cone with axle hole
- 1x2x1 half cylinder
- 1x2 plate with socket
- 1x1 half circle tile

SWIMMING LIMBS

It's time to build the flattened flippers of *Plesiosaurus*. Their diamond-like shape is created from wedge plates, and the front flippers are slightly larger than the hind ones. Also add the tail at this stage. It's built from two cones and two half cylinders—they all slide onto the LEGO Technic axle piece.

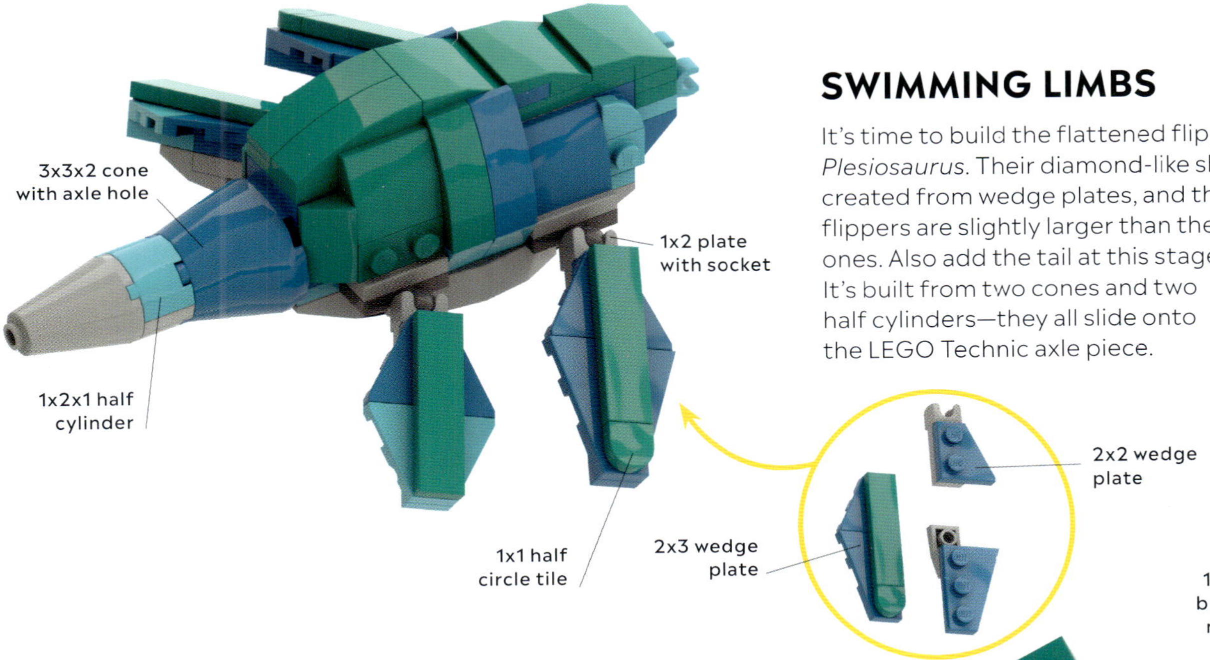

- 2x2 wedge plate
- 2x3 wedge plate
- 1x1/1x1 bracket mouth
- 1x2 plate with clip
- 1x2 plate with bar
- 2x4 double curved slopes give the neck a rounded shape
- Rounded plates underneath hold together the small plates on top

FLEXIBLE NECK

Without its long, slender neck, *Plesiosaurus* wouldn't have been quite so efficient at catching its prey. This model's neck is built in three small parts that connect together with clip and bar pieces. Add a small head to the top of the neck, then attach the lowest part of the neck to the body to make this ancient creature seaworthy!

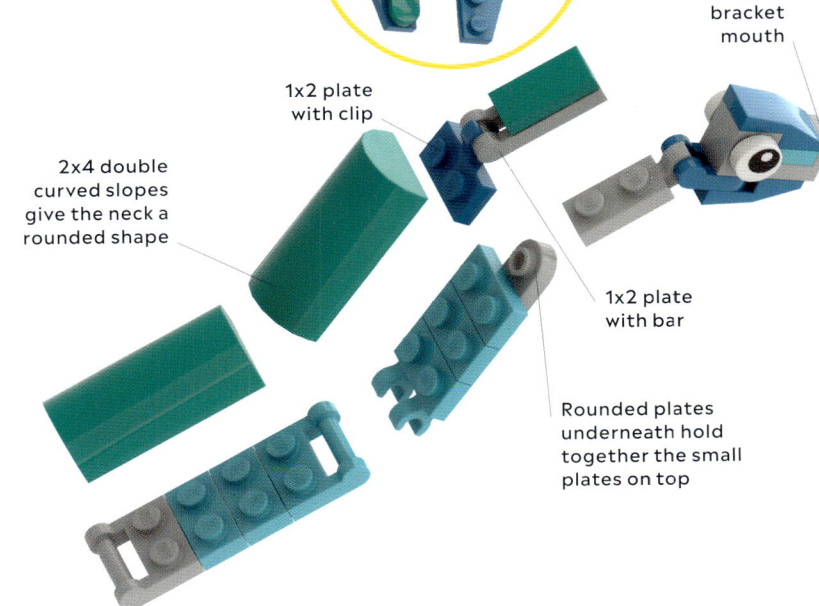

JURASSIC PERIOD

IGUANODON

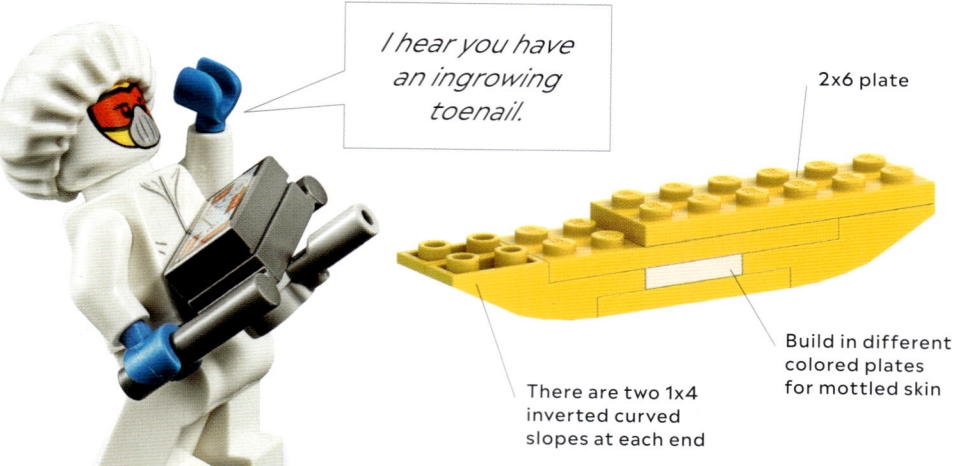

This heavyweight herbivore was one of the first dinosaurs ever to be discovered and named. It gets its name from its teeth—*Iguanodon* (ig-gwah-no-don) means "iguana-tooth." It was larger than an elephant and could move around on its two back legs, particularly when it needed to reach food in high places or defend itself.

- Bulky body with a tough hide
- Narrow head topped with curved slopes
- Sharp beak for gathering plants
- Muscular hind legs for rearing up
- Thumb spike is a horizontal tooth plate
- 1x2 grille slopes create toe claws

DINO FACT
A notable feature of *Iguanodon* was a sharp thumb spike on each hand. It may have used it as a weapon and to break open fruit and seeds.

I hear you have an ingrowing toenail.

- 2x6 plate
- Build in different colored plates for mottled skin
- There are two 1x4 inverted curved slopes at each end

FULL BELLY

Iguanodon spent its days chomping prehistoric plants, such as cycads and horsetails. Give your model a nice, long belly to fill using plates and inverted curved slopes.

SKILL LEVEL

267

STIFF TAIL

Build up the height of the barrel-shaped body with bricks and more plates. Some of the bricks have holes, which will be connection points for the limbs. Add a stiff tail at the rear—this will help your *Iguanodon* to balance on two legs.

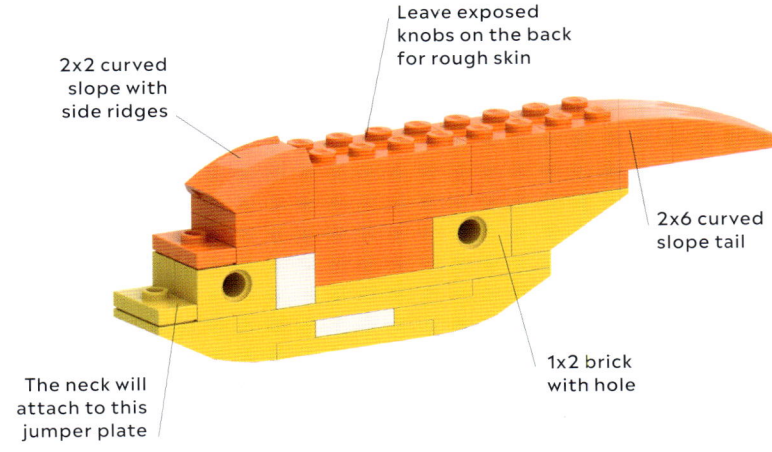

- 2x2 curved slope with side ridges
- Leave exposed knobs on the back for rough skin
- 2x6 curved slope tail
- 1x2 brick with hole
- The neck will attach to this jumper plate

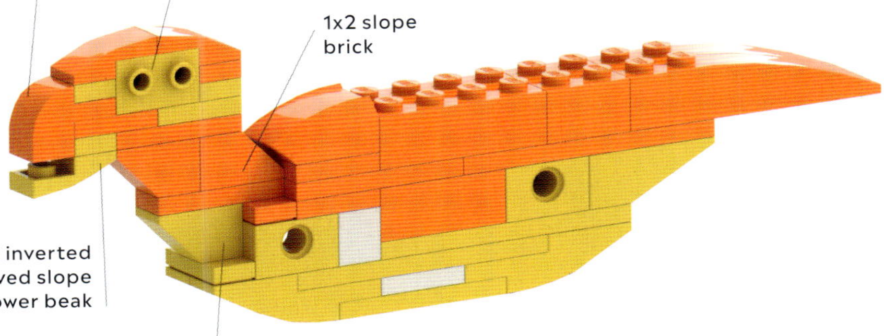

- 1x2x1⅓ slope brick
- Include a 1x2 brick with side knobs here for the eyes and cheek details
- 1x2 slope brick
- 1x2 inverted curved slope lower beak
- 1x2 inverted slope brick

IGUANA FEATURES

The vertical neck shape of *Iguanodon* is created from three slope bricks. Build the head on top using lots of plates and add a smooth-looking beak. *Iguanodon* had teeth like an iguana, but they would be too small to see at this scale so there's no need to add them.

- 1x1 printed eye tile
- 2x2 round tile is a chewing cheek
- LEGO Technic half pins connect to the legs

VERSATILE LIMBS

Iguanodon had four limbs supporting its hefty body, but its two bigger back legs could hold all of its weight when necessary. Both the front and hind legs are created using sideways building techniques.

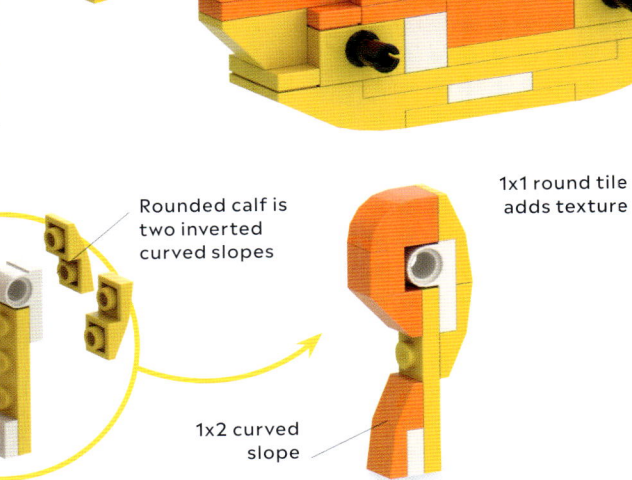

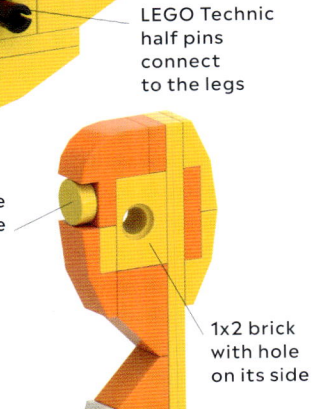

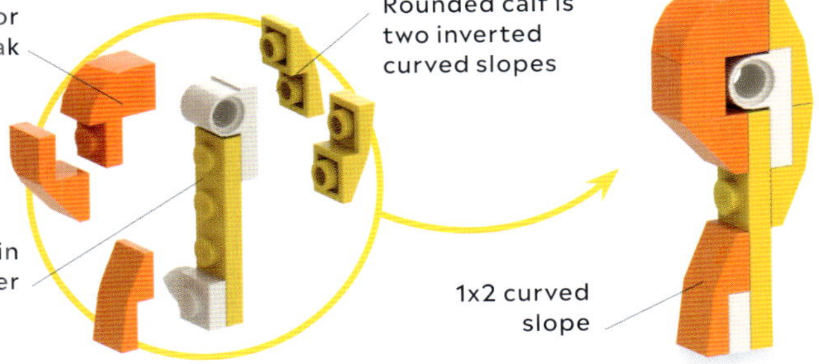

- This piece is also used for the beak
- Rounded calf is two inverted curved slopes
- 1x4 plate in the center
- 1x2 curved slope
- 1x1 round tile adds texture
- 1x2 brick with hole on its side

PLEISTOCENE PERIOD

WOOLLY MAMMOTH

This Ice Age animal was not a dinosaur—it was related to elephants that we see in the world today. A hulking herbivore, it was covered in thick, shaggy fur that kept it warm in freezing temperatures. It had smaller ears than a modern elephant (so it didn't get frostbite), an extra-long trunk, and enormous, curved tusks.

FUN FACT
The woolly mammoth was not alive at the time of the dinosaurs—they lived 66 million years earlier. But it did live alongside the first humans. Sadly, they hunted it so much that it became extinct.

- Ruffled head fur
- These curved slopes look like thick hair hanging down
- Bendy trunk for feeding and trumpeting
- Stacked 2x2 round brick legs
- Massive tusks for digging and fighting

They make great pets until they step on your toe!

BEHEMOTH BASE

This mega mammoth model's body is flat at the front and slopes down at the back. You can already see its distinctive shape at this early building stage. Start with a 4x8 base plate at the bottom and build bricks, plates, and slope bricks on top.

- 2x2x2 slope brick
- 4x8 plate
- Front of the body is wider than the back

SKILL LEVEL

LEG SUPPORT

After building up the curved top half of the body, create some wide legs to support the woolly mammoth's great weight. Also add circular foot pads at the bottom of the front legs—these helped the colossal creature walk on soft snow and ice.

These plates make the front of the body taller

1x4 curved slope rump

The head will attach here

Include bricks with side knobs for building out sideways

1x2 rounded plates make the front legs wider at the top

The tail will attach to this inverted bracket

1x2 plate adds height

The hind legs have large thighs

1x2 plate with ball

2x3 inverted slope

1x1 brick with side knob

1x1 round plate foot pad

1x1/1x1 inverted bracket

1x2 plate with rocks looks like tufts of hair

HEAD START

It's time to focus on the woolly mammoth's huge head. It's filled with connecting pieces with side knobs and clips that the eyes, ears, tusks, and trunk can attach to. Also build in a plate with a ball at the back so you can add the head to the body.

Two 1x2x1⅓ curved slopes form the forehead

The trunk attaches to this clip

1x2 plate with bar ear

1x1 slopes widen the cheeks

Darker brown back fur

Small 1x2 curved slope tail

1x2 plate with bar

Rounded trunk tip

TAIL, TRUNK, TUSKS

Add more furry finishes to your woolly mammoth and a tail at the back. The tail was smaller than a modern elephant's so it didn't get too cold in Ice Age temperatures. Finally, clip on a bendy trunk and sharp tusks.

Tusk bases are upside-down candle pieces clipped to bars

Another 1x2 plate with rocks here adds to the shaggy look!

CRETACEOUS PERIOD

ANKYLOSAURUS

This is one tough dinosaur! The armored *Ankylosaurus* (an-kie-lo-sore-us) had spikes and plates down the length of its body and a bone-shattering, club-like tail that was always poised to swing at potential predators. All that body armor was heavy to lumber around—*Ankylosaurus* weighed as much as three rhinoceroses!

DINO FACT
The plates on the neck, back, and tail of *Ankylosaurus* are called osteoderms. They are bones that grow on the skin. The dinosaur's dangerous tail club was four large osteoderms fused together.

Upright back plates are 1x1 tiles with bars

Shoulder plates are 1x1 round tiles

Clubbed tail tip made from 2x2 dome pieces

Wide legs and feet

Four head horns help protect the brain

My ankles are sore as well!

GIGANTIC GUT

Ankylosaurus had a very large digestive system because it ate tough plants and leaves that needed to be broken down inside its body, so begin your *Ankylosaurus* model with a big, broad gut! This model's is made from layers of wide plates and bricks placed horizontally.

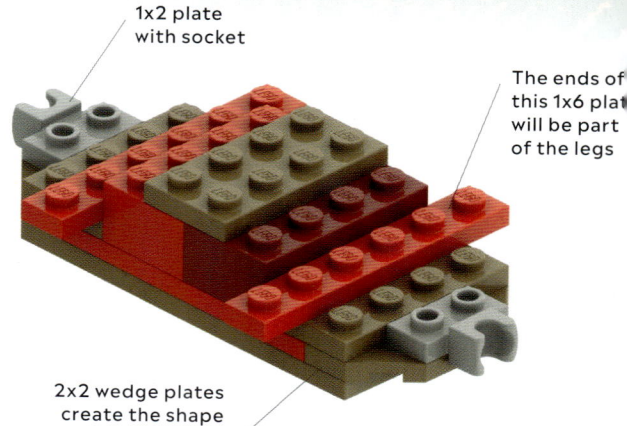

1x2 plate with socket

The ends of this 1x6 plate will be part of the legs

2x2 wedge plates create the shape of the shoulder

SKILL LEVEL

271

The rear of the body mirrors the front

1x2/1x2 inverted bracket

Smaller 1x1/1x1 inverted bracket

1x1 pyramid slope

Another layer of wedge plates

The tops of the legs are now in place too

SIDE SPIKES

After building up the height of the soft belly—which is the only unprotected part of *Ankylosaurus*—it's time to build some spiky sections. The pyramid slope spikes on the sides of the body attach sideways to inverted bracket pieces.

Two 1x3 curved slopes

1x2 plate with ball

HARD HEAD

Once the bumpy back plates are in place, build the triangle-shaped head. The head of *Ankylosaurus* was covered in a thick layer of bone, with four spikes for added protection. It even had armored eyelids, though this model's eyes don't look quite so durable—they're made from doughnut-printed tiles!

2x2 inverted slope thigh

The belly hangs low to the ground

Wide feet are 2x2 slope bricks

1x1 doughnut tile

2x2x^2/$_3$ wedge tile forms the pointed snout

Bar with ball

This 1x2 plate has a ball at one end and a socket at the other

The larger plate underneath holds the small plates together

JOIN THE CLUB

All *Ankylosaurus* needs now is its slashing club tail, which is built in three parts. There's a thicker, two-knob wide section that connects to a narrow, one-knob wide part. The tail tip is a bar with a ball—the dome pieces that form the bony club fit onto this.

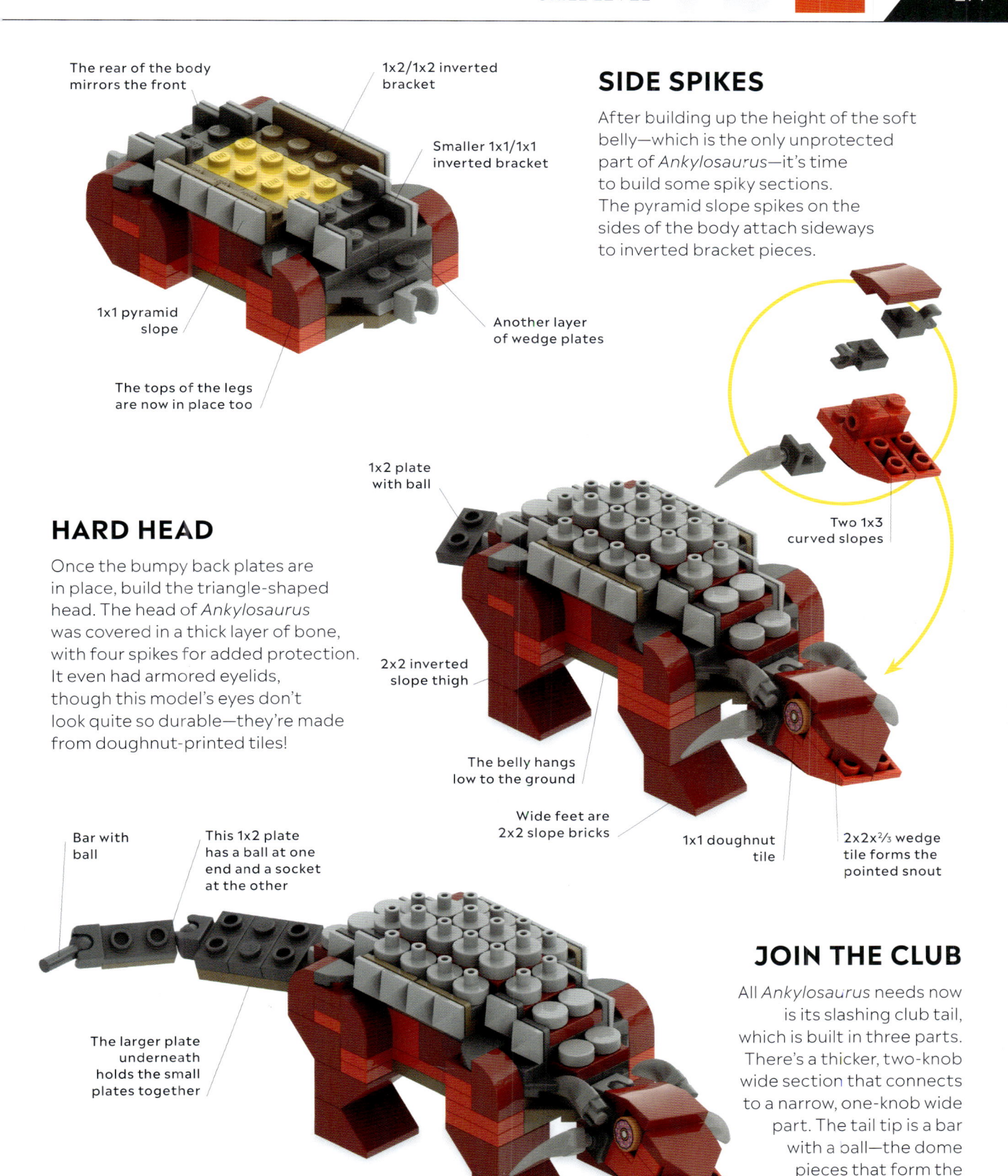

MECHA BIRD
PAGES 334–335

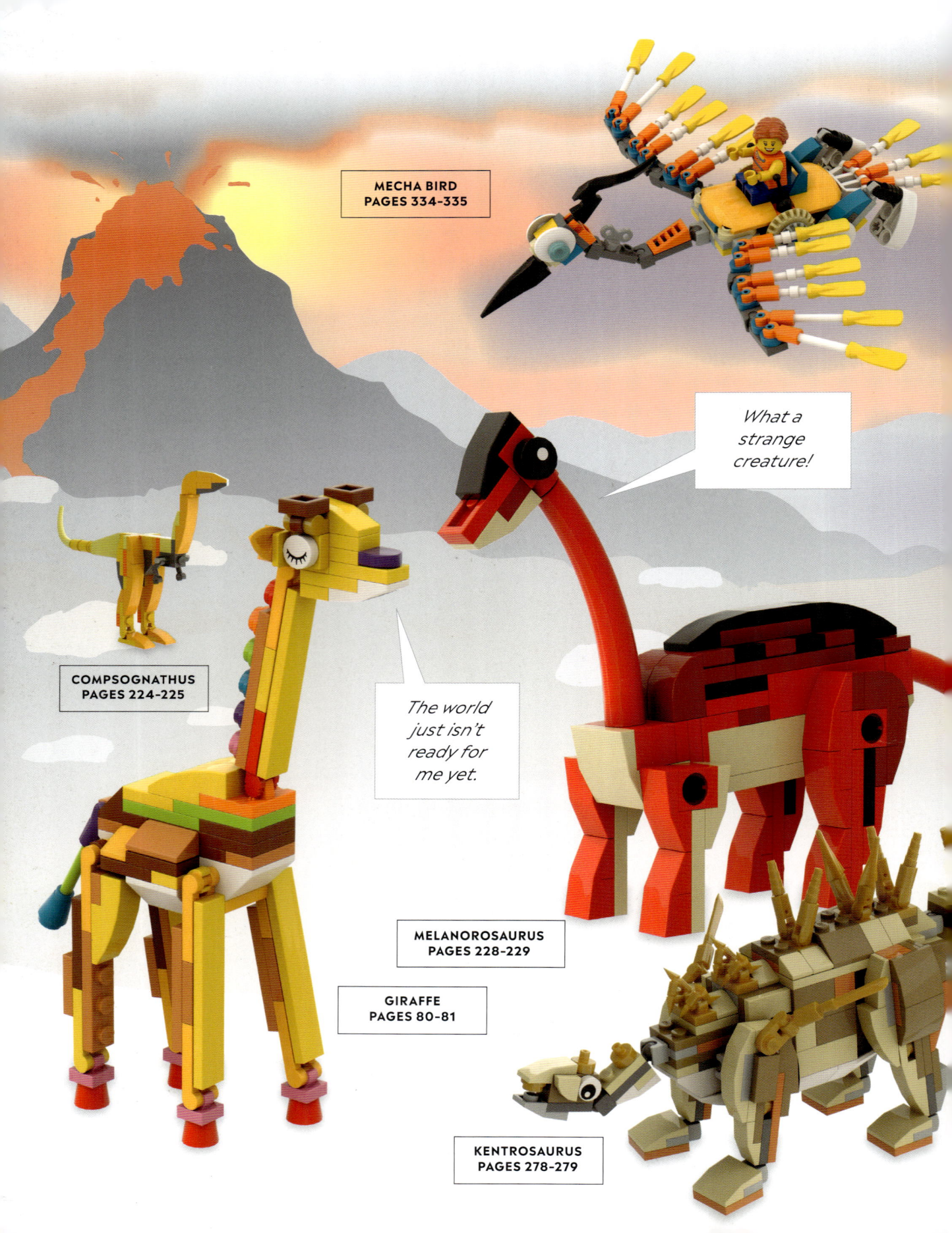

What a strange creature!

COMPSOGNATHUS
PAGES 224–225

The world just isn't ready for me yet.

MELANOROSAURUS
PAGES 228–229

GIRAFFE
PAGES 80–81

KENTROSAURUS
PAGES 278–279

MODEL MASH-UP

DIMORPHODON
PAGES 226–227

BARYONYX
PAGES 244–245

SIMPLE HOUSE
PAGES 10–11

Can you dinos keep the noise down?

CITY CAR
PAGES 166–167

ANKYLOSAURUS
PAGES 270–271

DIPLODOCUS

JURASSIC PERIOD

With a body almost the length of three school buses, *Diplodocus* (dip-lod-oh-cus) was one long land animal! We know a lot about this dinosaur from the many fossils that have been found. This model is based on a cast of a skeleton that stood in the Natural History Museum in London, UK, for more than 100 years. Its name is "Dippy."

DINO FACT
Paleontologists originally thought that the neck and tail of *Diplodocus* hung downward. In the 1960s, they discovered that this sauropod held them at a horizontal angle, so they changed Dippy's stance.

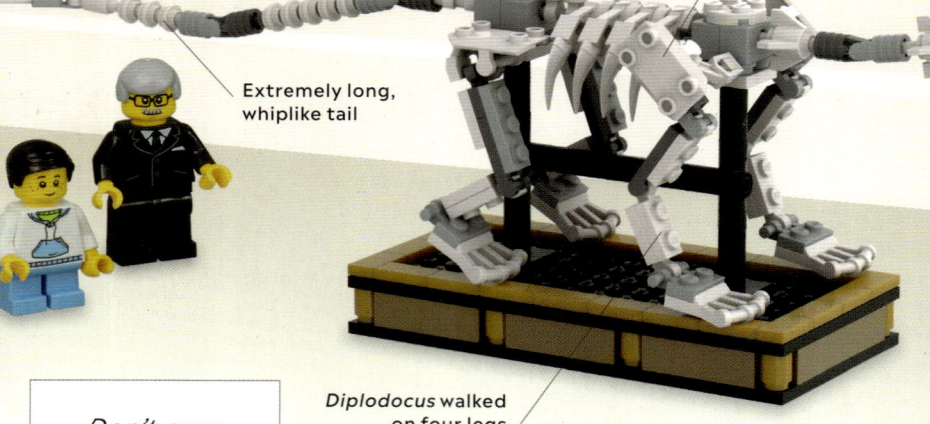

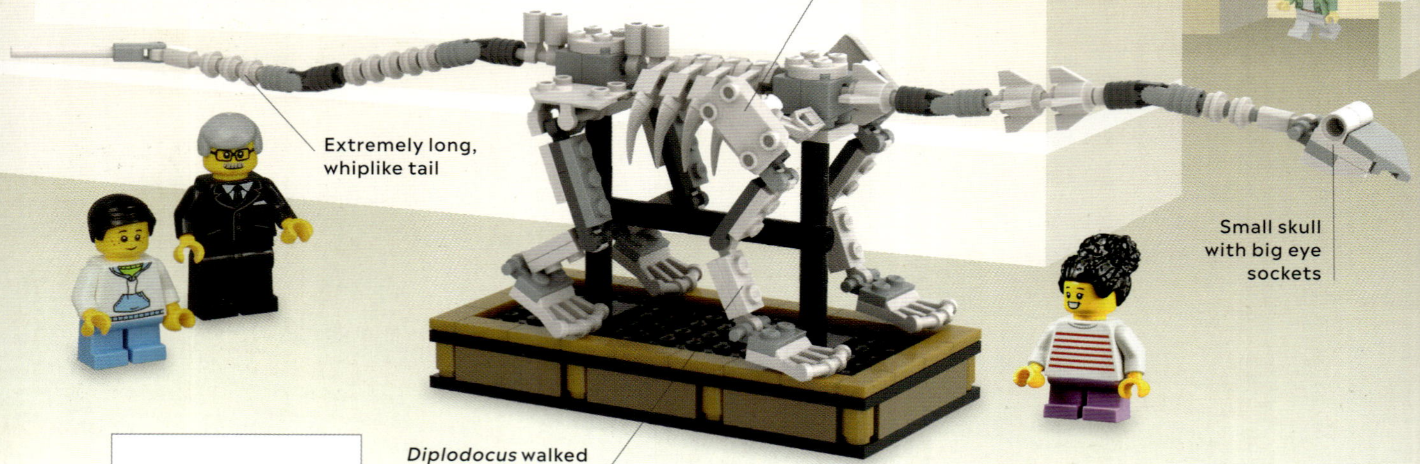

- Shoulder blade
- Extremely long, whiplike tail
- Small skull with big eye sockets
- *Diplodocus* walked on four legs

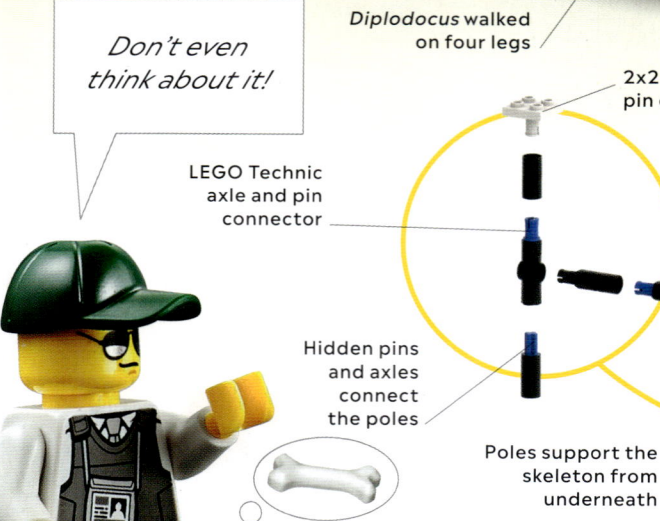

Don't even think about it!

- LEGO Technic axle and pin connector
- 2x2 plate with pin on bottom
- Hidden pins and axles connect the poles
- Poles support the skeleton from underneath

SOMEWHERE TO STAND

Before you start piecing together your Dippy skeleton, build a suitably sturdy display stand for it. This stand looks similar to the one that the real Dippy rests on. It's built up from an 8x16 plate, which is wide enough for the feet to fit onto.

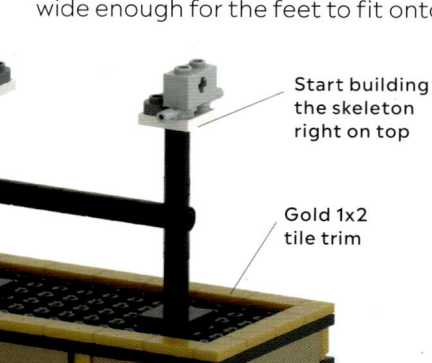

- Start building the skeleton right on top
- Gold 1x2 tile trim
- 8x16 base plate
- Gold 1x1 round bricks add a fancy finish

SKILL LEVEL

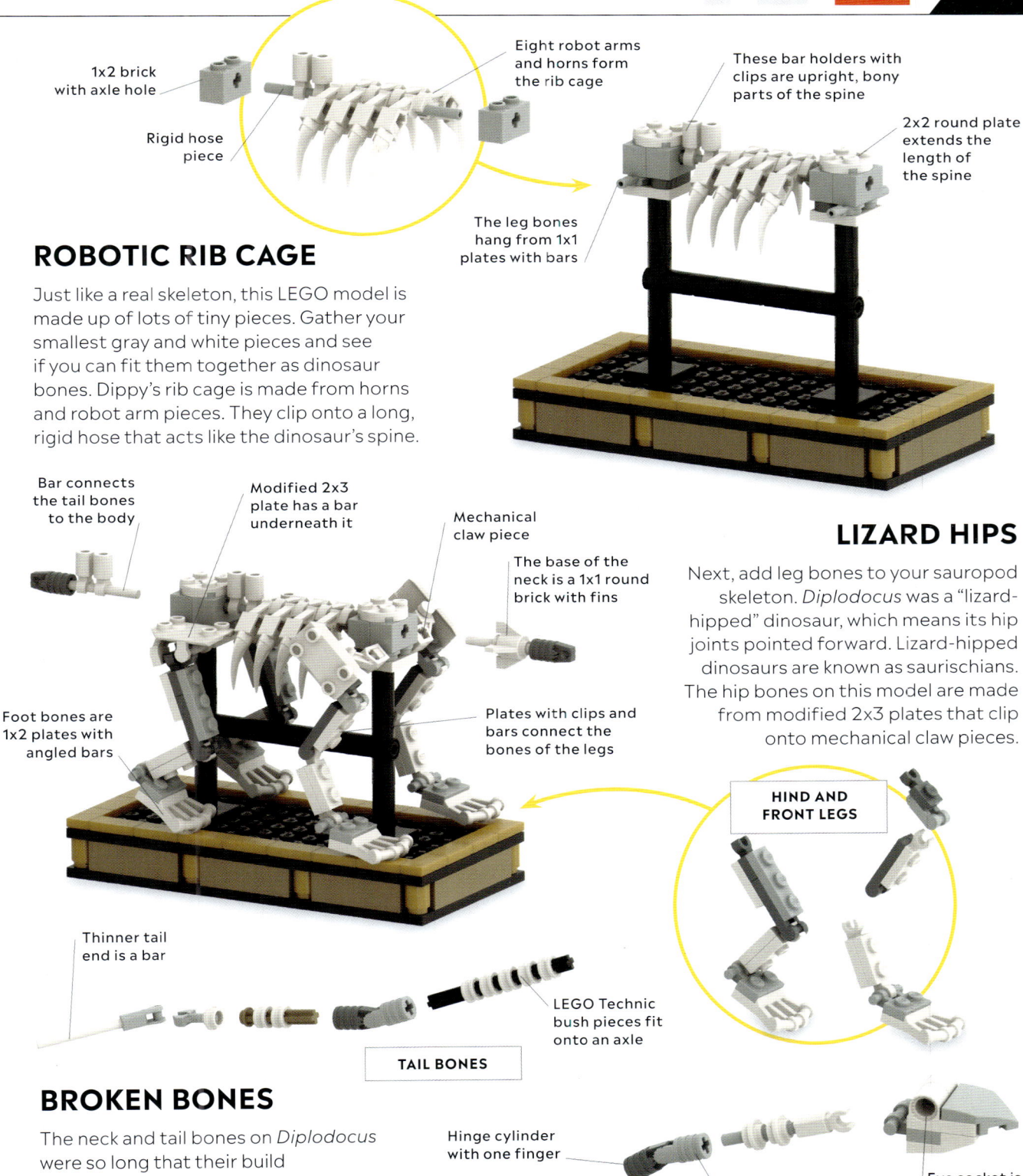

ROBOTIC RIB CAGE

- 1x2 brick with axle hole
- Rigid hose piece
- Eight robot arms and horns form the rib cage
- These bar holders with clips are upright, bony parts of the spine
- 2x2 round plate extends the length of the spine
- The leg bones hang from 1x1 plates with bars

Just like a real skeleton, this LEGO model is made up of lots of tiny pieces. Gather your smallest gray and white pieces and see if you can fit them together as dinosaur bones. Dippy's rib cage is made from horns and robot arm pieces. They clip onto a long, rigid hose that acts like the dinosaur's spine.

- Bar connects the tail bones to the body
- Modified 2x3 plate has a bar underneath it
- Mechanical claw piece
- The base of the neck is a 1x1 round brick with fins
- Foot bones are 1x2 plates with angled bars
- Plates with clips and bars connect the bones of the legs

LIZARD HIPS

Next, add leg bones to your sauropod skeleton. *Diplodocus* was a "lizard-hipped" dinosaur, which means its hip joints pointed forward. Lizard-hipped dinosaurs are known as saurischians. The hip bones on this model are made from modified 2x3 plates that clip onto mechanical claw pieces.

HIND AND FRONT LEGS

- Thinner tail end is a bar
- LEGO Technic bush pieces fit onto an axle

TAIL BONES

BROKEN BONES

The neck and tail bones on *Diplodocus* were so long that their build breakdowns can barely fit on this page! Attach these intricate skeletal structures to the completed body from the previous building stage, and Dippy is ready to wow millions of minifigure museum-goers.

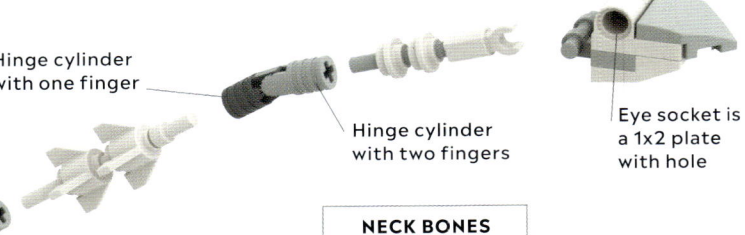

- Hinge cylinder with one finger
- Hinge cylinder with two fingers
- Eye socket is a 1x2 plate with hole

NECK BONES

SMILODON

PLEISTOCENE PERIOD

Look at the terrifying teeth on this not-so-cuddly cat! Often referred to as the saber-toothed cat or tiger, *Smilodon* (*smile-oh-don*) is best known for its two dagger-like canines that extended past its lower jaw. Like its relatives, the big cats of today, *Smilodon* was immensely strong, and an expert hunter and killer.

FUN FACT
Smilodon prowled around forests and grasslands in a pack, hunting for larger but slower animals to eat. You could build a whole pack of these ferocious felines ... if you dare!

- Thick, muscular neck and shoulders
- Smooth ears are upside-down plates with clips
- 1x1 quarter tile nostrils
- Small 2x2 curved slope tail
- Extremely long canine teeth
- Sharp claws are a 1x2 plate with three teeth
- Powerful front legs for pinning down prey

STRONG START

The powerful body of *Smilodon* was strong enough to bring down huge animals, so give yours a sturdy base. This model's belly is built around a 2x6 plate. Four inverted curved slopes and a double inverted slope fit underneath the plate.

- 1x4 inverted curved slope
- 2x6 plate
- 4x4 double inverted slope

BROAD BODY

At twice the weight of a lion, *Smilodon* was bulkily built. Attach bricks horizontally across the belly base to broaden it.

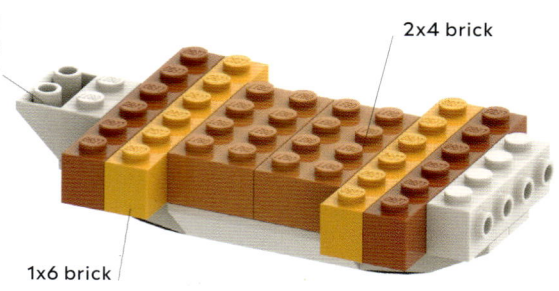

- This inverted slope brick makes the body longer
- 2x4 brick
- 1x6 brick

SKILL LEVEL

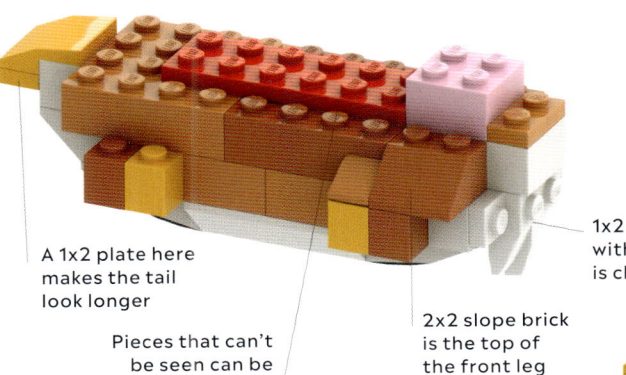

277

FURRY FEATURES

Now *Smilodon* has some soft-looking tufts of chest fur and a small tail, which is much shorter than a lion's tail. The body is also taller and the tops of the legs are starting to take shape.

- A 1x2 plate here makes the tail look longer
- Pieces that can't be seen can be any color
- 1x2 plate with rocks is chest fur
- 2x2 slope brick is the top of the front leg
- This rocky 4x4 wedge looks like ruffled back fur

MUSCULAR LIMBS

It's time to give this creature some super-strong legs to chase down its prey with. The lower part of each leg is built in exactly the same way, but the thighs on the hind legs are extra wide.

- Small gold plates and bricks add texture to the fur
- 2x2 inverted slope brick
- This 2x2 round brick looks like a pad on the paw
- 1x3 inverted slope brick rear thigh

FEARSOME HEAD

Finish off your *Smilodon* with a huge head and those trademark teeth. Give it a lower jaw that can open wide by using connecting pieces that move together, such as clips and bars. This model's head can also swivel around thanks to a ball and socket connection at the neck.

- Leave space here for the back of the head to move freely
- Headlight brick will hold the eye tile
- 1x1 slope creates a smooth nose
- 1x2 plate with bar
- 1x2 plate with socket
- 2x2 inverted curved slope
- The large canine teeth attach here
- 1x1 rounded plate with bar
- 1x3 curved slope adds a curve around the hips

UNDERSIDE VIEW

KENTROSAURUS

JURASSIC PERIOD

This plant-eating dinosaur had a very striking look, in more ways than one! *Kentrosaurus* (*ken-troh-saw-russ*) had two long shoulder spikes, more spikes along its back and on its tail, and pairs of protective plates that ran down its neck and back. It could also use its flexible tail like a weapon, whipping it from side to side at high speed to strike down any predators.

- Sharp tail spike is a spear tip piece
- This pair of plates is made from unicorn horns
- Flexible neck with smaller spikes
- Broad hips support the heavy body
- Long hind legs
- Sharp beak for gathering leafy food
- Extra-long shoulder spike made from a minifigure machete (knife)

DINO FACT
Kentrosaurus had a very small skull that housed a tiny brain. Scientists think its brain was about the size of a plum. Who needs brains when you have all those impressive spikes?

Urgh. Those spikes remind me of stakes!

UNDERBELLY

The rounded underbelly of the heavy body is made from inverted curved wedges and inverted slope bricks. They attach to the underside of a layer of bricks.

- 1x3 brick
- 2x8 brick
- 2x2 inverted slope bricks continue the curve
- 4x4 inverted wedge
- 4x6 triple inverted curved wedge

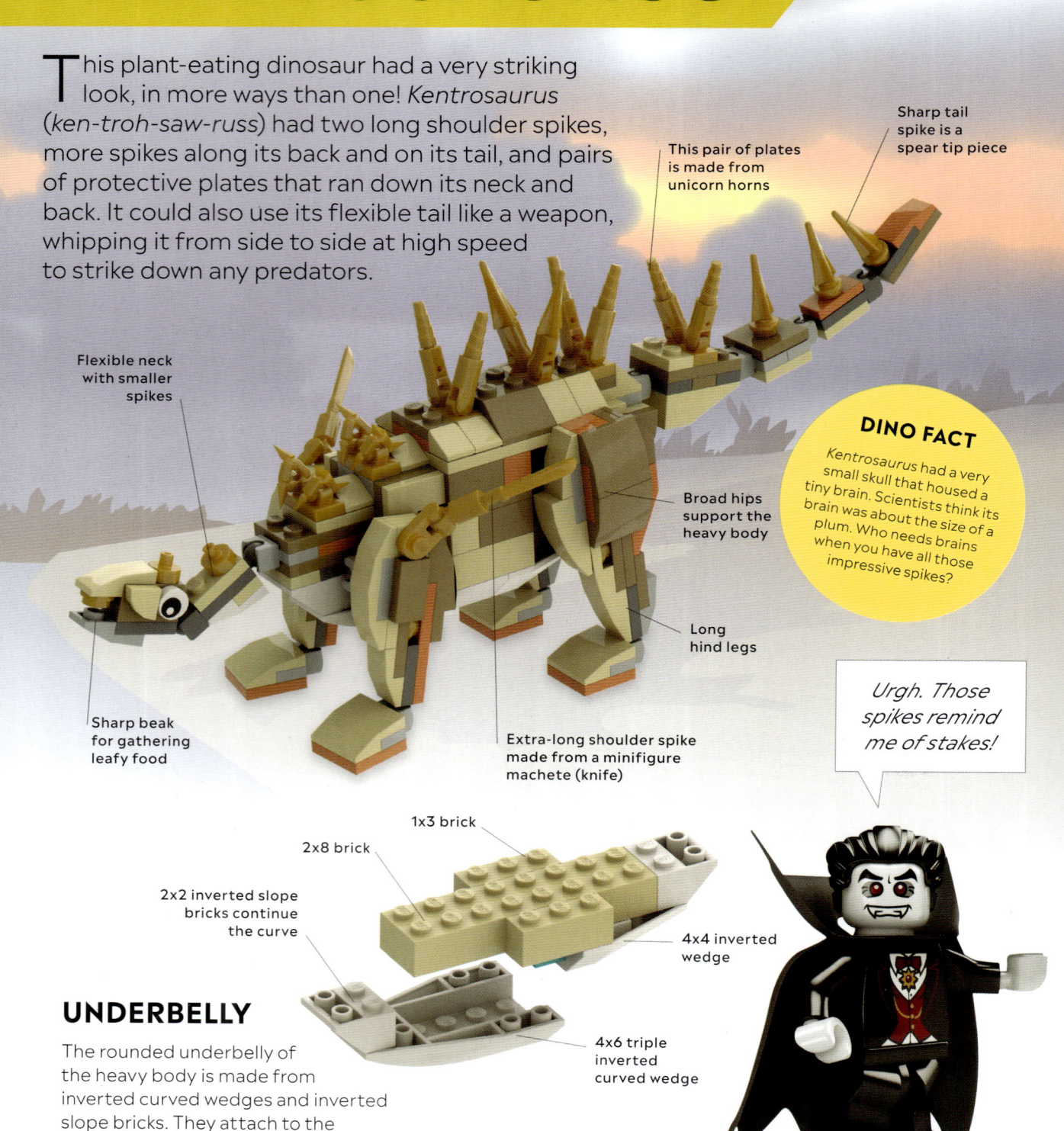

SKILL LEVEL

BUILDING UP

Kentrosaurus was a smaller relative of *Stegosaurus* (pages 248–249), but it still had a hefty body. Build up the height of the body with more bricks and plates, including some with connecting parts where the limbs will attach later.

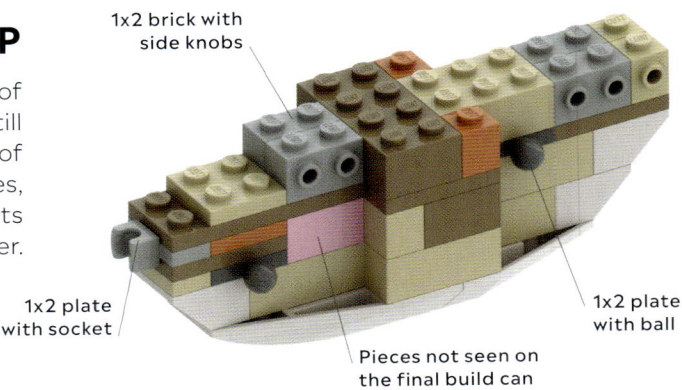

1x2 brick with side knobs

1x2 plate with socket

1x2 plate with ball

Pieces not seen on the final build can be any color

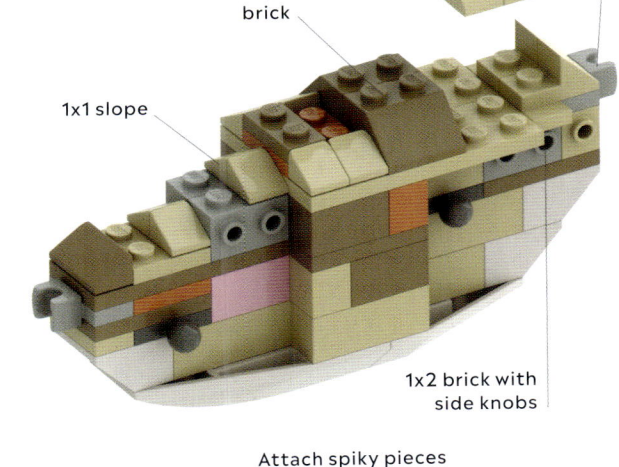

1x2 slope with cutout

2x2 slope brick

1x1 slope

The tail will connect here

BUILDING ROUND

Once you've reached a height you're happy with, top the body with sloped pieces to give it a rounded look. Then add curved slopes and tiles to the sides of the body to make them smoother and chunkier.

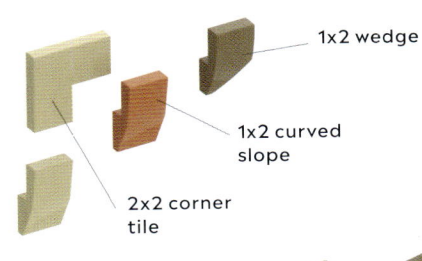

1x2 wedge

1x2 curved slope

2x2 corner tile

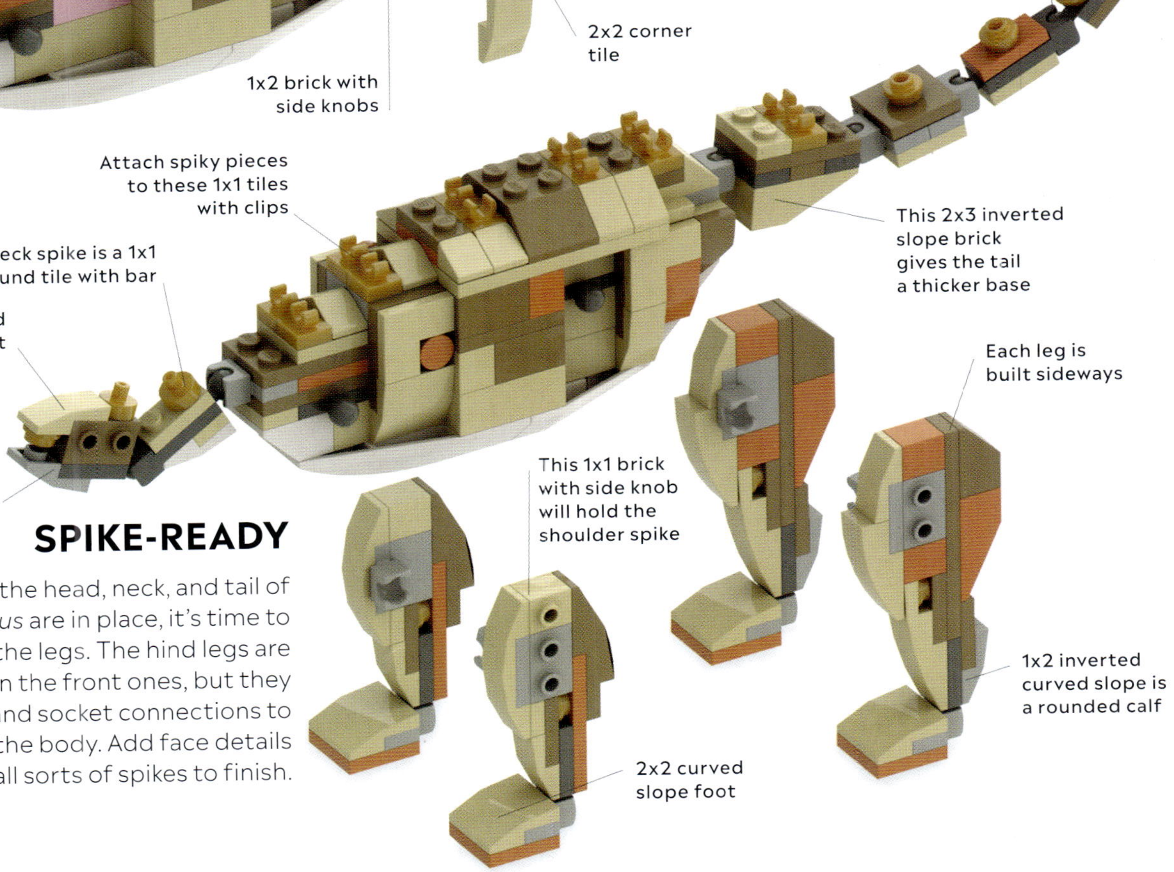

Four of these ball connections make the tail really flexible

1x2 brick with side knobs

Attach spiky pieces to these 1x1 tiles with clips

Neck spike is a 1x1 round tile with bar

1x2 curved slope snout

1x2 inverted curved slope jaw

This 2x3 inverted slope brick gives the tail a thicker base

SPIKE-READY

Now that the head, neck, and tail of *Kentrosaurus* are in place, it's time to attach the legs. The hind legs are longer than the front ones, but they all use ball and socket connections to attach to the body. Add face details and all sorts of spikes to finish.

This 1x1 brick with side knob will hold the shoulder spike

Each leg is built sideways

1x2 inverted curved slope is a rounded calf

2x2 curved slope foot

279

CRETACEOUS PERIOD

TRICERATOPS

This huge herbivore is perhaps one of the most recognizable dinosaurs. The bulkily built *Triceratops* (try-seh-ra-tops) had a tough, bony frill covering its neck and three sharp horns on its head. It needed them to protect itself against meat-eating predators, such as the fearsome *Tyrannosaurus* (pages 282–283).

- Shield-like frill protects the neck and shoulders
- Large rump made from two 2x3x2 round corner bricks
- Shorter snout horn
- Toothless beak for munching on plants
- 1x2 slope toes
- Short, wide legs support its heavy body
- Flexible, three-part tail
- Two 1x4 plates form the chest
- 2x6 plate
- The tail will attach to this 1x2 plate with ball
- 1x4 inverted curved slope
- This 2x2 round plate provides another connection point
- 4x4 double inverted slope with cutout

This hat helps me channel my inner Triceratops.

BULKY BASE

Begin your *Triceratops* by building the lowest part of its heavyweight body. On this model, the belly is built around a 2x6 plate. Four inverted curved slopes and a double inverted slope attach underneath the plate.

SKILL LEVEL

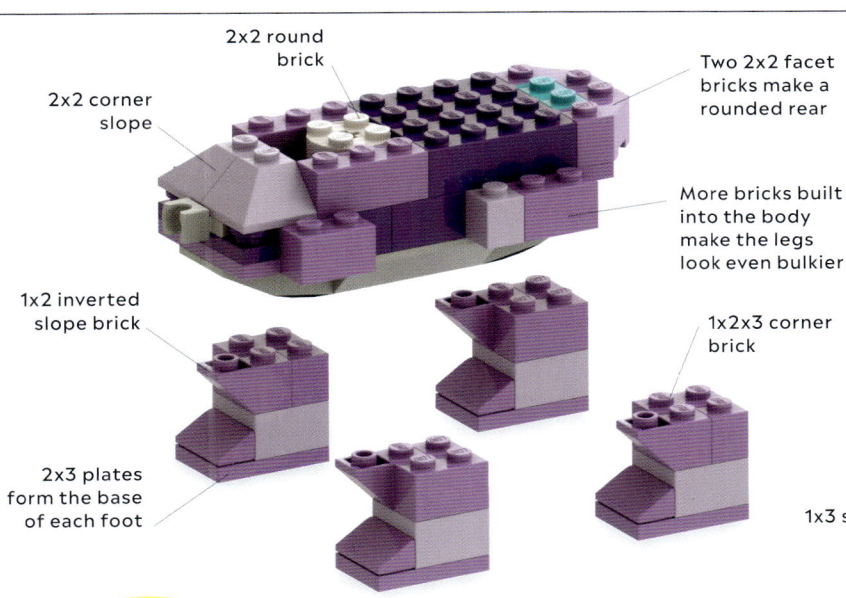

281

2x2 round brick

2x2 corner slope

Two 2x2 facet bricks make a rounded rear

STRONG LIMBS

Bulk up the body of *Triceratops*, then build sturdy legs and feet to support it. Each of its four limbs has a 2x3 plate base, with square bricks on top. Inverted slope bricks add width to the tops of the legs, and slope bricks form the toes.

More bricks built into the body make the legs look even bulkier

1x2 inverted slope brick

1x2x3 corner brick

2x3 plates form the base of each foot

You can also use layers of smaller bricks to make a rounded rear

1x3 slope brick

1x2 slope brick

UPPER BODY

The shoulders, back, and rump of this dinosaur are starting to take shape. The shoulders slope downward to make space for the neck frill at the front end. Round corner bricks add curves at the rear.

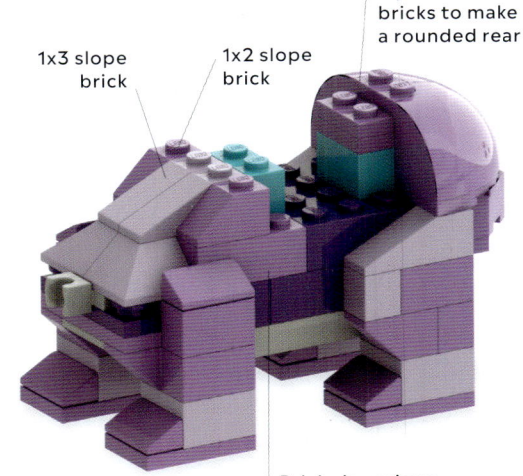

DINO FACT
Herbivorous Triceratops would have eaten lots of prehistoric plants, such as cycads and ferns. You could build some for your model to chomp on. Have a look at pictures online and get building!

Bricks in various sizes add height to the body

Forward-facing horns

6x6 round corner brick for the bony frill

1x4 curved slopes form the curves of the back

2x2 slope brick

2x3 slope brick for this wider tail segment

1x2 plate with ball

MOVING PARTS

Now that the top of the body is complete, it's time to add the distinctive head and thick tail. Both parts attach to the body using ball and socket connections. These connecting pieces allow *Triceratops* to swish its tail from side to side.

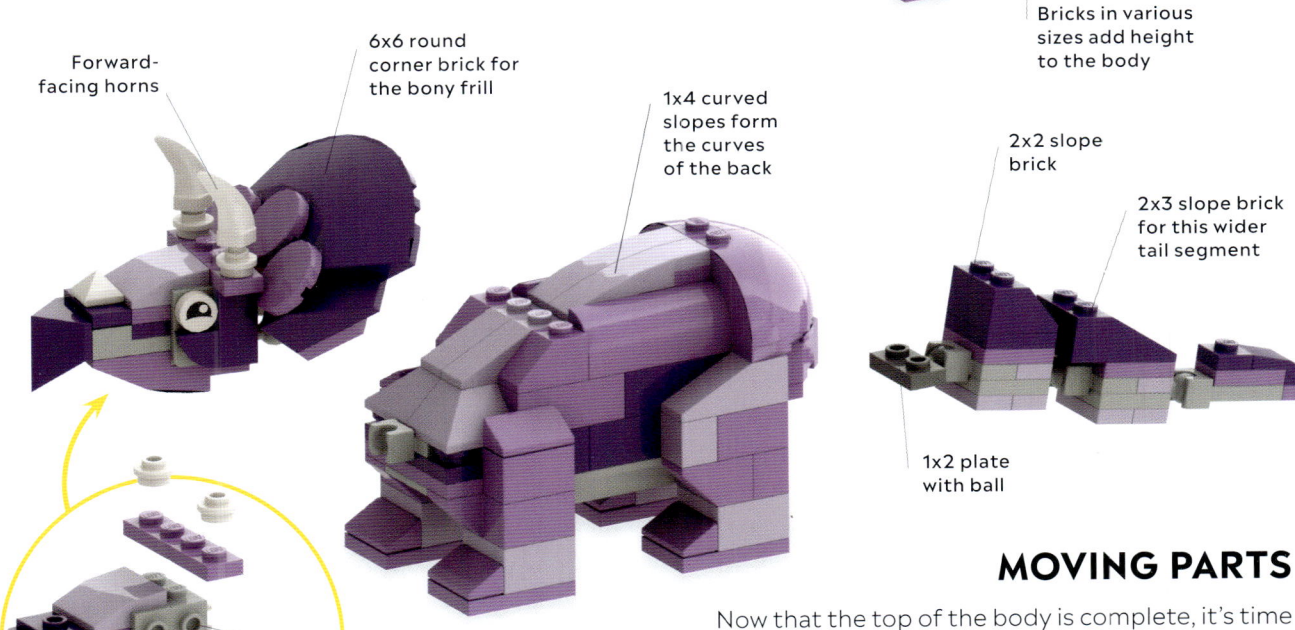

2x2 plate with ball

The face details attach to this 1x2/2x2 bracket piece

TYRANNOSAURUS

CRETACEOUS PERIOD

Make way for the intimidating king of the dinosaurs—*Tyrannosaurus* (tie-ran-oh-sore-us). This threatening theropod was one of the biggest predators ever to roam the planet! It could chase down prey on its two powerful legs and crunch through bones with its massive jaws and sharp teeth.

Huge neck muscles helped it tear through prey

Pointed fangs are 1x1 plates with vertical teeth

Long tail held out for balance on two legs

Long legs with powerful thighs

1x4 curved slope shin

Clawed foot made from a 1x2 plate with three teeth

He's actually quite sweet!

DINO FACT
Tyrannosaurus had two short arms at the front of its body, each with two sharp, curved claws. Scientists aren't sure what they were used for but they may have slashed at prey in close combat while the dinosaur bit down with its jaws.

Two 2x3 wedge plates make this shape

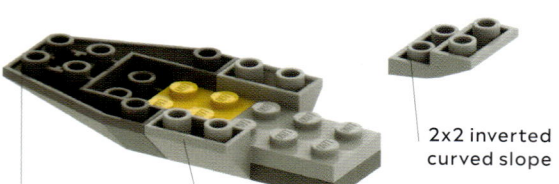

4x4 triple inverted wedge slope

2x4 double inverted slope

2x2 inverted curved slope

START SMALL

A real *Tyrannosaurus* was the size of two elephants, but a LEGO version can be any size you like! Begin the curved underside of this model with inverted slopes and plates in various sizes.

SKILL LEVEL

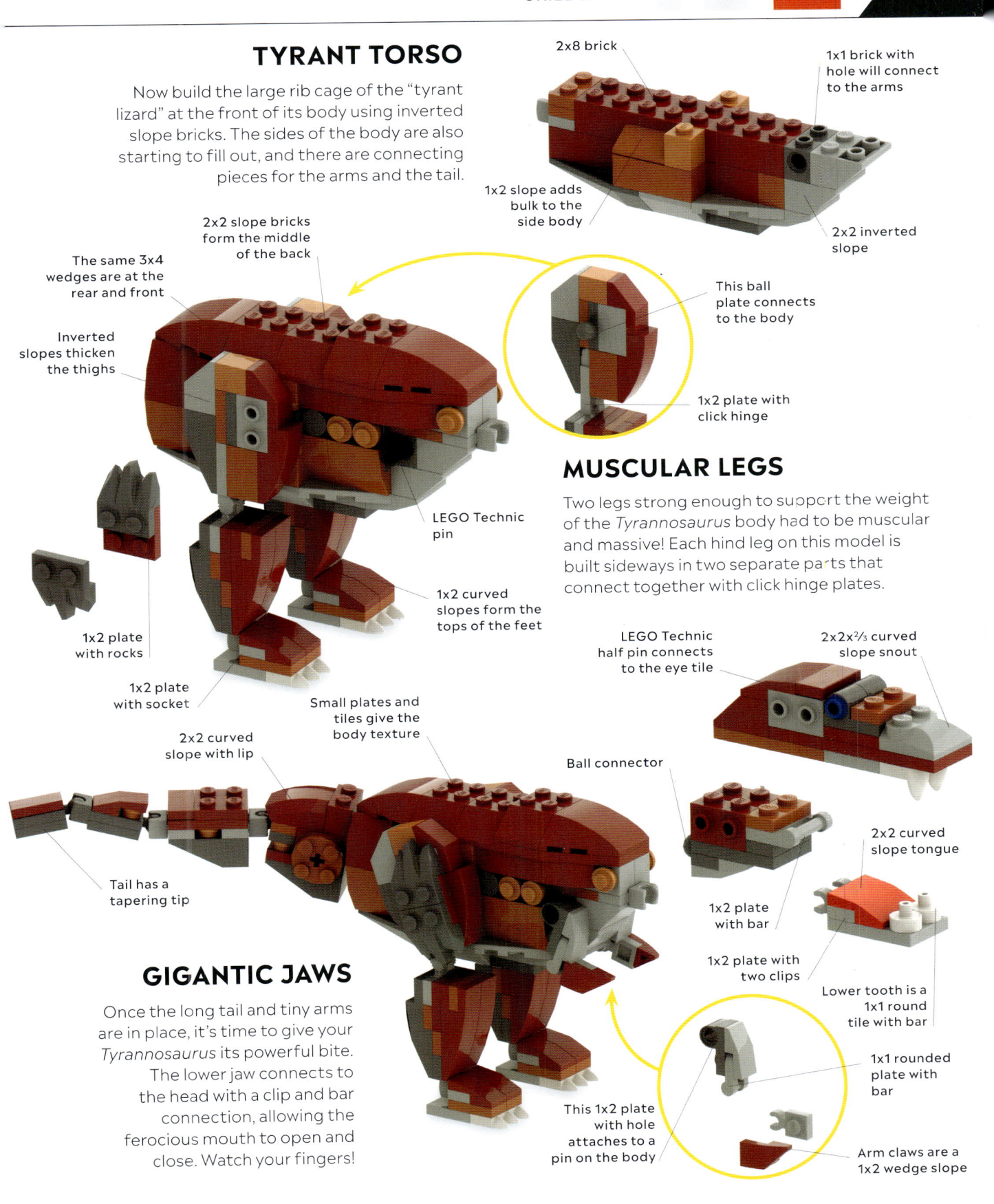

283

TYRANT TORSO

Now build the large rib cage of the "tyrant lizard" at the front of its body using inverted slope bricks. The sides of the body are also starting to fill out, and there are connecting pieces for the arms and the tail.

2x8 brick

1x1 brick with hole will connect to the arms

1x2 slope adds bulk to the side body

2x2 inverted slope

2x2 slope bricks form the middle of the back

The same 3x4 wedges are at the rear and front

Inverted slopes thicken the thighs

This ball plate connects to the body

1x2 plate with click hinge

MUSCULAR LEGS

Two legs strong enough to support the weight of the *Tyrannosaurus* body had to be muscular and massive! Each hind leg on this model is built sideways in two separate parts that connect together with click hinge plates.

LEGO Technic pin

1x2 curved slopes form the tops of the feet

LEGO Technic half pin connects to the eye tile

2x2x2/3 curved slope snout

1x2 plate with rocks

1x2 plate with socket

2x2 curved slope with lip

Small plates and tiles give the body texture

Ball connector

2x2 curved slope tongue

1x2 plate with bar

1x2 plate with two clips

Lower tooth is a 1x1 round tile with bar

Tail has a tapering tip

1x1 rounded plate with bar

GIGANTIC JAWS

Once the long tail and tiny arms are in place, it's time to give your *Tyrannosaurus* its powerful bite. The lower jaw connects to the head with a clip and bar connection, allowing the ferocious mouth to open and close. Watch your fingers!

This 1x2 plate with hole attaches to a pin on the body

Arm claws are a 1x2 wedge slope

Robots

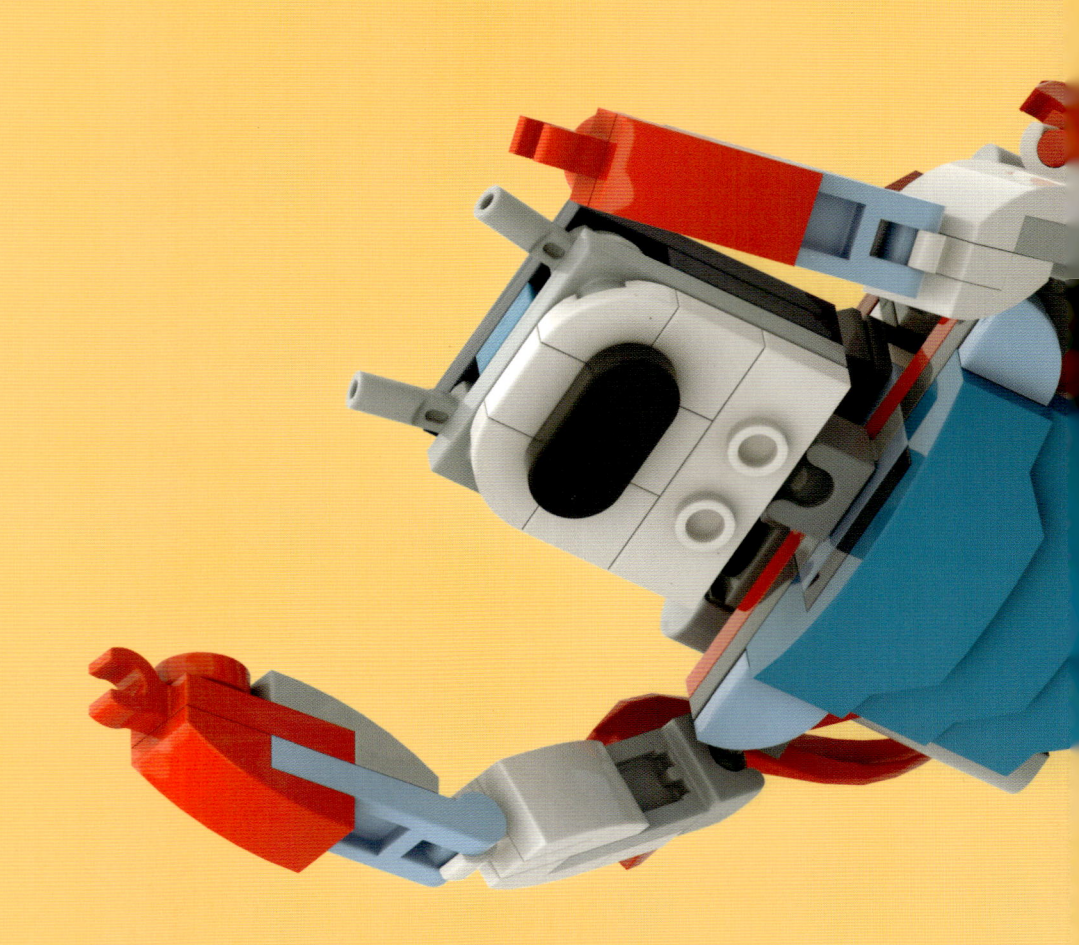

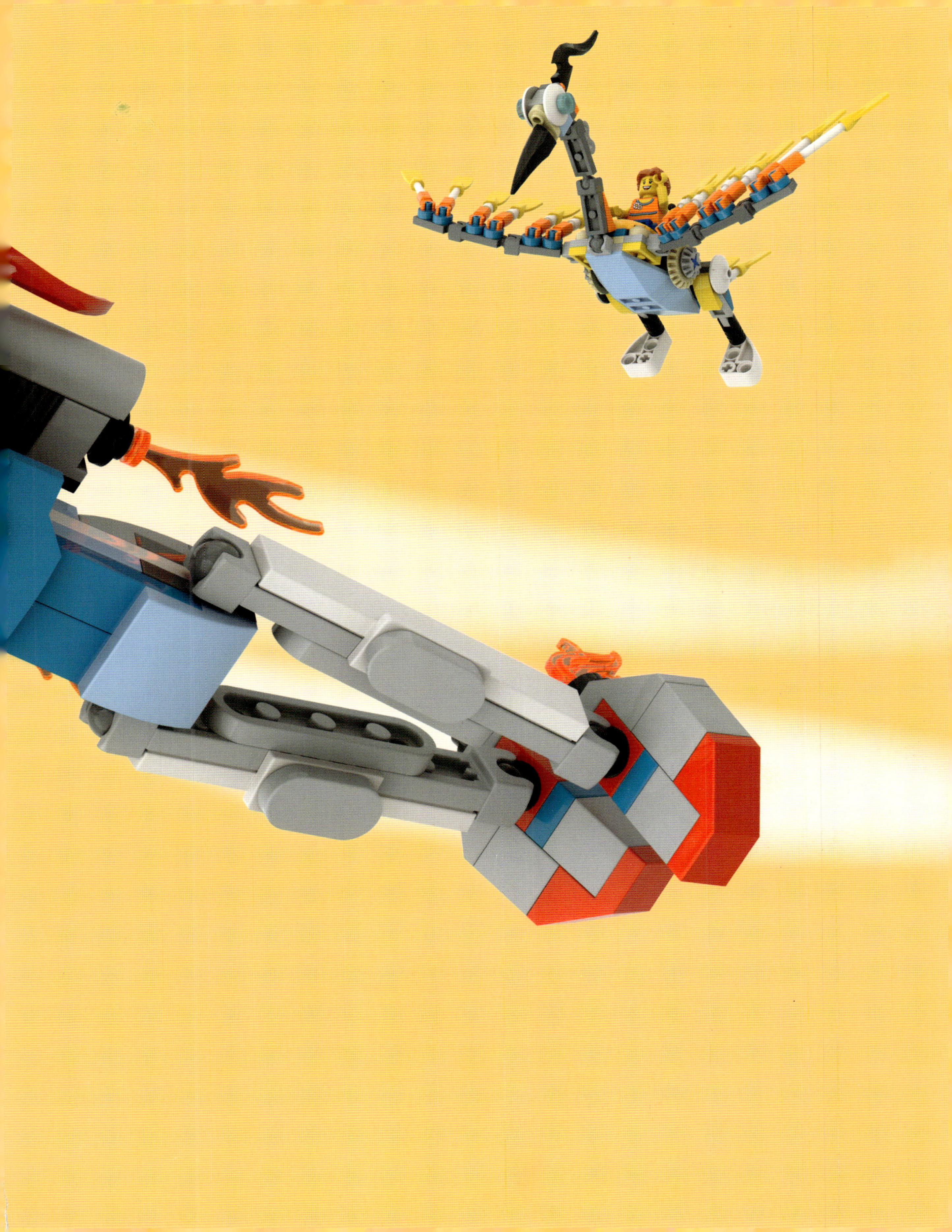

BOXY BOT

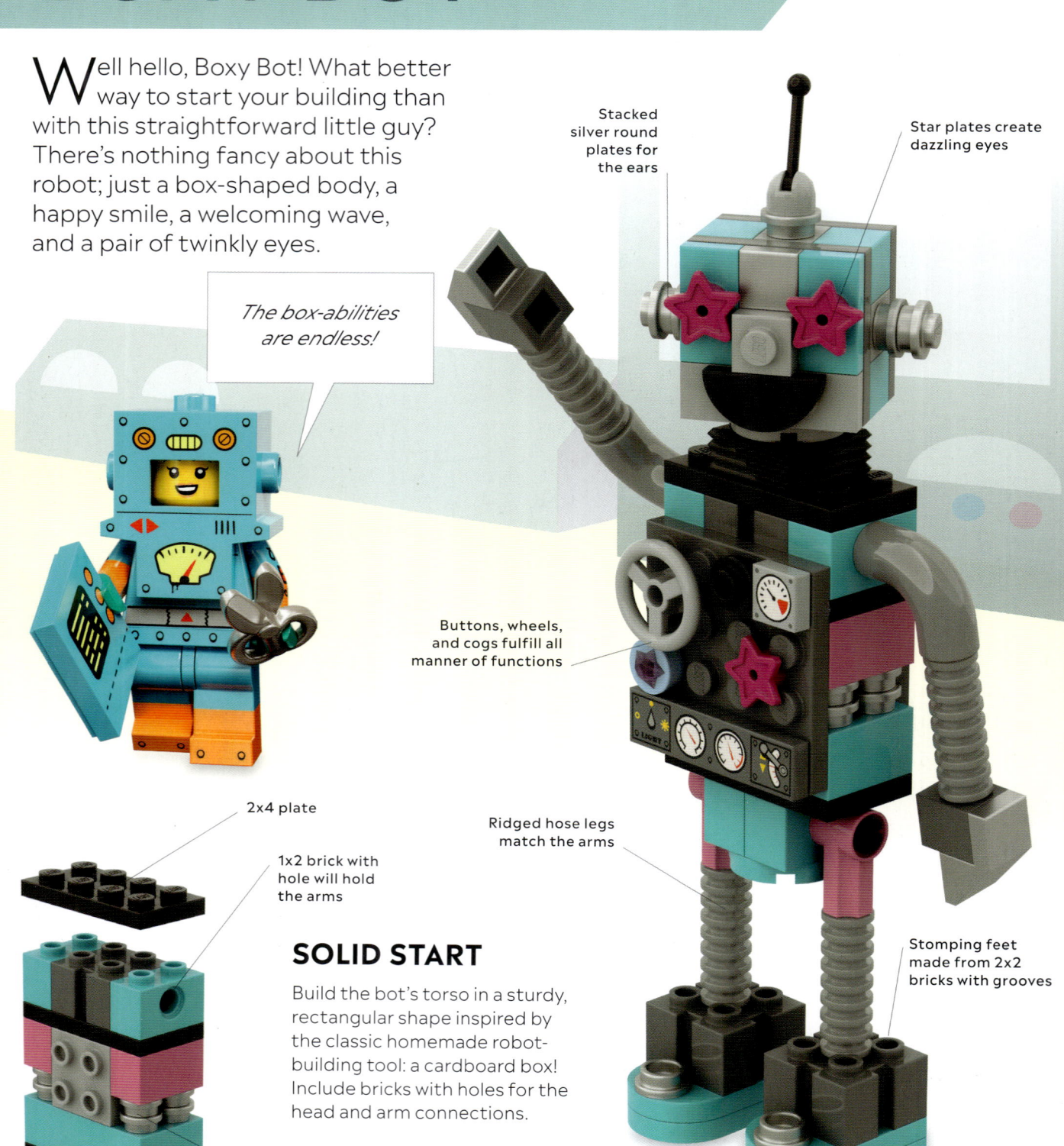

Well hello, Boxy Bot! What better way to start your building than with this straightforward little guy? There's nothing fancy about this robot; just a box-shaped body, a happy smile, a welcoming wave, and a pair of twinkly eyes.

The box-abilities are endless!

Stacked silver round plates for the ears

Star plates create dazzling eyes

Buttons, wheels, and cogs fulfill all manner of functions

Ridged hose legs match the arms

Stomping feet made from 2x2 bricks with grooves

2x4 plate

1x2 brick with hole will hold the arms

SOLID START

Build the bot's torso in a sturdy, rectangular shape inspired by the classic homemade robot-building tool: a cardboard box! Include bricks with holes for the head and arm connections.

Stacked 1x1 round plates look like machinery

SKILL LEVEL

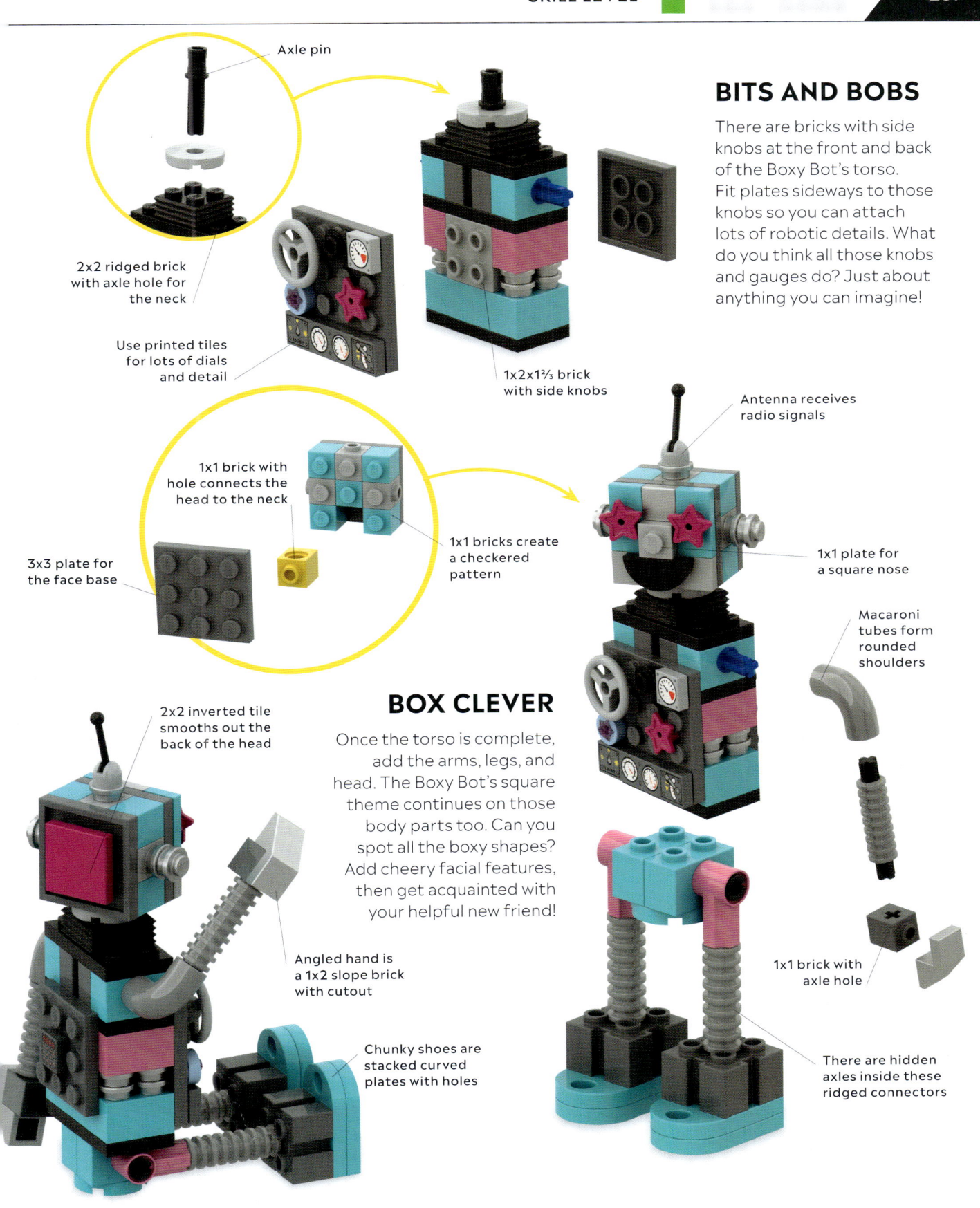

287

BITS AND BOBS

There are bricks with side knobs at the front and back of the Boxy Bot's torso. Fit plates sideways to those knobs so you can attach lots of robotic details. What do you think all those knobs and gauges do? Just about anything you can imagine!

Axle pin

2x2 ridged brick with axle hole for the neck

Use printed tiles for lots of dials and detail

1x2x1²/₃ brick with side knobs

1x1 brick with hole connects the head to the neck

1x1 bricks create a checkered pattern

3x3 plate for the face base

Antenna receives radio signals

1x1 plate for a square nose

Macaroni tubes form rounded shoulders

2x2 inverted tile smooths out the back of the head

BOX CLEVER

Once the torso is complete, add the arms, legs, and head. The Boxy Bot's square theme continues on those body parts too. Can you spot all the boxy shapes? Add cheery facial features, then get acquainted with your helpful new friend!

Angled hand is a 1x2 slope brick with cutout

Chunky shoes are stacked curved plates with holes

1x1 brick with axle hole

There are hidden axles inside these ridged connectors

ROBOT HELPER

Have you had a hard day? This eager Robot Helper is waiting to roll forward and greet you with a drink on a tray. One of its flexible arms is free, so perhaps there is something else it could bring you too. A copy of *Robot News*?

Don't fall in the pool again, Botty.

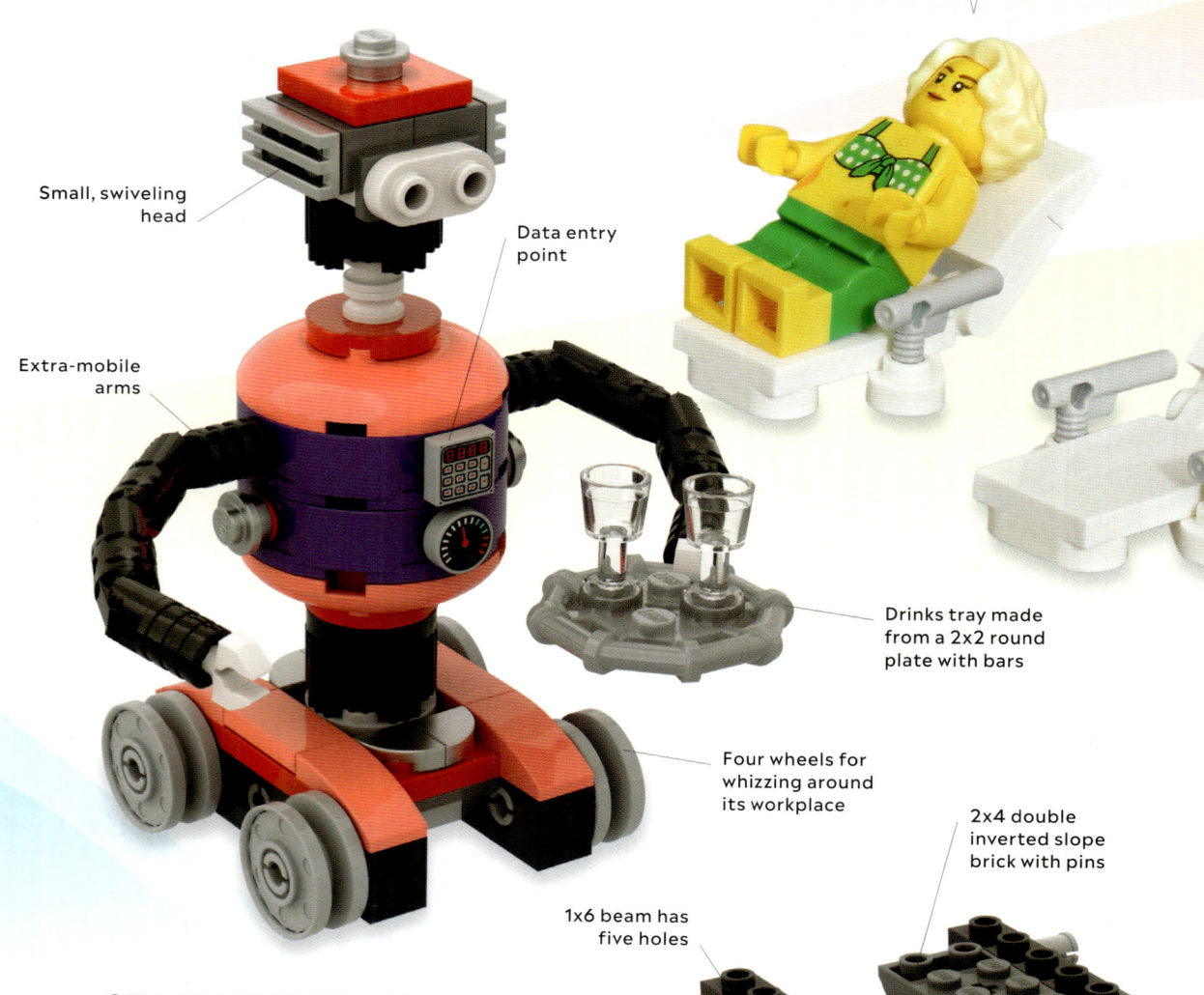

- Small, swiveling head
- Data entry point
- Extra-mobile arms
- Drinks tray made from a 2x2 round plate with bars
- Four wheels for whizzing around its workplace
- 2x4 double inverted slope brick with pins
- 1x6 beam has five holes
- Pin will later hold a wheel

START THE TANK

With its wide, rolling wheels, the Robot Helper's lower body looks a bit like an army tank. Begin that section by sandwiching a double inverted slope brick between two beams. Don't add wheels yet, as they have a habit of rolling away!

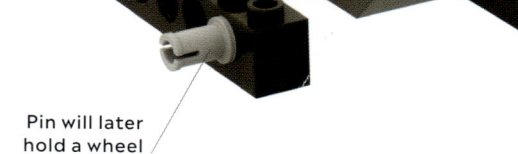

SKILL LEVEL

289

This 2x2 round plate has an axle hole

4x4 frame plate with 2x2 square cutout

BIGGER BASE

Next, build more plates onto the lower body to make it a stable base for the busy bot. A 4x4 plate with a square hole in the middle fits neatly across the wheel beams.

This ridged round brick is the lower part of the head

Tiny wheel pieces form the neck

4x4 round brick with holes

Long axle

Upside-down 4x4 dome

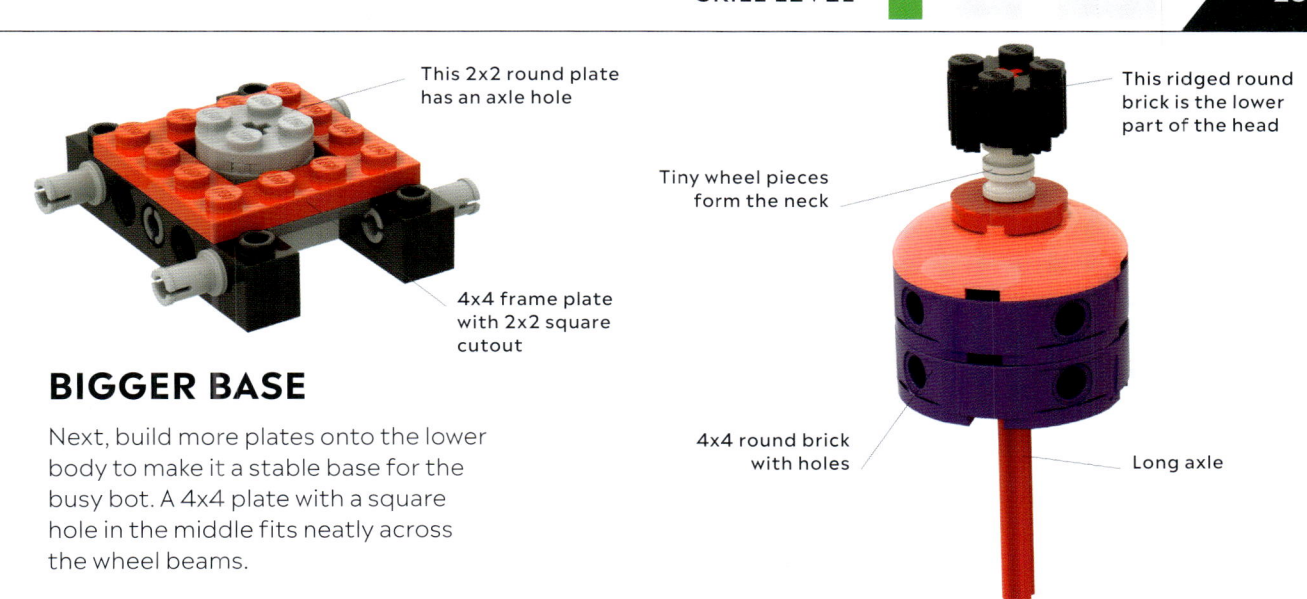

UP AND DOWN

Keep building up and up until you reach the Robot Helper's head. A long, red axle threads right through the bot, from the lower body to the face, and holds the whole body together. Can you spot the two pieces that fit onto the robot upside down?

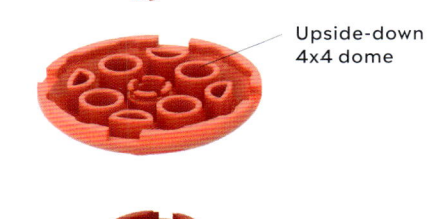

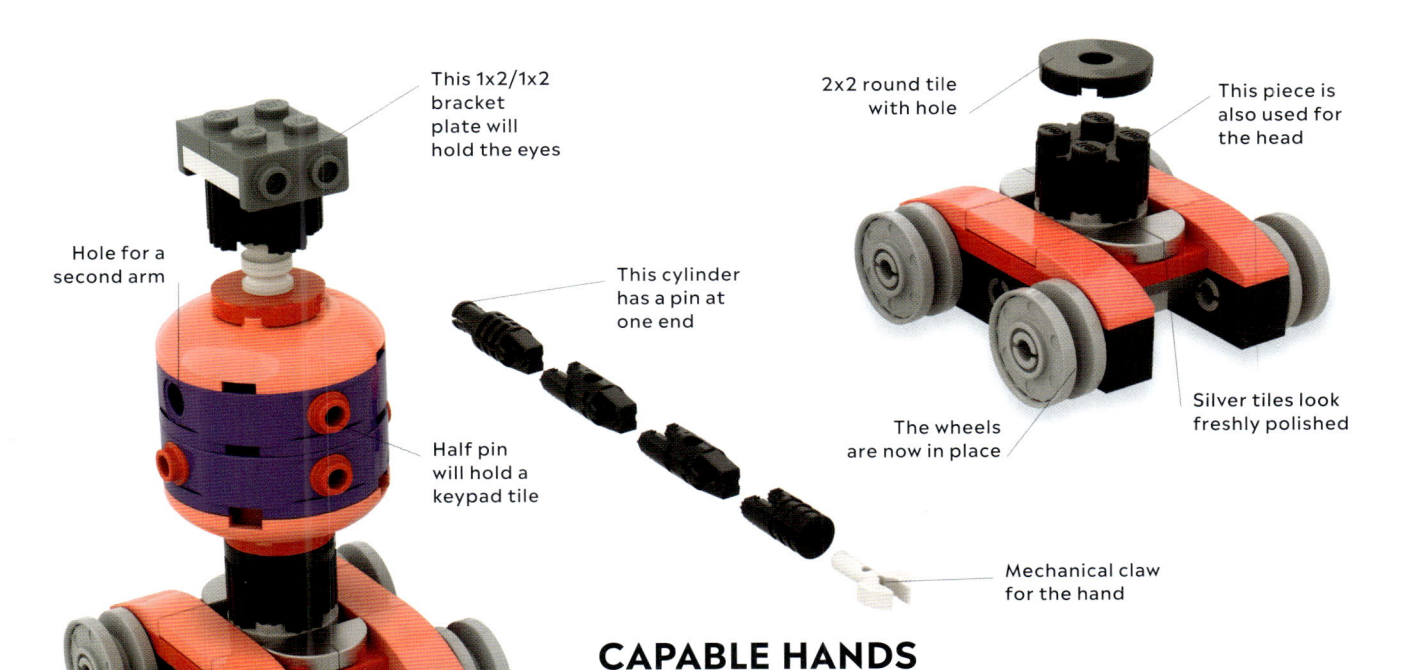

Hole for a second arm

This 1x2/1x2 bracket plate will hold the eyes

This cylinder has a pin at one end

Half pin will hold a keypad tile

2x2 round tile with hole

This piece is also used for the head

The wheels are now in place

Silver tiles look freshly polished

Mechanical claw for the hand

CAPABLE HANDS

LEGO® Technic click hinge cylinders link together to form the Robot Helper's extra-long, flexible arms. Its helping hands are made from mechanical claw pieces. They look like the pristine white-gloved hands of a butler. This little bot is ready to serve!

ROBO RECYCLER

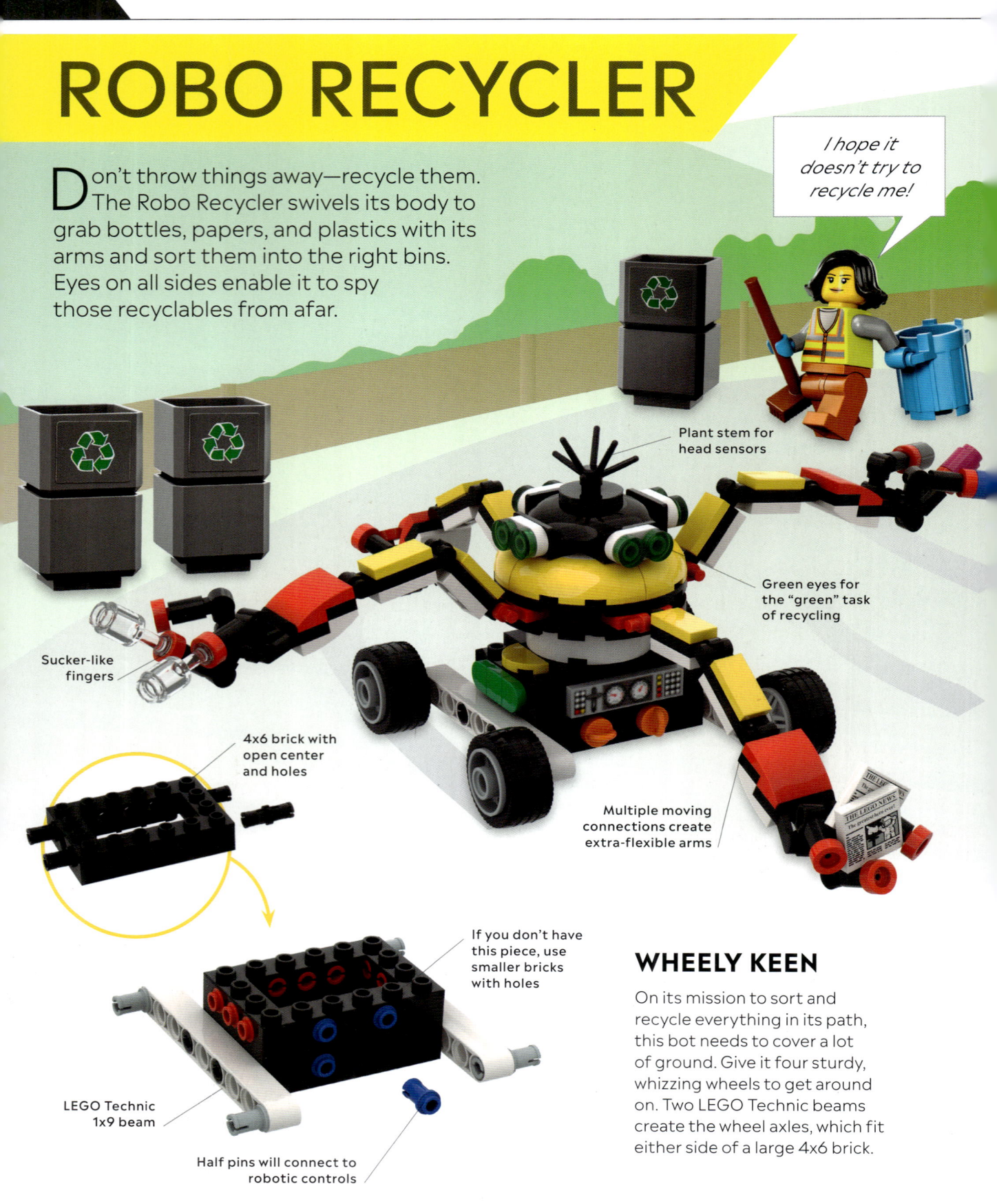

Don't throw things away—recycle them. The Robo Recycler swivels its body to grab bottles, papers, and plastics with its arms and sort them into the right bins. Eyes on all sides enable it to spy those recyclables from afar.

I hope it doesn't try to recycle me!

Plant stem for head sensors

Green eyes for the "green" task of recycling

Multiple moving connections create extra-flexible arms

Sucker-like fingers

4x6 brick with open center and holes

If you don't have this piece, use smaller bricks with holes

LEGO Technic 1x9 beam

Half pins will connect to robotic controls

WHEELY KEEN

On its mission to sort and recycle everything in its path, this bot needs to cover a lot of ground. Give it four sturdy, whizzing wheels to get around on. Two LEGO Technic beams create the wheel axles, which fit either side of a large 4x6 brick.

SKILL LEVEL

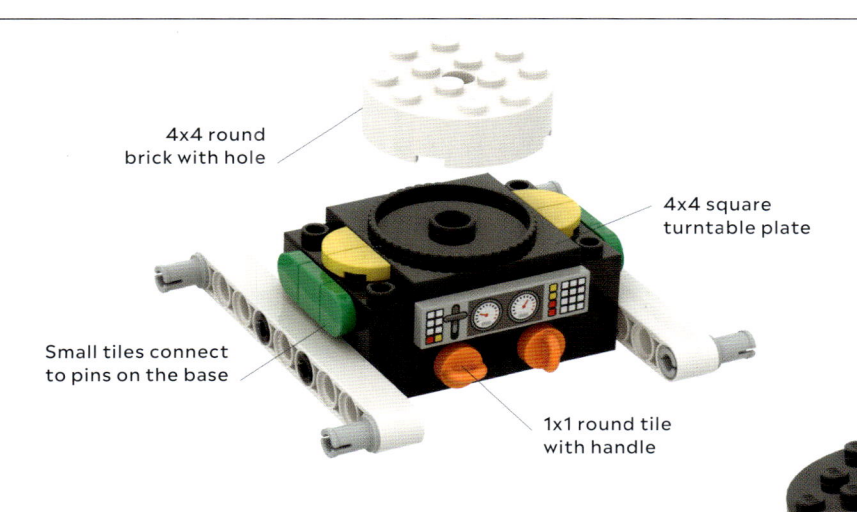

TWISTS AND TURNS

Add knobs, buttons, and dials to the control center base of the bot, then top that section with a turntable piece. This will allow the bot's upper body, including the arms, to swivel around in two directions.

- 4x4 round brick with hole
- 4x4 square turntable plate
- Small tiles connect to pins on the base
- 1x1 round tile with handle

IN A SPIN

Now the spinning section of the body is coming together. Sandwiched between two black round plates are plates with bars for the arm connections. Small transparent round plates at this level look like a strip of red light.

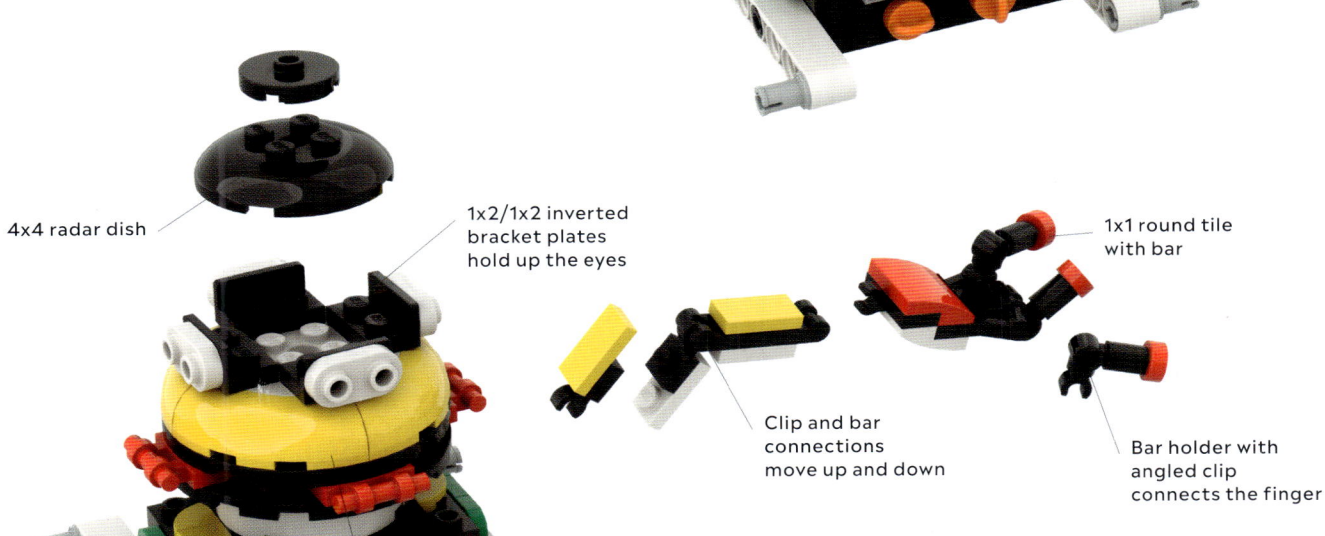

- 6x6 round plate
- 1x2 plate with bar
- This round plate supports the center
- 1x1 round plate can be seen through the gap

- 4x4 radar dish
- 1x2/1x2 inverted bracket plates hold up the eyes
- Clip and bar connections move up and down
- 1x1 round tile with bar
- Bar holder with angled clip connects the finger
- Round corner bricks create upper body curves

ARMED AND READY

This bot would be nowhere near as helpful without its four long, articulated arms and garbage-grabbing hands. Attach them before you finish, along with the multi-eyed head. Finally, add the wheels and just watch this green machine go!

DETAILS AND GREEBLES

LEGO® robots should look scientifically convincing, but you don't have to explain (even to yourself) what each little detail is meant to be. Cover your builds with any tiny elements that look like vital components but are really just decorative.

CIRCUIT BOARD

Have you ever heard the LEGO term "greebling"? This circuit board is a good example of it. It means to group together or layer up lots of small details to create an interesting, mechanical appearance. Give greebling a try on your robotic creations!

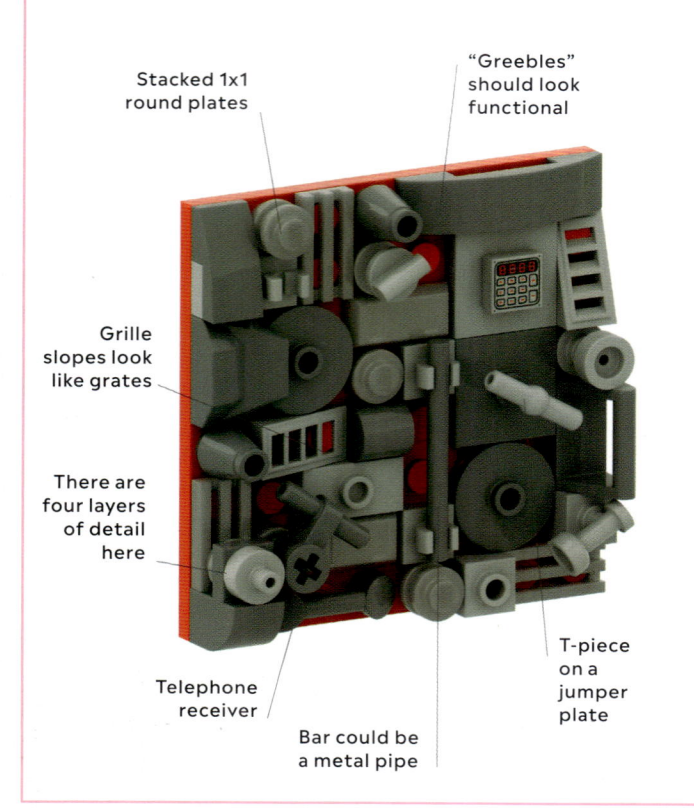

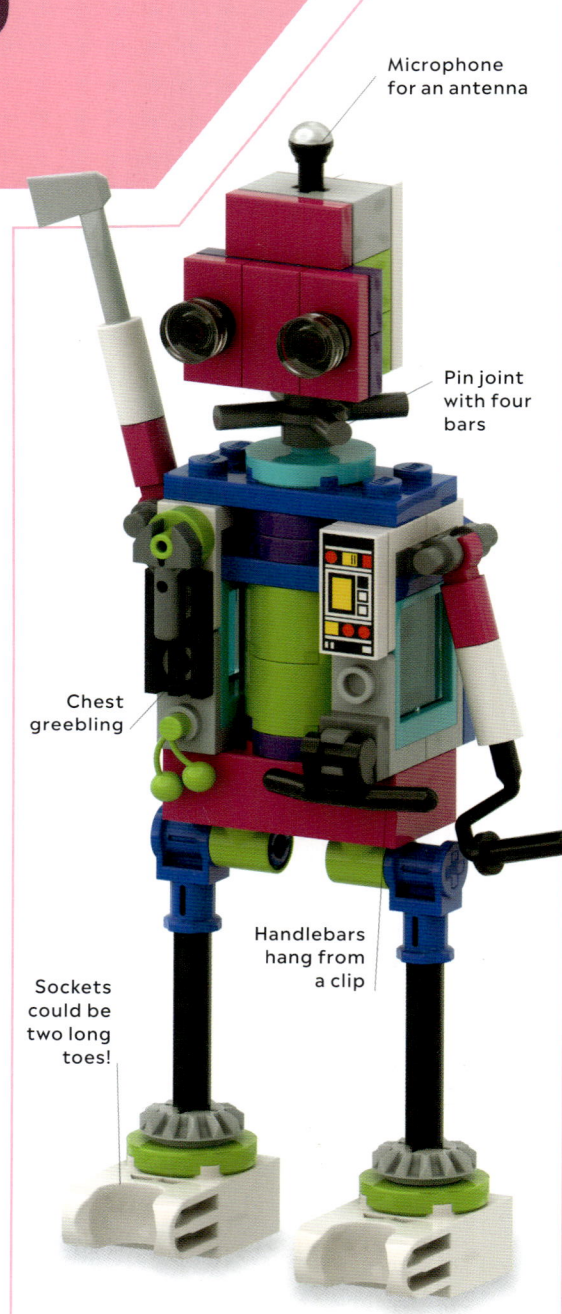

- Stacked 1x1 round plates
- "Greebles" should look functional
- Grille slopes look like grates
- There are four layers of detail here
- Telephone receiver
- Bar could be a metal pipe
- T-piece on a jumper plate

- Microphone for an antenna
- Pin joint with four bars
- Chest greebling
- Sockets could be two long toes!
- Handlebars hang from a clip

EVEN JUNKIER BOT

This is the Junk Robot from pages 340–341 reimagined. Its core body parts are the same, but it now has a paint roller and an ax for hands. Can you spot what else has changed? You really can make something out of nothing!

IDEAS GALLERY 293

DIFFERENT PARTS

The tiniest details can completely change the look of a LEGO robot part. These mechanical arms are all created using the same basic elements and connections, but they look very different.

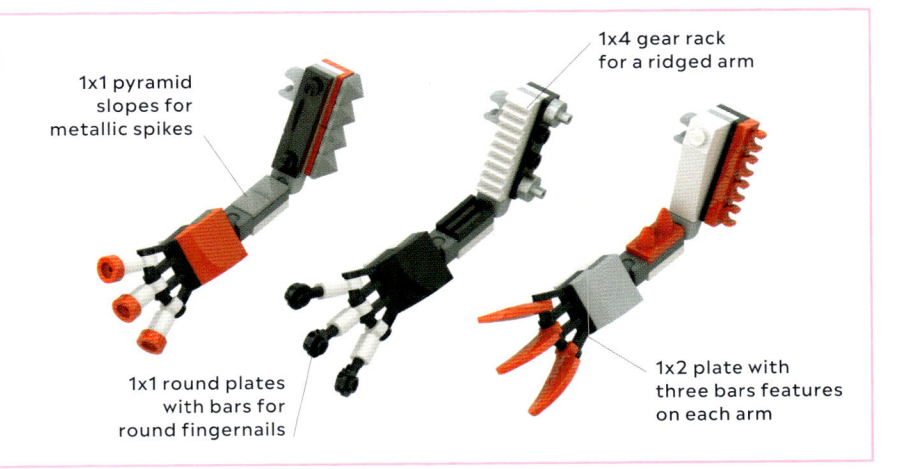

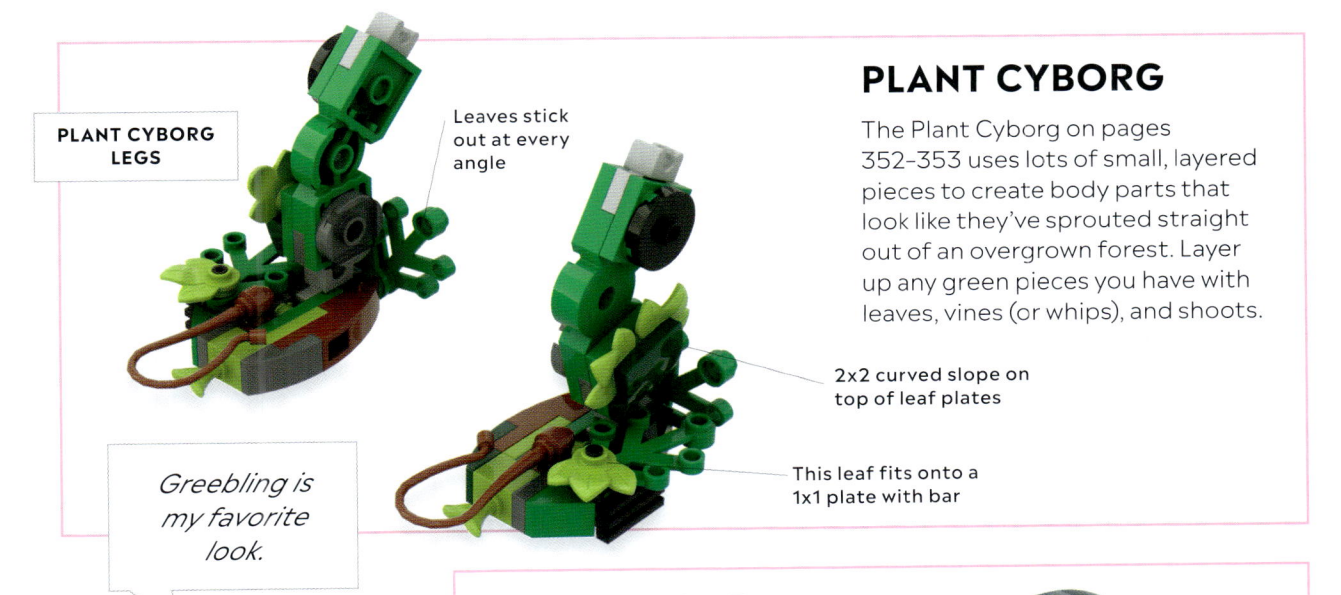

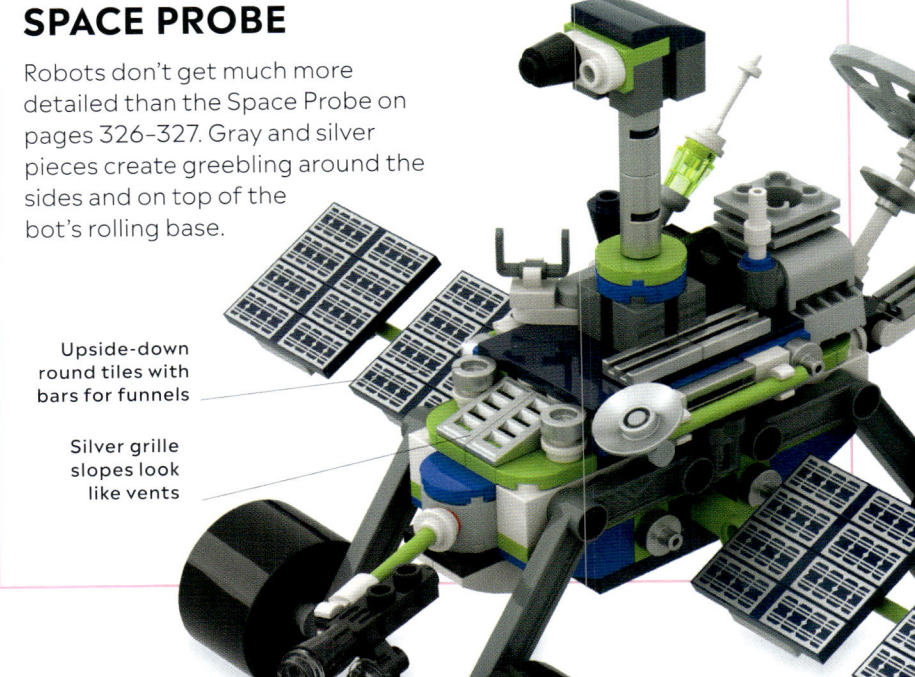

- 1x1 pyramid slopes for metallic spikes
- 1x4 gear rack for a ridged arm
- 1x1 round plates with bars for round fingernails
- 1x2 plate with three bars features on each arm

PLANT CYBORG

The Plant Cyborg on pages 352–353 uses lots of small, layered pieces to create body parts that look like they've sprouted straight out of an overgrown forest. Layer up any green pieces you have with leaves, vines (or whips), and shoots.

PLANT CYBORG LEGS

- Leaves stick out at every angle
- 2x2 curved slope on top of leaf plates
- This leaf fits onto a 1x1 plate with bar

Greebling is my favorite look.

SPACE PROBE

Robots don't get much more detailed than the Space Probe on pages 326–327. Gray and silver pieces create greebling around the sides and on top of the bot's rolling base.

- Upside-down round tiles with bars for funnels
- Silver grille slopes look like vents

CHEF BOT

Yes, Chef! Don't upset Chef Bot, or it might flip. The hot-and-bothered bot is in a bit of a stew as it rolls around the kitchen in its white hat and apron. Perhaps it's worried about dropping that precious last egg.

Egg-cellent cooking, Chef.

Tall white chef's hat

Doughnut tiles for eyes

Flick this arm and the egg flips out of the pan

Skateboard wheels for kitchen whizzing

Recipe input device

3x3 plates fill out the middle

6x6 round plate base

2x2x1⅓ curved corner slope

1x1/1x2 bracket plate will hold techy details

1x3 brick adds height to the body

APRON ON

What's the first thing a chef should do in the kitchen? Put on their apron! (Then wash their hands ...) Build Chef Bot in the same order, placing white pieces on top of the circular base to form the classic cover-up. Use a different color for the pieces at the back of the body.

SKILL LEVEL

295

- 1x1 brick with hole makes the arm connection
- 2x1x1⅓ curved slope
- White plate for the apron string
- Layers of small plates

RISING UP

Build upward with more bricks and plates, then add curved slopes for the bot's smooth shoulders. Include two bricks with holes in the upper body for the arms to attach to at the next stage.

- Skateboard wheels on a 2x2 round plate with wheel holder

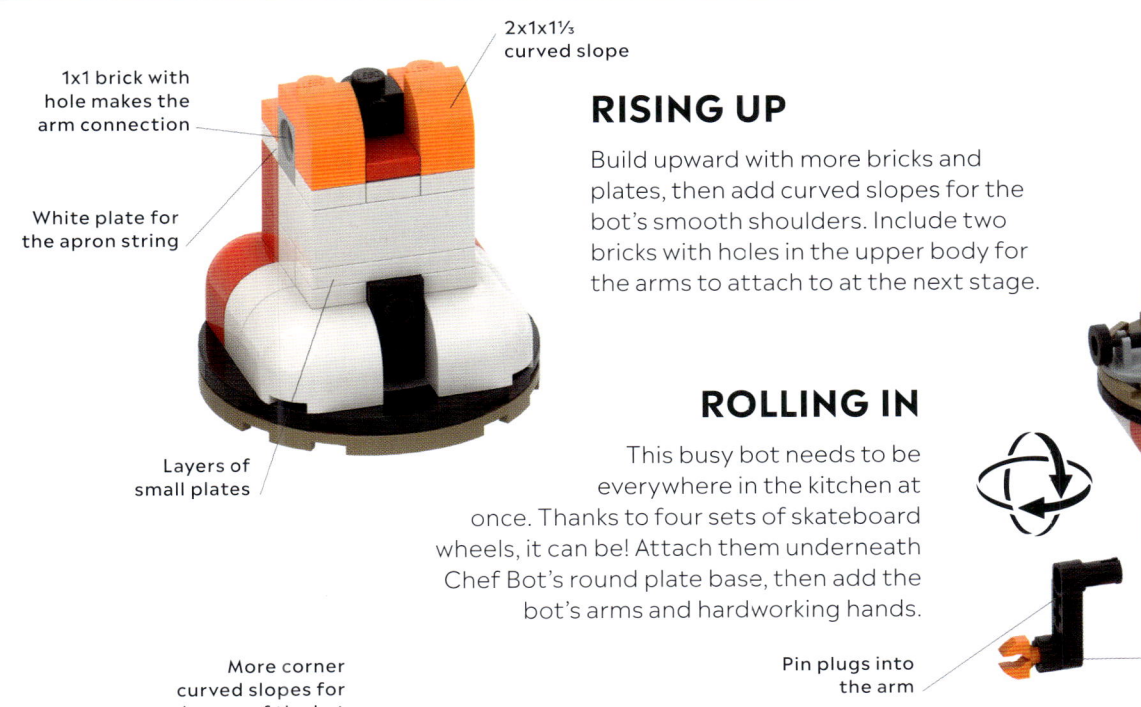

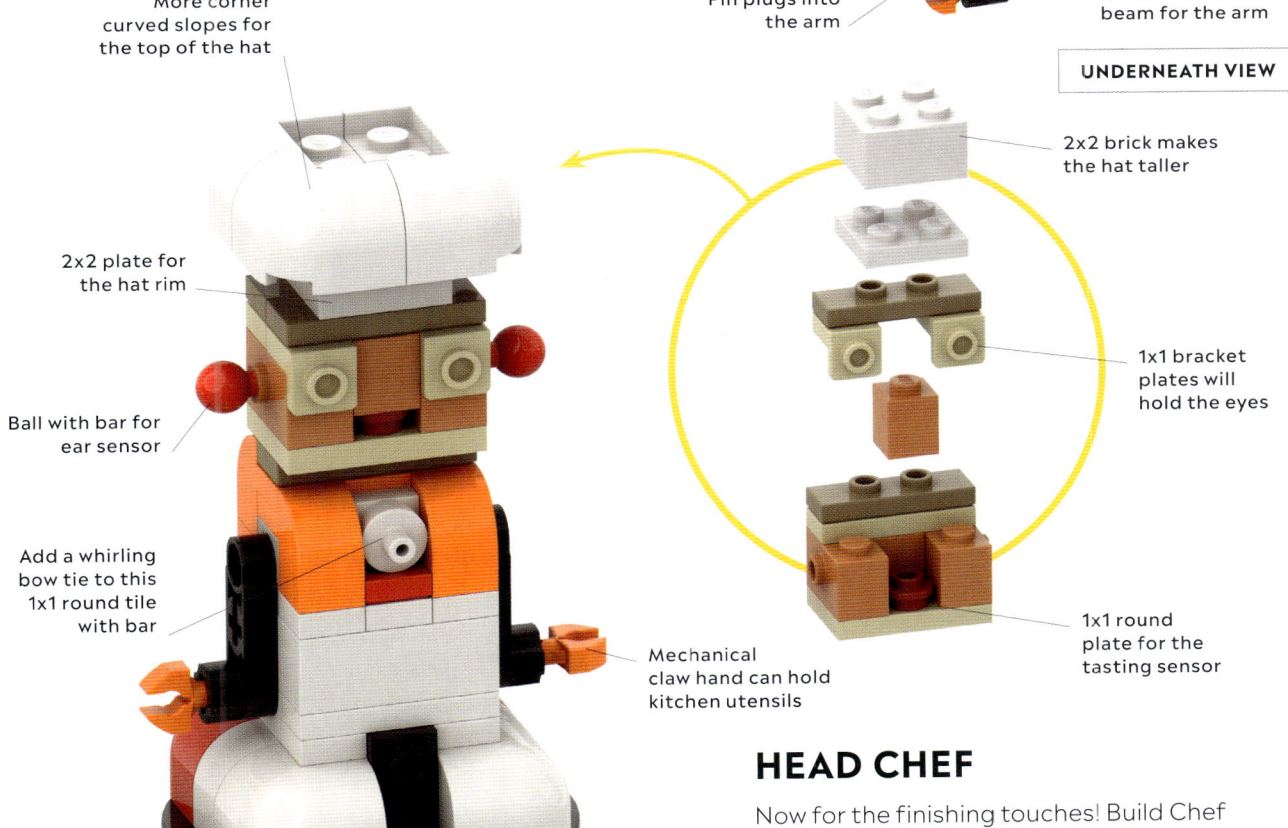

ROLLING IN

This busy bot needs to be everywhere in the kitchen at once. Thanks to four sets of skateboard wheels, it can be! Attach them underneath Chef Bot's round plate base, then add the bot's arms and hardworking hands.

- Pin plugs into the arm
- LEGO Technic beam for the arm

UNDERNEATH VIEW

- More corner curved slopes for the top of the hat
- 2x2 plate for the hat rim
- Ball with bar for ear sensor
- Add a whirling bow tie to this 1x1 round tile with bar
- Mechanical claw hand can hold kitchen utensils
- 2x2 brick makes the hat taller
- 1x1 bracket plates will hold the eyes
- 1x1 round plate for the tasting sensor

HEAD CHEF

Now for the finishing touches! Build Chef Bot's stock-cube shaped head, including mouth and ear sensors for added functionality. Top it with a classic white chef's hat, or "toque," and this chef is ready to take the heat in any kitchen.

SHOPPER BOT

This busy little Shopper Bot thinks supermarkets are super fun. It whizzes happily up and down the aisles, gathering groceries with its grabber arm. That box of crackers at the back of the shelf? No problem. Next stop, the checkout.

I'm programed to look for promotions.

Be Ha-Pea!

Roomy cart for a shopping spree

Giant grabber picks up produce

Eye sensors scan supermarket shelves

Small wheels whiz across tile floors

2x2 tiles smooth out the cart base

2x2 brick with grooves and axle hole

1x4 plate with pins

CART START

The Shopper Bot's base is like a small car chassis. It's built around a long 2x8 plate, with plates with pins underneath that will connect to the wheels. The gray brick with grooves has an axle hole in the middle that will hold the grabber arm.

SKILL LEVEL

SUPERMARKET SWEEP

Yellow trapezoid pieces with lots of side knobs make shaping the Shopper Bot a piece of cake. If you don't have any, you could build a similar shape using plates in different sizes plus bracket plates or bricks with side knobs.

This trapezoid piece is also found on LEGO dump trucks

2x2 round jumper plate

Half circle tile tail light

1x3 wedge plate (right)

1x8 tile

WARNING LIGHT

Four tiles make each smooth side of the Shopper Bot. A 3x6 wedge plate is just the right shape for the front of the cart, while two smaller wedge plates make a similar shape at the back. Connect a red tail light there to warn fellow shoppers away from the grabber arm.

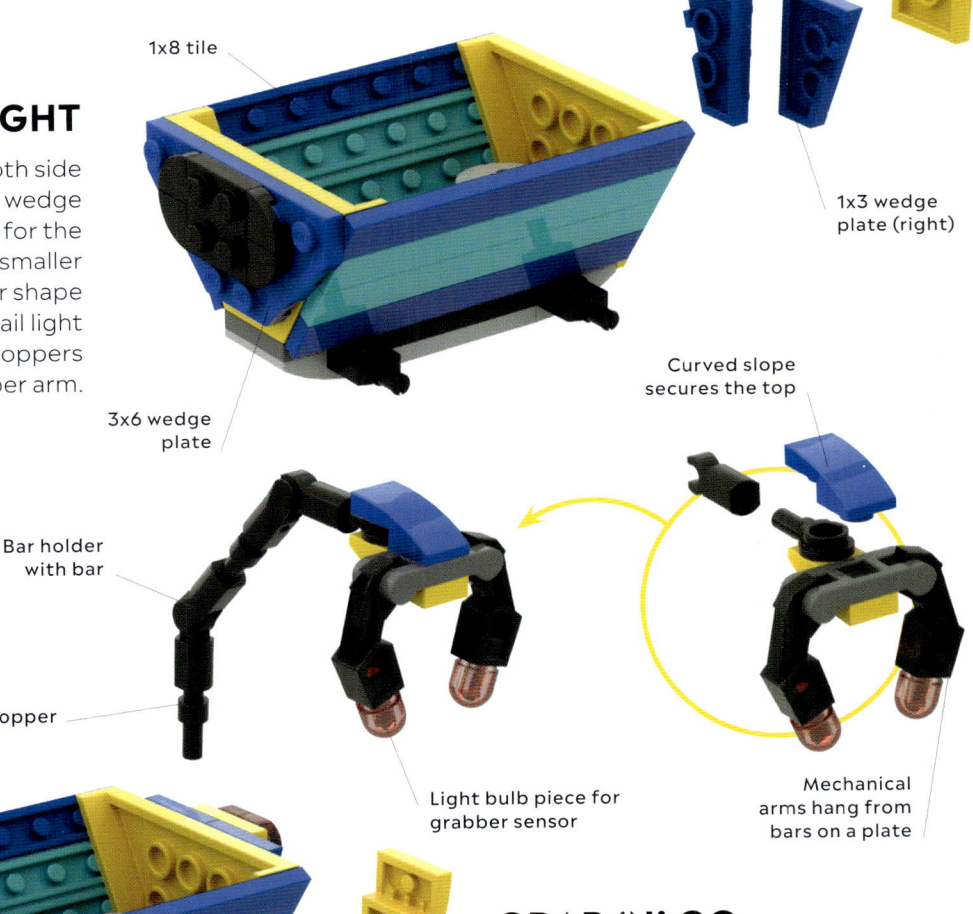

3x6 wedge plate

Curved slope secures the top

Bar holder with bar

Bar with stopper

Light bulb piece for grabber sensor

Mechanical arms hang from bars on a plate

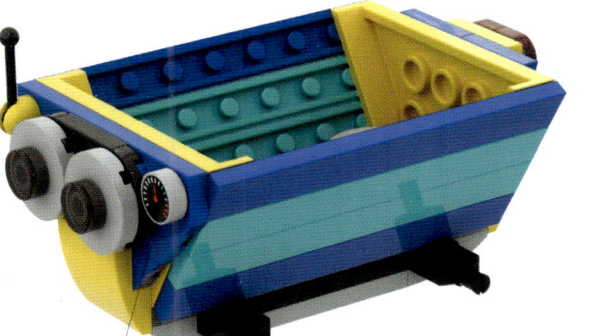

Gauge displays the weight of the shop

GRAB 'N' GO

Give your cart some cute computerized parts at the front, then build the grabber arm. This one is made from bars and bar holders. Clips and bars on the bar holders make the arm flexible. Build gripping pincers at the end of the arm and get grabbing! What's on your shopping list?

ROBO PUPPY

Lets' face it, real puppies can be a little messy. Not like Robo Puppy! Build this pristine pooch and you can forget about house training, dog hairs, and chewed slippers forever. You'll still have floppy ears, snuffly nose, and cute waggy tail, though.

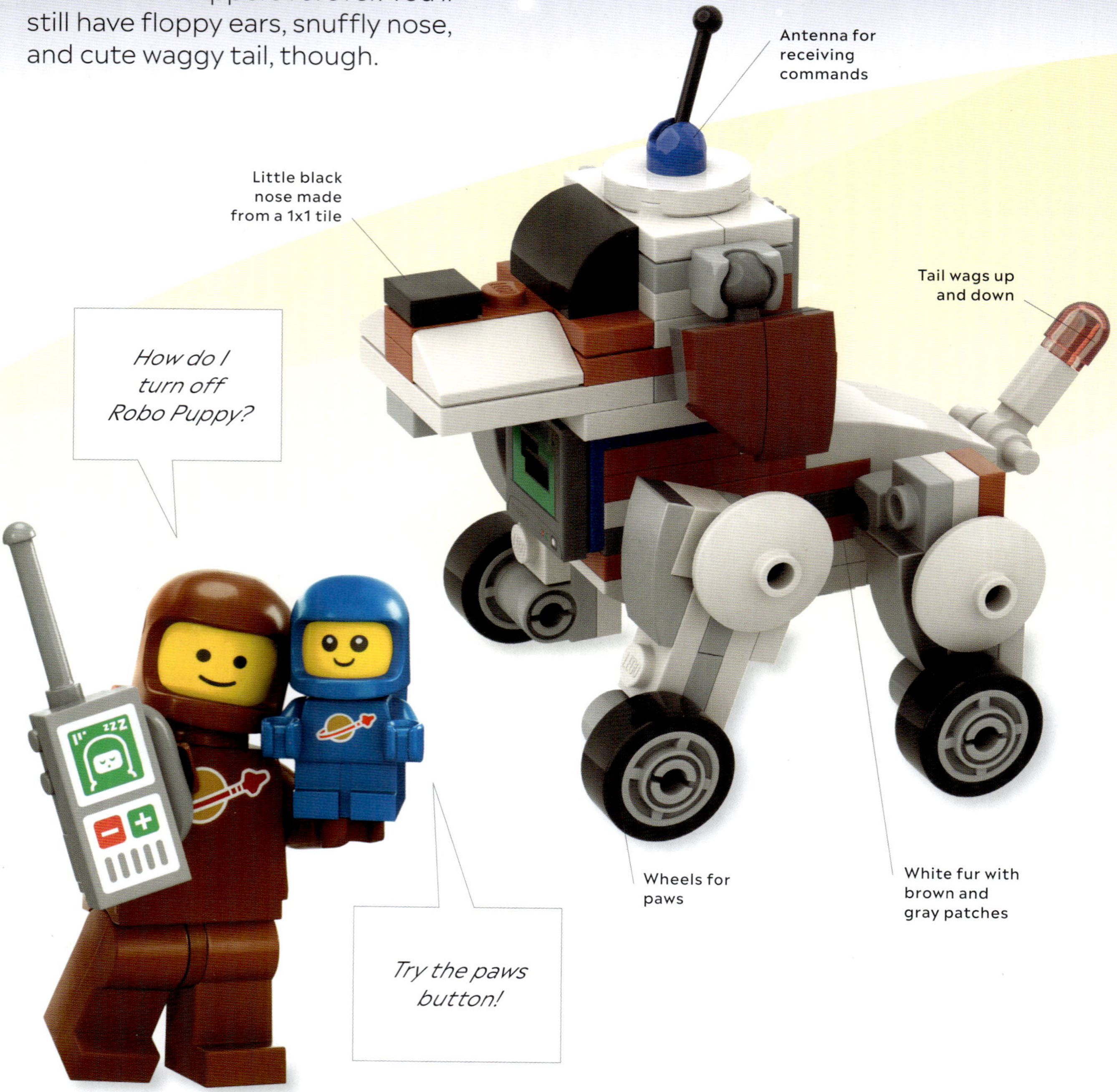

Antenna for receiving commands

Little black nose made from a 1x1 tile

Tail wags up and down

How do I turn off Robo Puppy?

Try the paws button!

Wheels for paws

White fur with brown and gray patches

SKILL LEVEL

DOG'S BODY

Build the body of your new furry friend first. This one is made from layers of small plates, with curved slopes on top and below to create a smooth shape. Those gray plates with balls in the middle of the body are where the four legs will attach.

1x2 plate for a blue collar

The tail will attach here

Brown pieces look like patches of fur

1x2 plate with ball

Hidden 1x1/1x2 angle plate

1x2 plate with socket

1x2 plate for the mouth

1x2 slope brick

Printed 2x2 tile looks like a computer screen

HEAD'S UP

Next, piece together the head on top of the blue collar pieces. Like the body, the head is made from lots of small plates, with white slope bricks to shape the snout and a black curved slope for the eye visor. The flapping ears fit to sockets in the head.

1x2 curved slope

2x2 round jumper tile for a curved head

Grille tile could be a speaker for woofing

1x2 plate with ball for the ear base

1x2 curved slope shapes the front leg

ON A ROLL

Now fit four legs to your fur ball. Each one is built sideways, but the front legs are slightly different to the rear legs. Can you spot the difference? Attach the wheels last, otherwise Robo Puppy might roll away before you finish!

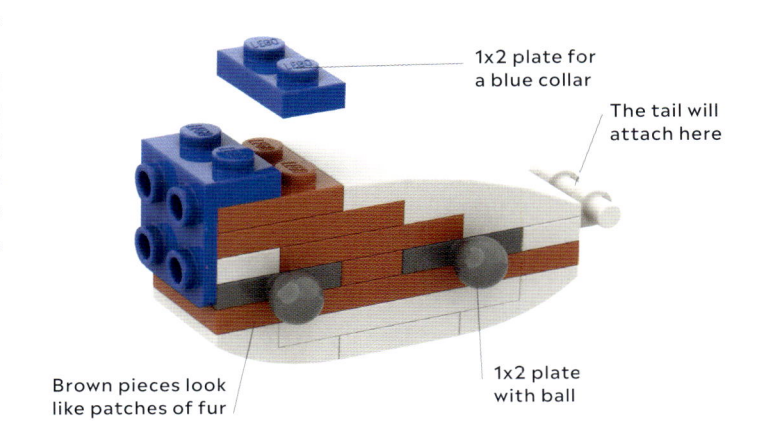

A pin fits here to connect a wheel

MONSTER MECHANIC

Has a bot had a breakdown? Don't panic! The Monster Mechanic will get to grips with any robot repairs. With an enthusiastic smile, fast-moving feet, and a big, swiveling eye, this monster is definitely more goofy than gruesome.

X-500, I hope you don't have a microchip on your shoulder!

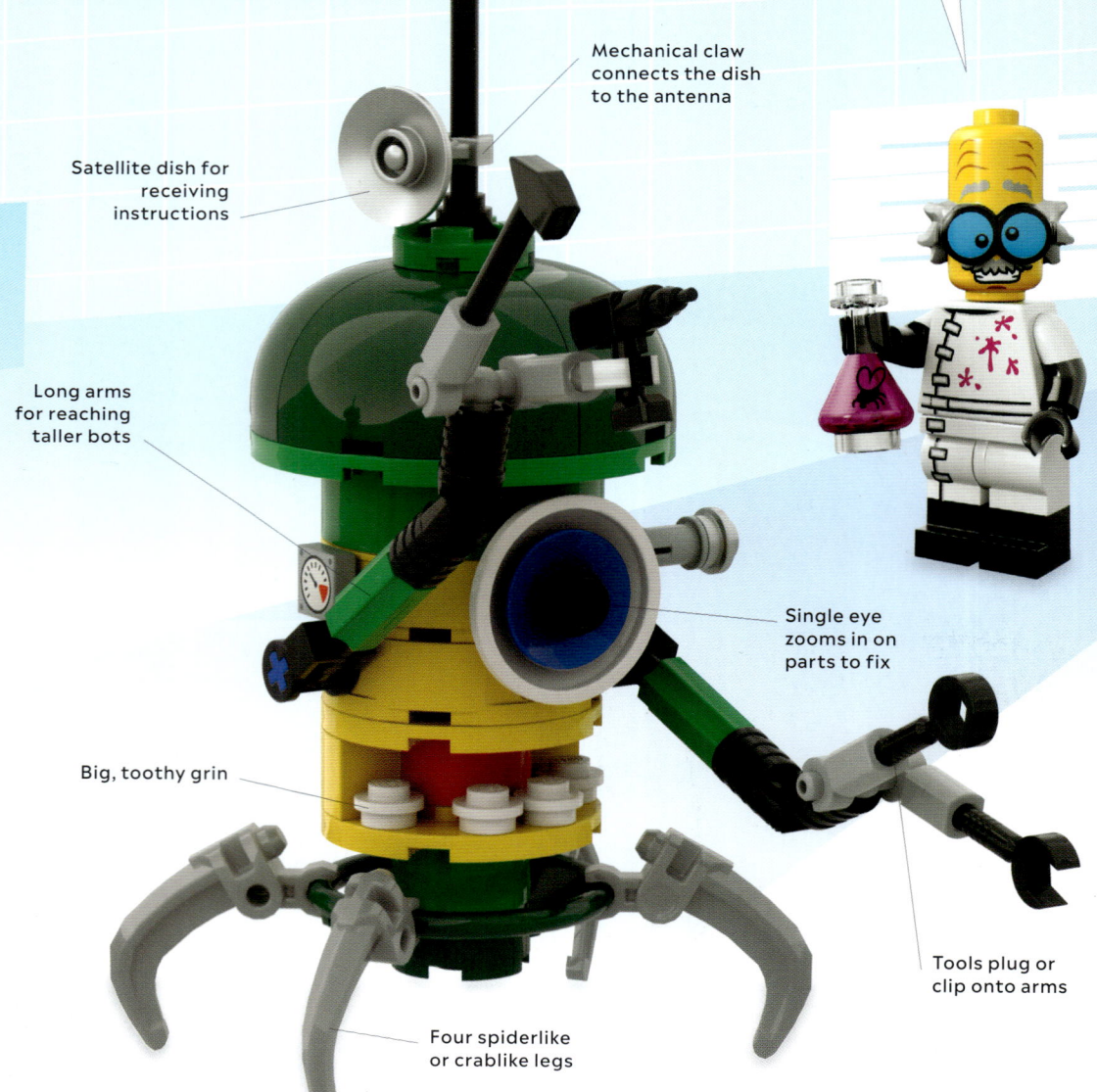

- Satellite dish for receiving instructions
- Mechanical claw connects the dish to the antenna
- Long arms for reaching taller bots
- Single eye zooms in on parts to fix
- Big, toothy grin
- Tools plug or clip onto arms
- Four spiderlike or crablike legs

SKILL LEVEL   301

STEERING START

Right in the middle of the mechanic monster's body is a steering wheel. It forms the base of the body, and it's the piece the legs will attach to in a future step.

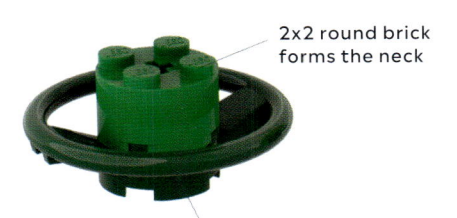

4x4 round brick for the top of the face

4x4 round plate starts off the face

2x2 round brick forms the neck

4x4 brick with holes

Large steering wheel piece

BEGIN THE GRIN

With a smile that wide, this robot must enjoy its job! Four 1x1 round plates form the robot's teeth, and a red round brick inside the mouth looks like a tongue. Once they're in place, layer up the top section of the face.

Hidden axle holds the face together

Bar piece creates an antenna

3x3x2 round corner dome

2x2 round brick looks like a tongue

1x1 round plate

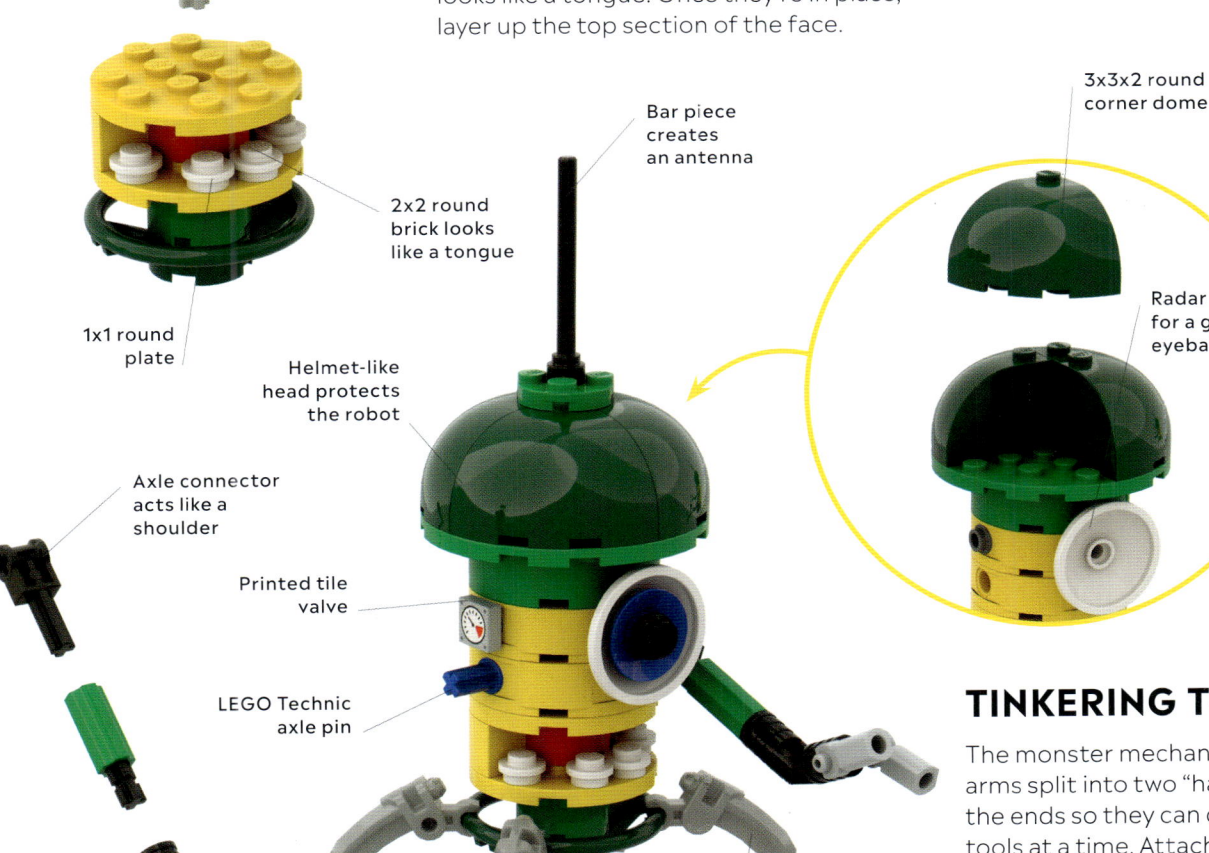

Radar dish for a giant eyeball

Helmet-like head protects the robot

Axle connector acts like a shoulder

Printed tile valve

LEGO Technic axle pin

TINKERING TOOLS

The monster mechanic's long arms split into two "hands" at the ends so they can carry four tools at a time. Attach the arms to axles on the body, finish the head, and add scuttling legs to complete the job.

Click hinge cylinders look like an elbow joint

Large claw for a leg

Clip with bar holder makes a hand

DELIVERY BOT

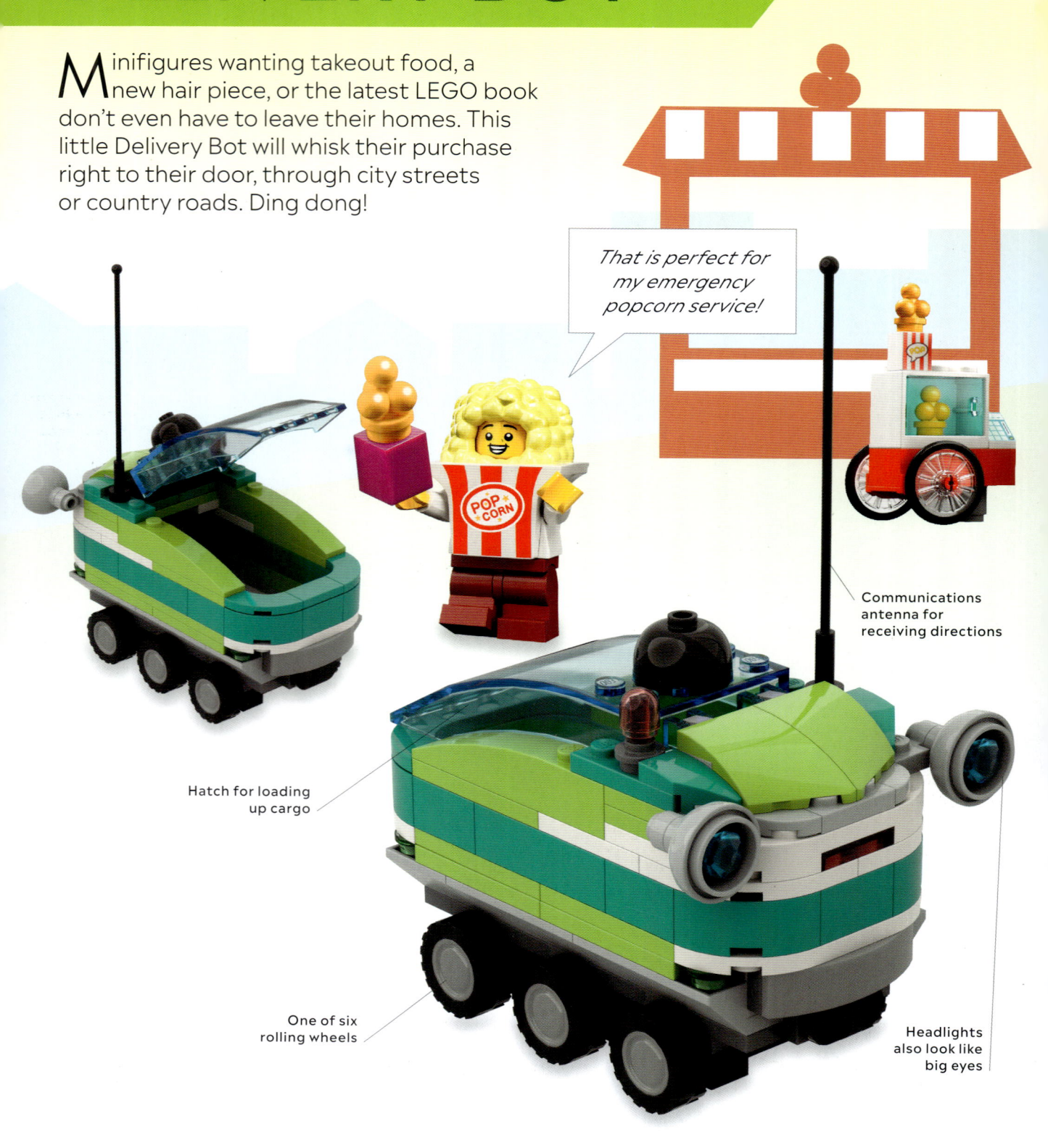

Minifigures wanting takeout food, a new hair piece, or the latest LEGO book don't even have to leave their homes. This little Delivery Bot will whisk their purchase right to their door, through city streets or country roads. Ding dong!

That is perfect for my emergency popcorn service!

Communications antenna for receiving directions

Hatch for loading up cargo

One of six rolling wheels

Headlights also look like big eyes

SKILL LEVEL   303

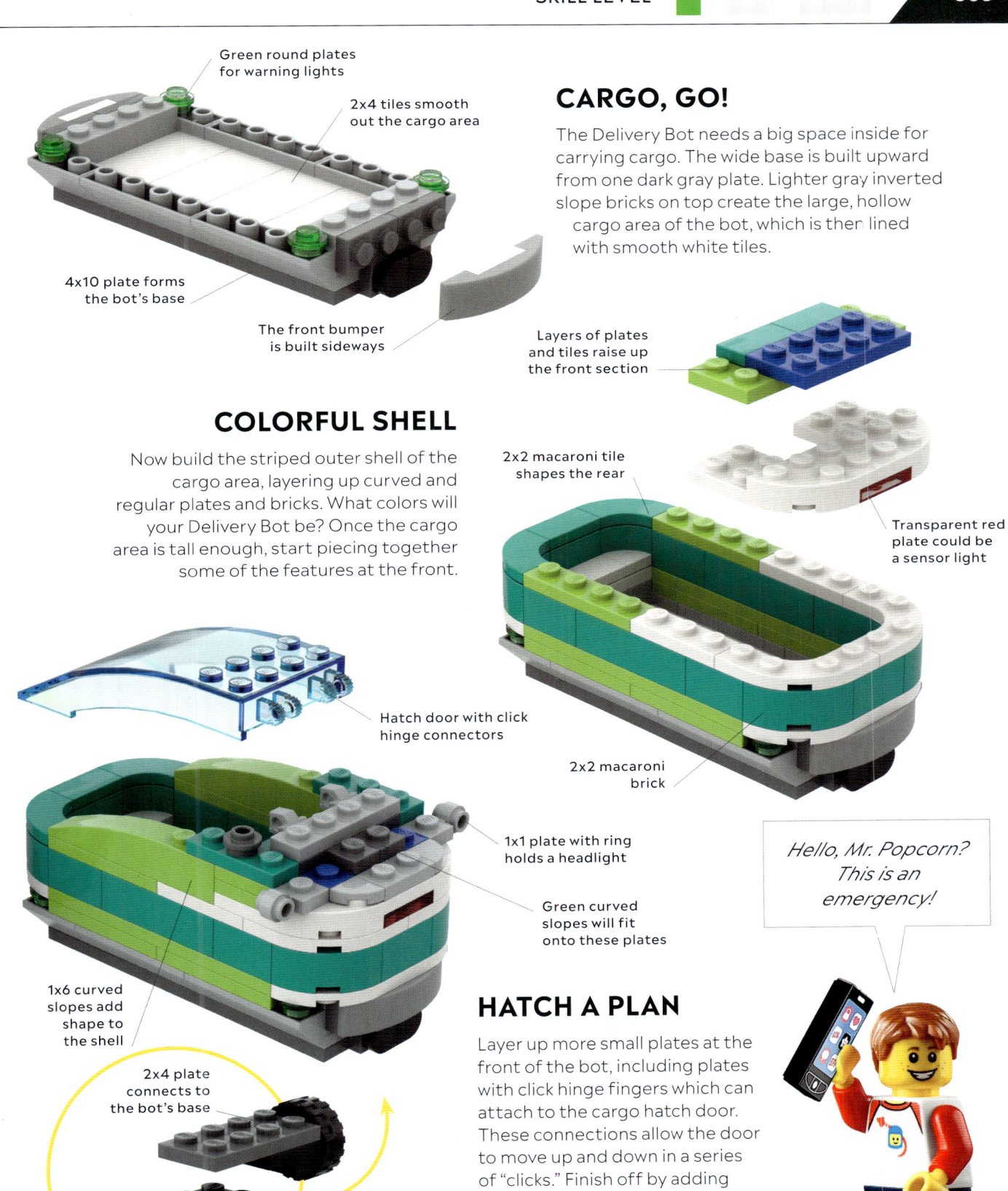

Green round plates
for warning lights

2x4 tiles smooth
out the cargo area

CARGO, GO!

The Delivery Bot needs a big space inside for carrying cargo. The wide space is built upward from one dark gray plate. Lighter gray inverted slope bricks on top create the large, hollow cargo area of the bot, which is then lined with smooth white tiles.

4x10 plate forms
the bot's base

The front bumper
is built sideways

Layers of plates
and tiles raise up
the front section

COLORFUL SHELL

Now build the striped outer shell of the cargo area, layering up curved and regular plates and bricks. What colors will your Delivery Bot be? Once the cargo area is tall enough, start piecing together some of the features at the front.

2x2 macaroni tile
shapes the rear

Transparent red
plate could be
a sensor light

Hatch door with click
hinge connectors

2x2 macaroni
brick

1x1 plate with ring
holds a headlight

Green curved
slopes will fit
onto these plates

1x6 curved
slopes add
shape to
the shell

HATCH A PLAN

Layer up more small plates at the front of the bot, including plates with click hinge fingers which can attach to the cargo hatch door. These connections allow the door to move up and down in a series of "clicks." Finish off by adding headlights, smooth curved slopes, and six whizzing wheels.

2x4 plate
connects to
the bot's base

2x2 plate with
pin holds a wheel

Hello, Mr. Popcorn? This is an emergency!

DINOBOT TRANSPORTER

Make prehistory with this Dinobot Transporter. The minifigure passengers are having a wild time as they ride their robot reptile with its big, stomping feet. The Dinobot is based on a giant plant-eater, so no need to mind your fingers while you build.

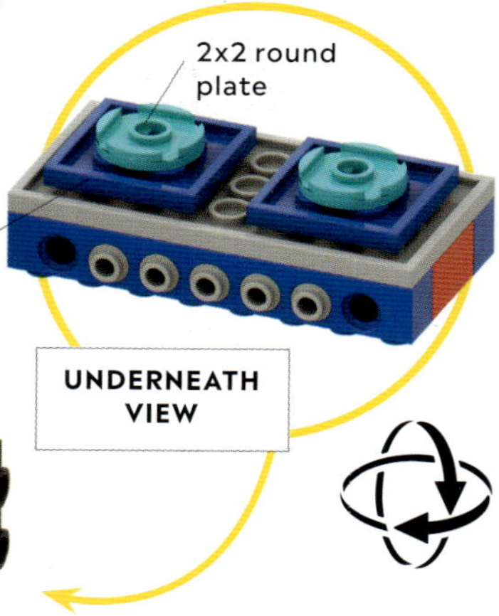

This is dino-riffic!

- Saddle basket for two riders
- Start/stop lever
- Air vents keep the Dinobot cool
- Small head, like a diplodocus's
- Wide feet prevent it from toppling over
- These pieces make the belly look heavier
- 2x2 round plate
- 1x8 beam
- Axle pin
- 1x2/2x2 bracket plate

UNDERNEATH VIEW

WIDE LOAD

This is one big bot! Its broad mid-section is the perfect perch for minifigure riders. Start it off by building a belly using a 4x8 plate and LEGO Technic beams. Then add pieces that can create side knobs and connection points for building outward.

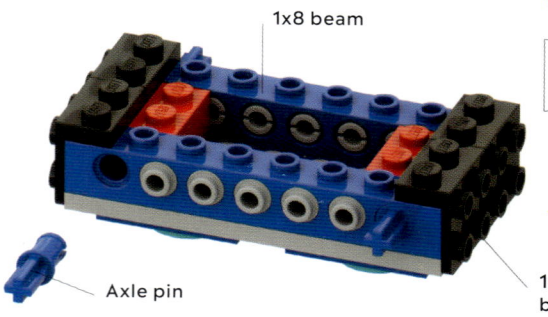

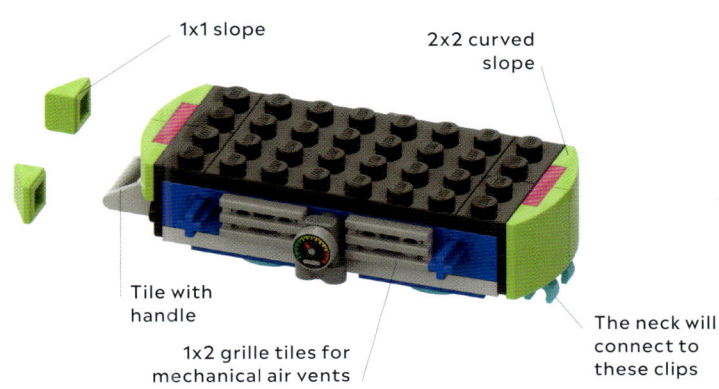

SKILL LEVEL

305

BUILT-UP BODY

Now add lots of elements sideways to the Dinobot's body, plus a layer of plates on top. Curved slopes create a smooth shape at the front and rear of the body. There are now connection points for the neck and the tail too.

- 1x1 slope
- 2x2 curved slope
- Tile with handle
- 1x2 grille tiles for mechanical air vents
- The neck will connect to these clips

SWIVEL AND STOMP

Once the back is complete—with curved pieces, and a turntable plate for the saddle basket to swivel on—start on the legs. They're each made the same way, with long axles reaching down to the wide ankles and feet.

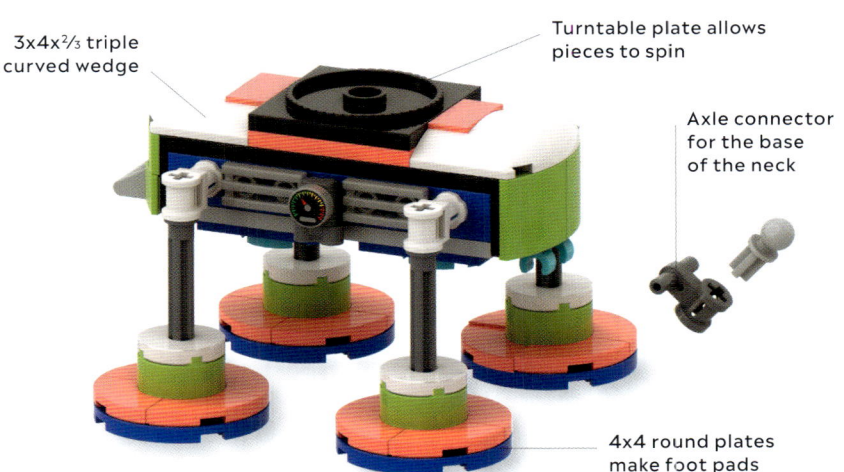

- 3x4x²⁄₃ triple curved wedge
- Turntable plate allows pieces to spin
- Axle connector for the base of the neck
- 4x4 round plates make foot pads

TIP TO TIP

Finish off the Dinobot with the two end points of its body: the tail and the neck. Both parts have interesting mechanical details. Let's hope the Dinobot doesn't swish its tail too hard and accidentally power down!

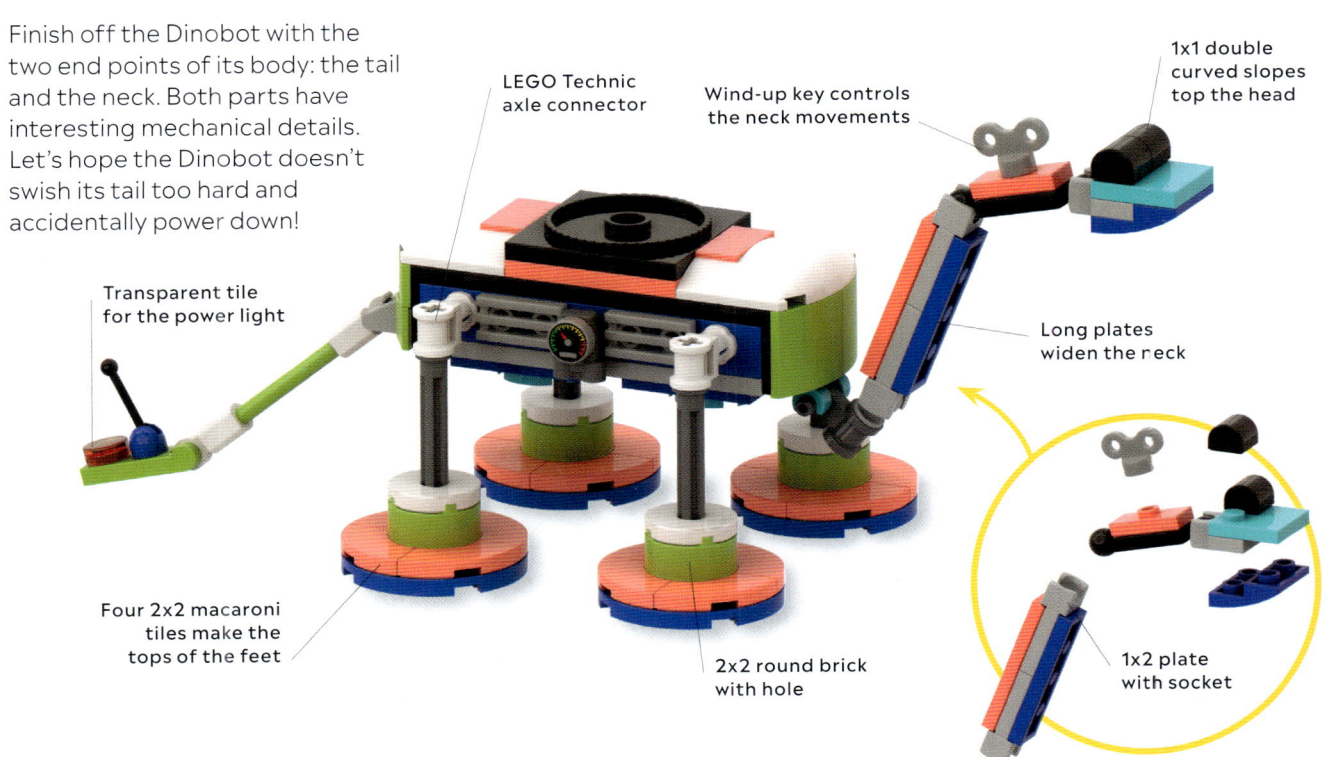

- Transparent tile for the power light
- LEGO Technic axle connector
- Wind-up key controls the neck movements
- 1x1 double curved slopes top the head
- Long plates widen the neck
- Four 2x2 macaroni tiles make the tops of the feet
- 2x2 round brick with hole
- 1x2 plate with socket

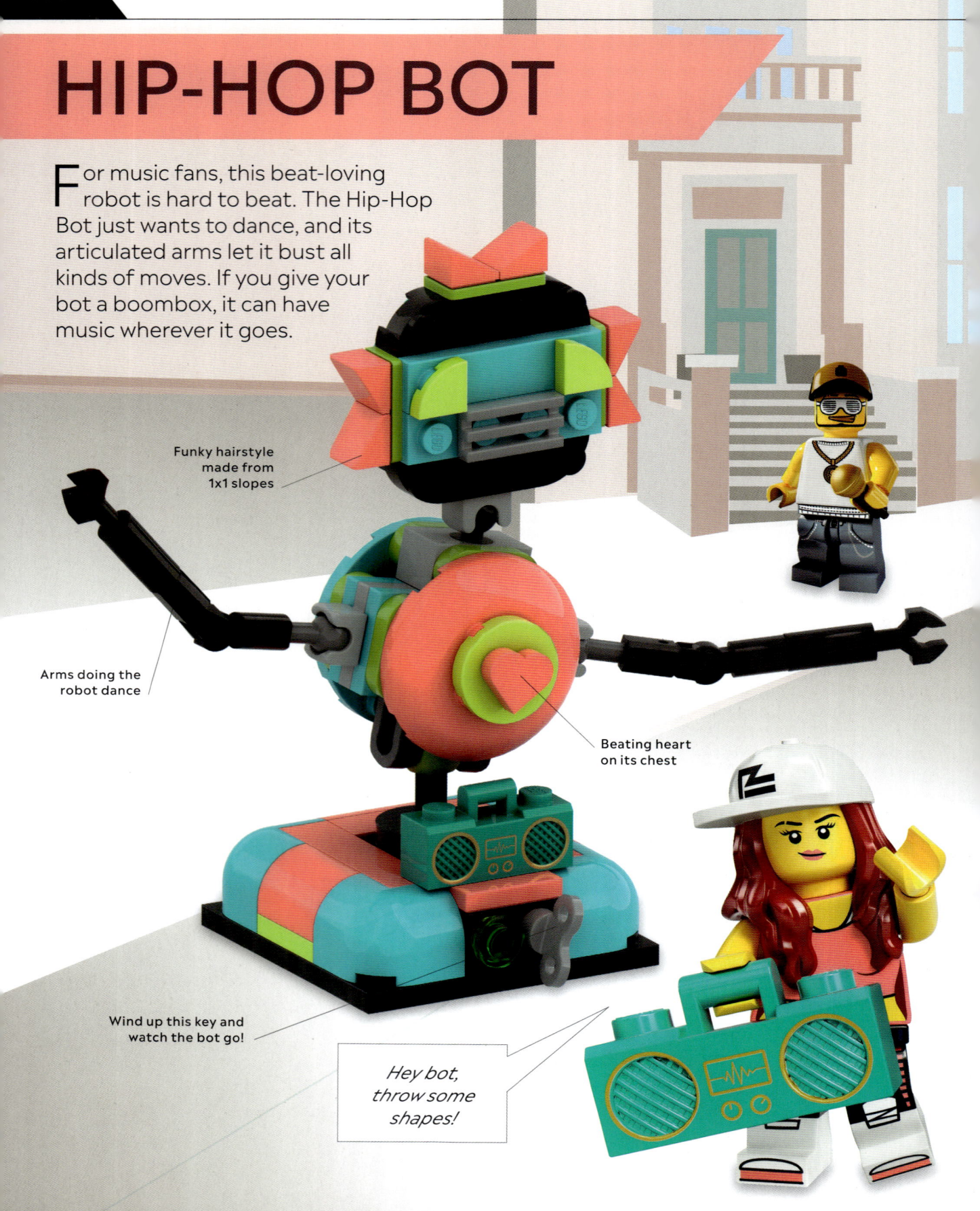

HIP-HOP BOT

For music fans, this beat-loving robot is hard to beat. The Hip-Hop Bot just wants to dance, and its articulated arms let it bust all kinds of moves. If you give your bot a boombox, it can have music wherever it goes.

Funky hairstyle made from 1x1 slopes

Arms doing the robot dance

Beating heart on its chest

Wind up this key and watch the bot go!

Hey bot, throw some shapes!

SKILL LEVEL

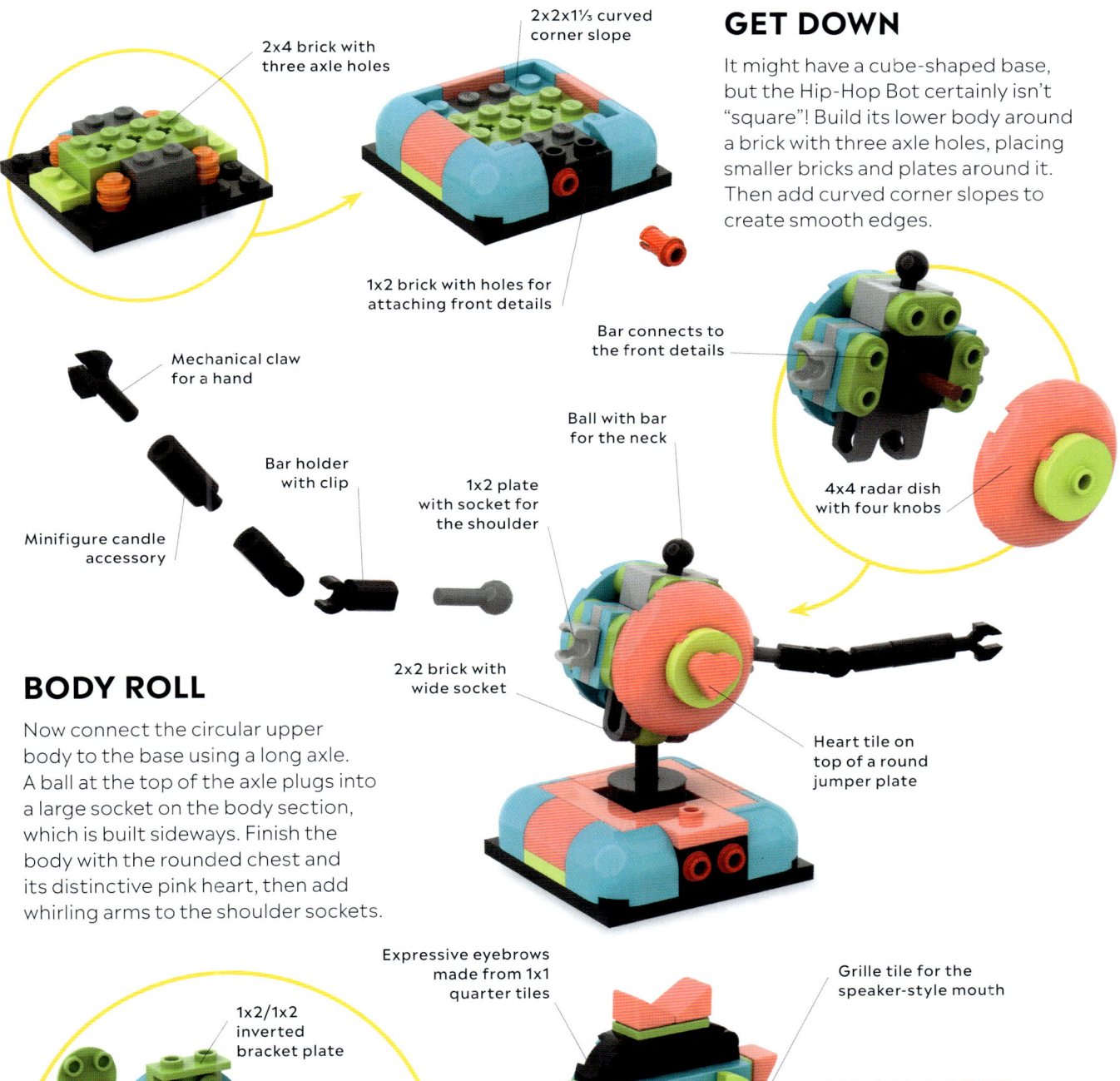

GET DOWN

It might have a cube-shaped base, but the Hip-Hop Bot certainly isn't "square"! Build its lower body around a brick with three axle holes, placing smaller bricks and plates around it. Then add curved corner slopes to create smooth edges.

- 2x4 brick with three axle holes
- 2x2x1⅓ curved corner slope
- 1x2 brick with holes for attaching front details
- Bar connects to the front details
- Ball with bar for the neck
- 1x2 plate with socket for the shoulder
- 2x2 brick with wide socket
- 4x4 radar dish with four knobs
- Heart tile on top of a round jumper plate
- Mechanical claw for a hand
- Bar holder with clip
- Minifigure candle accessory

BODY ROLL

Now connect the circular upper body to the base using a long axle. A ball at the top of the axle plugs into a large socket on the body section, which is built sideways. Finish the body with the rounded chest and its distinctive pink heart, then add whirling arms to the shoulder sockets.

- 1x2/1x2 inverted bracket plate
- 4x4x⅔ plate
- Expressive eyebrows made from 1x1 quarter tiles
- Grille tile for the speaker-style mouth
- Exposed knobs look like eyes

HEAD SLIDE

Just like the body, the Hip-Hop Bot's bobbing head is built sideways. The first layer is a 4x4 round plate, followed by small plates that connect to the neck and hair spikes. Add another plate and fun face details, then this bot is ready to bop!

TIDY BOT

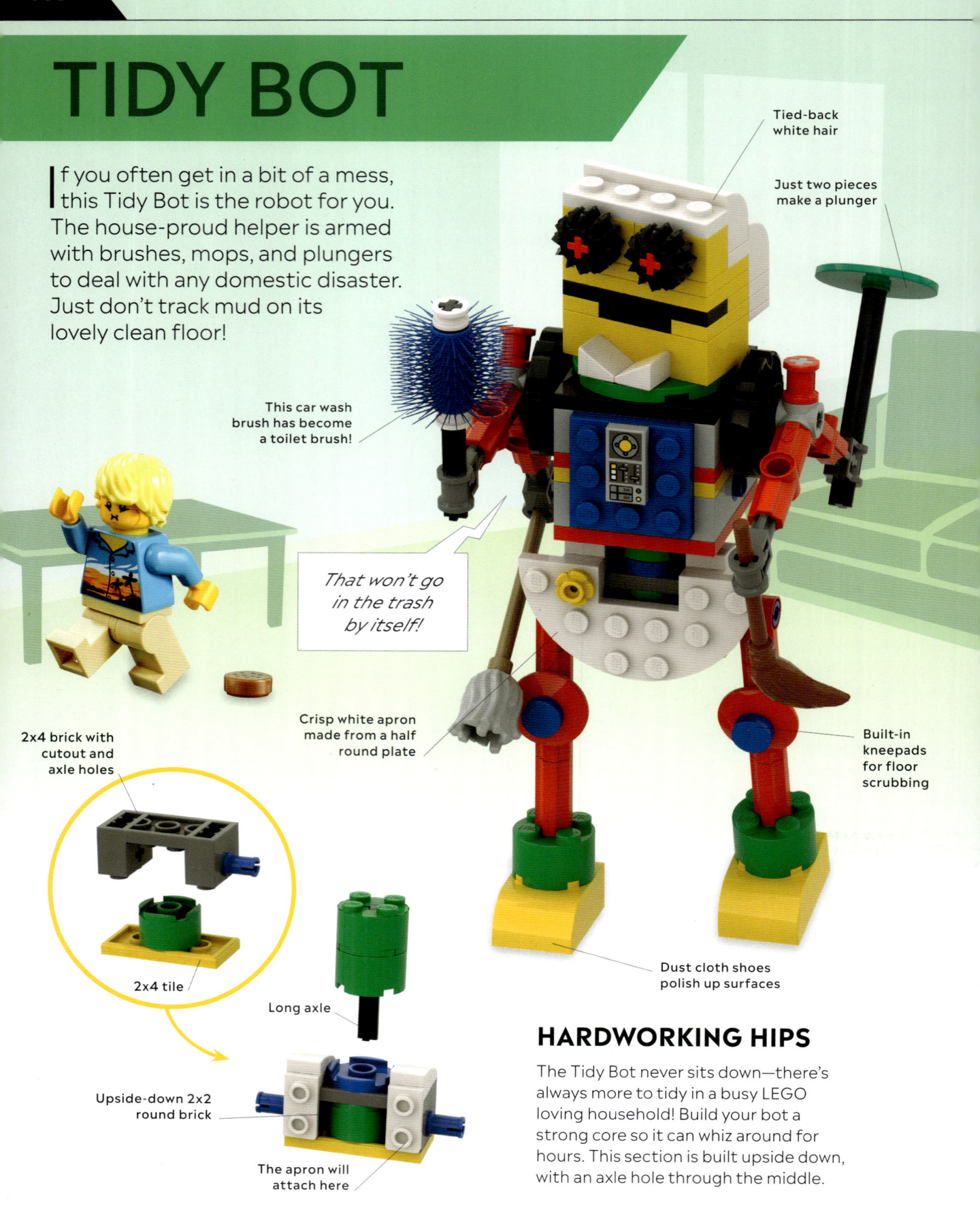

If you often get in a bit of a mess, this Tidy Bot is the robot for you. The house-proud helper is armed with brushes, mops, and plungers to deal with any domestic disaster. Just don't track mud on its lovely clean floor!

Tied-back white hair

Just two pieces make a plunger

This car wash brush has become a toilet brush!

That won't go in the trash by itself!

Crisp white apron made from a half round plate

Built-in kneepads for floor scrubbing

2x4 brick with cutout and axle holes

2x4 tile

Long axle

Upside-down 2x2 round brick

The apron will attach here

Dust cloth shoes polish up surfaces

HARDWORKING HIPS

The Tidy Bot never sits down—there's always more to tidy in a busy LEGO loving household! Build your bot a strong core so it can whiz around for hours. This section is built upside down, with an axle hole through the middle.

SKILL LEVEL

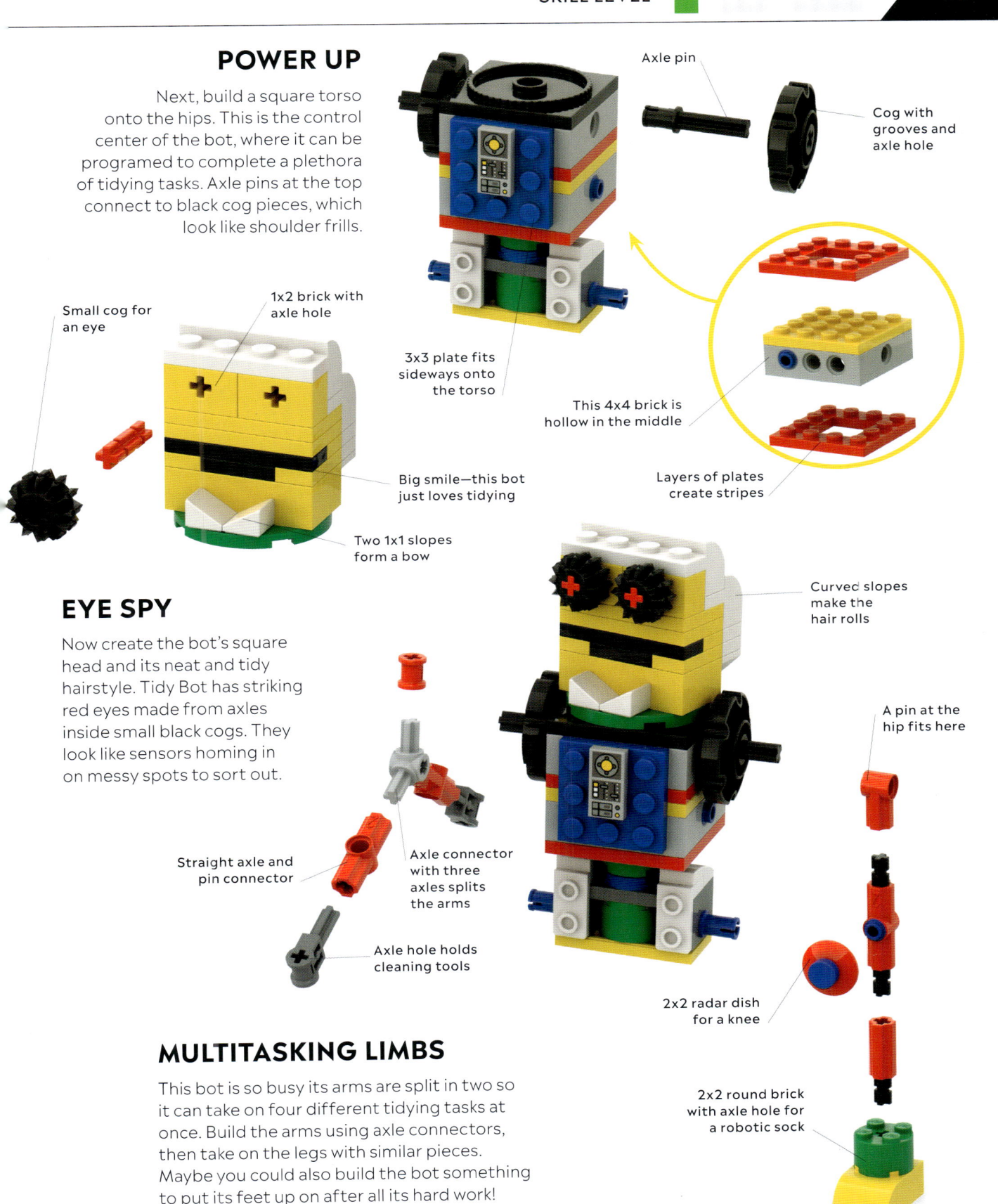

309

POWER UP

Next, build a square torso onto the hips. This is the control center of the bot, where it can be programed to complete a plethora of tidying tasks. Axle pins at the top connect to black cog pieces, which look like shoulder frills.

Axle pin

Cog with grooves and axle hole

1x2 brick with axle hole

Small cog for an eye

3x3 plate fits sideways onto the torso

This 4x4 brick is hollow in the middle

Layers of plates create stripes

Big smile—this bot just loves tidying

Two 1x1 slopes form a bow

EYE SPY

Now create the bot's square head and its neat and tidy hairstyle. Tidy Bot has striking red eyes made from axles inside small black cogs. They look like sensors homing in on messy spots to sort out.

Curved slopes make the hair rolls

A pin at the hip fits here

Straight axle and pin connector

Axle connector with three axles splits the arms

Axle hole holds cleaning tools

2x2 radar dish for a knee

MULTITASKING LIMBS

This bot is so busy its arms are split in two so it can take on four different tidying tasks at once. Build the arms using axle connectors, then take on the legs with similar pieces. Maybe you could also build the bot something to put its feet up on after all its hard work!

2x2 round brick with axle hole for a robotic sock

GARDENER ROBOT

Forget batteries; this Gardener Robot runs solely on sunshine. Its hothouse head has a cloche that lifts to reveal a beautiful bevy of blooms. Don't forget to give your robot a spade so it can keep its garden well tended.

It's a bot-anical marvel!

Clear dome floods the bot with sunlight

Growing and blooming flowers

Freshly picked blooms

Four macaroni tubes make pronged feet

2x2 ridged brick

2x2 turntable

1x2x1⅔ brick with eight side knobs

This 1x2 plate with hole will connect to a leg

TALL TRUNK

Grow this bot outward from its middle, which looks a little like a sturdy tree trunk. Right at the center of the trunk are four dark gray bricks with side knobs. Add small plates (sideways) to those, in all four directions.

SKILL LEVEL

311

BULBOUS BODY

The Gardener Robot's body is now shaped a little like a flower bulb. Find some curved slopes to get the same effect. They attach to the four sides of the trunk along with dark pink plates and tiles. At this stage, add some techy details and four arched legs.

1x2 tile

2x2 curved slope

Gardening controls and data bank

There's a hidden axle that connects this tube to a hole

This piece is often used for cockpits on LEGO spaceships

1x1 round plates with leaves

1x4 tile with click hinge connections

1x2/1x2 bracket plate

2x2 jumper plate

4x4 round plate

2x2 curved slope for the sides

Eyes made from jam sandwich cookie tiles

TOP SOIL

Keep climbing upward, creating the bot's eccentric eyes and that hard-to-miss hothouse head. The latter gets its round shape with some clever sideways building involving bracket plates. Add a brown 4x4 plate layer of soil, then let the plants grow!

The head fits onto this piece

Clip looks like a grabber hand

The arms can bend at the elbows too

2x2 curved slope for the shoulder

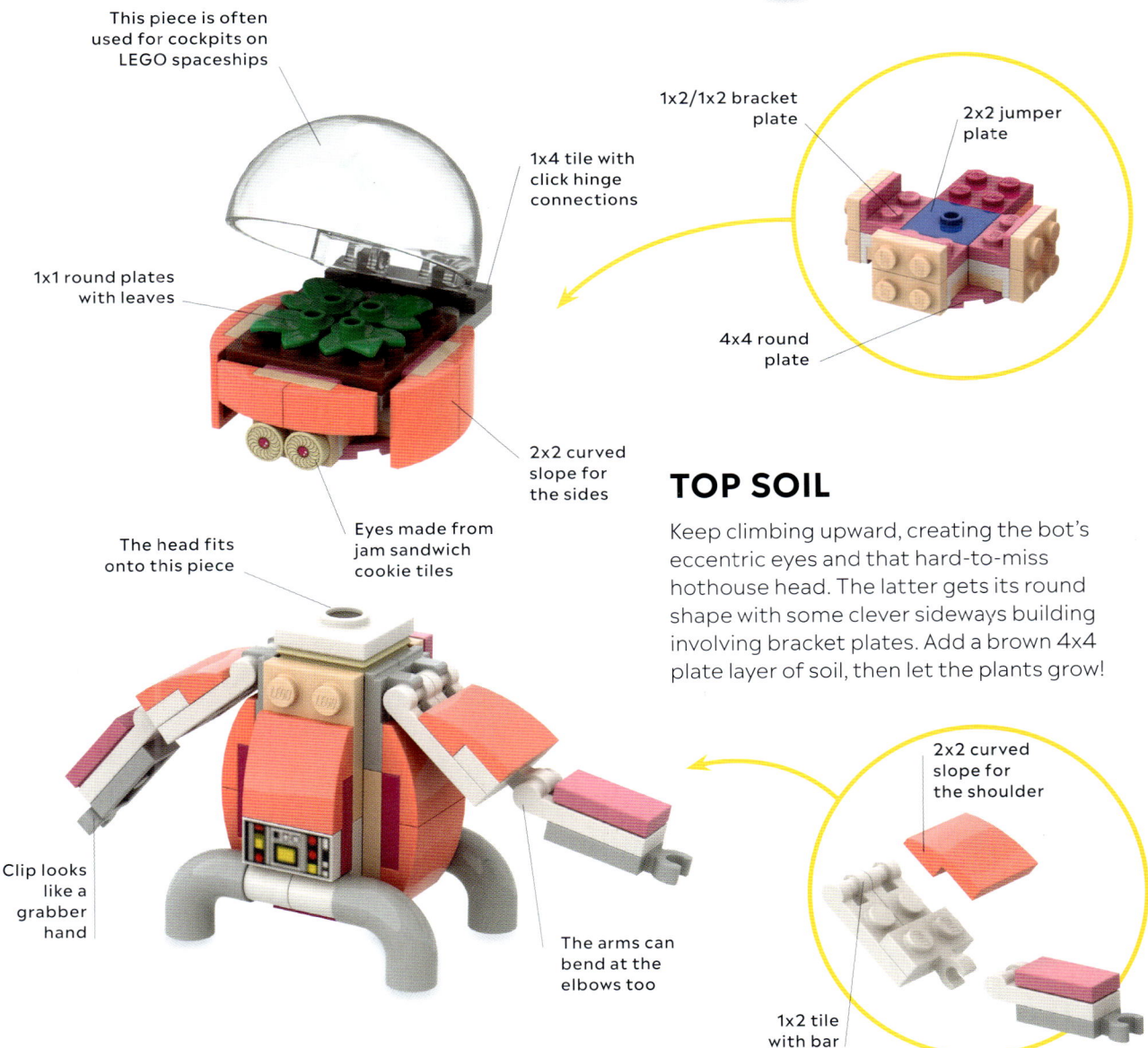

1x2 tile with bar

DRIVERLESS HOVERCAR

These minifigure friends never argue about who is driving. They just hop into their Driverless Hovercar, set their destination, and they're off. The futuristic pod-style car has a roof antenna and an all-around bumper in case of bumps in the road.

Lever piece for a circling antenna

Did you know, a hundred years ago, cars used to have wheels?

That's so weird!

Rear windshield is as wide as the front

Clear headlights at the front

Passenger step for easy entry

Black bumper wraps around the car

2x2 slide plates let the base slide

6x10 plate

UNDERNEATH VIEW

1x2 plate

2x2x⅔ modified plate with side knobs

SPACIOUS BASE

One big plate forms the base. Since there's no driver, there's no need to leave space for any controls, so your minifigures will have lots of leg room! Add small plates with side knobs to the big plate's perimeter for the bumper to attach to later.

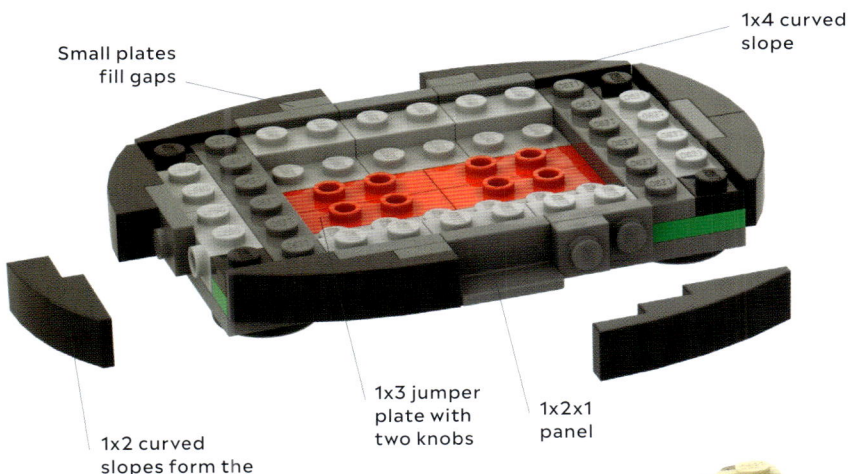

SKILL LEVEL

BUMPER TO BUMPER

Next, use black curved slopes to create a protective circular bumper all around the Driverless Hovercar. On the long sides of the bumper, add in panel pieces for footboards. Deck out the car's interior too, with a bright red carpet that matches the car's shell.

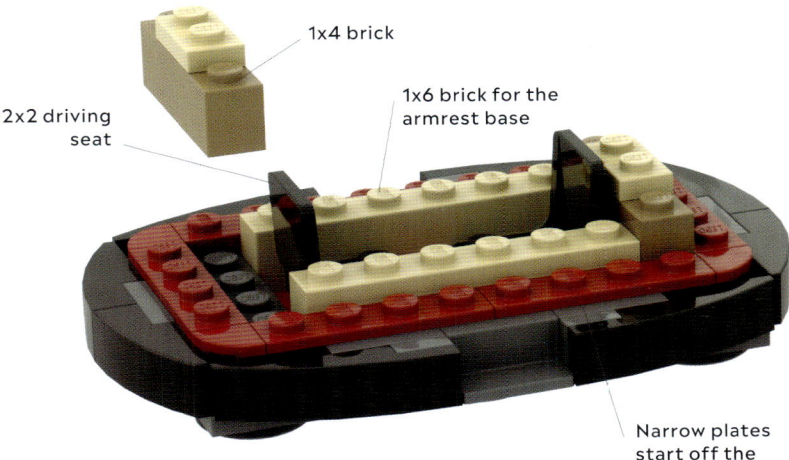

Small plates fill gaps

1x4 curved slope

1x3 jumper plate with two knobs

1x2x1 panel

1x2 curved slopes form the shorter sides

1x4 brick

1x6 brick for the armrest base

2x2 driving seat

SOCIABLE SEATING

In a driverless car no one needs to watch the road, so the passenger seats can face each other. Place the seats on the red interior carpet and add bricks and plates around them to start building up the car's outer shell.

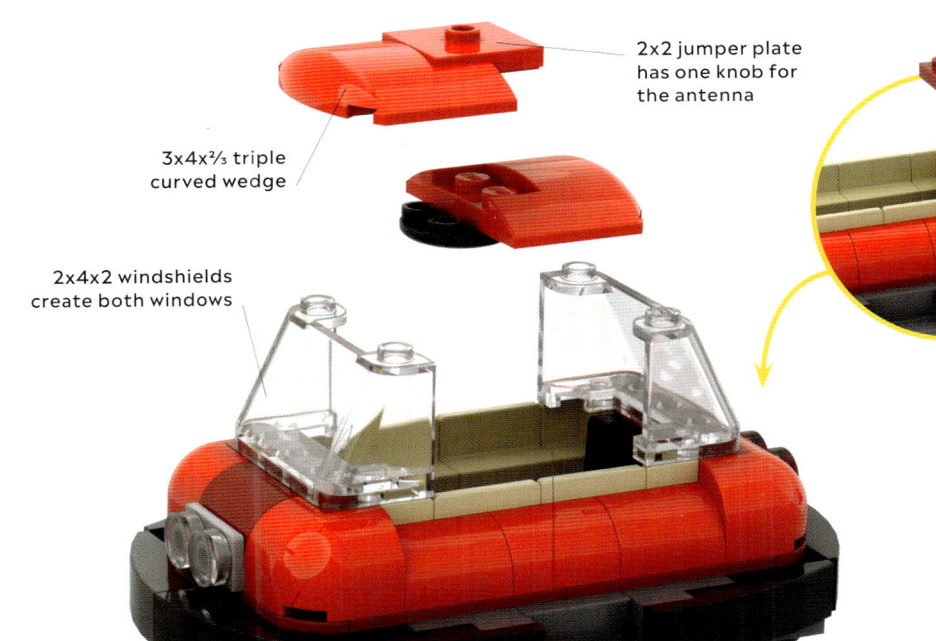

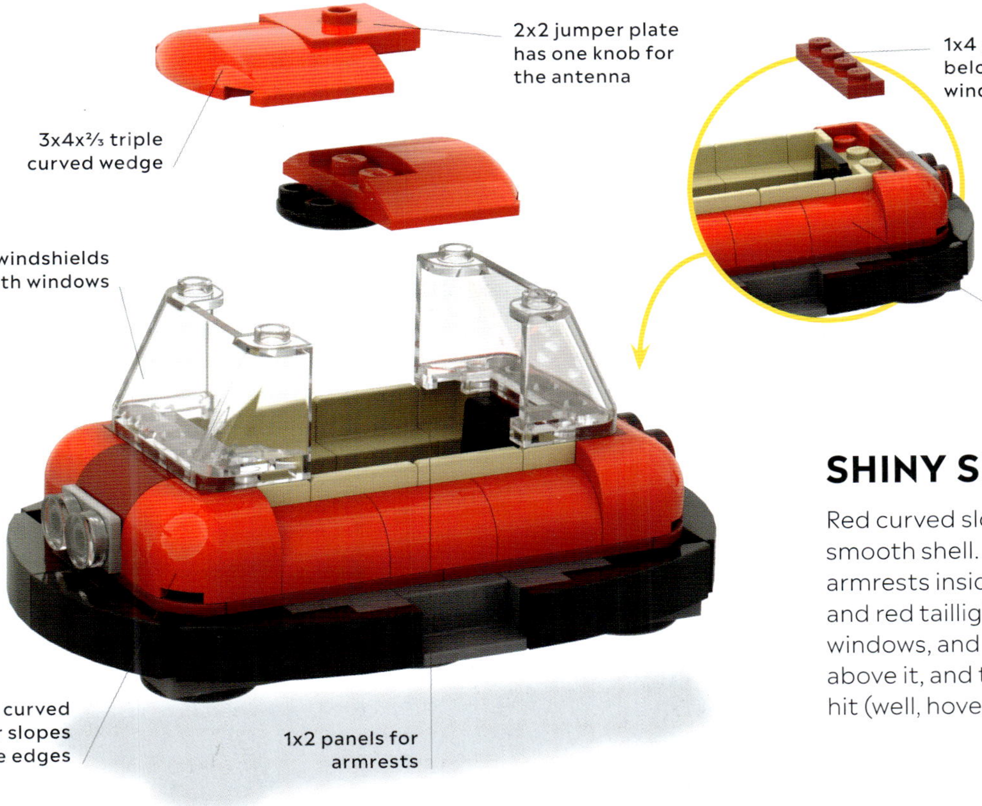

Narrow plates start off the car's shell

2x2 jumper plate has one knob for the antenna

1x4 plate fits below the windshield

3x4x²⁄₃ triple curved wedge

2x4x2 windshields create both windows

1x2 curved slope

SHINY SHELL

Red curved slopes create the smooth shell. Add panel pieces for armrests inside it, clear headlights and red taillights outside it, large windows, and a matching red roof above it, and this car is ready to hit (well, hover over) the road!

2x1x1⅓ curved corner slopes for the edges

1x2 panels for armrests

ALIEN ROBOT

What kind of robot would aliens build? A little green robot, perhaps. This Alien Robot has tall antennas, suction pad feet, and arms like ray guns. Yours could look different, depending on its home planet (and what LEGO pieces you have).

How do I tell it to invade Earth?

Antennas sense the robot's environment

Silver tiles with bars look like a neck bolt

Pink tiles with clips for tiny grabber hands

What could flow through these hoses?

Broad suction pad feet support the tall body

3x3 dome looks like a round belly

This 2x2 plate allows the dome to attach

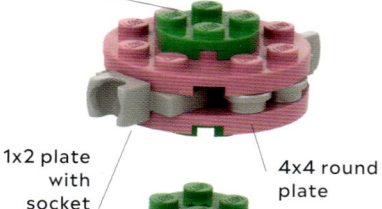

1x2 plate with socket

4x4 round plate

HIP, HIP

Start off your extraterrestrial pal right in the middle of its body. The robot's wide pink hips are made from two 4x4 round plates. They have a layer of smaller pieces between them, including plates with sockets that connect to the movable legs.

SKILL LEVEL   315

Upside-down 4x4 dome

Inverted curved slopes shape the shoulders

2x2 plate with ball will connect to an arm

Bar fits through the pieces above and below

HIDDEN CONNECTION

The top half of this robot's techy torso is wide like the bottom half, and it mirrors the shape with a gray dome. A hidden bar piece connects the two parts of the torso. Can you see that the shoulder section is built upside down?

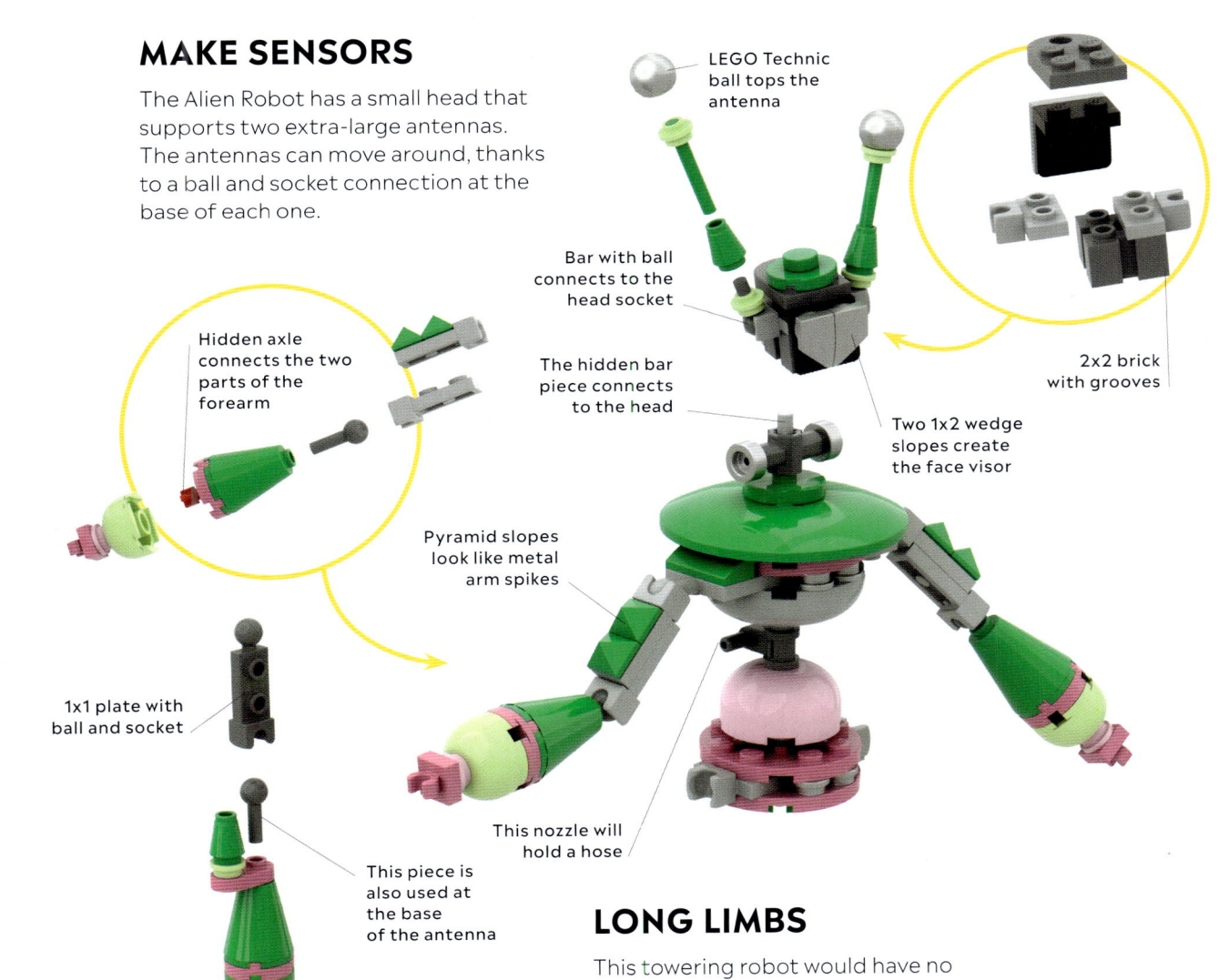

MAKE SENSORS

The Alien Robot has a small head that supports two extra-large antennas. The antennas can move around, thanks to a ball and socket connection at the base of each one.

LEGO Technic ball tops the antenna

Hidden axle connects the two parts of the forearm

Bar with ball connects to the head socket

The hidden bar piece connects to the head

2x2 brick with grooves

Two 1x2 wedge slopes create the face visor

Pyramid slopes look like metal arm spikes

1x1 plate with ball and socket

This nozzle will hold a hose

This piece is also used at the base of the antenna

LONG LIMBS

This towering robot would have no problem reaching items on the top shelf of an alien kitchen! Its lengthy arms and legs can move in two places—at the top and in the middle—making them extra mobile.

LEFT LEG

4x4 dome foot

SPACE PROBE
PAGES 326–327

ALIEN ROBOT
PAGES 314–315

"Your Highness, we have received a scan of an Earthling. Take a look."

"Is that to scale?"

MODEL MASH-UP

**TYRANNOSAURUS
PAGES 282–283**

Errr... let's invade Mars instead.

BATH TIME BOT

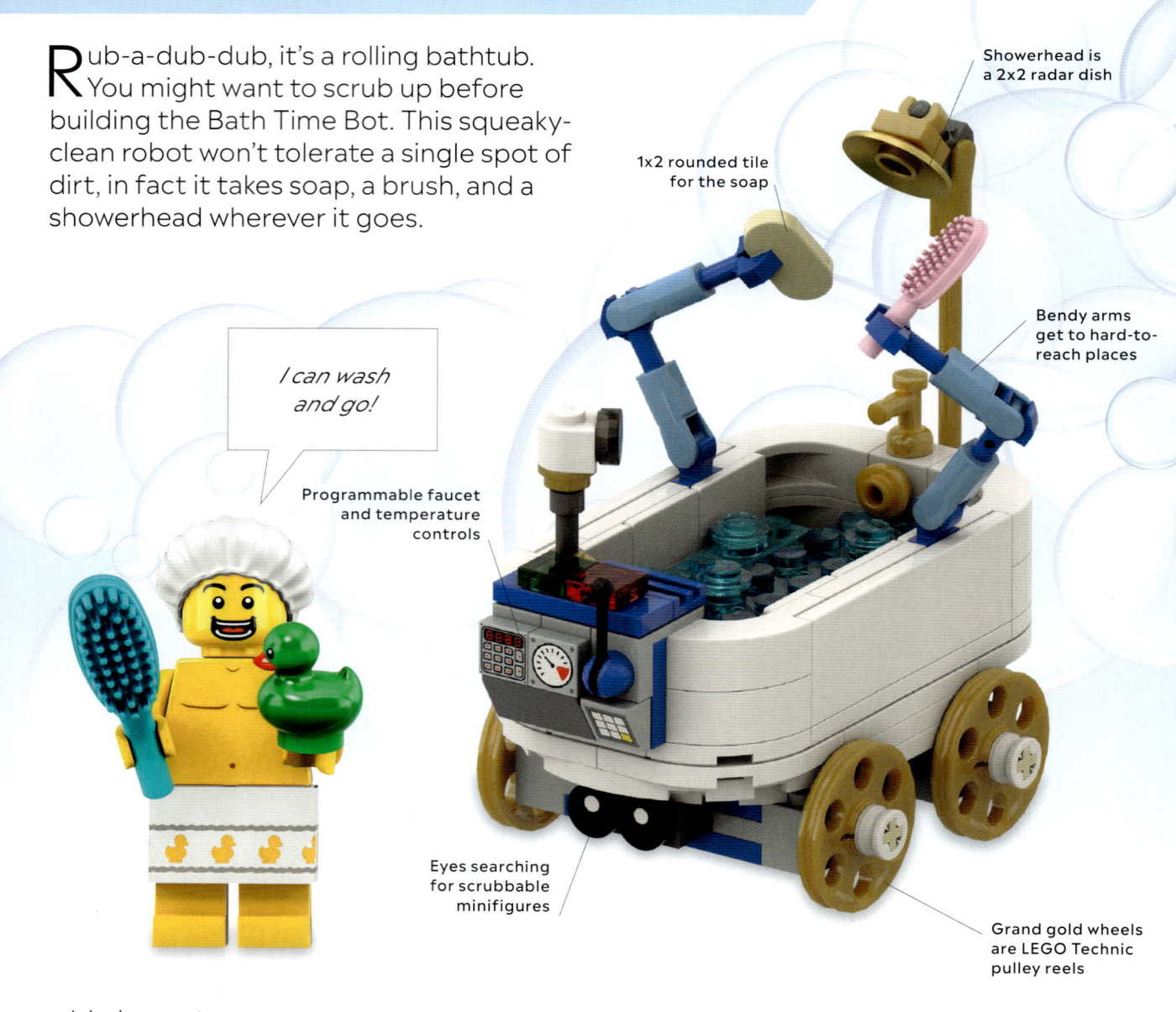

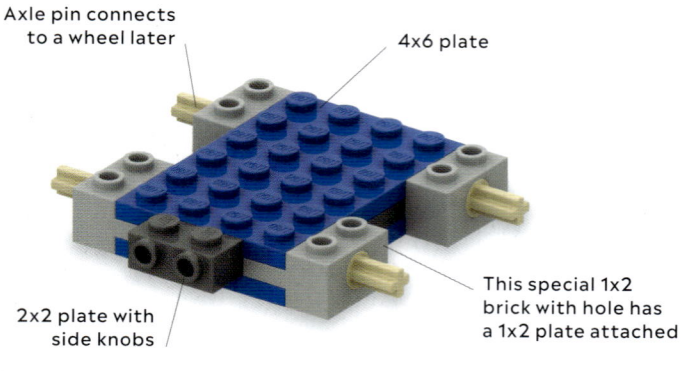

Rub-a-dub-dub, it's a rolling bathtub. You might want to scrub up before building the Bath Time Bot. This squeaky-clean robot won't tolerate a single spot of dirt, in fact it takes soap, a brush, and a showerhead wherever it goes.

Showerhead is a 2x2 radar dish

1x2 rounded tile for the soap

Bendy arms get to hard-to-reach places

I can wash and go!

Programmable faucet and temperature controls

Eyes searching for scrubbable minifigures

Grand gold wheels are LEGO Technic pulley reels

Axle pin connects to a wheel later

4x6 plate

2x2 plate with side knobs

This special 1x2 brick with hole has a 1x2 plate attached

WHEEL BASE

Begin your Bath Time Bot with a broad base for the tub to sit on, made from blue 4x6 plates. Those plates fit above and below the pieces that will connect to the bot's wheels at the sides and eyes at the front. Add axle pins for the wheels, but don't add the wheels just yet.

SKILL LEVEL

TUB BOTTOM

Now use white pieces to start building up the bathtub. (Or use another color if you prefer a brighter bathroom!) The tub's curved ends are made from half round plates. There's another, smaller layer of white plates below the top one here.

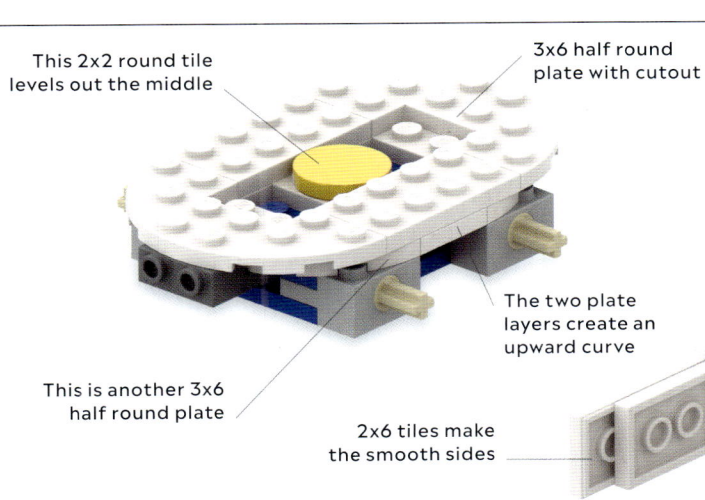

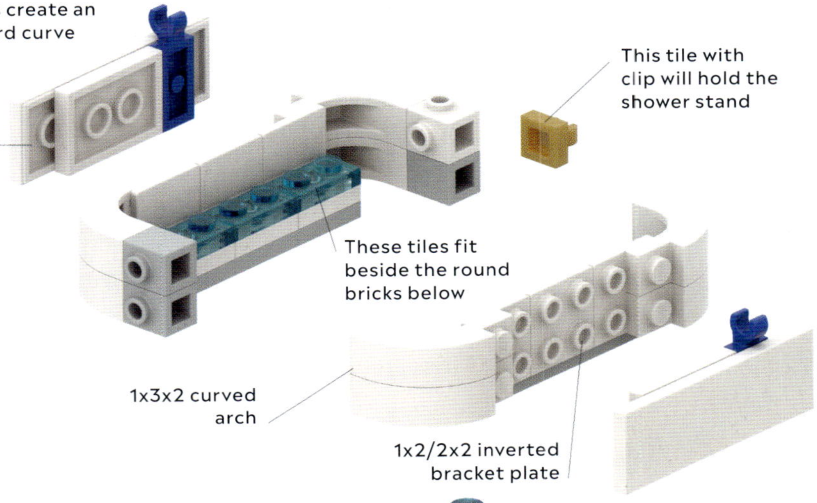

- This 2x2 round tile levels out the middle
- 3x6 half round plate with cutout
- The two plate layers create an upward curve
- This is another 3x6 half round plate
- 2x6 tiles make the smooth sides
- This tile with clip will hold the shower stand
- These tiles fit beside the round bricks below
- 1x3x2 curved arch
- 1x2/2x2 inverted bracket plate
- Transparent blue 2x2 round brick
- 2x2 macaroni tiles follow the curve of the plates below

RUN THE BATH

Add smooth tiles around the circumference of the tub, then pour in some watery-looking pieces. Use blue ones for water, or green if you prefer a slime bath! Next, create the bathtub's sides to hold in all that liquid. They're mostly made from a mixture of curved arch bricks, bracket plates, and tiles.

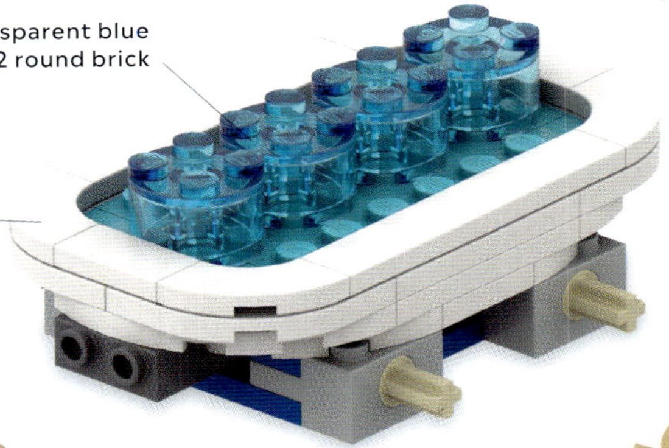

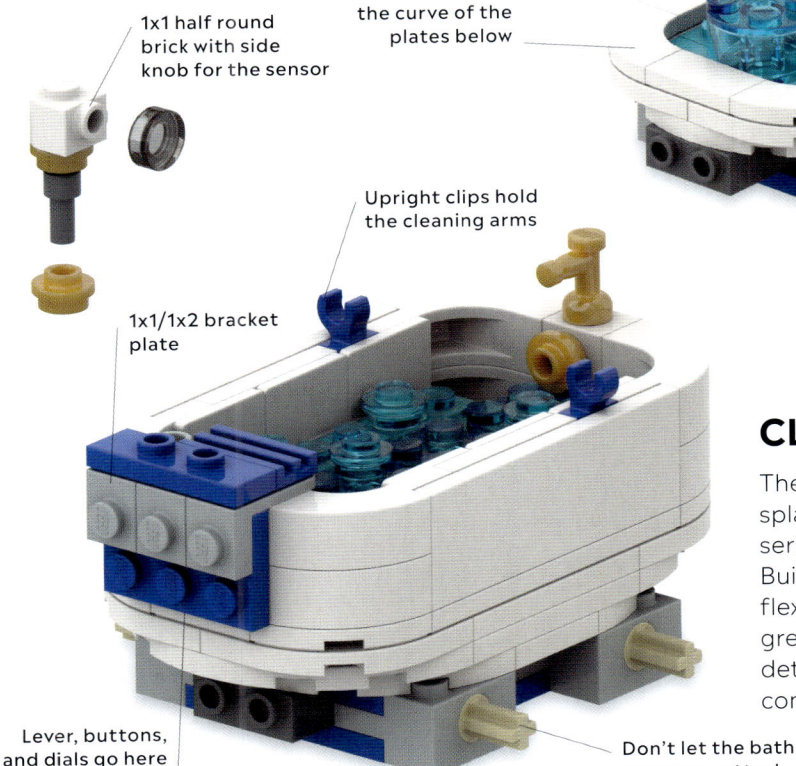

- 1x1 half round brick with side knob for the sensor
- Upright clips hold the cleaning arms
- 1x1/1x2 bracket plate
- Radar dish knob fits into the back of a tile with clip

CLEANING TOOLS

The Bath Time Bot is not just for splashing around—it's designed for seriously synchronized scrubbing! Build it a shower for blasting water, flexible cleaning arms for elbow grease, and an upright sensor to determine when cleaning is complete. What a brilliant bot!

- Lever, buttons, and dials go here
- Don't let the bath run away—attach the wheels last!

SHOWER

ROBO-ON-THE-GO

Stand clear—robot reversing. Lifting and shifting is what Robo-On-The-Go does. Its pallet stacker rises and falls to move bricks, tiles, and other LEGO parts. This one is waving its arms wildly. Perhaps there's a minifigure in its way.

- Heavy lifting pallet stacker
- One long brick with holes creates the mast
- L-shaped beam
- 1x8 brick with holes for the main chassis
- 1x4 brick with holes strengthen the chassis
- Deep tread tires for indoor and outdoor work

You have great shelf-awareness.

STRONG SUPPORT

Robo-On-The-Go has a super strong base to support and balance out the long lift arm. Two long bricks with holes create the bot's chassis. They fit together at the front with blue pins. More long pins on the sides of the chassis connect to other pieces with holes to make it even stronger.

SKILL LEVEL

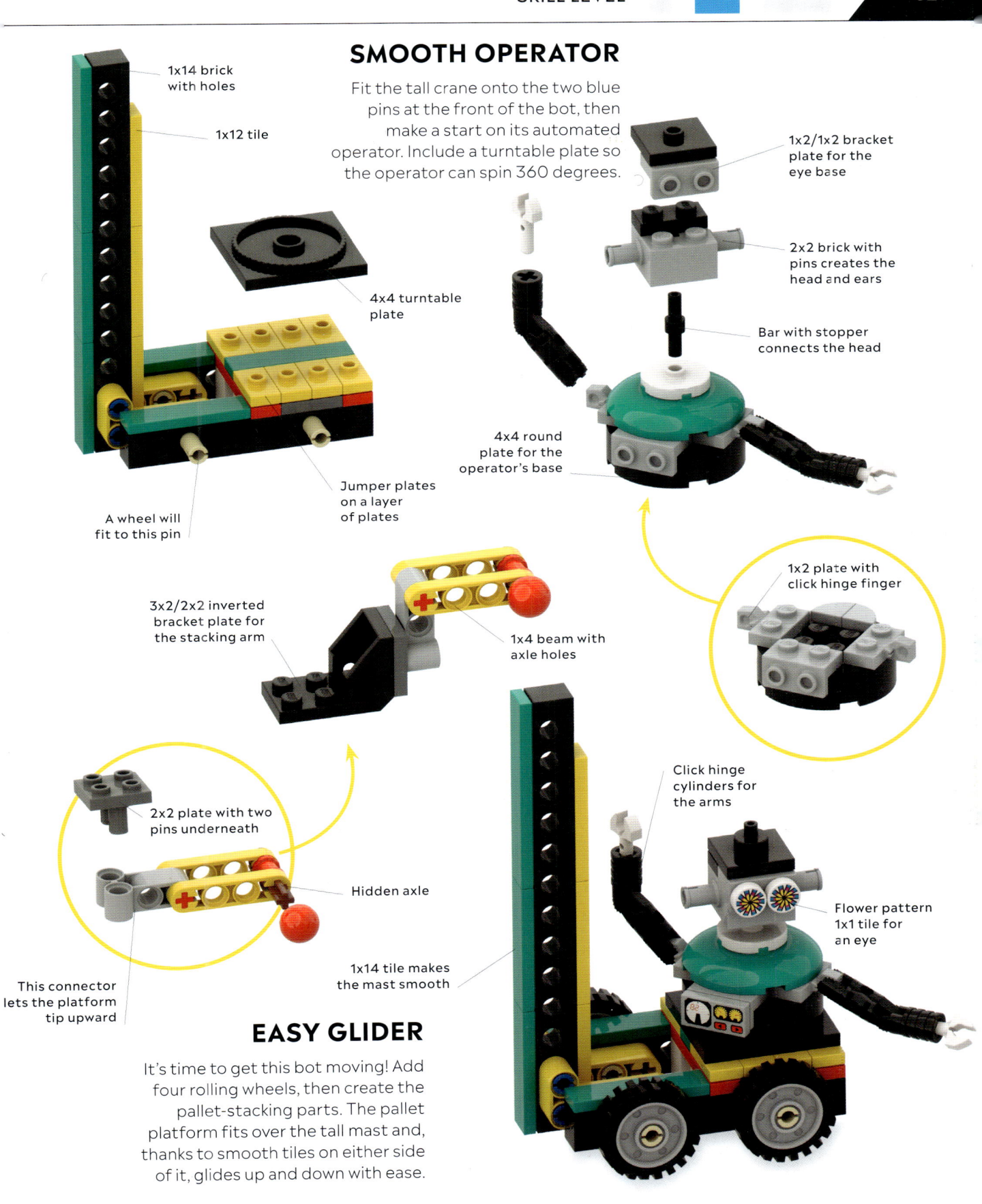

321

SMOOTH OPERATOR

Fit the tall crane onto the two blue pins at the front of the bot, then make a start on its automated operator. Include a turntable plate so the operator can spin 360 degrees.

- 1x14 brick with holes
- 1x12 tile
- 4x4 turntable plate
- Jumper plates on a layer of plates
- A wheel will fit to this pin
- 1x2/1x2 bracket plate for the eye base
- 2x2 brick with pins creates the head and ears
- Bar with stopper connects the head
- 4x4 round plate for the operator's base
- 1x2 plate with click hinge finger
- 3x2/2x2 inverted bracket plate for the stacking arm
- 1x4 beam with axle holes
- 2x2 plate with two pins underneath
- Hidden axle
- This connector lets the platform tip upward
- Click hinge cylinders for the arms
- Flower pattern 1x1 tile for an eye
- 1x14 tile makes the mast smooth

EASY GLIDER

It's time to get this bot moving! Add four rolling wheels, then create the pallet-stacking parts. The pallet platform fits over the tall mast and, thanks to smooth tiles on either side of it, glides up and down with ease.

SUPERSONIC ROBOT

Supersonic means faster than the speed of sound. With a jet pack on its back and boosters on its heels, this Supersonic Robot looks more than ready to break the sound barrier. Get ready for the sonic boom!

Arms can raise up in a flying pose

Layered chest armor is all one piece

Whoa! What was that?

Narrow legs are light in the air

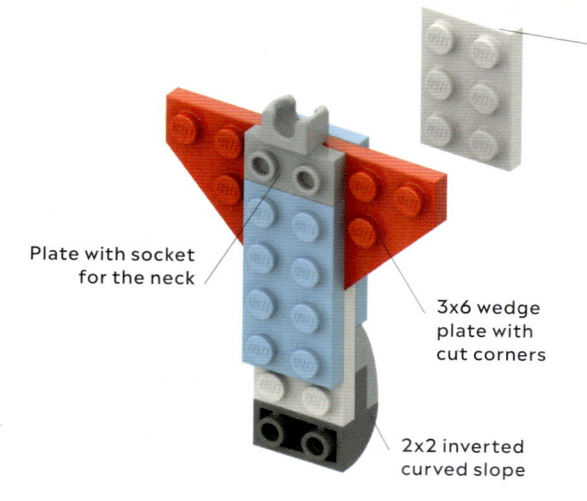

2x3 plate

Plate with socket for the neck

3x6 wedge plate with cut corners

2x2 inverted curved slope

WINGED WONDER

Start your engines and get this speedy bot off the ground with plates in the shape of a jet plane! These plates all fit together sideways to form the base of the triangular torso. At the bottom, two curved slopes form the robot's rounded rear.

SKILL LEVEL

BULKY BODY

Add more layers of plates to the torso, including plates with balls for the arms and legs to attach to at the next step. The tiles and plates at the front create the right shape for the armor to be added.

- 1x1 round plate
- 1x2 plate with ball
- 1x4 tile with two knobs
- Ball with pin
- Curved flame with pin
- 1x1 round plate with hole fits onto a knob
- Socket connects to the body

FAST FEATURES

This robot is humanoid, which means it resembles a human in shape. Build two humanoid arms and legs with added superhero-like supersonic features: blasts of fire at the heels and flame adornments on the shoulders.

- 1x2 slope with cutout shapes the toes

RIGHT LEG

LEFT ARM

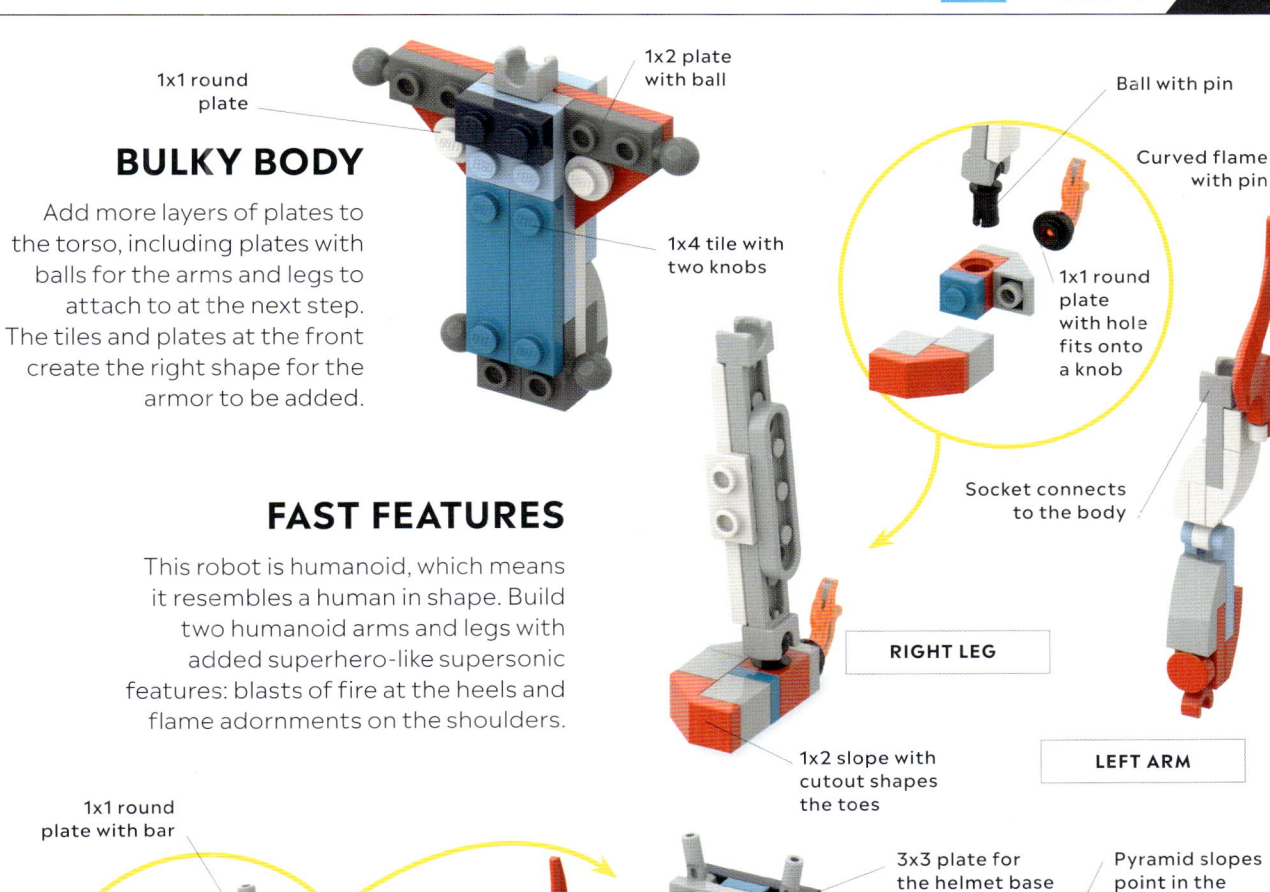

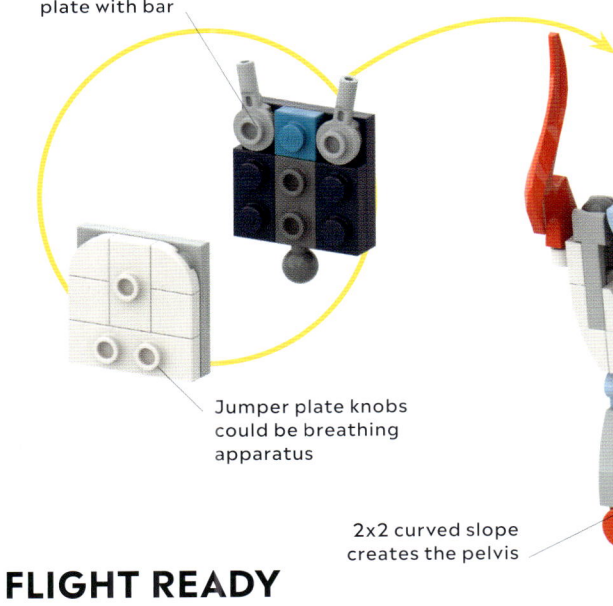

- 1x1 round plate with bar
- Jumper plate knobs could be breathing apparatus
- 3x3 plate for the helmet base
- Pyramid slopes point in the direction of travel
- 2x2 curved slope creates the pelvis
- 1x1 round tile and clip for a hand
- Rocket flames fit into round plates with open knobs
- Long 1x6 tiles smooth out the sides of the legs

FLIGHT READY

Now build the mysterious head of the robot, in its white flight helmet and black visor. Two round plates with bars facing upward create the helmet antennas for ground-to-air radio communications. Then all this bot needs is a powerful jet pack to send it anywhere, FAST.

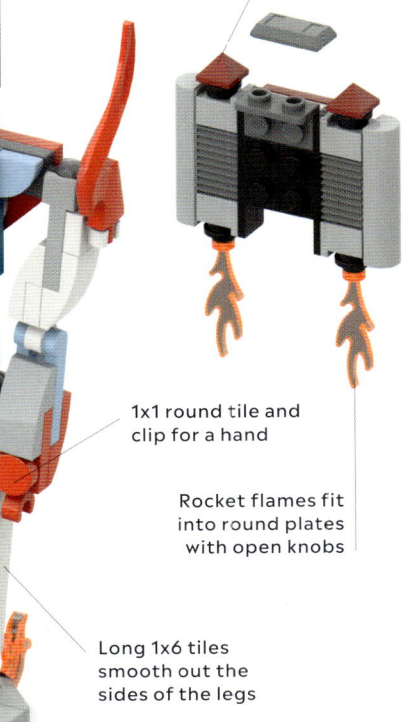

UNDERWATER EXPLORER

I think we need a can opener!

Scientists learn about the deep ocean by sending down underwater explorer robots. Why not build your own, with a spinning tail fin and claws for grabbing samples? This one has a swiveling camera too, for snapping deep-sea snapper fish.

- Jagged 1x8 panels cut through water
- Jumper plate knobs look like tiny portholes
- 2x2 double curved slopes give the head a round shape
- Round observation window
- Grippers help the arms grab samples
- 1x4 curved slopes for rear end
- 1x2/1x2 bracket plate
- Line of 1x1 slopes
- 2x6 base plate

LOBSTER BODY

The Underwater Explorer blends in well with its nautical environment, because it resembles a lobster! Build it a long body, moving upward from a yellow base plate. The small slopes that line the sides give the body a rounded shape at the bottom.

SKILL LEVEL

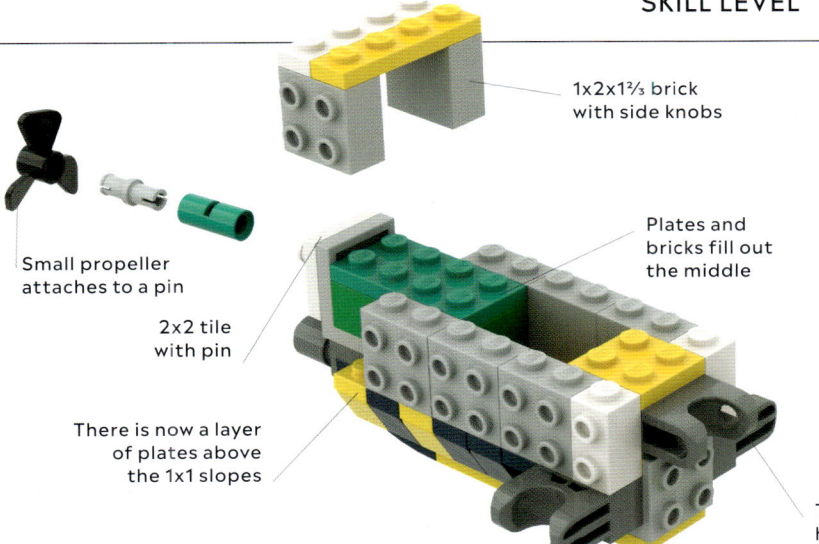

PROPELLER POWER

Now the bot's body is much taller, thanks to lots of bricks with side knobs. At the back there's a whirling propeller that fits (via a pin and connector) onto a sideways tile with pin. At the front there are connection points for the pincer arms and head.

- 1x2x2/3 brick with side knobs
- Small propeller attaches to a pin
- 2x2 tile with pin
- Plates and bricks fill out the middle
- There is now a layer of plates above the 1x1 slopes
- This big socket holds the head

SMOOTH SURFACES

Connect smooth tiles and curved slopes to the sides of the bot's body so it can glide with ease through water. There's now another layer of plates on top of the body and a slope brick for shape at the back.

- 1x2 slope with cutout
- These plates continue the colors of the lower body
- 1x2 jumper plate
- 2x8 curved slope for the smooth back
- 2x2x2/3 curved slope with point
- Transparent 2x2 radar dish for a round window
- Build the camera and stand onto these clips
- These tiles continue the bot's color scheme
- 2x2 brick with two ball joints
- Bar holder with angled clip
- Black light bulb for a gripper
- Two 1x2 wedge slopes curve around the face
- 1x2/2x2 bracket plate for the face base

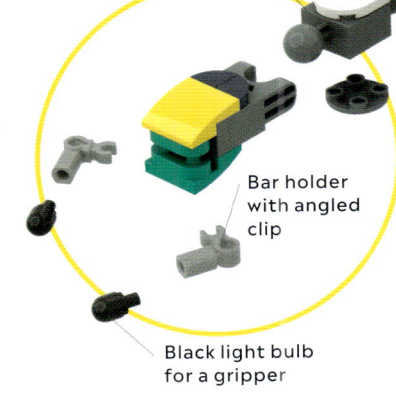

EXPLORING CLAWS

This bot is almost ready to take to the water. Last but not least, build an inquisitive round head with probing eyes and clawlike, grasping arms for collecting interesting finds.

325

SPACE PROBE

If there's life on Mars—or other far-flung planets—the robot Space Probe is out to find it. It scuttles over the planet's surface, taking photographs and collecting samples. Solar panels power the plucky probe.

Fez looks like a probing eye

Star scanner surveys its position in space

Solar panels power the probe

Camera records where the probe travels

2x2 turntable plate

Axle will connect to solar panels

1x2/1x2 inverted bracket plate

1x10 brick with holes

2x2 round brick with hole

TURNING TOOLS

A space-bound probe needs to be solidly built. Give yours a stable base to start with, made from layers of bricks and plates. At either end there are turntable plates—these will hold movable tools at the front and rear of the probe.

PROBE PROGRESS

Continue building upward to finish off the probe's robotic body. Long bricks with holes hold pins that will later connect to the roving wheels. There's now a round brick with hole at the front for the camera arm base.

SKILL LEVEL

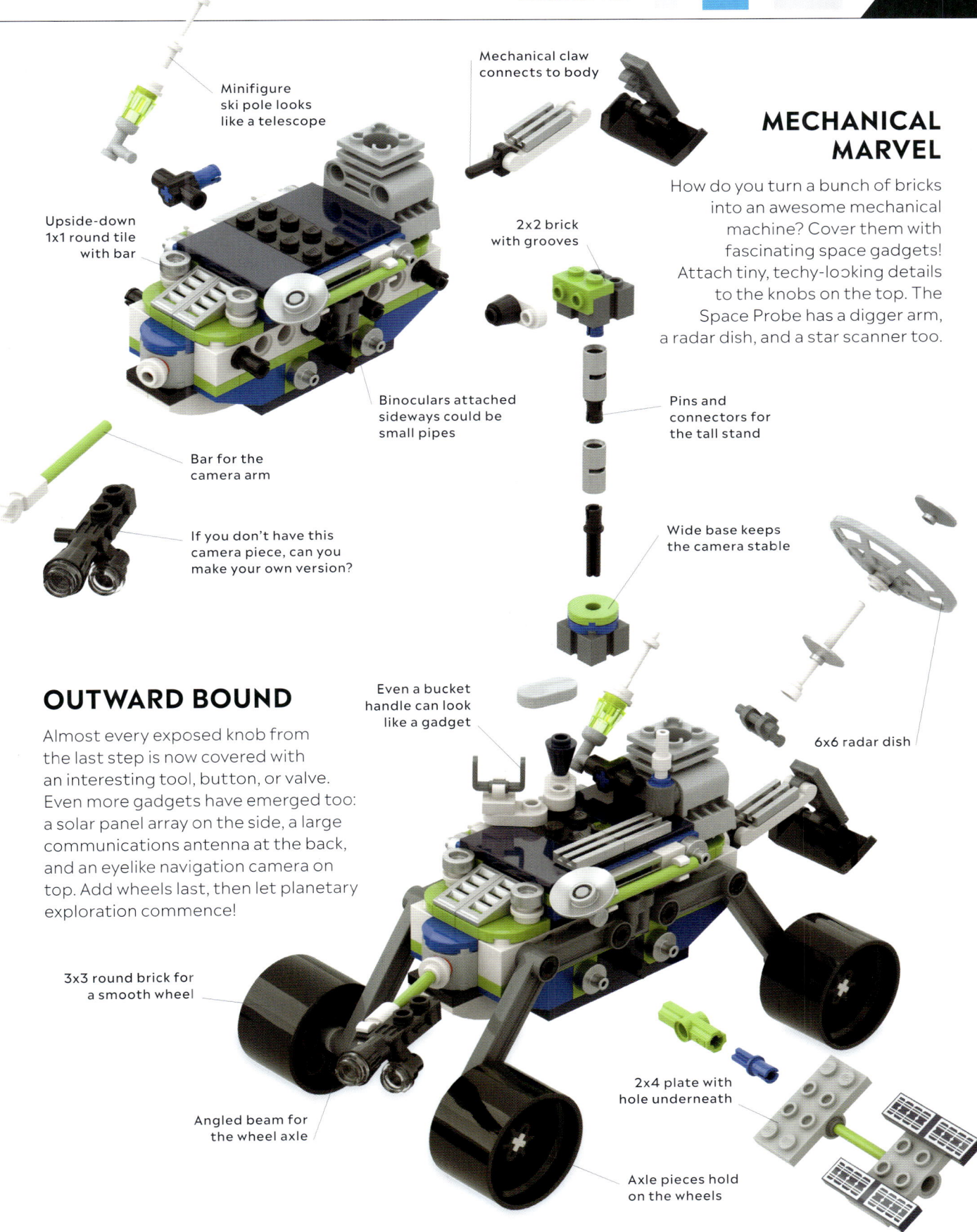

MECHANICAL MARVEL

How do you turn a bunch of bricks into an awesome mechanical machine? Cover them with fascinating space gadgets! Attach tiny, techy-looking details to the knobs on the top. The Space Probe has a digger arm, a radar dish, and a star scanner too.

- Minifigure ski pole looks like a telescope
- Mechanical claw connects to body
- Upside-down 1x1 round tile with bar
- 2x2 brick with grooves
- Binoculars attached sideways could be small pipes
- Pins and connectors for the tall stand
- Bar for the camera arm
- If you don't have this camera piece, can you make your own version?
- Wide base keeps the camera stable
- 6x6 radar dish

OUTWARD BOUND

Almost every exposed knob from the last step is now covered with an interesting tool, button, or valve. Even more gadgets have emerged too: a solar panel array on the side, a large communications antenna at the back, and an eyelike navigation camera on top. Add wheels last, then let planetary exploration commence!

- Even a bucket handle can look like a gadget
- 3x3 round brick for a smooth wheel
- Angled beam for the wheel axle
- 2x4 plate with hole underneath
- Axle pieces hold on the wheels

BIRTHDAY BOT

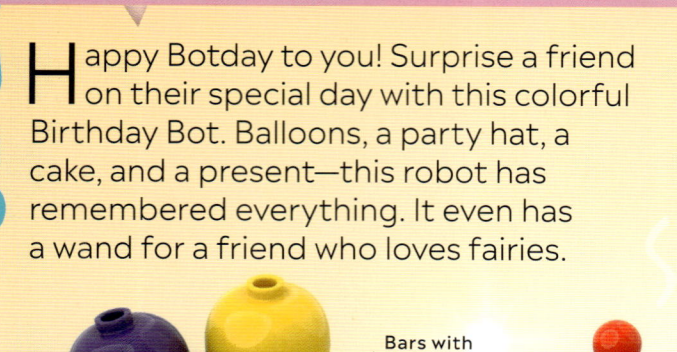

Happy Botday to you! Surprise a friend on their special day with this colorful Birthday Bot. Balloons, a party hat, a cake, and a present—this robot has remembered everything. It even has a wand for a friend who loves fairies.

I'm a walking party!

Bars with stoppers hold up the balloons

Multiple eyes for managing parties full of people

Freshly lit birthday candle

Clown-like ruff collar

Gift lists can be programed in here

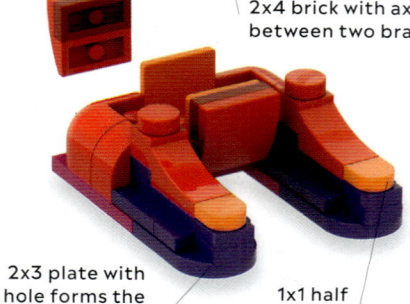

2x4 brick with axle holes fits between two bracket plates

2x3 plate with hole forms the foot base

1x1 half tile toes

PUT BOTH FEET IN ...

Without feet, this robot wouldn't be able to do the birthday Hokey Pokey! Build them first to give the Birthday Bot a sturdy base. These long feet resemble extra-large clown shoes.

SKILL LEVEL 329

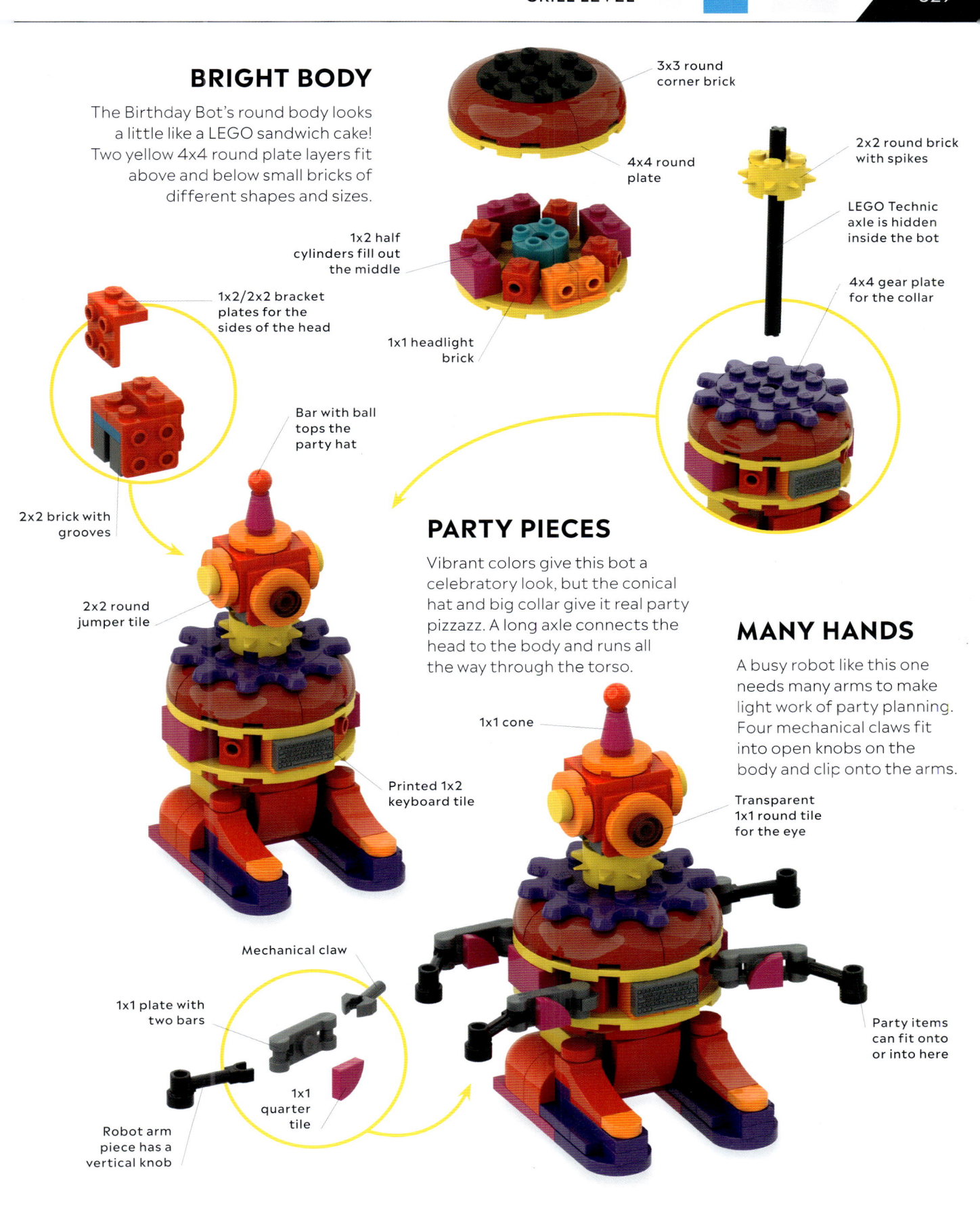

BRIGHT BODY

The Birthday Bot's round body looks a little like a LEGO sandwich cake! Two yellow 4x4 round plate layers fit above and below small bricks of different shapes and sizes.

3x3 round corner brick

4x4 round plate

1x2 half cylinders fill out the middle

1x1 headlight brick

2x2 round brick with spikes

LEGO Technic axle is hidden inside the bot

4x4 gear plate for the collar

1x2/2x2 bracket plates for the sides of the head

2x2 brick with grooves

2x2 round jumper tile

Bar with ball tops the party hat

PARTY PIECES

Vibrant colors give this bot a celebratory look, but the conical hat and big collar give it real party pizzazz. A long axle connects the head to the body and runs all the way through the torso.

1x1 cone

Printed 1x2 keyboard tile

MANY HANDS

A busy robot like this one needs many arms to make light work of party planning. Four mechanical claws fit into open knobs on the body and clip onto the arms.

Transparent 1x1 round tile for the eye

Mechanical claw

1x1 plate with two bars

1x1 quarter tile

Robot arm piece has a vertical knob

Party items can fit onto or into here

BOTS IN THE COMMUNITY

Have no fear, the robots are here. In this LEGO town, robots perform all kinds of helpful tasks. Cleaning up after messy pups? Yes! Seeing pedestrians safely across the road? For sure! Try to think of other tasks a community robot could do.

Flowers thrive in this greenhouse environment

CROSSING GUARD BOT

With its stop sign and a red light for a face, there's no mistaking this bot's role in the community. LEGO car drivers who see it stomp out into the road at a crosswalk must stop to let minifigures cross safely.

I just want to talk, Crossing Guard Bot!

GREEN-FINGERED BOT

The Gardener Robot on pages 310–311 is a breath of fresh air in any community. It not only grows plants in its hothouse head, it also tends to any flower beds it passes and delivers fresh flowers to brighten your day.

Red and orange pieces make this bot easy to spot

Paddle and red round tile for the stop sign

Outstretched arm is a bar holder with bar

2x2 jumper plates for the crosswalk

IDEAS GALLERY

331

Grabber arm twists in all directions

See more of this bot on pages 296-297

SHOPPER BOT

This happy shopper not only carries all your groceries for you, it also picks them off shelves with its grabber arm. You can even program it to carry your shopping home from the supermarket. Can you build LEGO groceries to go into it?

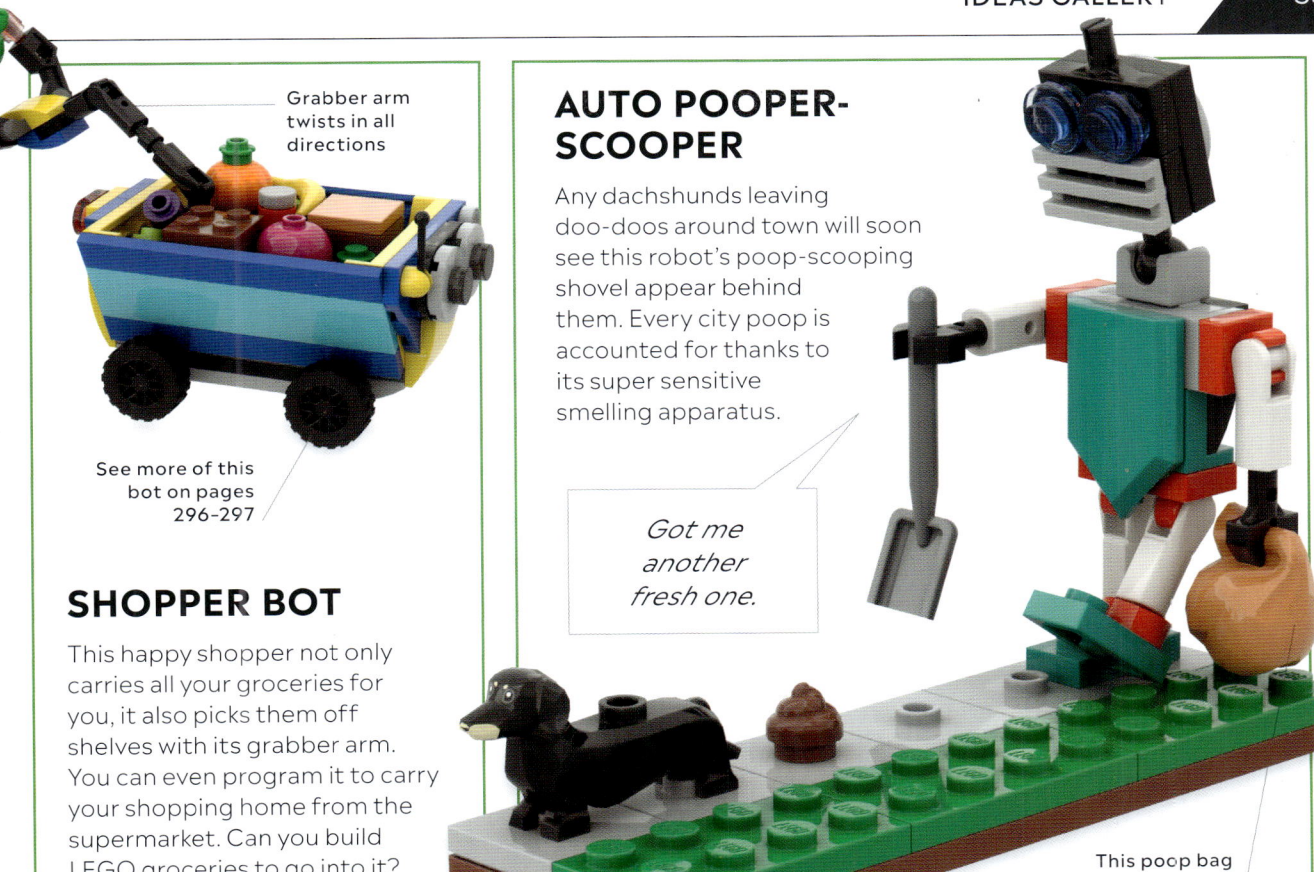

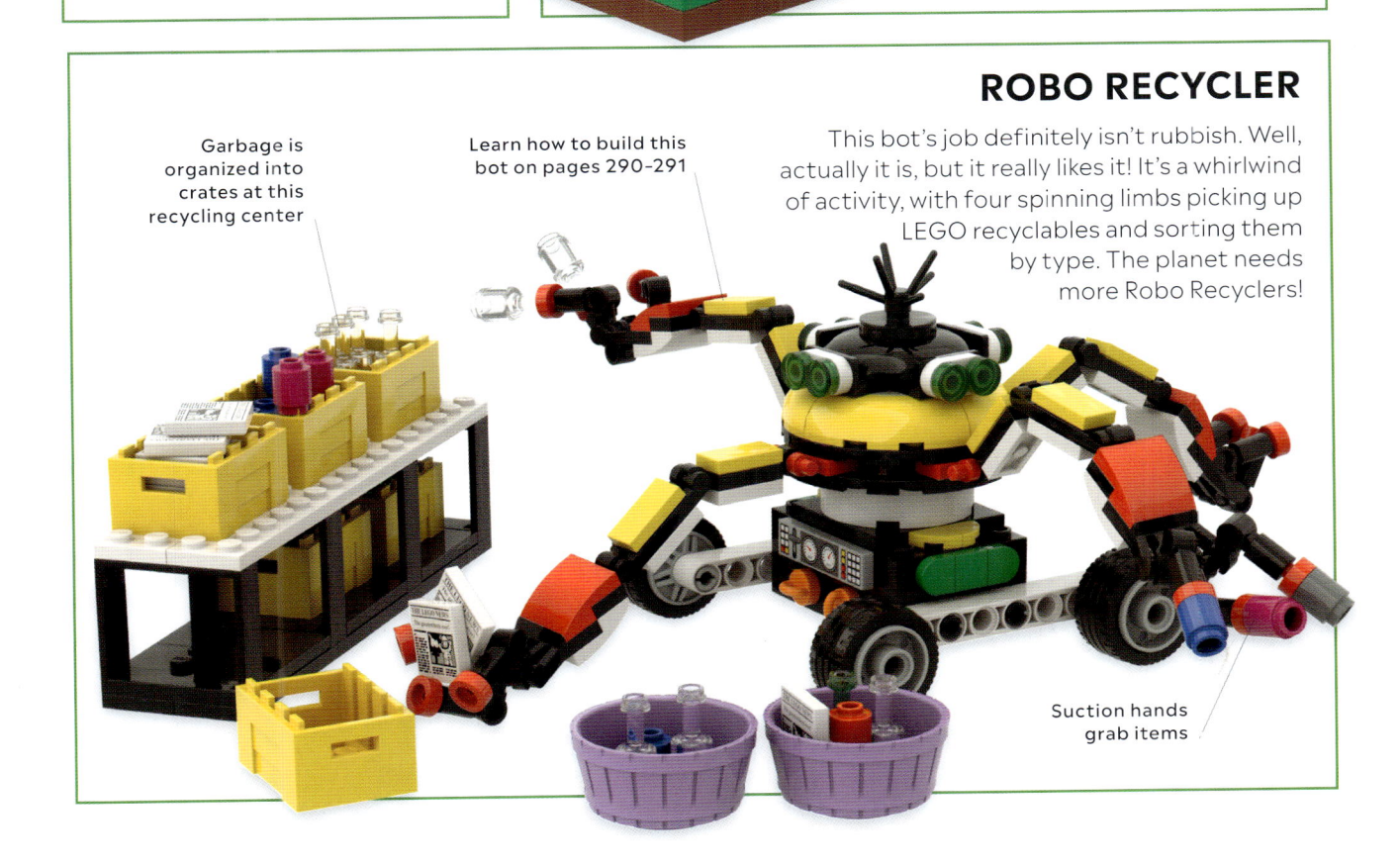

AUTO POOPER-SCOOPER

Any dachshunds leaving doo-doos around town will soon see this robot's poop-scooping shovel appear behind them. Every city poop is accounted for thanks to its super sensitive smelling apparatus.

Got me another fresh one.

This poop bag looks pretty full

Garbage is organized into crates at this recycling center

Learn how to build this bot on pages 290-291

ROBO RECYCLER

This bot's job definitely isn't rubbish. Well, actually it is, but it really likes it! It's a whirlwind of activity, with four spinning limbs picking up LEGO recyclables and sorting them by type. The planet needs more Robo Recyclers!

Suction hands grab items

ONE-BOT BAND

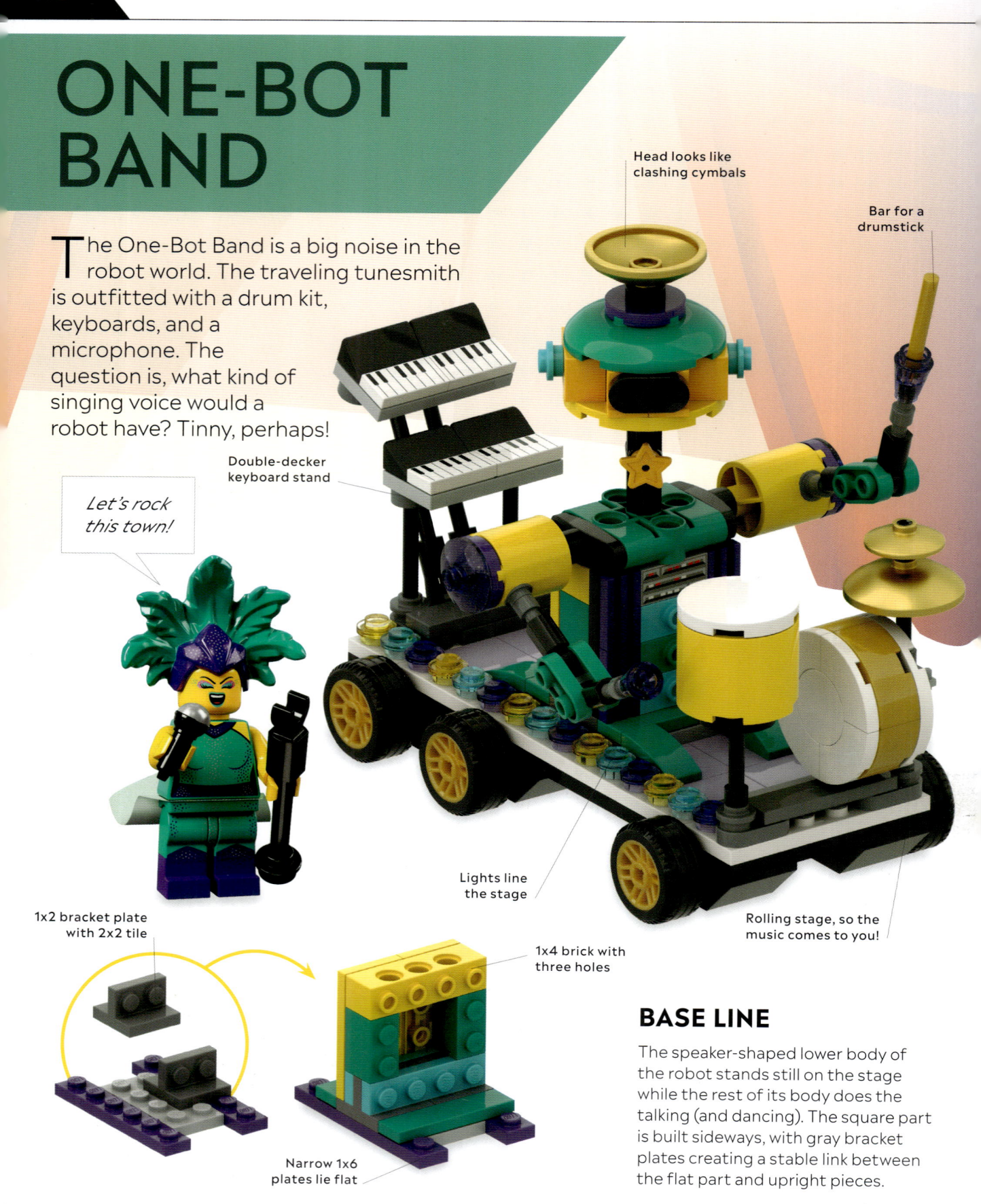

The One-Bot Band is a big noise in the robot world. The traveling tunesmith is outfitted with a drum kit, keyboards, and a microphone. The question is, what kind of singing voice would a robot have? Tinny, perhaps!

Head looks like clashing cymbals

Bar for a drumstick

Double-decker keyboard stand

"Let's rock this town!"

Lights line the stage

Rolling stage, so the music comes to you!

1x2 bracket plate with 2x2 tile

1x4 brick with three holes

Narrow 1x6 plates lie flat

BASE LINE

The speaker-shaped lower body of the robot stands still on the stage while the rest of its body does the talking (and dancing). The square part is built sideways, with gray bracket plates creating a stable link between the flat part and upright pieces.

SKILL LEVEL

GROOVY MOVER

Now the base is stable, add a swiveling shoulder section so the bot can switch between instruments quickly. At this stage, add the quick-moving arms and hands, and the One-Bot Band's turning head too.

- 4x4 round plate for the head base
- This axle holds the head securely
- 1x3x3 square pin connector
- 1x2 round brick with open center
- Curved slopes and tiles make the body base smooth
- 2x2x2 round brick with hole for the shoulder
- Axle pin lets the neck swivel
- Pin connector with two pins

ON STAGE

The One-Bot Band knows how to fill a stage, so build it a wide one! This stage has a 6x16 base plate, with tiles on top that make a smooth platform for the bot to perform on. Plates with pins underneath will hold the rotating wheels.

- Four knobs hold or the bot
- 2x2 plate with pin

MUSICAL MAESTRO

Give the stage some disco flair with flashing lights, then create some instruments. There's a three-piece drum kit at one end, and double keyboards at the other. What would your bot play? Add wheels, then get this show on the road!

- Bar holders connect to clips
- Bar through the center of this 3x3x2 round brick
- 2x2 round brick
- 1x2 piano keys printed tile
- Bars create the stand
- Tiles and round plates connect the drum kit at the base

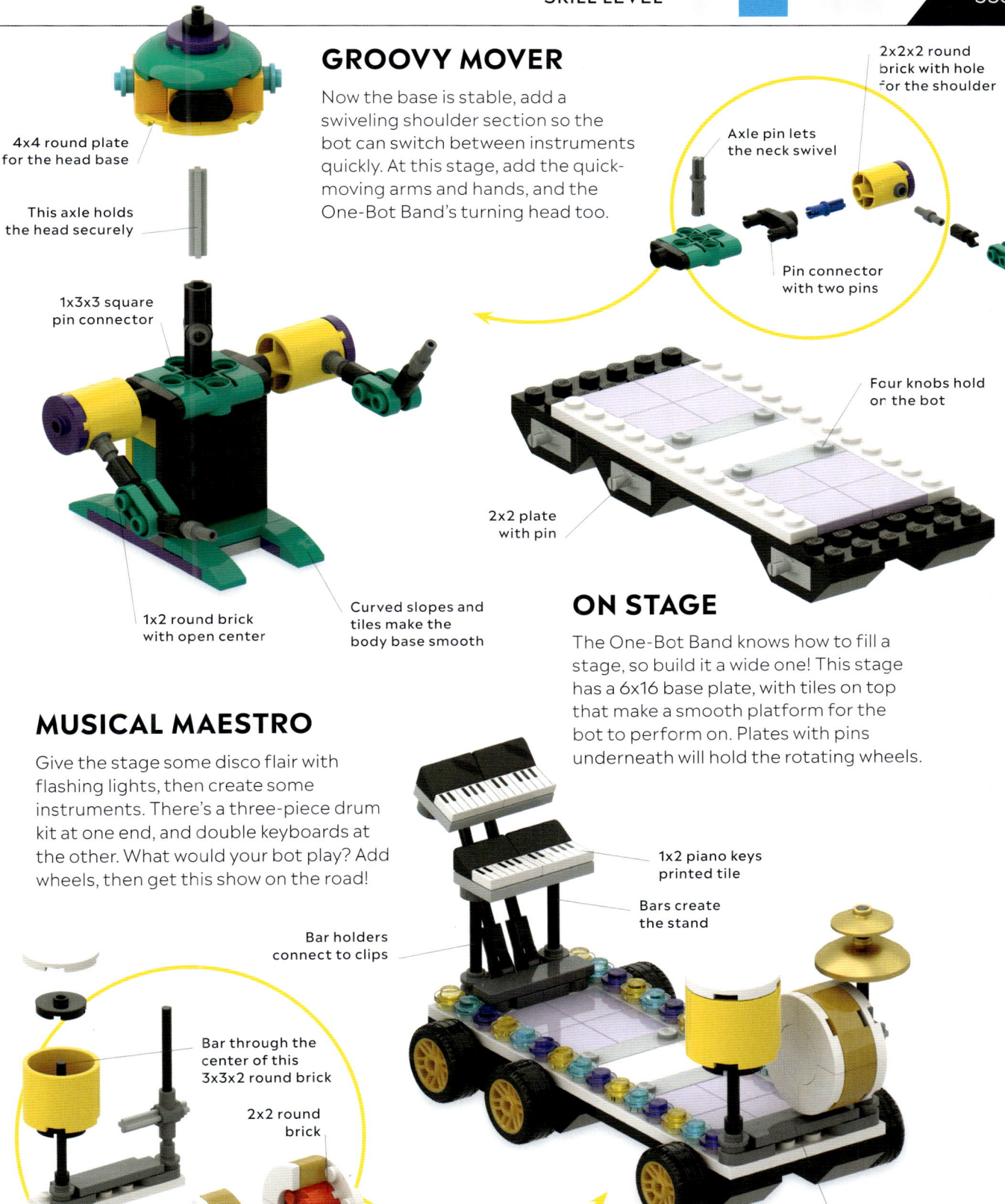

333

MECHA BIRD

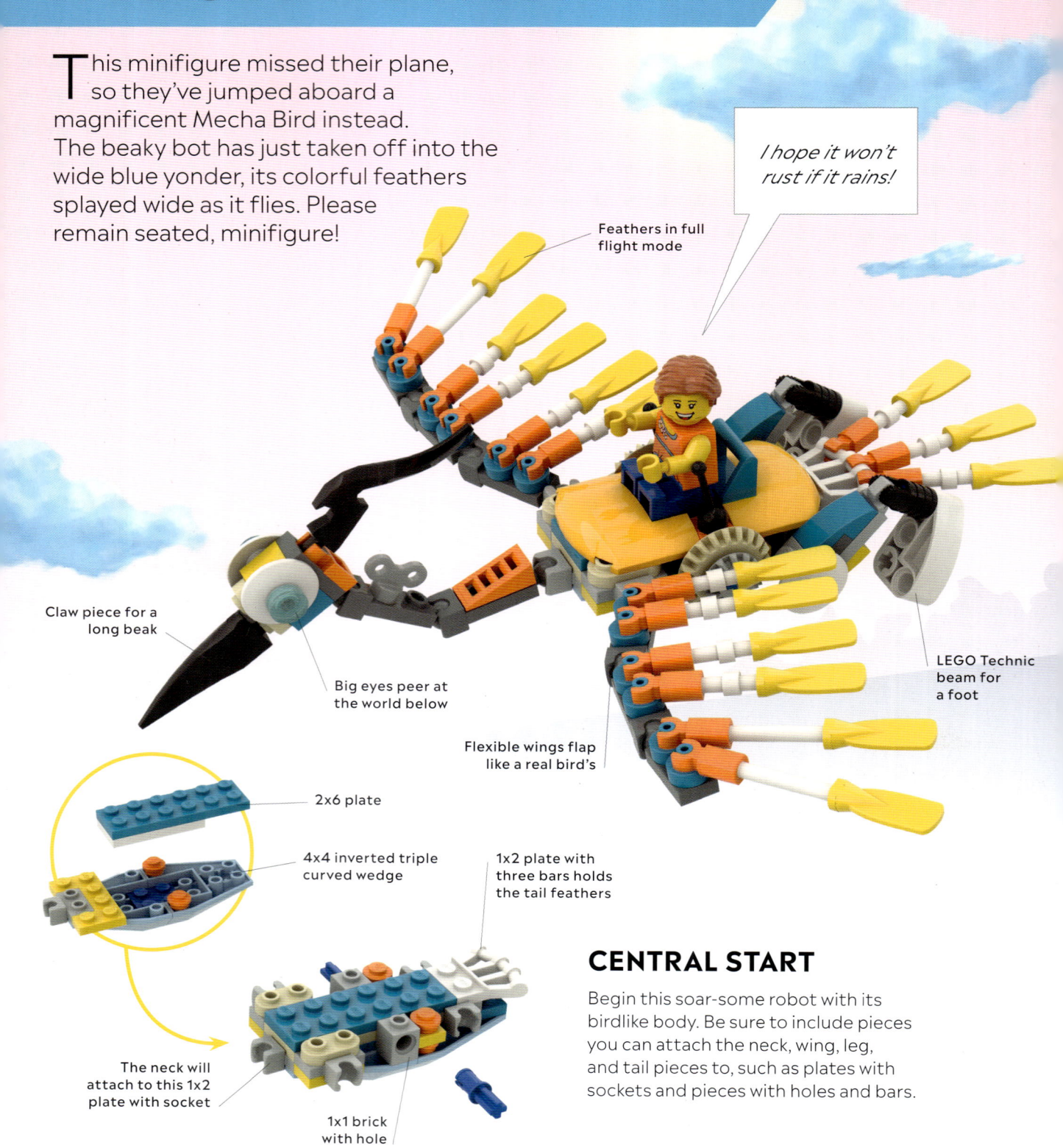

This minifigure missed their plane, so they've jumped aboard a magnificent Mecha Bird instead. The beaky bot has just taken off into the wide blue yonder, its colorful feathers splayed wide as it flies. Please remain seated, minifigure!

I hope it won't rust if it rains!

Feathers in full flight mode

Claw piece for a long beak

Big eyes peer at the world below

Flexible wings flap like a real bird's

LEGO Technic beam for a foot

2x6 plate

4x4 inverted triple curved wedge

1x2 plate with three bars holds the tail feathers

The neck will attach to this 1x2 plate with socket

1x1 brick with hole

CENTRAL START

Begin this soar-some robot with its birdlike body. Be sure to include pieces you can attach the neck, wing, leg, and tail pieces to, such as plates with sockets and pieces with holes and bars.

SKILL LEVEL 335

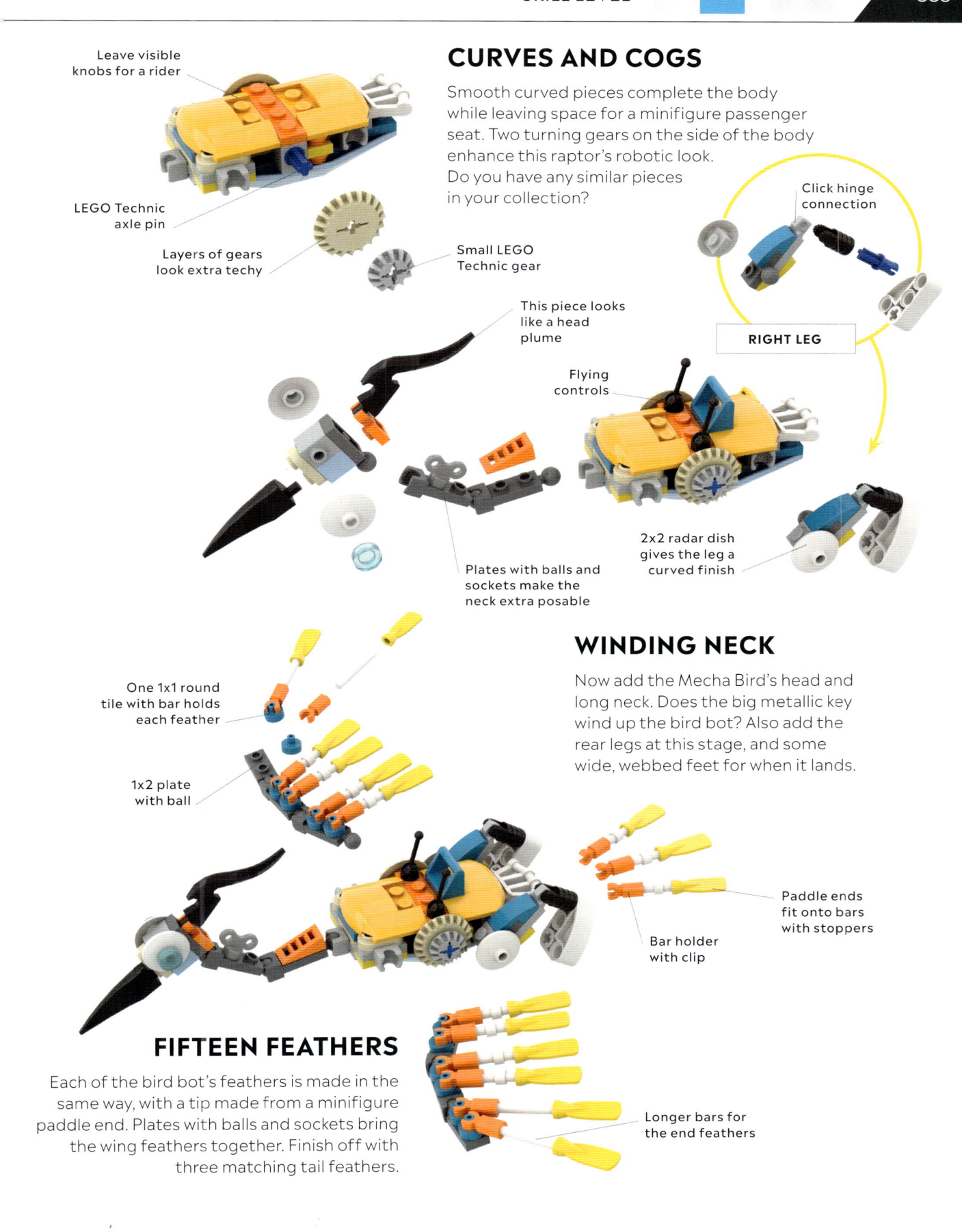

CURVES AND COGS

Smooth curved pieces complete the body while leaving space for a minifigure passenger seat. Two turning gears on the side of the body enhance this raptor's robotic look. Do you have any similar pieces in your collection?

- Leave visible knobs for a rider
- LEGO Technic axle pin
- Layers of gears look extra techy
- Small LEGO Technic gear
- This piece looks like a head plume
- Click hinge connection
- Flying controls
- Plates with balls and sockets make the neck extra posable
- 2x2 radar dish gives the leg a curved finish

RIGHT LEG

WINDING NECK

Now add the Mecha Bird's head and long neck. Does the big metallic key wind up the bird bot? Also add the rear legs at this stage, and some wide, webbed feet for when it lands.

- One 1x1 round tile with bar holds each feather
- 1x2 plate with ball
- Paddle ends fit onto bars with stoppers
- Bar holder with clip

FIFTEEN FEATHERS

Each of the bird bot's feathers is made in the same way, with a tip made from a minifigure paddle end. Plates with balls and sockets bring the wing feathers together. Finish off with three matching tail feathers.

- Longer bars for the end feathers

QUACKBOT

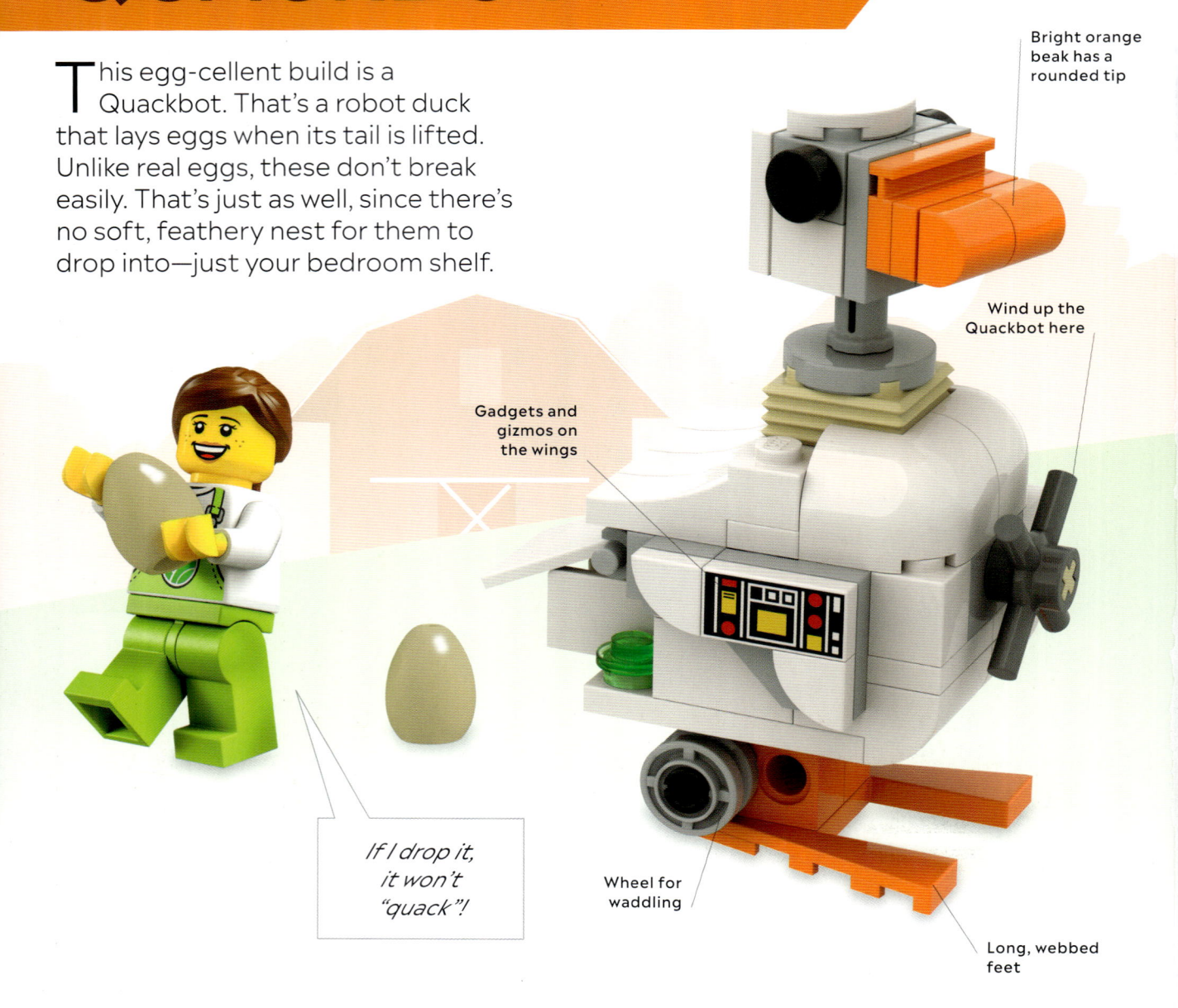

This egg-cellent build is a Quackbot. That's a robot duck that lays eggs when its tail is lifted. Unlike real eggs, these don't break easily. That's just as well, since there's no soft, feathery nest for them to drop into—just your bedroom shelf.

Bright orange beak has a rounded tip

Wind up the Quackbot here

Gadgets and gizmos on the wings

If I drop it, it won't "quack"!

Wheel for waddling

Long, webbed feet

SITTING DUCK

The Quackbot's rounded lower body is the starting point for this build. A 4x4 plate forms the base, and inverted curved slopes create the smooth shape of the feathers. Create a little tunnel for an egg to sit in at this stage too. This one is made from two panel pieces.

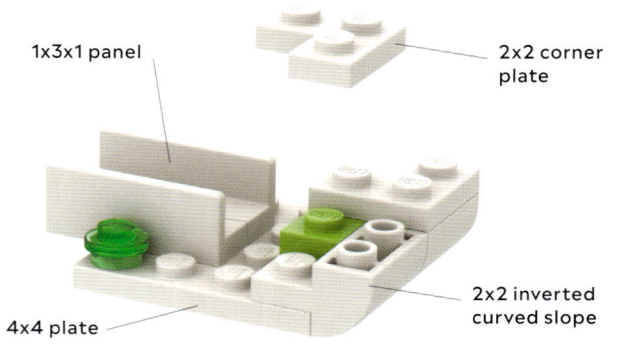

1x3x1 panel

2x2 corner plate

4x4 plate

2x2 inverted curved slope

SKILL LEVEL

337

1x2 brick with hole

LEGO Technic axle pin

Side knobs will hold the wing

KEY DETAILS

Build up the top part of the Quackbot's body, adding some pieces with side knobs and holes for attaching extra details. The axle pin on the front will become part of the wind-up key, which will actually turn.

Bar

2x2 inverted tile

This piece lies sideways

1x1 double curved slopes for the beak

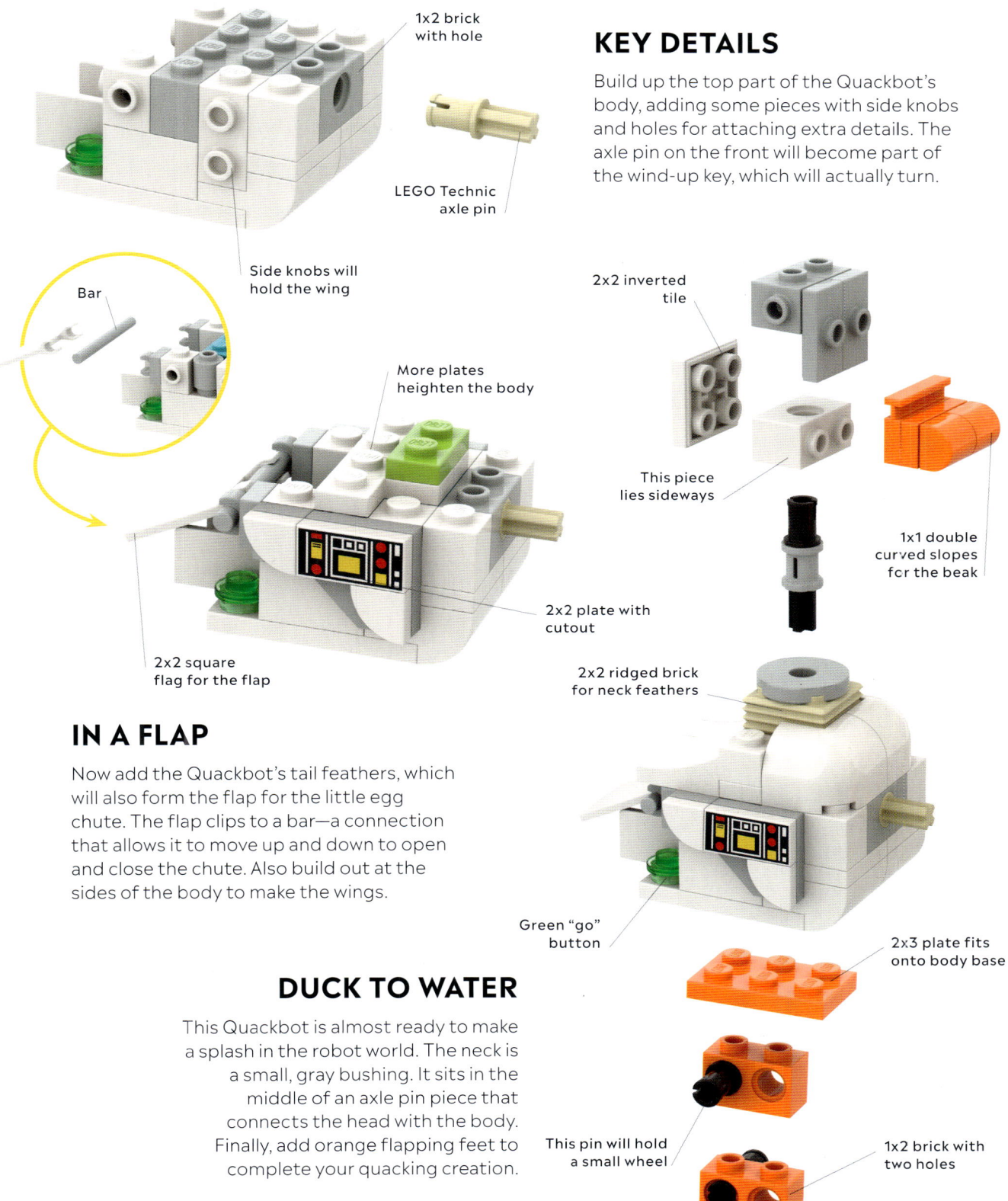

More plates heighten the body

2x2 plate with cutout

2x2 square flag for the flap

2x2 ridged brick for neck feathers

IN A FLAP

Now add the Quackbot's tail feathers, which will also form the flap for the little egg chute. The flap clips to a bar—a connection that allows it to move up and down to open and close the chute. Also build out at the sides of the body to make the wings.

Green "go" button

2x3 plate fits onto body base

DUCK TO WATER

This Quackbot is almost ready to make a splash in the robot world. The neck is a small, gray bushing. It sits in the middle of an axle pin piece that connects the head with the body. Finally, add orange flapping feet to complete your quacking creation.

This pin will hold a small wheel

1x2 brick with two holes

Webbed feet are one 4x4 wedge plate

ODD BOT

There's something adorably amiss about this robot. None of its parts go together, from the antenna on its head to its tanklike treads. Use colorful, random, or unexpected elements to create a unique Odd Bot of your own.

To be different is to be unique!

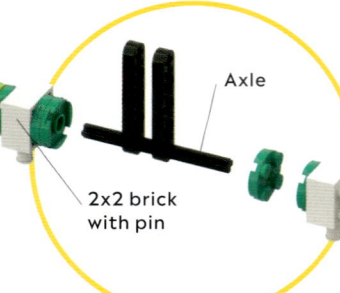

- Three claws make big grabber hands
- Rotating head antenna
- Shoulder satellite dishes send out strange signals
- Solar panels provide power
- Central computer outputs garbled data
- Rolling treads balance the unwieldy body

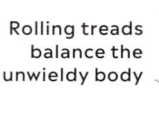

- Axle
- The hip pieces all have axle holes
- 2x2 brick with pin
- 1x5 beam with axle holes
- Pin connector tube
- Pin will connect to a wheel
- 1x3x3 square pin connector

BEAM BUILDING

The Odd Bot's bulky base is made from several LEGO Technic beams. Add pins to the bottom two beams to attach the tread tires to later, then start building upward toward the hips. They're held together with a hidden axle.

SKILL LEVEL

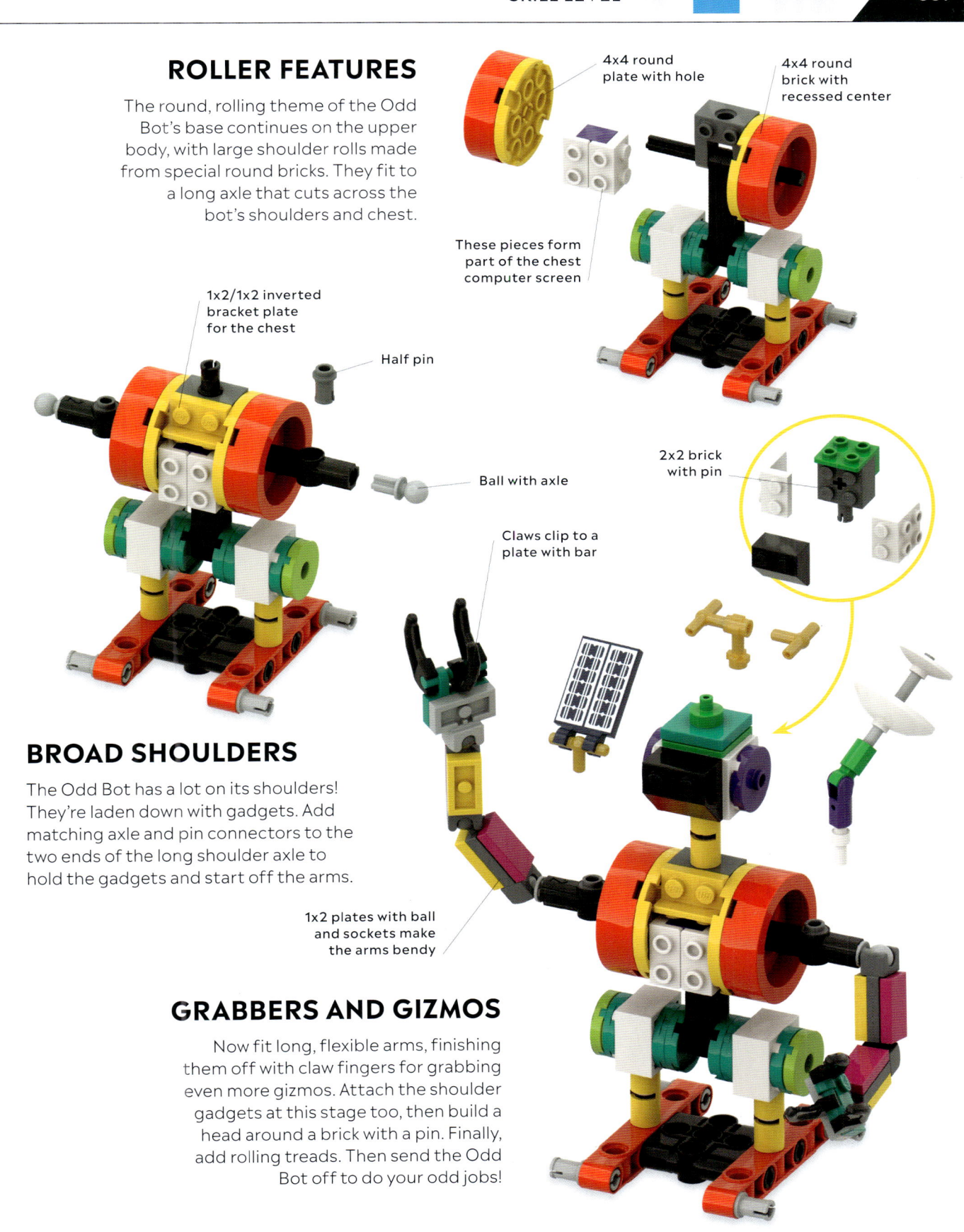

ROLLER FEATURES

The round, rolling theme of the Odd Bot's base continues on the upper body, with large shoulder rolls made from special round bricks. They fit to a long axle that cuts across the bot's shoulders and chest.

4x4 round plate with hole

4x4 round brick with recessed center

These pieces form part of the chest computer screen

1x2/1x2 inverted bracket plate for the chest

Half pin

Ball with axle

2x2 brick with pin

Claws clip to a plate with bar

BROAD SHOULDERS

The Odd Bot has a lot on its shoulders! They're laden down with gadgets. Add matching axle and pin connectors to the two ends of the long shoulder axle to hold the gadgets and start off the arms.

1x2 plates with ball and sockets make the arms bendy

GRABBERS AND GIZMOS

Now fit long, flexible arms, finishing them off with claw fingers for grabbing even more gizmos. Attach the shoulder gadgets at this stage too, then build a head around a brick with a pin. Finally, add rolling treads. Then send the Odd Bot off to do your odd jobs!

JUNK ROBOT

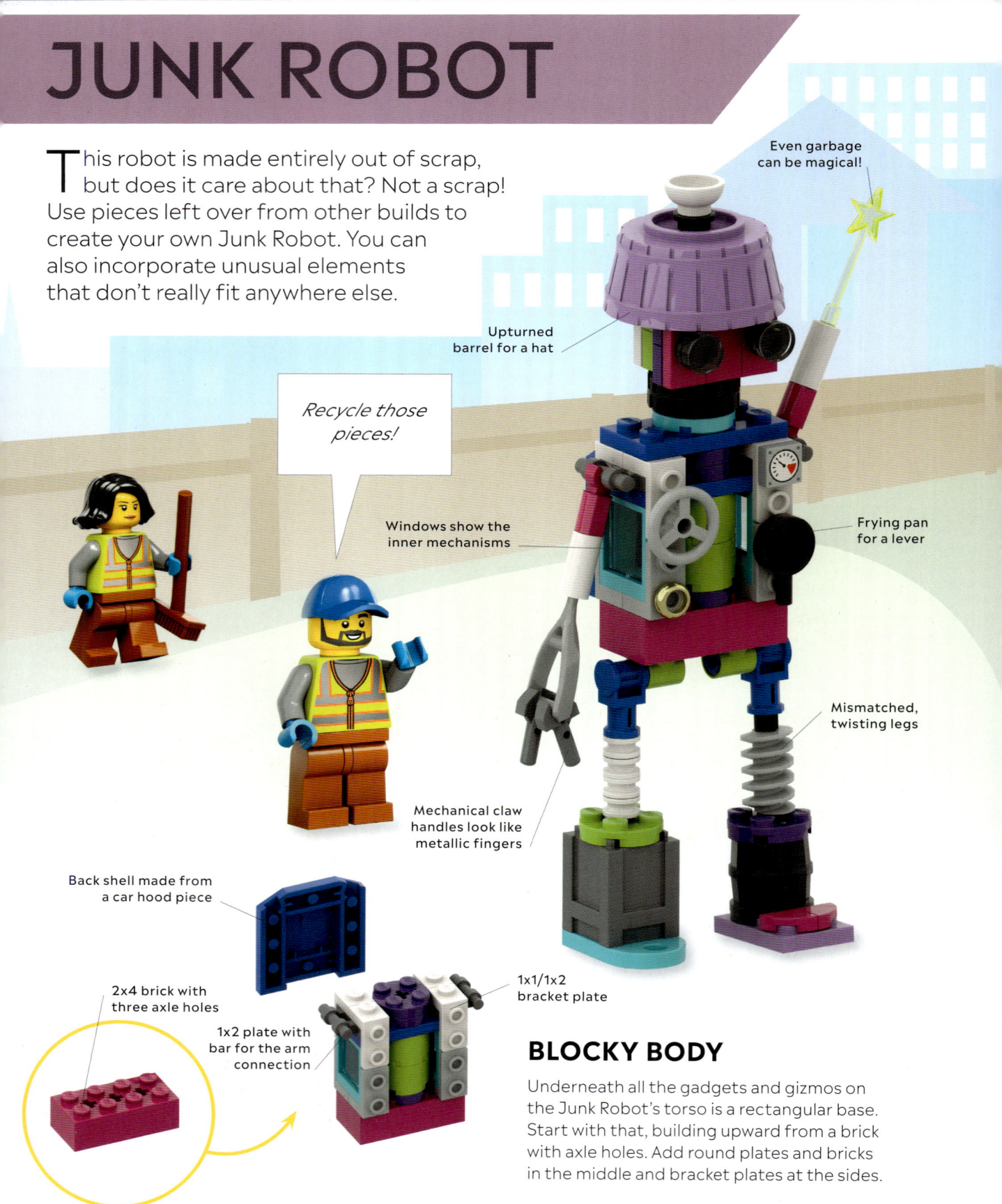

This robot is made entirely out of scrap, but does it care about that? Not a scrap! Use pieces left over from other builds to create your own Junk Robot. You can also incorporate unusual elements that don't really fit anywhere else.

Recycle those pieces!

Even garbage can be magical!

Upturned barrel for a hat

Windows show the inner mechanisms

Frying pan for a lever

Mechanical claw handles look like metallic fingers

Mismatched, twisting legs

Back shell made from a car hood piece

2x4 brick with three axle holes

1x2 plate with bar for the arm connection

1x1/1x2 bracket plate

BLOCKY BODY

Underneath all the gadgets and gizmos on the Junk Robot's torso is a rectangular base. Start with that, building upward from a brick with axle holes. Add round plates and bricks in the middle and bracket plates at the sides.

SKILL LEVEL

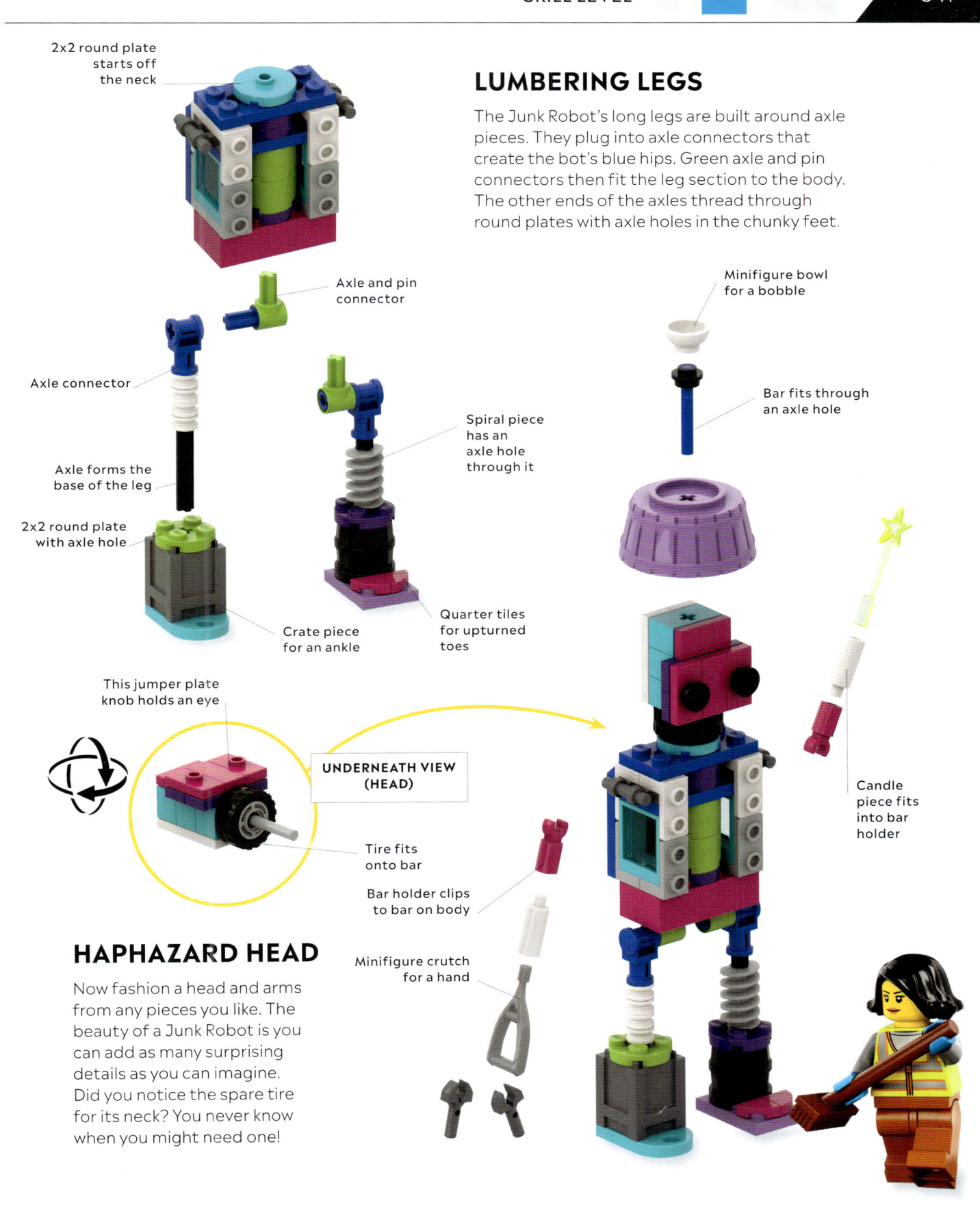

341

2x2 round plate starts off the neck

LUMBERING LEGS

The Junk Robot's long legs are built around axle pieces. They plug into axle connectors that create the bot's blue hips. Green axle and pin connectors then fit the leg section to the body. The other ends of the axles thread through round plates with axle holes in the chunky feet.

Axle and pin connector

Minifigure bowl for a bobble

Axle connector

Bar fits through an axle hole

Axle forms the base of the leg

Spiral piece has an axle hole through it

2x2 round plate with axle hole

Crate piece for an ankle

Quarter tiles for upturned toes

This jumper plate knob holds an eye

UNDERNEATH VIEW (HEAD)

Tire fits onto bar

Bar holder clips to bar on body

Candle piece fits into bar holder

HAPHAZARD HEAD

Now fashion a head and arms from any pieces you like. The beauty of a Junk Robot is you can add as many surprising details as you can imagine. Did you notice the spare tire for its neck? You never know when you might need one!

Minifigure crutch for a hand

WARRIOR MECH

With its chunky armor and towering stature, this Warrior Mech is built to take on any battle, any place, with any other robot. There's a minifigure pilot in the cockpit controlling its movements and two different weapons—a samurai sword and a futuristic ray gun.

Minifigure operator is safe inside the cockpit

A beehive hat forms part of the ray gun

No one will cut in front of me at the supermarket!

Gold details for a grand look

Bent knees, poised for battle

Sharp unicorn horn talons

1x2/1x4 inverted bracket

2x2 slope brick

Gap for the cockpit

Plate with socket for the arm connection

Wide, stomping feet support the heavy frame

1x2 slope with cutout

WARRIOR WAIST

The Warrior Mech's color scheme and armor shape resemble a samurai, which is an ancient Japanese warrior. Samurai wore leg armor that tied around the waist, cinching the waist in. The slope bricks on this bot's waist create a similar effect.

SKILL LEVEL   343

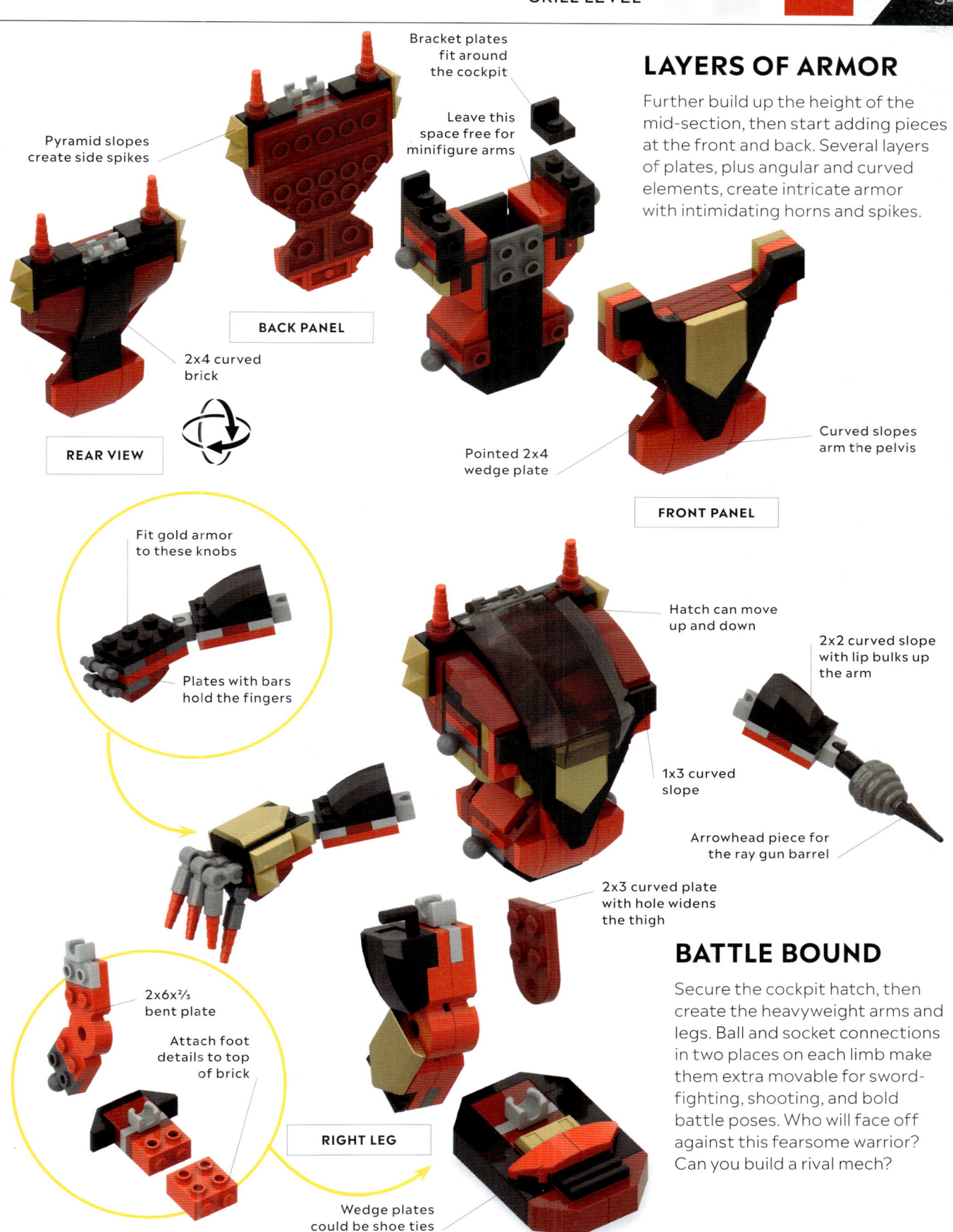

LAYERS OF ARMOR

Further build up the height of the mid-section, then start adding pieces at the front and back. Several layers of plates, plus angular and curved elements, create intricate armor with intimidating horns and spikes.

- Pyramid slopes create side spikes
- Bracket plates fit around the cockpit
- Leave this space free for minifigure arms
- 2x4 curved brick
- **BACK PANEL**
- **REAR VIEW**
- Pointed 2x4 wedge plate
- Curved slopes arm the pelvis
- **FRONT PANEL**

- Fit gold armor to these knobs
- Plates with bars hold the fingers
- Hatch can move up and down
- 2x2 curved slope with lip bulks up the arm
- 1x3 curved slope
- Arrowhead piece for the ray gun barrel
- 2x3 curved plate with hole widens the thigh

- 2x6x²⁄₃ bent plate
- Attach foot details to top of brick
- **RIGHT LEG**
- Wedge plates could be shoe ties

BATTLE BOUND

Secure the cockpit hatch, then create the heavyweight arms and legs. Ball and socket connections in two places on each limb make them extra movable for sword-fighting, shooting, and bold battle poses. Who will face off against this fearsome warrior? Can you build a rival mech?

ROBOT HORSE

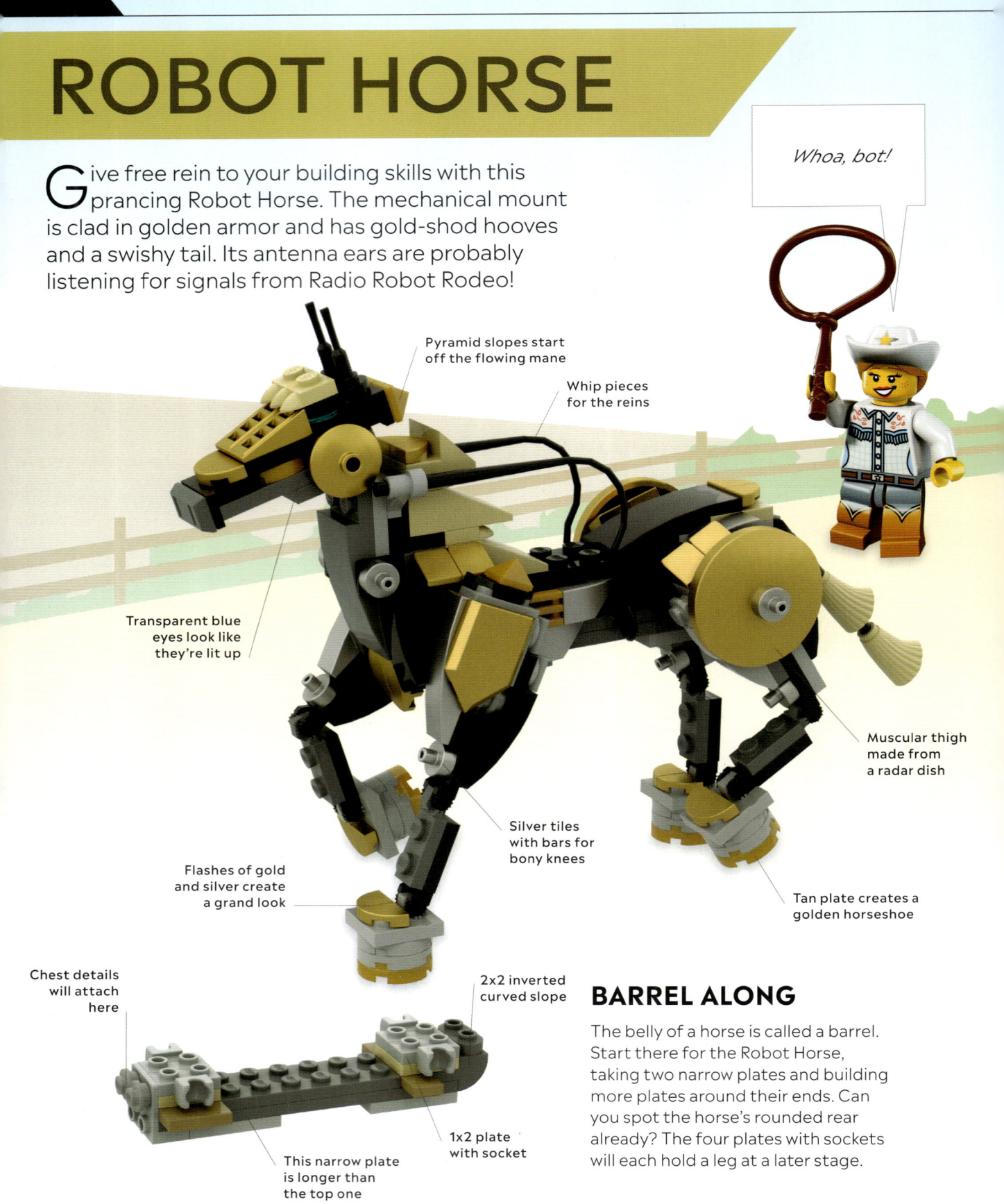

Give free rein to your building skills with this prancing Robot Horse. The mechanical mount is clad in golden armor and has gold-shod hooves and a swishy tail. Its antenna ears are probably listening for signals from Radio Robot Rodeo!

Whoa, bot!

Pyramid slopes start off the flowing mane

Whip pieces for the reins

Transparent blue eyes look like they're lit up

Muscular thigh made from a radar dish

Silver tiles with bars for bony knees

Flashes of gold and silver create a grand look

Tan plate creates a golden horseshoe

Chest details will attach here

2x2 inverted curved slope

1x2 plate with socket

This narrow plate is longer than the top one

BARREL ALONG

The belly of a horse is called a barrel. Start there for the Robot Horse, taking two narrow plates and building more plates around their ends. Can you spot the horse's rounded rear already? The four plates with sockets will each hold a leg at a later stage.

SKILL LEVEL

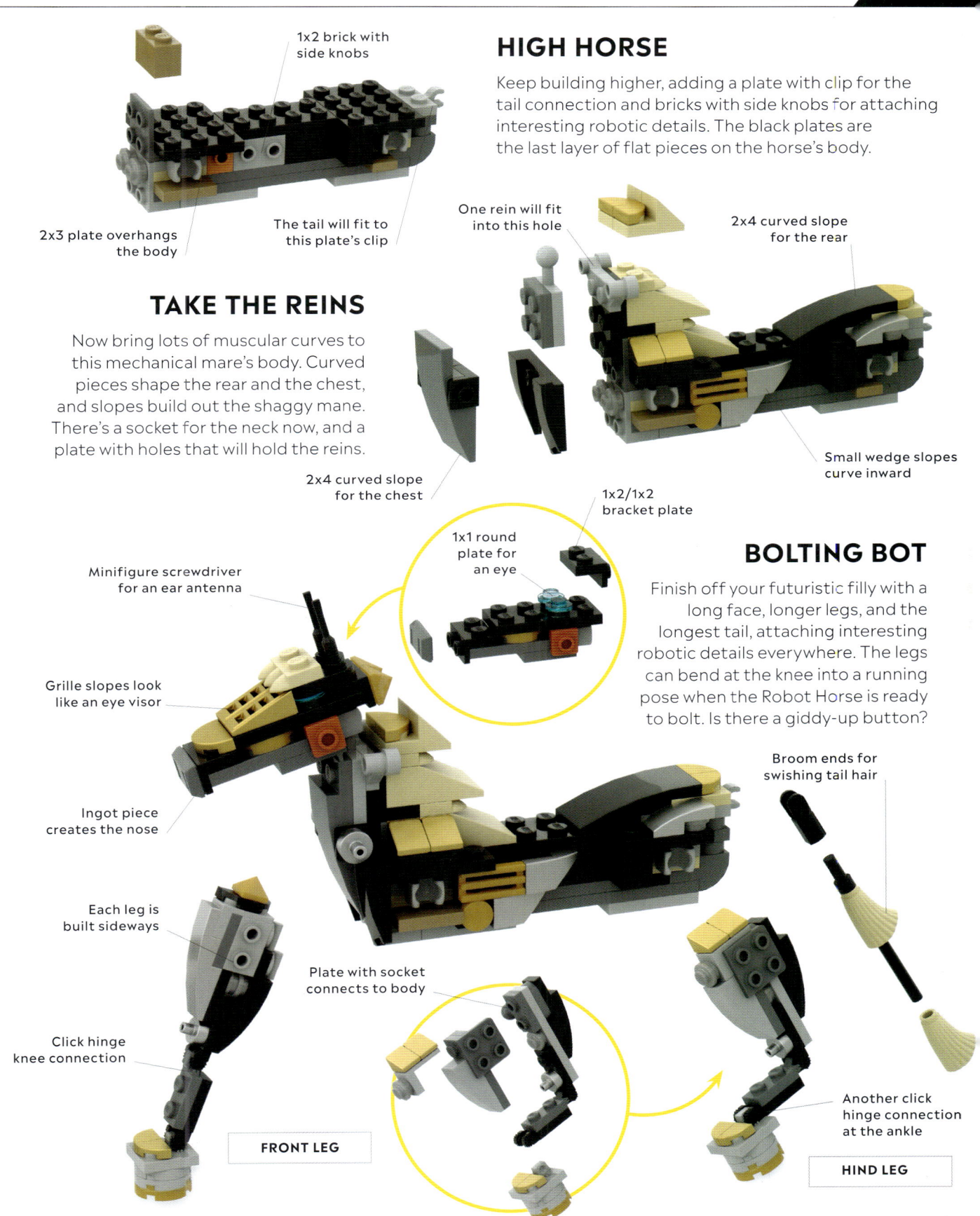

345

HIGH HORSE

Keep building higher, adding a plate with clip for the tail connection and bricks with side knobs for attaching interesting robotic details. The black plates are the last layer of flat pieces on the horse's body.

1x2 brick with side knobs

2x3 plate overhangs the body

The tail will fit to this plate's clip

One rein will fit into this hole

2x4 curved slope for the rear

Small wedge slopes curve inward

TAKE THE REINS

Now bring lots of muscular curves to this mechanical mare's body. Curved pieces shape the rear and the chest, and slopes build out the shaggy mane. There's a socket for the neck now, and a plate with holes that will hold the reins.

2x4 curved slope for the chest

1x2/1x2 bracket plate

1x1 round plate for an eye

BOLTING BOT

Finish off your futuristic filly with a long face, longer legs, and the longest tail, attaching interesting robotic details everywhere. The legs can bend at the knee into a running pose when the Robot Horse is ready to bolt. Is there a giddy-up button?

Minifigure screwdriver for an ear antenna

Grille slopes look like an eye visor

Ingot piece creates the nose

Broom ends for swishing tail hair

Each leg is built sideways

Plate with socket connects to body

Click hinge knee connection

Another click hinge connection at the ankle

FRONT LEG

HIND LEG

ROBO CASHIER

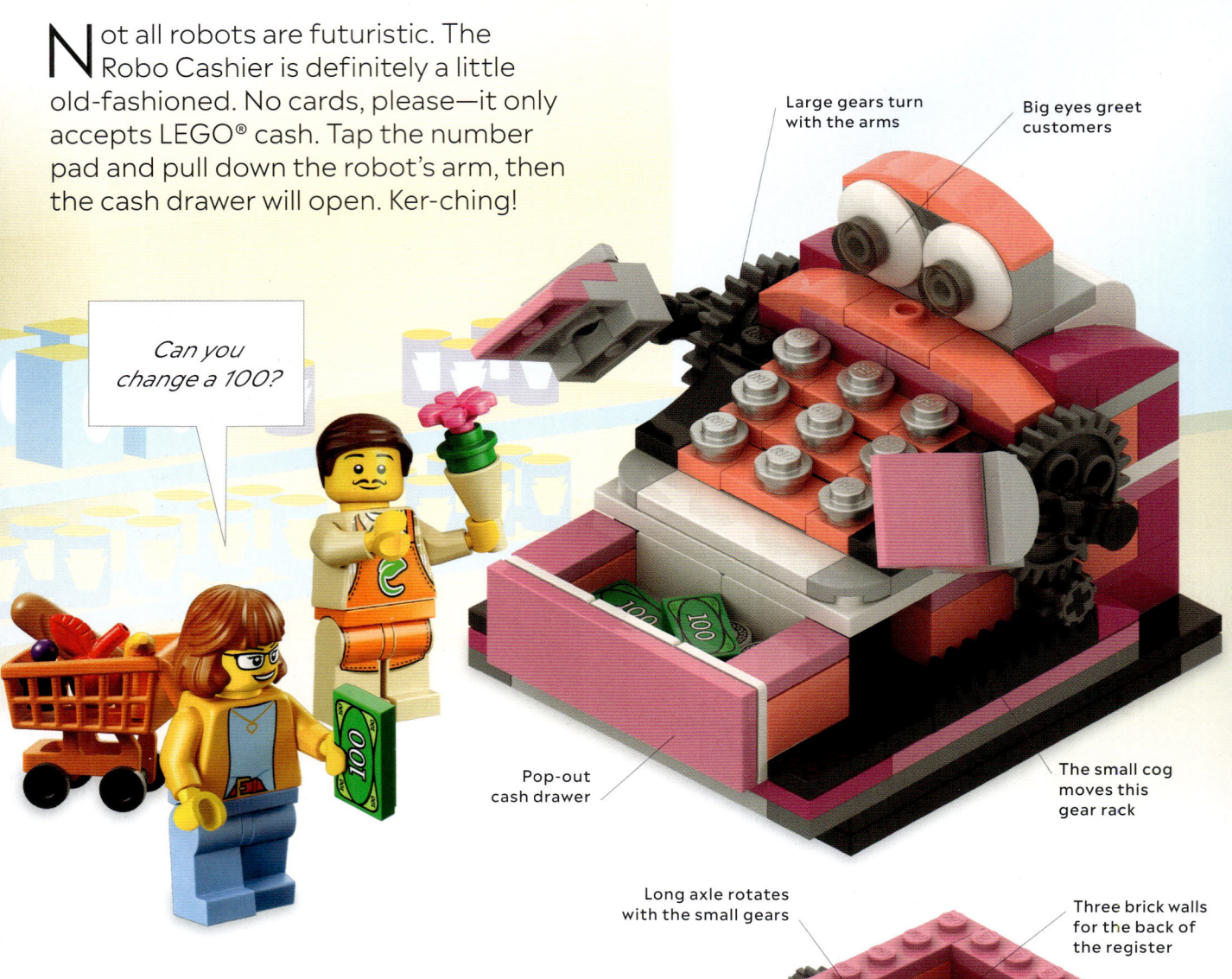

Not all robots are futuristic. The Robo Cashier is definitely a little old-fashioned. No cards, please—it only accepts LEGO® cash. Tap the number pad and pull down the robot's arm, then the cash drawer will open. Ker-ching!

Can you change a 100?

Large gears turn with the arms

Big eyes greet customers

Pop-out cash drawer

The small cog moves this gear rack

Long axle rotates with the small gears

Three brick walls for the back of the register

2x4 tile

SLIDE AND GLIDE

The Robo Cashier is built upward from a first layer of plates. There are smooth tiles on top of that for the cash drawer to glide over in a later step. Create the back of the register on top of the base, including an axle and small gears that will be part of the drawer mechanism.

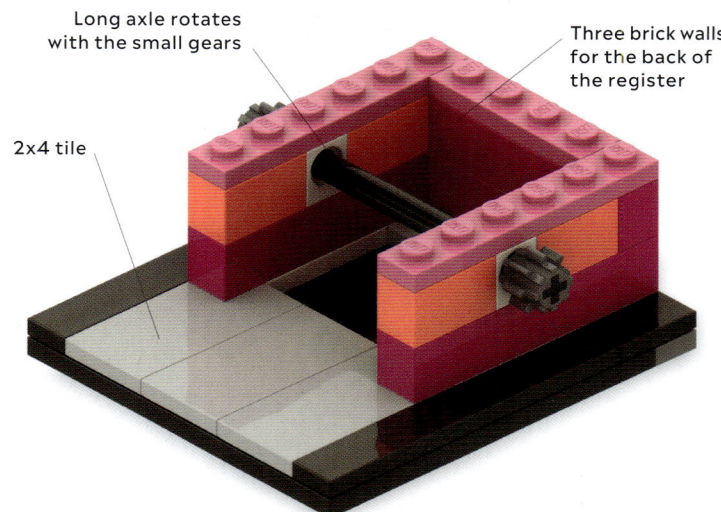

SKILL LEVEL

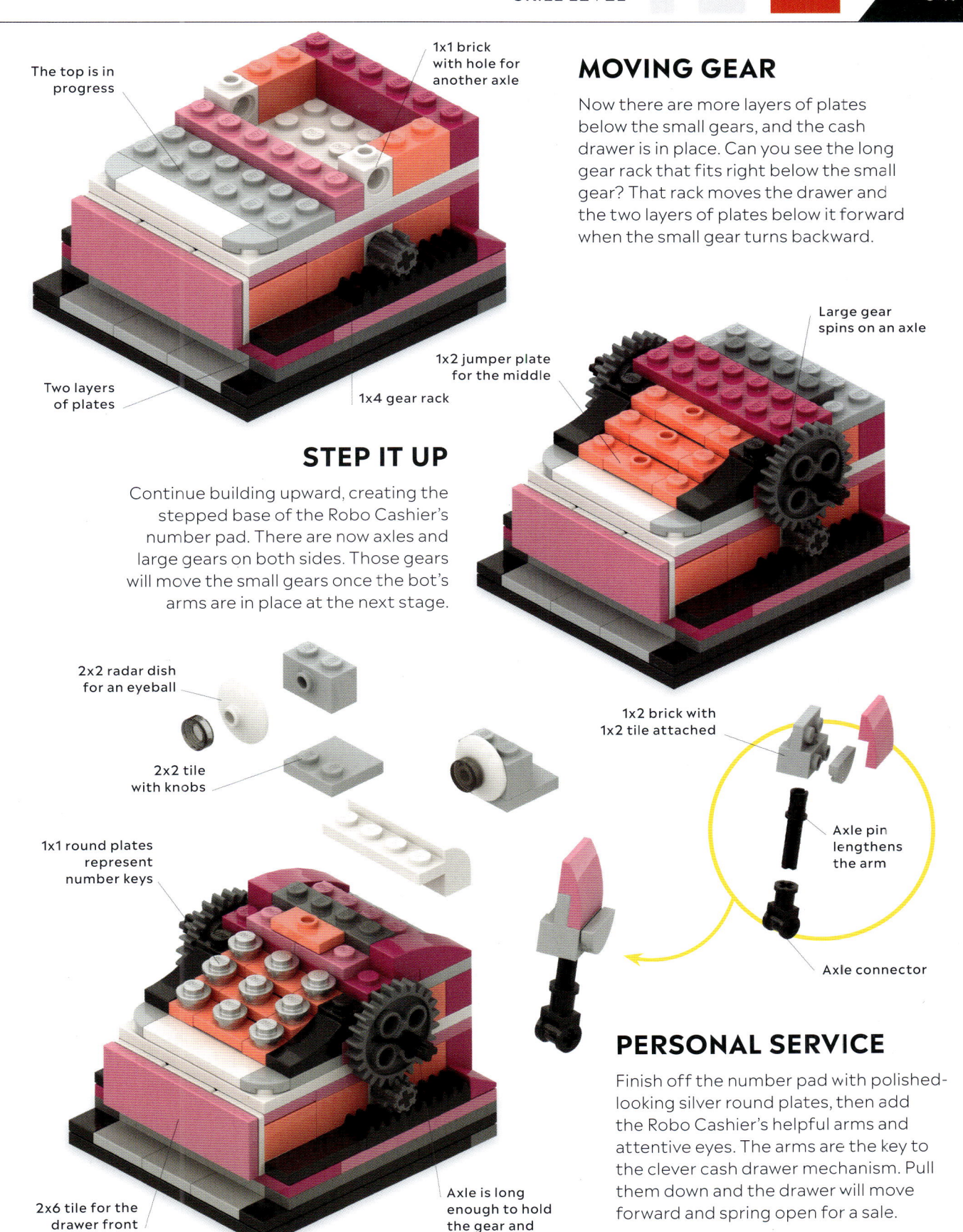

347

Labels (top-left build):
- The top is in progress
- 1x1 brick with hole for another axle
- Two layers of plates
- 1x2 jumper plate for the middle
- 1x4 gear rack

MOVING GEAR

Now there are more layers of plates below the small gears, and the cash drawer is in place. Can you see the long gear rack that fits right below the small gear? That rack moves the drawer and the two layers of plates below it forward when the small gear turns backward.

STEP IT UP

Continue building upward, creating the stepped base of the Robo Cashier's number pad. There are now axles and large gears on both sides. Those gears will move the small gears once the bot's arms are in place at the next stage.

Label (middle-right build): Large gear spins on an axle

Labels (parts):
- 2x2 radar dish for an eyeball
- 2x2 tile with knobs
- 1x2 brick with 1x2 tile attached
- Axle pin lengthens the arm
- Axle connector

Labels (bottom-left build):
- 1x1 round plates represent number keys
- 2x6 tile for the drawer front
- Axle is long enough to hold the gear and the arm

PERSONAL SERVICE

Finish off the number pad with polished-looking silver round plates, then add the Robo Cashier's helpful arms and attentive eyes. The arms are the key to the clever cash drawer mechanism. Pull them down and the drawer will move forward and spring open for a sale.

ROBOT SPA

At the Robot Spa, robots get the pampering of their lives. Check out that hot tub (perfect for flushing fluff out of a robot's filters) and the battery recharger chair. And how about a bracing brush-down followed by a relaxing oiling? Aaaahhhh!

Axle holds up the beam for the brush

SCRUB STATION

Robotic spa users wind up thoroughly scrubbed and polished to perfection after an encounter with this fast-moving facility. Robots lie on the scrubbing table as the two brushes leave no screw, wire, or gizmo unturned.

Scrubbing table has a 2x6 plate base

Checkered tile floor made from 1x1 tiles

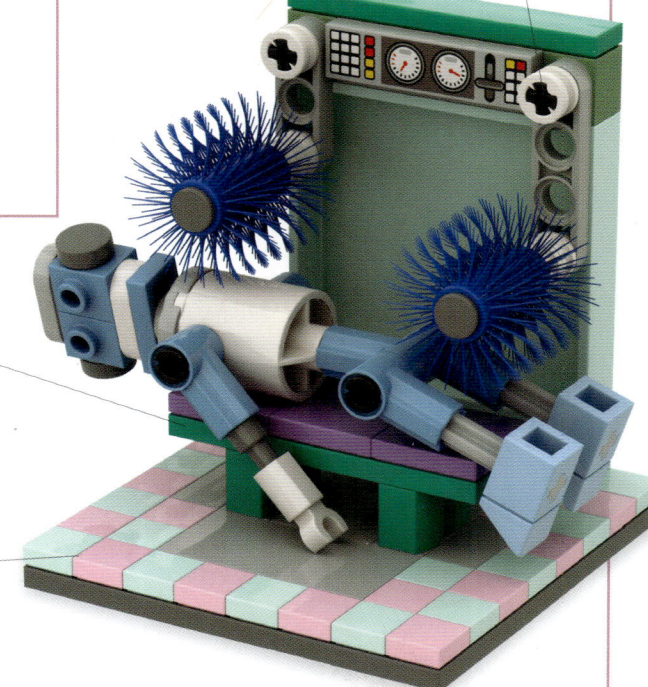

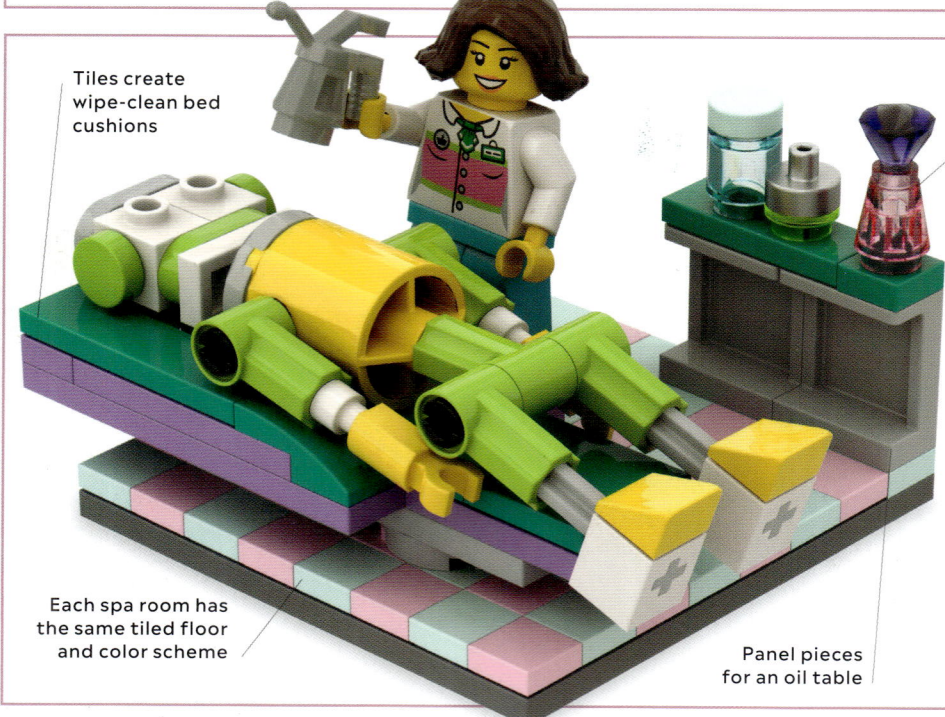

Tiles create wipe-clean bed cushions

A pink cone and a crystal make this fancy oil

OIL BED

Robots can lie back, relax, and let nourishing oils sink into their squeakiest parts in the Robot Spa's oil room. They can choose from a selection of oils that ease all types of rust and other robotic ailments.

Each spa room has the same tiled floor and color scheme

Panel pieces for an oil table

IDEAS GALLERY | 349

I can feel my springs unwinding.

RECHARGE ROOM

The busyness of daily robot life can drain a robot's batteries over time. This recharge room offers express battery charging in a relaxing environment. Robots just plug themselves into the solar-powered charging station, then hit the green button. They're guaranteed to leave feeling renewed.

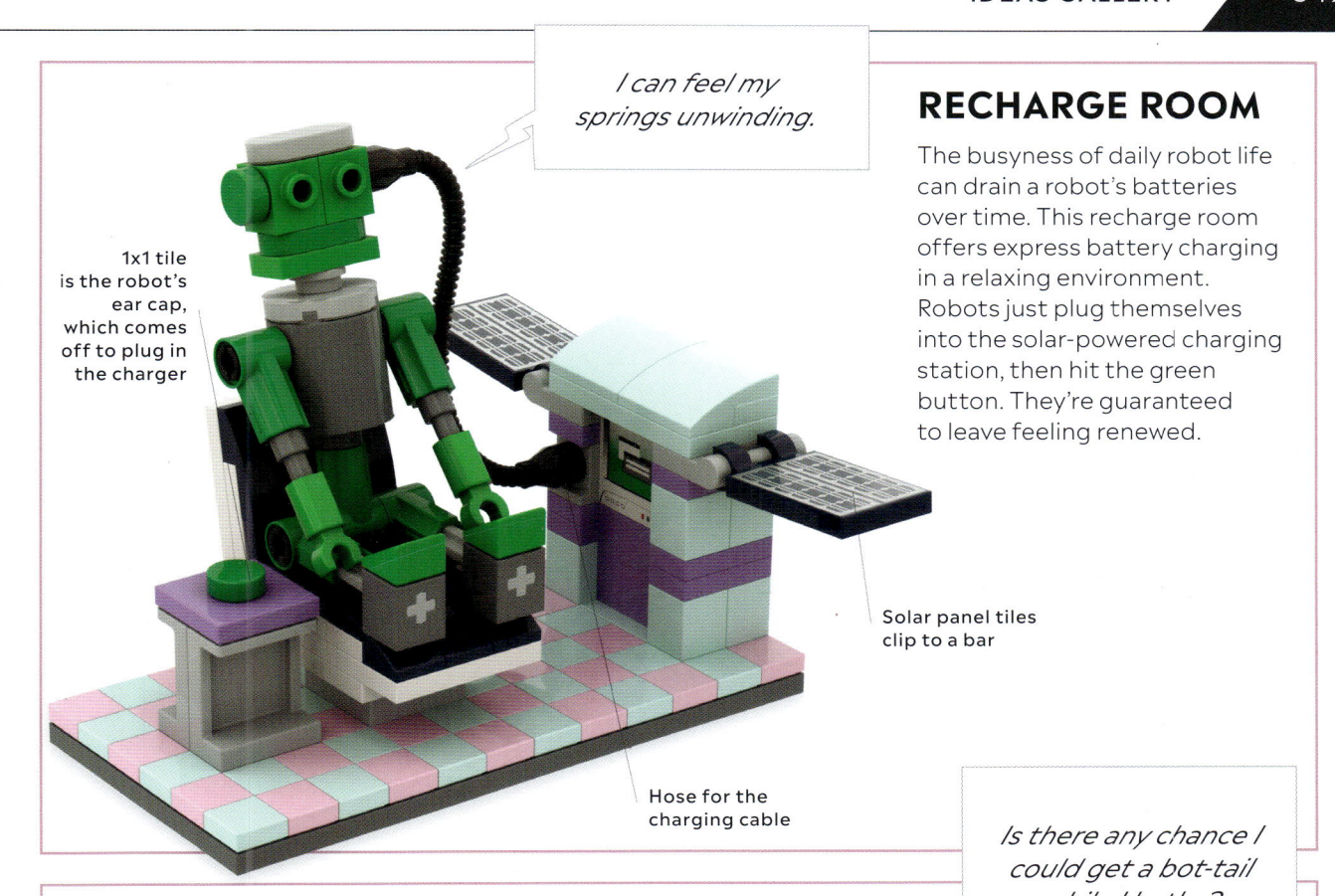

1x1 tile is the robot's ear cap, which comes off to plug in the charger

Solar panel tiles clip to a bar

Hose for the charging cable

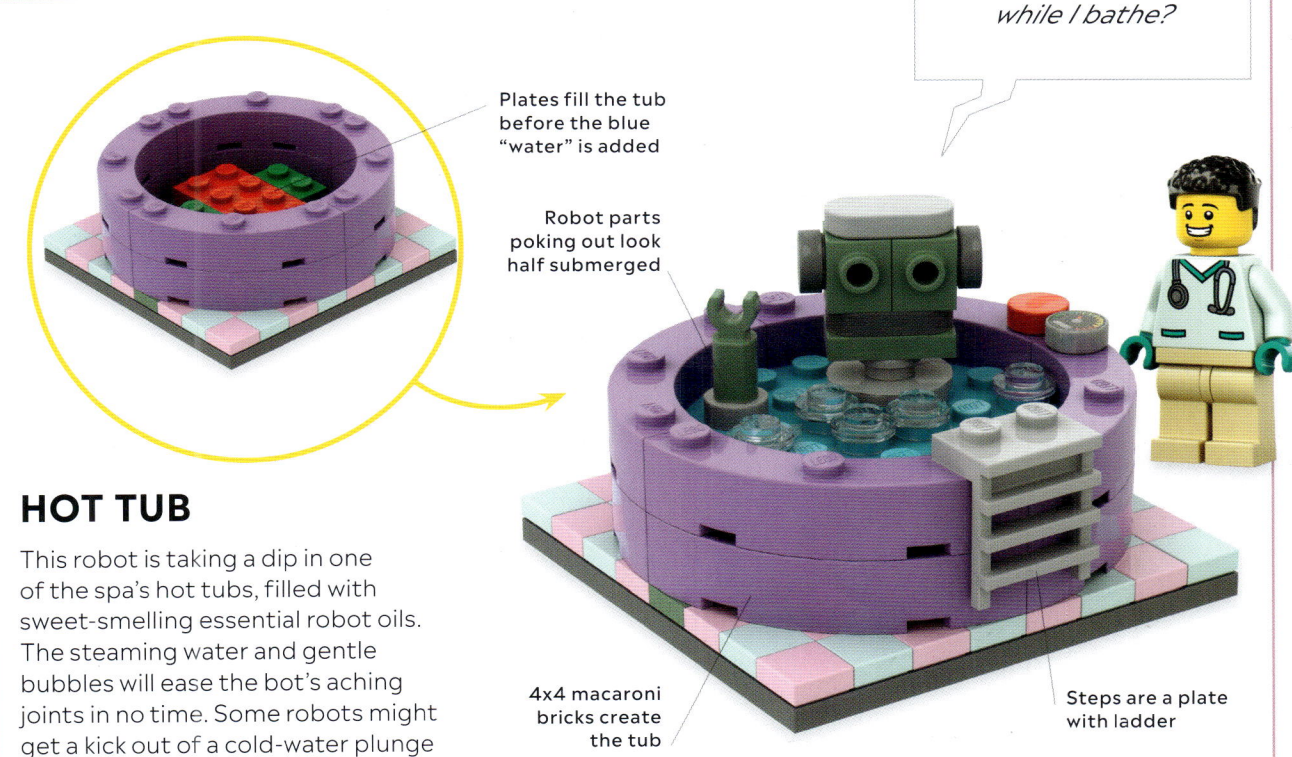

Plates fill the tub before the blue "water" is added

Robot parts poking out look half submerged

Is there any chance I could get a bot-tail while I bathe?

HOT TUB

This robot is taking a dip in one of the spa's hot tubs, filled with sweet-smelling essential robot oils. The steaming water and gentle bubbles will ease the bot's aching joints in no time. Some robots might get a kick out of a cold-water plunge pool instead—can you build one?

4x4 macaroni bricks create the tub

Steps are a plate with ladder

ANDROID

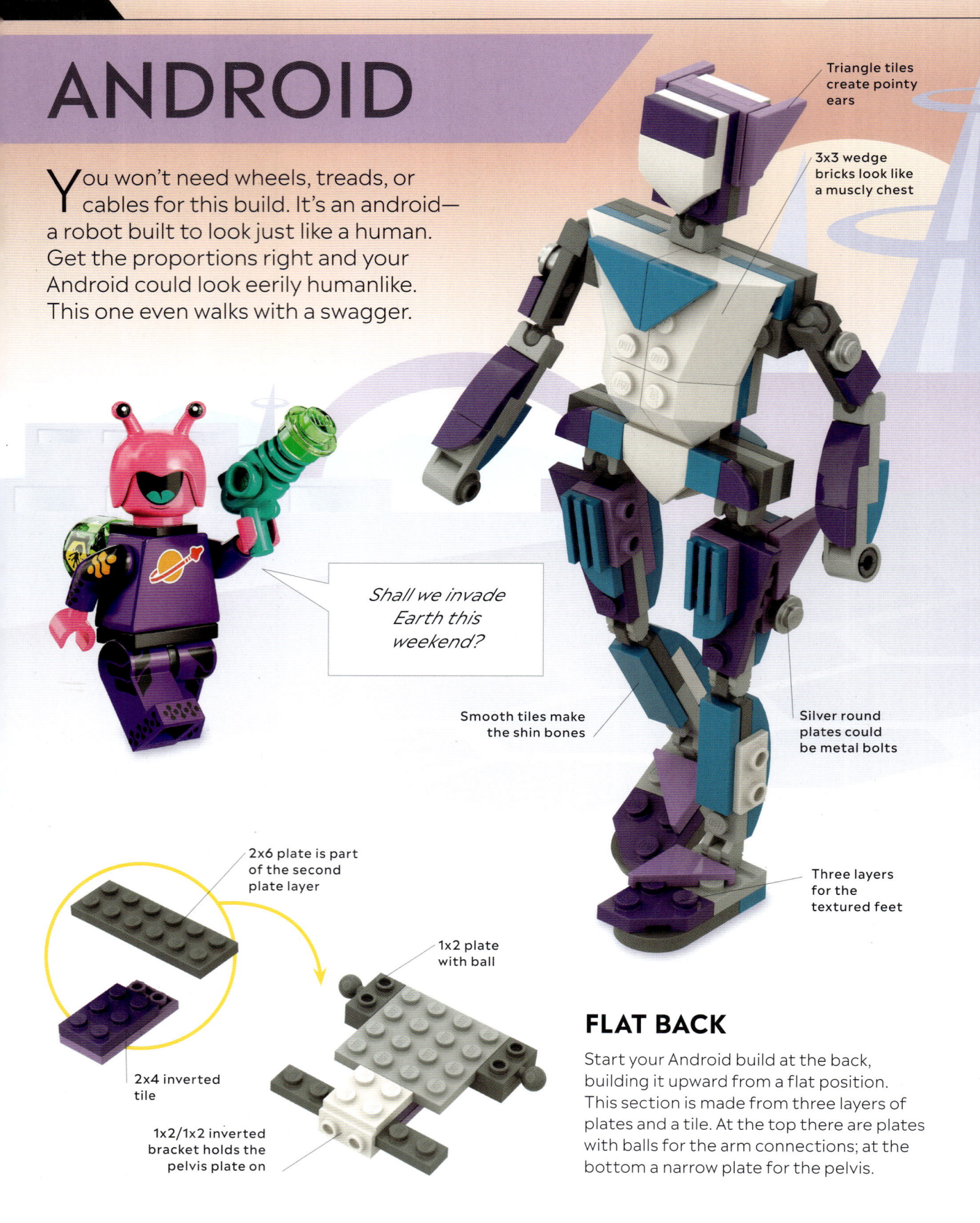

You won't need wheels, treads, or cables for this build. It's an android—a robot built to look just like a human. Get the proportions right and your Android could look eerily humanlike. This one even walks with a swagger.

Shall we invade Earth this weekend?

- Triangle tiles create pointy ears
- 3x3 wedge bricks look like a muscly chest
- Smooth tiles make the shin bones
- Silver round plates could be metal bolts
- Three layers for the textured feet

- 2x6 plate is part of the second plate layer
- 1x2 plate with ball
- 2x4 inverted tile
- 1x2/1x2 inverted bracket holds the pelvis plate on

FLAT BACK

Start your Android build at the back, building it upward from a flat position. This section is made from three layers of plates and a tile. At the top there are plates with balls for the arm connections; at the bottom a narrow plate for the pelvis.

SKILL LEVEL

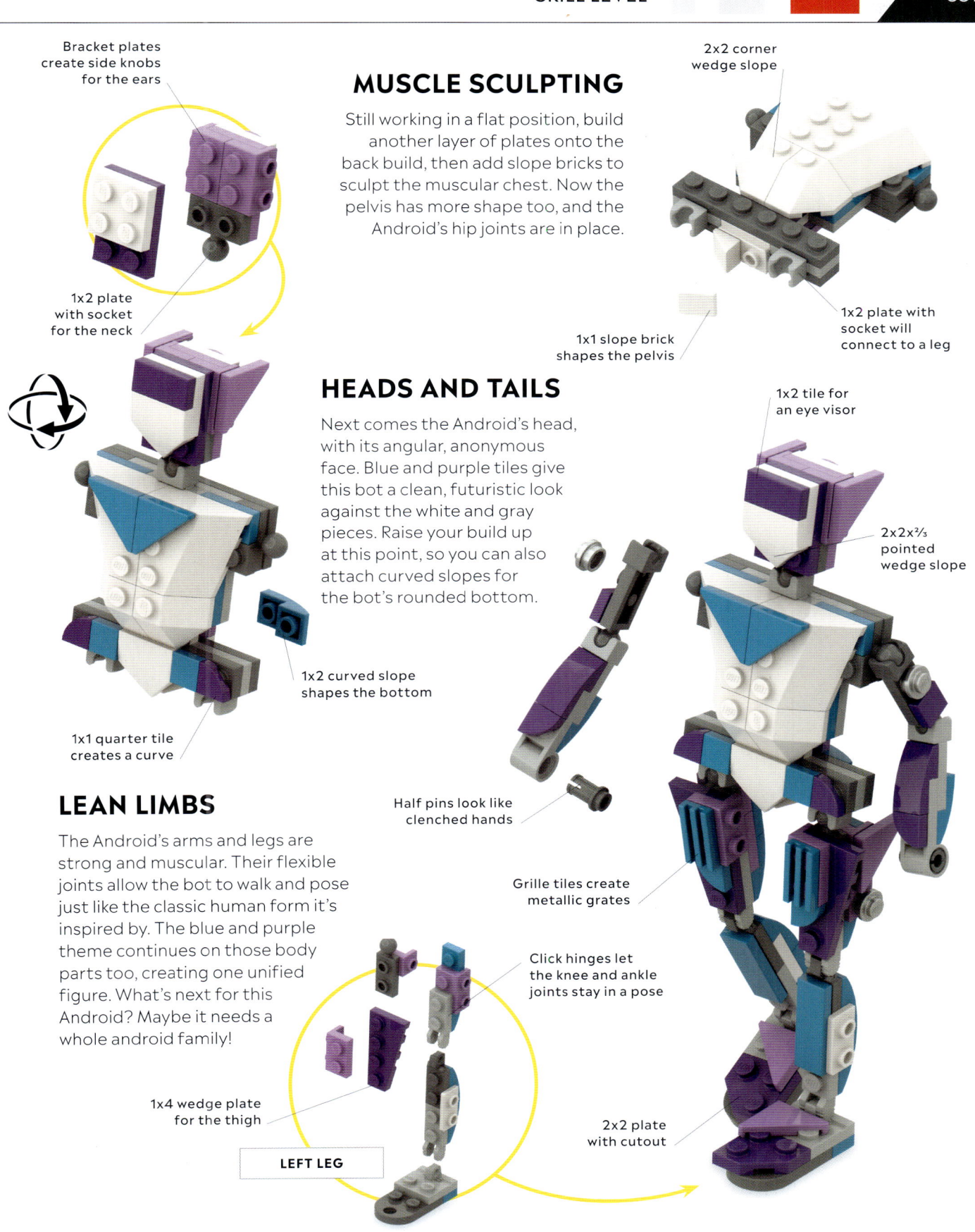

MUSCLE SCULPTING

Still working in a flat position, build another layer of plates onto the back build, then add slope bricks to sculpt the muscular chest. Now the pelvis has more shape too, and the Android's hip joints are in place.

Bracket plates create side knobs for the ears

1x2 plate with socket for the neck

2x2 corner wedge slope

1x2 plate with socket will connect to a leg

1x1 slope brick shapes the pelvis

HEADS AND TAILS

Next comes the Android's head, with its angular, anonymous face. Blue and purple tiles give this bot a clean, futuristic look against the white and gray pieces. Raise your build up at this point, so you can also attach curved slopes for the bot's rounded bottom.

1x2 tile for an eye visor

2x2x2/3 pointed wedge slope

1x2 curved slope shapes the bottom

1x1 quarter tile creates a curve

LEAN LIMBS

The Android's arms and legs are strong and muscular. Their flexible joints allow the bot to walk and pose just like the classic human form it's inspired by. The blue and purple theme continues on those body parts too, creating one unified figure. What's next for this Android? Maybe it needs a whole android family!

Half pins look like clenched hands

Grille tiles create metallic grates

Click hinges let the knee and ankle joints stay in a pose

1x4 wedge plate for the thigh

2x2 plate with cutout

LEFT LEG

PLANT CYBORG

Your friends won't be-leaf their eyes when they see your Plant Cyborg build. It's a walking tangle of cables, vines, bricks, and leaves that brings science and nature together. Just don't leave it outside. You might never find it again.

My new sidekick is growing on me.

Round plates with petals for eyes

Silver round tiles with bars for metallic details

Armored chest

Whip looks like a twisting foot root

Layers of plants and leaves create a wild look

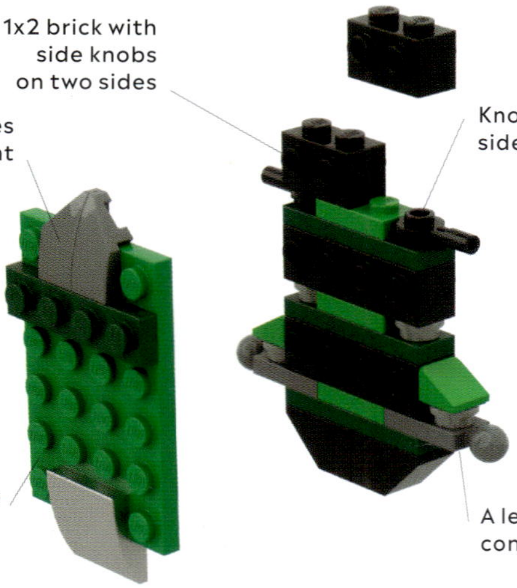

Two wedge slopes make a point

1x2 brick with side knobs on two sides

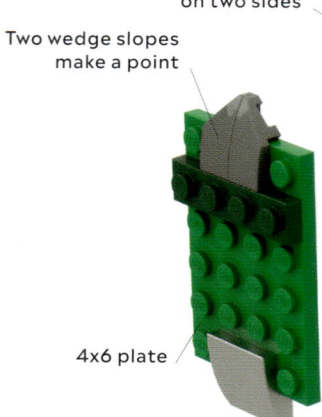

Knobs on both sides

4x6 plate

A leg will later connect here

GROWTH SPURT

Grow the Plant Cyborg outward from its armored chest. Its bottom layer is made from lots of bricks with side knobs, and the second layer is a 4x6 plate that attaches sideways to them. The gray pieces for the chest armor then fit on top of the plate.

SKILL LEVEL

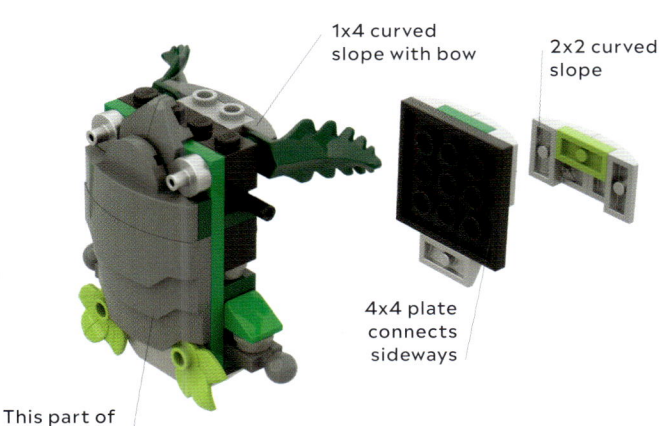

ROBOTIC ROOTS

The Plant Cyborg may be covered in soft-looking leaves, but it's still a bot at heart. Its back is a metallic shell made from four gray curved slopes. They fit onto a square black plate that connects to knobs on the chest pieces.

- 1x4 curved slope with bow
- 2x2 curved slope
- 4x4 plate connects sideways
- This part of the chest is one big piece

WILD STYLE

Shape the shoulders next, layering up plates and curved slopes. Include plates with balls at either side for the arms to fit to. Then move up to the head with its large, leafy headdress for both camouflage and forest fashion.

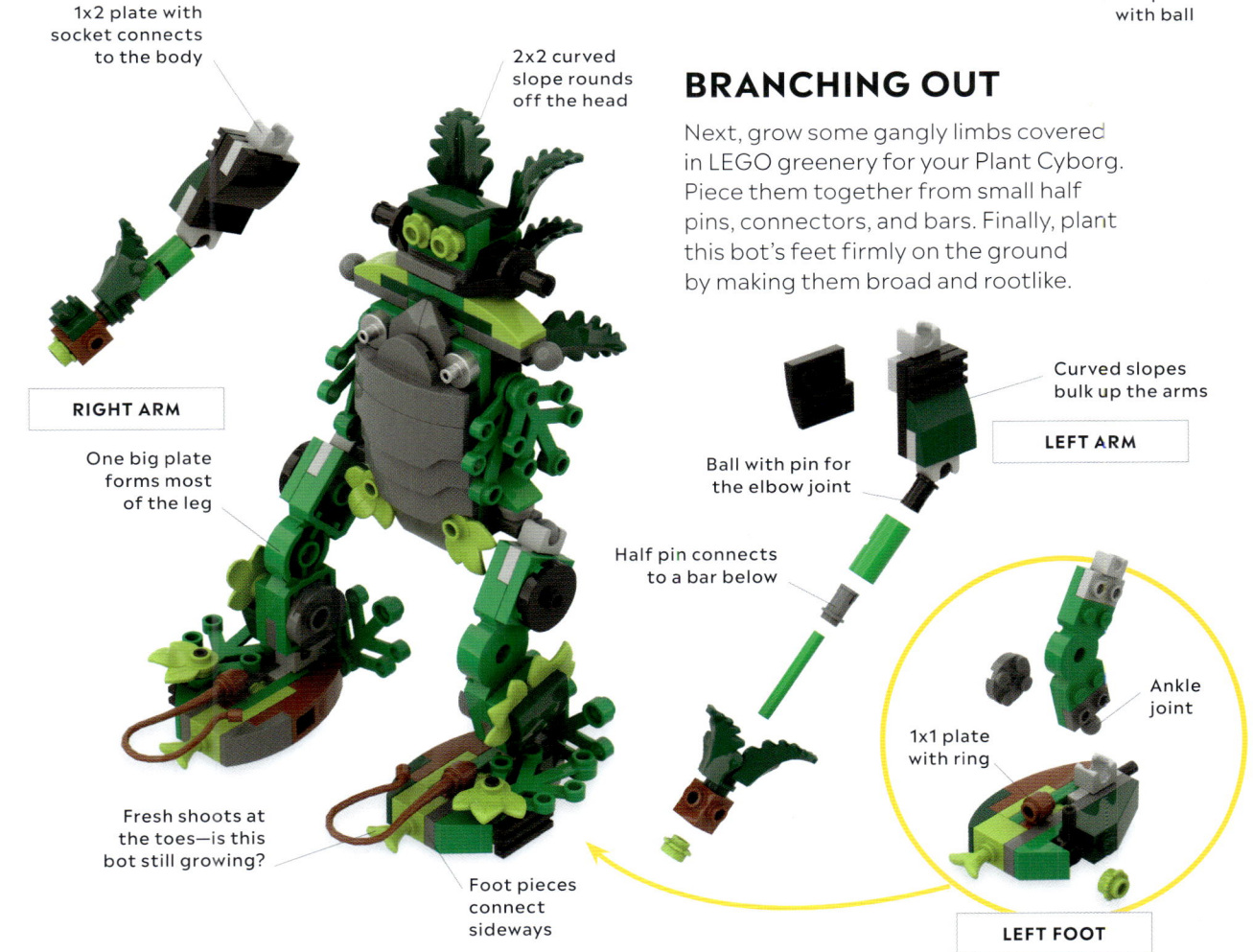

- Antenna ears are one car chassis piece
- 1x1 round plate with upright leaf
- 1x2 plate with ball

BRANCHING OUT

Next, grow some gangly limbs covered in LEGO greenery for your Plant Cyborg. Piece them together from small half pins, connectors, and bars. Finally, plant this bot's feet firmly on the ground by making them broad and rootlike.

- 1x2 plate with socket connects to the body
- 2x2 curved slope rounds off the head
- Curved slopes bulk up the arms
- Ball with pin for the elbow joint
- Half pin connects to a bar below
- One big plate forms most of the leg
- Fresh shoots at the toes—is this bot still growing?
- Foot pieces connect sideways
- 1x1 plate with ring
- Ankle joint

RIGHT ARM

LEFT ARM

LEFT FOOT

353

EXTRA BITS AND PIECES

GLOSSARY
This information will help you understand interesting words and terms used in some of the chapters.

ANIMAL DICTIONARY

blubber
A layer of fat underneath the skin that keeps sea mammals warm.

carnivore
An animal that mostly eats meat.

habitat
The environment where an animal lives.

herbivore
An animal that only eats plants, fruit, and vegetables.

mammal
A warm-blooded animal that gives birth to live young (not eggs). It may have fur.

marsupial
A type of mammal whose females have pouches to carry their young.

nocturnal
When an animal is active during the night.

omnivore
An animal that eats both meat and plants.

reptile
A cold-blooded animal that lays eggs. It may be covered in scales or plates.

DINOSAUR DICTIONARY

anatomy
The body or structure of a creature or plant.

ceratopsians
A group of horned dinosaurs with frills on their heads.

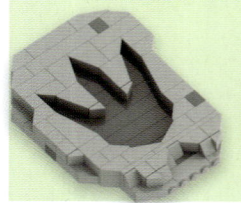

crest
A comb or tuft of feathers on the head.

fossil
The remains or traces of a once-living animal or plant.

paleontologist
A scientist who studies dinosaurs and other fossils.

predator
A dinosaur or creature that naturally preys on others.

prehistoric
An ancient time before recorded history.

sauropods
A group of dinosaurs with long necks and tails.

theropods
A group of mostly meat-eating dinosaurs that stood on two legs.

PREHISTORIC TIME PERIODS

The prehistoric creatures in the dinosaur chapter of this book existed at a specific point in time. Each creature's time period is flagged at the top of its page.

Permian
299–252 million years ago

Triassic
252–201 million years ago

Jurassic
201–145 million years ago

Cretaceous
145–66 million years ago

Pleistocene (or Ice Age)
2.6 million –11,700 years ago

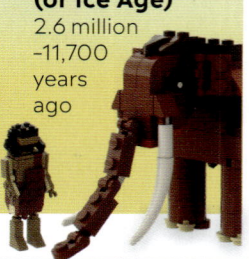

EXTRA BITS AND PIECES 355

ROBOT DICTIONARY

antenna
A metal device that can send or receive radio signals.

articulated limbs
Body parts with lots of joints that allow them to move easily.

chassis
The flat base of a robot that rolls on wheels.

gauge
A device that shows the pressure or temperature inside a robot.

mechanism
A piece of moving machinery that is part of a robot or other machine.

pincer
A tool with two grasping claws for grabbing things.

sensor
A part of a robot that senses its environment.

vent
A hole or grille on a machine that lets air in and heat or fumes out.

visor
A protective panel on a robot's face where its eye sensors are.

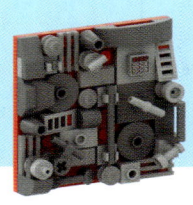

MEET THE BUILDERS

Jessica Farrell and Nate Dias designed and created the models in this book. Jessica built all of the houses, animals, robots, and half of the dinosaurs. She is a professional brick artist whose models are shown around the world. Nate built all of the cars and the other half of the dinosaurs. He is a science teacher by day and a LEGO® master builder by night!

We hope you enjoy our designs!

Which model is your favorite?

JESSICA'S MODERN HOUSE, PAGES 32-33

NATE'S BUMPER CAR, PAGES 158-159

DK | Penguin Random House

Project Editors **Lisa Stock**, **Nicole Reynolds**
Senior Editors **Helen Murray**,
Selina Wood, **Elizabeth Cook**
Senior US Editor **Megan Douglass**
Senior Designers **Anna Formanek**,
Lauren Adams, **Jenny Edwards**
Production Editors **Jennifer Murray**, **Siu Yin Chan**
Senior Production Controllers
Louise Minihane, **Lloyd Robertson**
Managing Editor **Paula Regan**
Managing Art Editor **Jo Connor**
Managing Director **Mark Searle**

Packaged for DK by **Plum Jam**
Editor **Hannah Dolan** Designer **Guy Harvey**

Models designed and created by
Jessica Farrell and **Nate Dias**

Animals consultant
Cathriona Hickey

Dinosaurs consultant
Dr. Dean Lomax

Dorling Kindersley would like to thank:
Randi Sørensen, **Heidi K. Jensen**, **Martin Leighton Lindhardt**, and **Nina Koopmann** at the LEGO Group; **Jessica Farrell** and **Nate Dias** for supplying all model images and breakdowns; **Julia March** for proofreading and additional text for the robots chapter.

First American Edition, 2025
Published in the United States by DK Publishing,
a division of Penguin Random House LLC
1745 Broadway, 20th Floor, New York, NY 10019

Page design copyright © 2025 Dorling Kindersley Limited
A Penguin Random House Company
LEGO, the LEGO logo, the Minifigure,
and the Brick and Knob configurations are
trademarks of the LEGO Group.
© 2025 The LEGO Group.

Contains content previously published in
How to Build LEGO® Houses (2021)
How to Build LEGO Cars (2021)
How to Build LEGO Dinosaurs (2022)
How to Build LEGO Animals (2023)
How to Build LEGO Robots (2024)

Manufactured by
Dorling Kindersley Limited
20 Vauxhall Bridge Road,
London SW1V 2SA
under license from the LEGO Group.

25 26 27 28 29 10 9 8 7 6 5 4 3 2 1
001–344896–Feb/2025

All rights reserved.
Without limiting the rights under the copyright reserved above, no part of this publication may be reproduced, stored in or introduced into a retrieval system, or transmitted, in any form, or by any means (electronic, mechanical, photocopying, recording, or otherwise), without the prior written permission of the copyright owner.
Published in Great Britain by Dorling Kindersley Limited

A catalog record for this book
is available from the Library of Congress.
ISBN: 978-0-5939-6587-0

Printed and bound in China

www.dk.com
www.LEGO.com

MIX
Paper | Supporting
responsible forestry
FSC™ C018179

This book was made with Forest Stewardship Council™ certified paper—one small step in DK's commitment to a sustainable future. Learn more at www.dk.com/uk/information/sustainability

Your opinion matters

Please scan this QR code to give feedback
to help us enhance your future experiences